T0212738

Lecture Notes in Physics

Founding Editors

Wolf Beiglböck, Heidelberg, Germany

Jürgen Ehlers, Potsdam, Germany

Klaus Hepp, Zürich, Switzerland

Hans-Arwed Weidenmüller, Heidelberg, Germany

Volume 992

Series Editors

Roberta Citro, Salerno, Italy

Peter Hänggi, Augsburg, Germany

Morten Hjorth-Jensen, Oslo, Norway

Maciej Lewenstein, Barcelona, Spain

Angel Rubio, Hamburg, Germany

Wolfgang Schleich, Ulm, Germany

Stefan Theisen, Potsdam, Germany

James D. Wells, Ann Arbor, MI, USA

Gary P. Zank, Huntsville, AL, USA

The Lecture Notes in Physics

The series Lecture Notes in Physics (LNP), founded in 1969, reports new developments in physics research and teaching - quickly and informally, but with a high quality and the explicit aim to summarize and communicate current knowledge in an accessible way. Books published in this series are conceived as bridging material between advanced graduate textbooks and the forefront of research and to serve three purposes:

- to be a compact and modern up-to-date source of reference on a well-defined topic;
- to serve as an accessible introduction to the field to postgraduate students and non-specialist researchers from related areas;
- to be a source of advanced teaching material for specialized seminars, courses and schools.

Both monographs and multi-author volumes will be considered for publication. Edited volumes should however consist of a very limited number of contributions only. Proceedings will not be considered for LNP.

Volumes published in LNP are disseminated both in print and in electronic formats, the electronic archive being available at springerlink.com. The series content is indexed, abstracted and referenced by many abstracting and information services, bibliographic networks, subscription agencies, library networks, and consortia.

Proposals should be sent to a member of the Editorial Board, or directly to the responsible editor at Springer:

Dr Lisa Scalone
Springer Nature
Physics
Tiergartenstrasse 17
69121 Heidelberg, Germany
lisa.scalone@springernature.com

More information about this series at http://www.springer.com/series/5304

Andreas Wipf

Statistical Approach to Quantum Field Theory

An Introduction

Second Edition

 Springer

Andreas Wipf
Theoretical Physics
Friedrich Schiller University Jena
Jena, Germany

ISSN 0075-8450　　　　　　　　　ISSN 1616-6361　(electronic)
Lecture Notes in Physics
ISBN 978-3-030-83262-9　　　　　ISBN 978-3-030-83263-6　(eBook)
https://doi.org/10.1007/978-3-030-83263-6

This Springer imprint is published by the registered company Springer Nature Switzerland AG.
The registered company address is: Gewerbestrasse 11, 6330 Cham, Switzerland

To Ingrid, Leonie, Severin, and Valentin

Preface to the Second Edition

This new expanded second edition has been totally revised and corrected. The reader finds two completely new chapters. One covers the exact solution of the finite temperature Schwinger model with periodic boundary conditions. This simple model supports instanton solutions—similarly as QCD—and allows for a detailed discussion of topological sectors in gauge theories, the anomaly-induced breaking of chiral symmetry, and the intriguing role of fermionic zero modes. The other new chapter is devoted to interacting fermions at finite fermion density and finite temperature. Such low-dimensional models are used to describe long-energy properties of Dirac-type materials in condensed matter physics. The large-N solutions of the Gross-Neveu, Nambu-Jona-Lasinio, and Thirring models are presented in great detail, where N denotes the number of fermion flavors. Towards the end of the book, corrections to the large-N solution and simulation results of a finite number of fermion flavors are presented. Further problems are added at the end of each chapter in order to guide the reader to a deeper understanding of the presented topics. This book is aimed at advanced students and young researchers who want to acquire the necessary tools and experience to produce research results in the statistical approach to quantum field theory.

It is a great pleasure to thank again the many collaborators, teachers, and colleagues already mentioned in the first edition of this book. In addition, I would like to thank the group in Frankfurt (Laurin Pannullo, Marc Wagner, and Marc Winstel) and my more recent PhD students and postdocs for a fruitful collaboration on interacting Fermi systems in the continuum and on the lattice. Several new sections in this second edition are based on an early collaboration with I. Sachs and an ongoing collaboration with the group in Frankfurt and with J. Lenz, M. Mandl, D. Schmidt, and B. Wellegehausen. I would like to thank Holger Gies and Felix Karbstein for many inspiring discussions about interacting Fermions, and Julian Lenz, Michael Mandl, and Ingrid Wipf for proofreading the new chapters.

Jena, Germany
June 2021

Andreas Wipf

Preface to the First Edition

Statistical field theory deals with the behavior of classical or quantum systems consisting of an enormous number of degrees of freedom in and out of equilibrium. Quantum field theory provides a theoretical framework for constructing quantum mechanical models of systems with an infinite number of degrees of freedom. It is the natural language of particle physics and condensed matter physics. In the past decades the powerful methods in statistical physics and Euclidean quantum field theory have come closer and closer, with common tools based on the use of path integrals. The interpretation of Euclidean field theories as particular systems of statistical physics opened up new avenues to understand strongly coupled quantum systems or quantum field theories at zero or finite temperature. The powerful methods of statistical physics and stochastics can be applied to study for example the vacuum sector, effective action, thermodynamic potentials, correlation functions, finite size effects, nature of phase transitions or critical behavior of quantum systems.

The first chapters of this book contain a self contained introduction to path integrals in Euclidean quantum mechanics and statistical mechanics. The resulting high-dimensional integrals can be estimated with the help of Monte-Carlo simulations based on Markov processes. The method is first introduced and then applied to ordinary integrals and to quantum mechanical systems. Thereby the most commonly used algorithms are explained in detail. Equipped with theses stochastic methods we may use high performance computers as an "experimental" tool for a new brand of theoretical physics.

The book contains several chapters devoted to an introduction into simple lattice field theories and a variety of spin systems with discrete and continuous spins. An ideal guide to the fascinating area of phase transitions is provided by the ubiquitous Ising model. Despite its simplicity the model is often used to illustrate the key features of statistical systems and the methods available to understand these features. The Ising model has always played an important role in statistical physics, both at pedagogical and methodological levels. Almost all chapters in the middle part of the book begin with introducing methods, approximations, expansions or rigorous results by first considering the Ising model. In a next step we generalize from the Ising model to other lattice systems, for example Potts models, O(N) models, scalar field theories, gauge theories and fermionic theories. For spin models and field theories on a lattice it is often possible to derive rigorous results or bounds. Important examples are the bounds provided by the mean field approximation, inequalities between correlation functions of ferromagnetic systems and the proofs that there exist spontaneously broken phases at low temperature or the duality transformations for Abelian models which relate the weak coupling and strong coupling regions or the low temperature and high temperature phases. All these interesting results are derived and discussed with great care.

As an alternative to the lattice formulation of quantum field theories one may use a variant of the flexible renormalization group methods. For example, implementing (spacetime) symmetries is not so much an issue for a functional renormalization group method as it sometimes is for a lattice regularization and hence the method is somehow complementary to the ab initio lattice approach. In cases where a lattice regularization based on a positive Boltzmann factor fails, for example for gauge theories at finite density, the functional method may work. Thus it is often a good strategy to consider both methods when it comes to properties of strongly coupled systems under extreme conditions. Knowledge of the renormalization group method and in particular the flow of scale dependent functionals from the microscopic to the macroscopic world is a key part of modern physics and thus we have devoted two chapters to this method.

According to present day knowledge all fundamental interactions in nature are described by gauge theories. Gauge theories can be formulated on a finite spacetime lattice without spoiling the important local gauge invariance. Thereby the functional integral turns into a finite-dimensional integral which can be handled by stochastic means. Problems arise when one considers gauge fields in interaction with fermions at finite temperature and non-zero baryon density. A lot of efforts have gone into solving or at least circumventing these problems to simulate quantum chromodynamics, the microscopic gauge theory underlying the strong interaction between quarks and gluons. The last chapters of the book deal with gauge theories without and with matter.

This book is based on an elaboration of lecture notes of the course Quantum Field Theory II given by the author at the Friedrich-Schiller-University Jena. It is designed for advanced undergraduate and beginning graduate students in physics and applied mathematics. For this reason, its style is greatly pedagogical; it assumes only some basics of mathematics, statistical physics, and quantum field theory. But the book contains some more sophisticated concepts which may be useful to researchers in the field as well. Although many textbooks on statistic physics and quantum field theory are already available, they largely differ in contents from the present book. Beginning with the path integral in quantum mechanics and with numerical methods to calculate ordinary integrals we bridge the gap to lattice gauge theories with dynamical fermions. Each chapter ends with some problems which should be useful for a better understanding of the material presented in the main text. At the end of many chapters you also find listings of computer programs, either written in C or in the freely available Matlab-clone Octave. Not only because of the restricted size of the book I did not want to include lengthy simulation programs for gauge theories.

Acknowledgments

Over the years I have had the pleasure of collaborating and discussing many of the themes of this book with several of my teachers, colleagues, and friends. First of all, I would like to especially thank the late Lochlain O'Raifeartaigh for the long and profitable collaboration on effective potentials, anomalies, and two-dimensional field theories, and for sharing his deep understanding of many aspects of symmetries and field theories. I would like to use this opportunity to warmly thank the academic teachers who have influenced me most—Jürg Fröhlich, Res Jost, John Lewis, Konrad Osterwalder, Eduard Stiefel, and especially Norbert Straumann. I assume that their influence on my way of thinking about quantum field theory and statistical physics might be visible in some parts of this book.

I have been fortunate in having the benefit of collaborations and discussions with many colleagues and friends and in particular with Manuel Asorey, Pierre van Baal, Janos Balog, Steven Blau, Jens Braun, Fred Cooper, Stefan Durr, Chris Ford, Lazlo Feher, Thomas Filk, Peter Forgacs, Christof Gattringer, Holger Gies, Tom Heinzl, Karl Jansen, Claus Kiefer, Kurt Langfeld, Axel Maas, Emil Mottola, Renato Musto, Jan Pawlowski, Ivo Sachs, Lorenz von Smekal, Thomas Strobl, Torsten Tok, Izumi Tsutsui, Sebastian Uhlmann, Matt Visser, Christian Wiesendanger, and Hiroshi Yoneyma. On several topics covered in the second and more advanced part of the book, I collaborated intensively with my present and former PhD students Georg Bergner, Falk Bruckmann, Leander Dittman, Marianne Heilmann, Tobias Kästner, Andreas Kirchberg, Daniel Körner, Dominque Länge, Franziska Synatschke-Czerwonka, Bjoern Wellegehausen, and Christian Wozar. Last but not least, I am indebted to Holger Gies and Kurt Langfeld for a critical reading of parts of the manuscript and Marianne Heilmann for translating the German lecture notes into English.

Jena, Germany   Andreas Wipf
June 2012

Contents

About the Author

Andreas Wipf is full professor at the Friedrich-Schiller-University (FSU) of Jena, Germany, where he teaches many courses related to theoretical physics. His research focuses on quantum field theory in the continuum and on lattice, structural aspects of quantum theory, quantum systems under extreme conditions, conformal and supersymmetric field theories, and the role of symmetries and symmetry breaking in theoretical physics. He obtained his diploma from the Department of Physics at ETH Zurich, and his PhD from the Institute for Theoretical Physics at the University of Zurich. He covered positions as postdoctoral fellow, assistant professor, and *Privatdozent* at the Dublin Institute for Advanced Studies, Los Alamos National Laboratory, Max-Planck-Institute for Physics (Werner-Heisenberg-Institute) in Munich, and at ETH Zurich where he habilitated in 1995. Then, he moved to Jena as full professor and has been director of the Institute for Theoretical Physics, dean of the Physics and Astronomy Faculty, spokesman of the research training group on gravity, quantum field theory, and mathematical physics. He co-authored the textbooks *Theoretische Physik* (Springer Spektrum, 2015), which presents all theoretical physics up to the bachelor's degree and beyond, and *Theoretische Physik 3 | Quantenmechanik* (Springer Spektrum, 2018).

Acronyms

A_μ	Gauge potential
$A(\omega, \omega')$	Acceptance rate for $\omega \to \omega'$
β, β_T	Inverse gauge coupling $\sim 1/g^2$ or inverse temperature $1/k_B T$
$D\!\!\!/, \partial\!\!\!/$	Dirac operator and free Dirac operator
$\mathscr{C}$	Closed path or loop
χ	Susceptibility
$\mathscr{D}\phi$	Formal path integral measure
D_μ	Covariant derivative in μ-direction
d_s	$= 2^{[d/2]}$ dimension of irreducible spinor in d dimensions
F, f	Free energy and free energy density
$F_{\mu\nu}$	Field strength tensor
ϕ_x, φ_x	Microscopic and macroscopic lattice field at lattice site x
$\hat{\phi}(x)$	Field operator at spacetime point x
$\hat{\phi}_E(x)$	Euclidian field operator at spacetime point x
Γ, Γ_k	Effective action, scale dependent effective action
$j(x), j_x$	External source at spacetime point x and at lattice site x
$\mathscr{L}$	Lagrangian density or Legendre transformation
Λ	Spacetime lattice
$H(\omega)$	Energy of spin configuration ω
$\hat{H}$	Hamiltonian operator of quantum system
$K(t), K(\tau)$	Propagator at real time t and imaginary time τ
N_t, N_s	Number of lattice points in time and space direction
$O, \hat{O}$	Observable in a classical statistical system and in quantum theory
$\mathscr{P}, P$	Polyakov loop and its gauge invariant trace
P_ω	Equilibrium probability for configuration ω
$\psi(x), \psi_x$	Fermionic field in continuum and on the lattice
$\hat{q}(t)$	Position operator at time t
s_x	Value of spin variable at lattice site x
S, S_E	Action and Euclidean action
S_B, S_F	Action for bosons and fermions
S_B	Shannon–Boltzmann entropy
T_c	Critical temperature
$T(\omega, \omega')$	Probability to test transition $\omega \to \omega'$
$U(\beta)$	Internal energy

$U_\ell, U_{x,\mu}$	Group valued link variable
$U_p, U_{x,\mu\nu}$	Plaquette variable
u, u_k	Effective potential and scale dependent effective potential
ω	Path, spin- or field configuration
Ω	Set of paths, spin- or field configurations
$\Omega(x), \Omega_x$	Gauge transformation at x in continuum and on the lattice
R_k	Regulator function in flow equation
$W(\omega, \omega')$	Transition probability $\omega \to \omega'$
W, W_k	Schwinger functional, scale dependent Schwinger functional
$W[\mathscr{C}]$	Wilson loop around closed path $\mathscr{C}$
$w(j)$	Schwinger function for constant j
$Z(\beta)$	Partition function
$\zeta_A(s)$	Zeta function of operator A

Introduction

1

A quantum field theory (QFT) is an extension of the principles of quantum mechanics to fields based on the wave properties of matter. It is generally accepted that QFT is an appropriate framework for describing the interaction between infinitely many degrees of freedom. It is the natural language of particle physics and condensed matter physics with applications ranging from the Standard Model of elementary particles and their interactions to the description of critical phenomena and phase transitions, such as in the theory of superconductivity.

A relativistic quantum field theory unifies the basic principles of quantum theory and special relativity in a consistent manner. The quantization of the electromagnetic field was outlined by M. Born and P. Jordan back in 1925 [1], and just a year later with Heisenberg, they showed how to quantize general systems with an infinite number of degrees of freedom [2]. The theory was further developed and applied to the photon–atom interactions by P. Dirac who studied the quantized photon field in interaction with atoms and calculated emission and absorption rates [3]. In a next step, P. Jordan, W. Pauli, and W. Heisenberg [4–6] completed the quantization of electrodynamics, compatible with special relativity. This was achieved by quantizing the Dirac and Maxwell fields in interaction. In particular, Heisenberg and Pauli emphasized the importance of the Lagrangian formulation in field theory and developed the quantization procedure which nowadays is called *canonical quantization*. To this day, their approach to QFT remains a popular one and is presented in textbooks.

In a perturbative approach, one first quantizes the non-interacting field and subsequently includes the interaction by a local interaction density. The direct application of this method leads to divergent expressions for physical quantities, e.g., an infinitely large self-energy. The solution of this problem has led to the renormalization method which originated in the early work and has been completed by Tomonaga, Schwinger, Feynman, and Dyson. The latter outlined a proof of renormalizability of QED [7], which was complemented by other authors in the 1950s and 1960s. For a renormalizable QFT, there exists a method consisting of

A. Wipf, *Statistical Approach to Quantum Field Theory*, Lecture Notes
in Physics 992, https://doi.org/10.1007/978-3-030-83263-6_1

a regularization and subsequent renormalization that gives finite and physically sensible results by absorbing the divergences into redefinitions of only a few coupling constants and the field. QED is the most studied and successful prototype of a renormalizable quantum field theory with a (local) gauge symmetry. The high precision of QED calculations is based upon the applicability of perturbation theory, where the dimensionless fine-structure constant $\alpha \sim 1/137$ serves as expansion parameter. The following years were dedicated to the formal improvement of field theory in general. The connection between spin and statistics was found, the CPT theorem was formulated, the representation theory of the (anti)commutation rules was developed, the Euclidean formulation of QFT' was investigated [8, 9], and symmetry principles came to the fore.

An exciting period followed which began with the advent of *non-Abelian gauge theories*, formulated by Yang and Mills to extend the concept of a local gauge symmetry from Abelian to non-Abelian groups [10], in order to describe the interactions between elementary particles. The efforts culminated in the model of S. Glashow, S. Weinberg, and A. Salam [11–13] for the electroweak interaction. In particular after G. 't Hooft proved the renormalizability of spontaneously broken gauge theories [14]—the weak interaction is described by such a theory—the interest in Yang-Mills theories continued to increase. Shortly after these developments, it became clear that the forces between strongly interacting particles is described by yet another non-Abelian gauge theory, namely, *quantum chromodynamics* (QCD) [15]. However, it is not easy to compare QCD with experiment. This is not only because the coupling is strong and not weak as in electrodynamics but also because of the related fact that the fundamental constituents of the theory, the quarks, and force-carrying gluons have never been seen directly and are generally believed to be unattainable because of the phenomenon of confinement. To prove confinement or related properties as chiral symmetry breaking in strongly coupled field theories, we must leave the realm of perturbation theory.

New insights into the renormalization procedure beyond perturbation theory are obtained with the path integral quantization of physical systems [16]. After a rotation of time to imaginary values, one obtains the Euclidean functional integral formulation of QFT. The integrand contains the exponential of the negative action of the classical field theory. When one approximates the continuous (Euclidean) spacetime by a finite lattice, the functional integral of a QFT turns into a well-defined finite dimensional integral. This means that the discretization regularizes the QFT without reference to perturbation theory. In a second step, one may renormalize the theory by performing the continuum limit in which the lattice spacing tends to zero. Equally important, the finite-dimensional integral is an ensemble average in classical equilibrium statistical mechanics. Thereby the classical action of the field theory discretized on a spacetime lattice becomes the energy function of a particular classical spin model. This far-reaching observation bridges the gap between two apparently unrelated branches of physics: quantum (field) theory and classical statistical physics. Actually, any QFT at finite temperature is described by an Euclidean functional integral on a cylinder with the imaginary time as periodic variable. It follows that after discretization the functional integral for the finite temperature QFT turns into an ensemble average of a classical spin model with periodic boundary conditions in one direction. In this context the formulation of gauge theories on a spacetime lattice was of utmost importance. Lattice gauge theories with discrete gauge groups have been investigated by F. Wegner [17], and 3 years later K. Wilson succeeded in putting non-Abelian gauge theories on a spacetime lattice [18]. Many recent non-perturbative results on gauge theories at zero and finite temperatures are based on this pioneering work.

The interrelation between quantum field theories at zero or finite temperature and classical spin models is extremely beneficial both for QFT and statistical physics. For example, many non-perturbative problems of interest in QFT can be handled with the powerful and well-established methods of classical statistical physics. For example, on may address difficult problems like mass generation, decay widths, symmetry breaking, phase transitions, or condensates, to name a few. Thereby many rigorous results, inequalities, dualities, and approximation methods in statistical physics can be put to use in lattice field theory. Shortly after the seminal work of K. Wilson, the first numerical simulations of lattice gauge theories were performed [19, 20]. Today observable quantities as, for example, particle masses, decay widths, condensates, thermodynamical potentials, and finite temperature phase diagrams can be calculated with the powerful Monte Carlo method. The following chapters contain an introduction into this exciting and active fields of research in theoretical physics. For a further reading, I included primary and secondary literature, including textbooks, at the end of each chapter. There are excellent books on the market which introduce into quantum field theory, for example, the texts [21–25].

References

1. M. Born, P. Jordan, Zur Quantenmechanik. Z. Phys. **34**, 858 (1925)
2. M. Born, W. Heisenberg, P. Jordan, Zur Quantenmechanik II. Z. Phys. **35**, 557 (1926)
3. P.A.M. Dirac, The quantum theory of emission and absorption of radiation. Roc. Roy. Soc. London **A 114**, 243 (1927)
4. P. Jordan, W. Pauli, Zur Quantenelektrodynamik. Z. Phys. **47**, 151 (1928)
5. W. Heisenberg, W. Pauli, Zur Quantendynamik der Wellenfelder I. Z. Phys. **56**, 1 (1929)
6. W. Heisenberg, W. Pauli, Zur Quantendynamik der Wellenfelder II. Z. Phys. **59**, 168 (1930)
7. F.J. Dyson, The S-matrix in quantum electrodynamics. Phys. Rev. **75**, 1736 (1949)
8. J. Schwinger, On the Euclidean structure of relativistic field theory. Proc. Natl. Acad. Sci. USA **44**, 956 (1958)
9. K. Symanzik, Euclidean quantum field theory, I. Equations for a scalar model. J. Math. Phys. **7**, 510 (1966)
10. C.N. Yang, R.L. Mills, Conservation of isotopic spin and isotopic gauge invariance. Phys. Rev. **96**, 191 (1954)
11. S.L. Glashow, Partial-symmetries of weak interaction. Nucl. Phys. **22**, 579 (1961)
12. S. Weinberg, A model of leptons. Phys. Rev. Lett. **19**, 1264 (1964)
13. A. Salam, Weak and electromagnetic interactions, in *Elementary Particle Theory* (Almquist and Wiksell, Stockholm, 1968)
14. G. 't Hooft, Renormalizable Lagrangians for massive Yang-Mills fields. Nucl. Phys. **B35**, 167 (1971)

15. H. Fritzsch, M. Gell-Mann, H. Leutwyler, Advantages of the color octet gluon picture. Phys. Lett. **B47**, 365 (1973)
16. R. Feynman, Spacetime approach to non-relativistic quantum mechanic. Rev. Mod. Phys. **20**, 267 (1948)
17. F.J. Wegner, Duality in generalized Ising models and phase transitions without local order parameters. J. Math. Phys. **10**, 2259 (1971)
18. K.G. Wilson, Confinement of quarks. Phys. Rev. **D10**, 2445 (1974)
19. M. Creutz, Confinement and the critical dimensionality of spacetime. Phys. Rev. Lett. **43**, 553 (1979)
20. M. Creutz, Monte Carlo simulations in lattice gauge theories. Phys. Rep. **95**, 201 (1983)
21. S. Weinberg, *The Quantum Theory of Fields, Volume 1: Foundations* (Cambridge University Press, Cambridge, 2005)
22. M. Maggiore, *A Modern Introduction to Quantum Field Theory* (Oxford University Press, Oxford, 2005)
23. G. Münster, *Von der Quantenfeldtheorie zum Standardmodell: Eine Einführung in die Teilchenphysik* (De Gruyter, Berlin, 2019)
24. J. Zinn-Justin, *Quantum Field Theory and Critical Phenomena*, 5th edn. (Oxford University Press, Oxford, 2021)
25. R. Shankar, *Quantum Field Theory and Condensed Matter: An Introduction* (Cambridge University Press, Cambridge, 2017)

Path Integrals in Quantum and Statistical Mechanics

There exist three apparently different formulations of quantum mechanics: Heisenberg's matrix mechanics, Schrödinger's wave mechanics, and Feynman's path integral approach. In contrast to matrix and wave mechanics, which are based on the Hamiltonian approach, the latter is based on the Lagrangian approach.

2.1 Summing Over All Paths

Already back in 1933, Dirac asked himself whether the classical Lagrangian and action are as significant in quantum mechanics as they are in classical mechanics [1,2]. He observed that for simple systems, the probability amplitude

$$K(t, q', q) = \langle q' | e^{-i\hat{H}t/\hbar} | q \rangle \tag{2.1}$$

for the propagation from a point with coordinate q to another point with coordinate q' in time t is given by

$$K(t, q', q) \propto e^{iS[q_{cl}]/\hbar}, \tag{2.2}$$

where q_{cl} denotes the classical trajectory from q to q'. In the exponent the action of this trajectory enters as a multiple of Planck's reduced constant $\hbar$. For a free particle with Lagrangian

$$L_0 = \frac{m}{2}\dot{q}^2 \tag{2.3}$$

© The Author(s), under exclusive license to Springer Nature Switzerland AG 2021
A. Wipf, *Statistical Approach to Quantum Field Theory*, Lecture Notes
in Physics 992, https://doi.org/10.1007/978-3-030-83263-6_2

the formula (2.2) is verified easily: A free particle moves with constant velocity $(q' - q)/t$ from q to q' and the action of the classical trajectory is

$$S[q_{cl}] = \int_0^t ds \, L_0 \, [q_{cl}(s)] = \frac{m}{2t} (q' - q)^2 \, .$$

The factor of proportionality in (2.2) is then uniquely fixed by the condition $e^{-i\hat{H}t/\hbar} \longrightarrow \mathbb{1}$ for $t \to 0$ which in position space reads

$$\lim_{t \to 0} K(t, q', q) = \delta(q', q) \, . \tag{2.4}$$

Alternatively, it is fixed by the property $e^{-i\hat{H}t/\hbar} e^{-i\hat{H}s/\hbar} = e^{-i\hat{H}(t+s)/\hbar}$ that takes the form

$$\int du \, K(t, q', u) K(s, u, q) = K(t + s, q', q) \tag{2.5}$$

in position space. Thus, the correct free particle propagator on a line is given by

$$K_0(t, q', q) = \left(\frac{m}{2\pi i \hbar t} \right)^{1/2} e^{im(q'-q)^2/2\hbar t} \, . \tag{2.6}$$

Similar results hold for the harmonic oscillator or systems for which $\langle \hat{q}(t) \rangle$ fulfills the classical equation of motion. For such systems $\langle V'(\hat{q}) \rangle = V'(\langle \hat{q} \rangle)$ holds true. However, for general systems, the simple formula (2.2) must be extended, and it was Feynman who discovered this extension back in 1948. He realized that *all paths* from q to q' (and not only the classical path) contribute to the propagator. This means that in quantum mechanics a particle can potentially move on any path $q(s)$ from the initial to the final destination,

$$q(0) = q \quad \text{and} \quad q(t) = q' \, . \tag{2.7}$$

The probability amplitude emerges as the superposition of contributions from all trajectories,

$$K(t, q', q) \sim \sum_{\text{all paths}} e^{iS[\text{path}]/\hbar} \, , \tag{2.8}$$

where a single path contributes a term $\sim \exp\left(iS[\text{path}]/\hbar\right)$.

In passing we note that already in 1923, Wiener introduced the sum over all paths in his studies of stochastic processes [3]. Thereby a single path was weighted with a real and positive probability and *not* with a complex amplitude as in (2.8). Wiener's path integral corresponds to Feynman's path integral for imaginary time and describes quantum systems in thermal equilibrium with a heat bath at fixed temperature. In this book we will explain this extraordinary result and apply it to interesting physical systems. Moreover, the path integral method allows for a uniform treatment of quantum mechanics, quantum field theory, and statistical mechanics and can be regarded as a basic tool in modern theoretical physics. It represents an alternative approach to the canonical quantization of classical systems and earned its first success in the 1950s. The path integral method is very beautifully and intelligibly presented in Feynman's original work [4] as well as in his book with Hibbs [5]. The latter reference contains many applications and is still recognized as a standard reference. Functional integrals have been developed further by outstanding mathematicians and physicists, especially by Kac [6]. An adequate reference for these developments is contained in the review article by Gelfand and Yaglom [7]. In the present chapter, we can only give a short *introduction* to path integrals. For a deeper understanding, the reader should consult more specialized books and review articles. Some of them are listed in the bibliography at the end of this chapter [8–15].

2.2 Recalling Quantum Mechanics

There are two well-established ways to quantize a classical system: *canonical quantization* and *path integral quantization*. For completeness and later use, we recall the main steps of canonical quantization both in Schrödinger's wave mechanics and Heisenberg's matrix mechanics.

A classical system is described by its coordinates $\{q^i\}$ and momenta $\{p_i\}$ on *phase space* Γ. An observable O is a real-valued function on Γ. Examples are the coordinates on phase space and the energy $H(q, p)$. We assume that phase space comes along with a symplectic structure and has local coordinates with Poisson brackets

$$\{q^i, p_j\} = \delta^i_j \ .\tag{2.9}$$

The brackets are extended to observables on through antisymmetry and the derivation rule $\{OP, Q\} = O\{P, Q\} + \{O, Q\}P$. The evolution in time of an observable is determined by

$$\dot{O} = \{O, H\}, \quad \text{e.g.} \quad \dot{q}^i = \{q^i, H\} \quad \text{and} \quad \dot{p}_i = \{p_i, H\} \ .\tag{2.10}$$

In the canonical quantization, functions on phase space are mapped to operators, and the Poisson brackets of two functions become commutators of the associated operators:

$$O(q, p) \to \hat{O}(\hat{q}, \hat{p}) \quad \text{and} \quad \{O, P\} \longrightarrow \frac{1}{i\hbar}[\hat{O}, \hat{P}] \ .\tag{2.11}$$

The time evolution of an (not explicitly time-dependent) observable is determined by *Heisenberg's equation*

$$\frac{\mathrm{d}\hat{O}}{\mathrm{d}t} = \frac{\mathrm{i}}{\hbar}[\hat{H}, \hat{O}] . \tag{2.12}$$

In particular the phase space coordinates (q^i, p_i) become operators with commutation relations $[\hat{q}^i, \hat{p}_j] = \mathrm{i}\hbar\delta^i{}_j$, and their time evolution is determined by

$$\frac{\mathrm{d}\hat{q}^i}{\mathrm{d}t} = \frac{\mathrm{i}}{\hbar}[\hat{H}, \hat{q}^i] \quad \text{and} \quad \frac{\mathrm{d}\hat{p}_i}{\mathrm{d}t} = \frac{\mathrm{i}}{\hbar}[\hat{H}, \hat{p}_i] .$$

For a system of non-relativistic and spinless particles, the Hamiltonian reads

$$\hat{H} = \hat{H}_0 + \hat{V} \quad \text{with} \quad \hat{H}_0 = \frac{1}{2m} \sum \hat{p}_i^2 , \tag{2.13}$$

and one arrives at Heisenberg's equations of motion

$$\frac{\mathrm{d}\hat{q}^i}{\mathrm{d}t} = \frac{\hat{p}_i}{m} \quad \text{and} \quad \frac{\mathrm{d}\hat{p}_i}{\mathrm{d}t} = -\hat{V}_{,i} . \tag{2.14}$$

Observables are represented by Hermitian operators on a Hilbert space $\mathscr{H}$, whose elements characterize the states of the system:

$$\hat{O}(\hat{q}, \hat{p}) : \mathcal{H} \longrightarrow \mathcal{H} . \tag{2.15}$$

Consider a particle confined to an endless wire. Its Hilbert space is $\mathcal{H} = L_2(\mathbb{R})$, and its position and momentum operator are represented in position space as

$$(\hat{q}\psi)(q) = q\psi(q) \quad \text{and} \quad (\hat{p}\psi)(q) = \frac{\hbar}{\mathrm{i}}\partial_q\psi(q) . \tag{2.16}$$

In experiments we can measure matrix elements of observables, represented by Hermitian operators, and in particular their expectation values in a state of the system. The time dependence of an expectation value $\langle\psi|\hat{O}(t)|\psi\rangle$ is determined by the Heisenberg equation (2.12).

The transition from the *Heisenberg picture* to the *Schrödinger picture* involves a time-dependent similarity transformation,

$$\hat{O}_\mathrm{s} = \mathrm{e}^{-\mathrm{i}t\hat{H}/\hbar}\,\hat{O}\,\mathrm{e}^{\mathrm{i}t\hat{H}/\hbar} \quad \text{and} \quad |\psi_\mathrm{s}\rangle = \mathrm{e}^{-\mathrm{i}t\hat{H}/\hbar}|\psi\rangle , \tag{2.17}$$

and leads to time-independent observables in the Schrödinger picture,

$$\frac{\mathrm{d}}{\mathrm{d}t}\hat{O}_s = \mathrm{e}^{-it\hat{H}/\hbar}\left(-\frac{i}{\hbar}[\hat{H},\,\hat{O}]+\frac{\mathrm{d}}{\mathrm{d}t}\hat{O}\right)\mathrm{e}^{it\hat{H}/\hbar} = 0\,.$$

Note that the Hamiltonian operator is the same in both pictures, $\hat{H}_s = \hat{H}$, and that all expectation values are left invariant by the similarity transformation,

$$\langle\psi|\hat{O}(t)|\psi\rangle = \langle\psi_s(t)|\hat{O}_s|\psi_s(t)\rangle\,. \tag{2.18}$$

A state vector in the Schrödinger picture $|\psi_s(t)\rangle$ fulfills the *Schrödinger equation*

$$i\hbar\frac{\mathrm{d}}{\mathrm{d}t}|\psi_s\rangle = \hat{H}|\psi_s\rangle \iff |\psi_s(t)\rangle = \mathrm{e}^{-it\hat{H}/\hbar}|\psi_s(0)\rangle\,. \tag{2.19}$$

In position space this formal solution of the evolution equation has the form

$$\psi_s(t,q') \equiv \langle q'|\psi_s(t)\rangle = \int \langle q'|\mathrm{e}^{-it\hat{H}/\hbar}|q\rangle\langle q|\psi_s(0)\rangle\mathrm{d}q$$

$$\equiv \int K(t,q',q)\psi_s(0,q)\mathrm{d}q\,, \tag{2.20}$$

where we inserted the resolution of the identity with $\hat{q}$-eigenstates,

$$\int \mathrm{d}q\ |q\rangle\langle q| = \mathbb{1}\,, \tag{2.21}$$

and introduced the kernel of the unitary time evolution operator

$$K(t,q',q) = \langle q'|\hat{K}(t)|q\rangle, \qquad \hat{K}(t) = \mathrm{e}^{-it\hat{H}/\hbar}\,. \tag{2.22}$$

The *propagator* $K(t,q',q)$ is interpreted as the probability amplitude for the propagation from q at time 0 to q' at time t. This is emphasized by the notation

$$K(t,q',q) \equiv \langle q',t|q,0\rangle\,. \tag{2.23}$$

The amplitude solves the time-dependent Schrödinger equation

$$i\hbar\frac{\mathrm{d}}{\mathrm{d}t}K(t,q',q) = \hat{H}K(t,q',q)\,, \tag{2.24}$$

where $\hat{H}$ acts on q' and fulfills the initial condition

$$\lim_{t\to 0} K(t,q',q) = \delta(q'-q)\,. \tag{2.25}$$

The conditions (2.24) and (2.25) uniquely define the propagator. In particular for a non-relativistic free particle with Hamiltonian $\hat{H}_0 = \hat{p}^2/2m$ in d dimensions, the solution reads

$$K_0(t, q', q) = \langle q'|e^{-it\hat{H}_0/\hbar}|q\rangle = \left(\frac{m}{2\pi i\hbar t}\right)^{d/2} e^{im(q'-q)^2/2\hbar t}, \quad q, q' \in \mathbb{R}^d .$$

$$(2.26)$$

In one dimension we recover the result (2.6). After this preliminaries we now turn to the path integral representation of the propagator.

2.3   Feynman–Kac Formula

We shall derive Feynman's path integral representation for the unitary time evolution operator $\exp(-i\hat{H}t)$ as well as Kac's path integral representation for the positive operator $\exp(-\hat{H}\tau)$. Thereby we shall utilize the product formula of Trotter. In case of matrices, this formula was already verified by Lie and has the form:

Theorem 2.1 (Lie's Theorem)   *For two matrices* A *and* B

$$e^{\mathsf{A}+\mathsf{B}} = \lim_{n\to\infty} \left(e^{\mathsf{A}/n}e^{\mathsf{B}/n}\right)^n .$$

To prove this theorem, we define for each n the two matrices $\mathsf{S}_n := \exp(\mathsf{A}/n+\mathsf{B}/n)$ and $\mathsf{T}_n := \exp(\mathsf{A}/n)\exp(\mathsf{B}/n)$ and telescope the difference of their n'th powers,

$$\mathsf{S}_n^n - \mathsf{T}_n^n = \mathsf{S}_n^{n-1}(\mathsf{S}_n - \mathsf{T}_n) + \mathsf{S}_n^{n-2}(\mathsf{S}_n - \mathsf{T}_n)\mathsf{T}_n + \cdots + (\mathsf{S}_n - \mathsf{T}_n)\mathsf{T}_n^{n-1} .$$

Now we choose any (sub-multiplicative) matrix norm, for example, the Frobenius norm. The triangle inequality together with $\|XY\| \leq \|X\|\,\|Y\|$ imply the inequality $\|\exp(X)\| \leq \exp(\|X\|)$ such that

$$\|\mathsf{S}_n\|, \|\mathsf{T}_n\| \leq a^{1/n} \quad \text{with} \quad a = e^{\|\mathsf{A}\|+\|\mathsf{B}\|} .$$

Thus we conclude

$$\left\|\mathsf{S}_n^n - \mathsf{T}_n^n\right\| \equiv \left\|e^{\mathsf{A}+\mathsf{B}} - \left(e^{\mathsf{A}/n}e^{\mathsf{B}/n}\right)^n\right\| \leq n \times a^{(n-1)/n} \left\|\mathsf{S}_n - \mathsf{T}_n\right\| .$$

Finally, using $\mathsf{S}_n - \mathsf{T}_n = -[\mathsf{A}, \mathsf{B}]/2n^2 + O(1/n^3)$, the product formula is verified for matrices. But the theorem also holds for self-adjoint operators.

Theorem 2.2 (Trotter's Theorem) *If $\hat{A}$ and $\hat{B}$ are self-adjoint operators and $\hat{A} + \hat{B}$ is essentially self-adjoint on the intersection $\mathscr{D}$ of their domains, then*

$$e^{-it(\hat{A}+\hat{B})} = \text{s-}\lim_{n\to\infty} \left(e^{-it\hat{A}/n} e^{-it\hat{B}/n} \right)^n . \tag{2.27}$$

If in addition $\hat{A}$ and $\hat{B}$ are bounded from below, then

$$e^{-\tau(\hat{A}+\hat{B})} = \text{s-}\lim_{n\to\infty} \left(e^{-\tau\hat{A}/n} e^{-\tau\hat{B}/n} \right)^n . \tag{2.28}$$

The operators need not be bounded and the convergence is with respect to the strong operator topology. For operators $\hat{A}_n$ and $\hat{A}$ on a common domain $\mathscr{D}$ in the Hilbert space, we have s-$\lim_{n\to\infty} \hat{A}_n = \hat{A}$ iff $\|\hat{A}_n\psi - \hat{A}\psi\| \to 0$ for all $\psi \in \mathscr{D}$. Formula (2.27) is used in quantum mechanics, and formula (2.28) finds its application in statistical physics and the Euclidean formulation of quantum mechanics [16].

Let us assume that $\hat{H}$ can be written as $\hat{H} = \hat{H}_0 + \hat{V}$ and apply the product formula to the evolution kernel in (2.22). With $\varepsilon = t/n$ and $\hbar = 1$, we obtain

$$K(t, q', q) = \lim_{n\to\infty} \langle q' | \left(e^{-i\varepsilon\hat{H}_0} e^{-i\varepsilon\hat{V}} \right)^n | q \rangle$$

$$= \lim_{n\to\infty} \int dq_1 \cdots dq_{n-1} \prod_{j=0}^{j=n-1} \langle q_{j+1} | e^{-i\varepsilon\hat{H}_0} e^{-i\varepsilon\hat{V}} | q_j \rangle , \tag{2.29}$$

where we repeatedly inserted the resolution of the identity (2.21) and denoted the initial and final point by $q_0 = q$ and $q_n = q'$, respectively. The potential $\hat{V}$ is diagonal in position space such that

$$\langle q_{j+1} | e^{-i\varepsilon\hat{H}_0} e^{-i\varepsilon\hat{V}} | q_j \rangle = \langle q_{j+1} | e^{-i\varepsilon\hat{H}_0} | q_j \rangle e^{-i\varepsilon V(q_j)} . \tag{2.30}$$

Here we insert the result (2.26) for the propagator of the free particle with Hamiltonian $\hat{H}_0$ and obtain

$$K(t, q', q) = \lim_{n\to\infty} \int dq_1 \cdots dq_{n-1} \left(\frac{m}{2\pi i\varepsilon} \right)^{n/2}$$

$$\times \exp \left\{ i\varepsilon \sum_{j=0}^{j=n-1} \left(\frac{m}{2} \left(\frac{q_{j+1} - q_j}{\varepsilon} \right)^2 - V(q_j) \right) \right\} . \tag{2.31}$$

This is the celebrated *Feynman-Kac formula* which provides the path integral representation for the propagator. The reader who wants to know more details about this formula and its applications in statistical physics, biology, and advanced engineering sciences may consult the text [17].

Fig. 2.1 Broken-line path entering the discretized path integral (2.31)

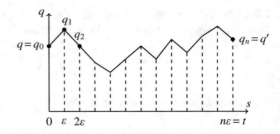

To make clear why it is called path integral, we divide the time interval $[0, t]$ into n subintervals of equal length $\varepsilon = t/n$ and identify q_k with $q(s = k\varepsilon)$. Now we connect the points

$$(0, q_0), \ (\varepsilon, q_1), \ \ldots, \ (t - \varepsilon, q_{n-1}), \ (t, q_n)$$

by straight line segments, which give rise to a broken-line path as depicted in Fig. 2.1. The exponent in (2.31) is just the Riemann integral for the action of a particle moving along the broken-line path,

$$\sum_{j=0}^{j=n-1} \varepsilon \left\{ \frac{m}{2} \left(\frac{q_{j+1} - q_j}{\varepsilon} \right)^2 - V(q_j) \right\} = \int_0^t ds \left\{ \frac{m}{2} \left(\frac{dq}{ds} \right)^2 - V(q(s)) \right\} . \tag{2.32}$$

The integral $\int dq_1 \ldots dq_{n-1}$ represents the sum over all broken-line paths from q to q'. Every continuous path can be approximated by a broken-line path if only ε is small enough. Next we perform the so-called continuum limit $\varepsilon \to 0$ or equivalently $n \to \infty$. In this limit the finite-dimensional integral (2.31) turns into an infinite-dimensional (formal) integral over all paths from q to q'. With the definition

$$\left(\frac{m}{2\pi i \varepsilon} \right)^{n/2} =: C_n \tag{2.33}$$

we arrive at the formal result

$$K(t, q', q) = C_\infty \int_{q(0)=q}^{q(t)=q'} \mathscr{D}q \, e^{iS[q]/\hbar} . \tag{2.34}$$

The "measure" $\mathscr{D}q$ is defined via the limiting process $n \to \infty$ in (2.31). Since the infinite product of Lebesgue measures does not exist, $\mathscr{D}$ has no precise mathematical meaning. Only after a continuation to imaginary time a measure on all paths can be rigorously defined.

The formal result (2.34) holds true for more general systems, for example, interacting particles moving in more than one dimension and in the presence of external fields. It also applies to mechanical systems with generalized coordinates $q^1, \ldots, q^N$ and to quantum field theories where one integrates over all fields instead of all paths. Further properties of the path integral as well as many examples and applications can be found in the references given at the end of this chapter.

2.4 Euclidean Path Integral

The oscillating integrand $\exp(iS/\hbar)$ entering the path integral (2.34) leads to distributions. If only we could suppress these oscillations, then it may be possible to construct a well-defined path integral. This may explain why most rigorous works on path integrals are based on imaginary time: for imaginary time it is indeed possible to construct a measure on all paths—the Wiener measure. The continuation from real to imaginary time is achieved by a *Wick rotation* and the continuation from imaginary time back to real time by an inverse Wick rotation. In practice, one replaces t by $-i\tau$ in the path integral (2.34), works with the resulting Euclidean path integral, and replaces τ by it in the final expressions.

2.4.1 Quantum Mechanics in Imaginary Time

The unitary time evolution operator has the spectral representation

$$\hat{K}(t) = e^{-i\hat{H}t} = \int e^{-iEt} d\hat{P}_E , \qquad (2.35)$$

where $\hat{P}_E$ is the spectral family of the Hamiltonian. If $\hat{H}$ has discrete spectrum, then $\hat{P}_E$ is the orthogonal projector onto the subspace of $\mathscr{H}$ spanned by all eigenfunctions with energies less than E. In the following we assume that the Hamiltonian operator is bounded from below. Then we can subtract its ground state energy to obtain a non-negative $\hat{H}$ for which the integration limits in (2.35) are 0 and ∞. With the substitution $t \to t - i\tau$, we obtain

$$e^{-(\tau+it)\hat{H}} = \int_0^\infty e^{-E(\tau+it)} d\hat{P}_E . \qquad (2.36)$$

This defines a holomorphic semigroup in the lower complex half-plane

$$\{z = t - i\tau \in \mathbb{C}, \ \tau \geq 0\} . \qquad (2.37)$$

If the operator (2.36) is known on the negative imaginary axis ($t = 0, \tau \geq 0$), one can perform an analytic continuation to the real axis ($t, \tau = 0$). The analytic continuation to complex time $t \to -i\tau$ corresponds to a transition from the Minkowski metric $ds^2 = dt^2 - dx^2 - dy^2 - dz^2$ to a metric with Euclidean signature. Hence a theory with imaginary time is called *Euclidean theory*.

The time evolution operator $\hat{K}(t)$ exists for real time and defines a *one-parametric unitary group*. It fulfills the Schrödinger equation

$$i\frac{d}{dt}\hat{K}(t) = \hat{H}\hat{K}(t)$$

with a complex and oscillating kernel $K(t,q',q) = \langle q'|\hat{K}(t)|q\rangle$. For imaginary time we have a Hermitian (and not unitary) evolution operator

$$\hat{K}(\tau) = e^{-\tau\hat{H}} \tag{2.38}$$

with positive spectrum. The $\hat{K}(\tau)$ exist for positive τ and form a *semi-group* only. For almost all initial data, evolution back into the "imaginary past" is impossible.

The evolution operator for imaginary time satisfies the *heat equation*

$$\frac{d}{d\tau}\hat{K}(\tau) = -\hat{H}\hat{K}(\tau)\,, \tag{2.39}$$

instead of the Schrödinger equation and has kernel

$$K(\tau,q',q) = \langle q'|e^{-\tau\hat{H}}|q\rangle, \quad K(0,q',q) = \delta(q',q)\,. \tag{2.40}$$

This kernel is real[1] for a real Hamiltonian. Furthermore it is strictly positive:

Theorem 2.3 *Let the potential V be continuous and bounded from below and $\hat{H} = -\Delta + \hat{V}$ be an essentially self-adjoint operator. Then*

$$\langle q'|e^{-\tau\hat{H}}|q\rangle > 0\,. \tag{2.41}$$

The reader may consult the textbook [18] for a proof of this theorem. As examples we consider the kernel of the *free particle* with mass m,

$$K_0(\tau,q',q) = \left(\frac{m}{2\pi\tau}\right)^{d/2} e^{-m(q'-q)^2/2\tau}\,, \tag{2.42}$$

and of the *harmonic oscillator* with frequency ω,

$$K_\omega(\tau,q',q) = \left(\frac{m\omega/(2\pi)}{\sinh\omega\tau}\right)^{d/2} \exp\left\{-\frac{m\omega}{2}\left((q'^2 + q^2)\coth\omega\tau - \frac{2q'q}{\sinh\omega\tau}\right)\right\}\,, \tag{2.43}$$

---

[1] If we couple the system to a magnetic field, $\hat{H}$ and $\hat{K}(\tau)$ become complex quantities.

both for imaginary time and in d dimensions. Both kernels are strictly positive. This positivity is essential for the far-reaching relation of Euclidean quantum theory and probability theory: The quantity

$$P_\tau(q) \equiv C\, K(\tau, q, 0) \tag{2.44}$$

can be interpreted as probability for the transition from point 0 to point q during the time interval τ.[2] The probability of ending somewhere should be 1,

$$\int dq\, P_\tau(q) = 1 , \tag{2.45}$$

and this requirement determines the constant C. For a free particle, we obtain

$$P_\tau(q) = \left(\frac{m}{2\pi\tau}\right)^{d/2} e^{-mq^2/2\tau} .$$

It represents the probability density for *Brownian motion* with diffusion coefficient inversely proportional to the mass, $D = 1/2m$.

In quantum field theory, vacuum expectation values of products of field operators at different spacetime points encode all information about the theory. They determine scattering amplitudes and spectral properties of the particles and hence play a distinguished role. In quantum mechanics these expectation values are given by

$$W^{(n)}(t_1, \ldots, t_n) = \langle 0| \hat{q}(t_1) \cdots \hat{q}(t_n) |0\rangle , \quad \hat{q}(t) = e^{it\hat{H}} \hat{q}\, e^{-it\hat{H}} . \tag{2.46}$$

These *Wightman functions* are not symmetric in their arguments $t_1, \ldots, t_n$ since the position operators at different times do not commute. Again we normalize the Hamiltonian such that the energy of the ground state $|0\rangle$ vanishes and perform an analytic continuation of the Wightman functions to complex times $z_i = t_i - i\tau_i$:

$$W^{(n)}(z_1, \ldots, z_n) = \langle 0| \hat{q}\, e^{-i(z_1-z_2)\hat{H}} \hat{q}\, e^{-i(z_2-z_3)\hat{H}} \hat{q} \cdots \hat{q}\, e^{-i(z_{n-1}-z_n)\hat{H}} \hat{q} |0\rangle . \tag{2.47}$$

We used that $\hat{H}$ annihilates the ground state or that $\exp(i\zeta\hat{H}) |0\rangle = |0\rangle$. The functions $W^{(n)}$ are well defined if the imaginary parts of their arguments z_k are ordered according to

$$im(z_k - z_{k+1}) \leq 0 .$$

---

[2] To keep the notation simple, we use q as the final point.

With $z_i = t_i - i\tau_i$, one ends up with analytic functions $W^{(n)}$ in the region

$$\tau_1 > \tau_2 \cdots > \tau_n \ . \tag{2.48}$$

The Wightman distributions for real time represent boundary values of the analytic Wightman functions with complex arguments:

$$W^{(n)}(t_1, \ldots, t_n) = \lim_{\substack{im z_i \to 0 \\ im(z_{k+1} - z_k) > 0}} W^{(n)}(z_1, \ldots, z_n) \ . \tag{2.49}$$

On the other hand, if the arguments are purely imaginary, then we obtain the *Schwinger functions*. For $\tau_1 > \tau_2 > \cdots > \tau_n$, they are given by

$$\begin{aligned}
S^{(n)}(\tau_1, \ldots, \tau_n) &= W^{(n)}(-i\tau_1, \ldots, -i\tau_n) \\
&= \langle 0| \hat{q} \, e^{-(\tau_1 - \tau_2)\hat{H}} \hat{q} e^{-(\tau_2 - \tau_3)\hat{H}} \hat{q} \cdots \hat{q} \, e^{-(\tau_{n-1} - \tau_n)\hat{H}} \hat{q} \, |0\rangle \ .
\end{aligned} \tag{2.50}$$

As an example we consider the harmonic oscillator with Hamiltonian

$$\hat{H} = \omega \hat{a}^\dagger \hat{a}$$

expressed in terms of the step operators $\hat{a}$, $\hat{a}^\dagger$ which obey the commutation relation $[\hat{a}, \hat{a}^\dagger] = 1$. The ground state $|0\rangle$ is annihilated by $\hat{a}$ and hence has zero energy. The first excited state $|1\rangle = \hat{a}^\dagger |0\rangle$ has energy ω. The two-point Wightman function depends on the time difference only,

$$\begin{aligned}
W^{(2)}(t_1 - t_2) &= \langle 0| \hat{q}(t_1)\hat{q}(t_2) |0\rangle = \frac{1}{2m\omega} \langle 0| (\hat{a} + \hat{a}^\dagger)e^{-i(t_1 - t_2)\hat{H}}(\hat{a} + \hat{a}^\dagger) |0\rangle \\
&= \frac{1}{2m\omega} \langle 1| e^{-i(t_1 - t_2)\omega \hat{a}^\dagger \hat{a}} |1\rangle = \frac{e^{-i\omega(t_1 - t_2)}}{2m\omega} \ .
\end{aligned}$$

The corresponding Schwinger function is given by

$$S^{(2)}(\tau_1 - \tau_2) = \frac{e^{-\omega(\tau_1 - \tau_2)}}{2m\omega} \quad (\tau_1 > \tau_2) \ . \tag{2.51}$$

In a relativistic quantum field theory, the Schwinger functions $S^{(n)}(x_1, \ldots, x_n)$ are invariant under Euclidean Lorentz transformation from the group SO(4). This invariance together with locality imply that the $S^{(n)}$ are symmetric functions of their arguments $x_i \in \mathbb{R}^4$. This is not necessarily true for the Schwinger functions in quantum mechanics.

2.4.2 Imaginary Time Path Integral

To formulate the path integral for imaginary time, we employ the product formula (2.28), which follows from the product formula (2.27) through the substitution of it by τ. For such systems the analog of (2.31) for Euclidean time τ is obtained by the substitution of $i\varepsilon$ by ε. Thus we find

$$
K(\tau, q', q) = \langle \hat{q}' | e^{-\tau \hat{H}/\hbar} | \hat{q} \rangle
$$

$$
= \lim_{n \to \infty} \int dq_1 \cdots dq_{n-1} \left(\frac{m}{2\pi \hbar \varepsilon} \right)^{n/2} e^{-S_E(q_0, q_1, \ldots, q_n)/\hbar} ,
$$

$$
S_E(\ldots) = \varepsilon \sum_{j=0}^{n-1} \left\{ \frac{m}{2} \left(\frac{q_{j+1} - q_j}{\varepsilon} \right)^2 + V(q_j) \right\} , \tag{2.52}
$$

where $q_0 = q$ and $q_n = q'$. The multidimensional integral represents the sum over all broken-line paths from q to q'. Interpreting S_E as Hamiltonian of a classical lattice model and $\hbar$ as temperature, it is (up to the fixed endpoints) the partition function of a one-dimensional lattice model on a lattice with $n + 1$ sites. The real-valued variable q_j defined on site j enters the action S_E which contains interactions between the variables q_j and q_{j+1} at neighboring sites. The values of the lattice field

$$
\{0, 1, \ldots, n-1, n\} \to \{q_0, q_1, \ldots, q_{n-1}, q_n\}
$$

are prescribed at the endpoints $q_0 = q$ and $q_n = q'$. Note that the classical limit $\hbar \to 0$ corresponds to the low-temperature limit of the lattice system.

The multidimensional integral (2.52) corresponds to the summation over all path on the time lattice. What happens to the finite-dimensional integral when we take the *continuum limit* $n \to \infty$? Then we obtain the Euclidean path integral representation for the positive kernel

$$
K(\tau, q', q) = \langle q' | e^{-\tau \hat{H}/\hbar} | q \rangle = C \int_{q(0)=q}^{q(\tau)=q'} \mathcal{D}q \; e^{-S_E[q]/\hbar} . \tag{2.53}
$$

The integrand contains the Euclidean action

$$
S_E[q] = \int_0^\tau d\sigma \left\{ \frac{m}{2} \dot{q}^2 + V(q(\sigma)) \right\} \tag{2.54}
$$

which for many physical systems is bounded from below.

2.5   Path Integral in Quantum Statistics

The Euclidean path integral formulation immediately leads to an interesting connection between quantum statistical mechanics and classical statistical physics. Indeed, if we set $\tau/\hbar \equiv \beta$ and integrate over $q = q'$ in (2.53), then we end up with the path integral representation for the canonical *partition function* of a quantum system with Hamiltonian $\hat{H}$ at inverse temperature $\beta = 1/k_B T$. More precisely, setting $q = q'$ and $\tau = \hbar\beta$ in the left-hand side of this formula, then the integral over q yields the trace of $\exp(-\beta\hat{H})$, which is just the canonical *partition function*,

$$\int dq \, K(\hbar\beta, q, q) = \mathrm{tr}\, e^{-\beta\hat{H}} = Z(\beta) = \sum e^{-\beta E_n} \quad \text{with} \quad \beta = \frac{1}{k_B T} \, .$$

(2.55)

Setting $q = q'$ in the Euclidean path integral in (2.53) means that we integrate over paths beginning and ending at q during the imaginary time interval $[0, \hbar\beta]$. The final integral over q leads to the path integral over *all periodic paths* with period $\hbar\beta$,

$$Z(\beta) = C \oint \mathscr{D}q \, e^{-S_E[q]/\hbar}, \quad q(\hbar\beta) = q(0) \, .$$

(2.56)

For example, the kernel of the harmonic oscillator in (2.43) on the diagonal is

$$K_\omega(\beta, q, q) = \sqrt{\frac{m\omega}{2\pi \, \sinh(\omega\beta)}} \exp\left\{-m\omega \tanh(\omega\beta/2)\, q^2\right\} \, ,$$

(2.57)

where we used units with $\hbar = 1$. The integral over q yields the partition function

$$Z(\beta) = \sqrt{\frac{m\omega}{2\pi \, \sinh(\omega\beta)}} \int dq \exp\left\{-m\omega \tanh(\omega\beta/2)\, q^2\right\}$$

$$= \frac{1}{2\sinh(\omega\beta/2)} = \frac{e^{-\omega\beta/2}}{1 - e^{-\omega\beta}} = e^{-\omega\beta/2} \sum_{n=0}^{\infty} e^{-n\omega\beta} \, ,$$

(2.58)

where we used $\sinh x = 2 \sinh x/2 \cosh x/2$. A comparison with the spectral sum over all energies in (2.55) yields the energies of the oscillator with (angular) frequency ω,

$$E_n = \omega\left(n + \frac{1}{2}\right), \qquad n = 0, 1, 2, \dots \, .$$

(2.59)

For large values of $\omega\beta$, i.e., for very low temperature, the spectral sum is dominated by the contribution of the ground state energy. Thus for cold systems, the *free energy* converges to the ground state energy

$$F(\beta) \equiv -\frac{1}{\beta} \log Z(\beta) \xrightarrow{\omega\beta \to \infty} E_0 . \qquad (2.60)$$

One often is interested in the energies and wave functions of excited states. We now discuss an elegant method to extract this information from the path integral.

2.5.1 Thermal Correlation Functions

The energies of excited states are encoded in the *thermal correlation functions*. These functions are expectation values of products of the position operator

$$\hat{q}_E(\tau) = e^{\tau \hat{H}/\hbar} \hat{q} \, e^{-\tau \hat{H}/\hbar}, \qquad \hat{q}_E(0) = \hat{q}(0) , \qquad (2.61)$$

at different imaginary times in the canonical ensemble,

$$\langle \hat{q}_E(\tau_1) \cdots \hat{q}_E(\tau_n) \rangle_\beta \equiv \frac{1}{Z(\beta)} \, \mathrm{tr} \left(e^{-\beta \hat{H}} \hat{q}_E(\tau_1) \cdots \hat{q}_E(\tau_n) \right) . \qquad (2.62)$$

The normalizing function $Z(\beta)$ is the partition function (2.56). From the thermal two-point function

$$\langle \hat{q}_E(\tau_1) \hat{q}_E(\tau_2) \rangle_\beta = \frac{1}{Z(\beta)} \mathrm{tr} \left(e^{-\beta \hat{H}} \hat{q}_E(\tau_1) \hat{q}_E(\tau_2) \right)$$

$$= \frac{1}{Z(\beta)} \mathrm{tr} \left(e^{-(\beta-\tau_1)\hat{H}} \hat{q} \, e^{-(\tau_1-\tau_2)\hat{H}} \hat{q} \, e^{-\tau_2 \hat{H}} \right) \qquad (2.63)$$

we can extract the *energy gap* between the ground state and the first excited state. For this purpose we use orthonormal energy eigenstates $|n\rangle$ to calculate the trace and in addition insert the resolution of the identity operator $\mathbb{1} = \sum |m\rangle\langle m|$. This yields

$$\langle \ldots \rangle_\beta = \frac{1}{Z(\beta)} \sum_{n,m} e^{-(\beta-\tau_1+\tau_2)E_n} e^{-(\tau_1-\tau_2)E_m} \langle n|\hat{q}|m\rangle\langle m|\hat{q}|n\rangle . \qquad (2.64)$$

Note that in the sum over n the contributions from the excited states are exponentially suppressed at low temperatures $\beta \to \infty$, implying that the thermal two-point function converges to the Schwinger function in this limit:

$$\langle \hat{q}_E(\tau_1) \hat{q}_E(\tau_2) \rangle_\beta \xrightarrow{\beta \to \infty} \sum_{m \geq 0} e^{-(\tau_1-\tau_2)(E_m-E_0)} |\langle 0|\hat{q}|m\rangle|^2 = \langle 0|\hat{q}_E(\tau_1)\hat{q}_E(\tau_2)|0\rangle .$$

$$(2.65)$$

In the first step, we used that for low temperature the partition function tends to $\exp(-\beta E_0)$. Likewise, we find for the one-point function the result

$$\lim_{\beta \to \infty} \langle \hat{q}_E(\tau) \rangle_\beta = \langle 0|\hat{q}|0 \rangle . \tag{2.66}$$

In the *connected two-point function*

$$\langle \hat{q}_E(\tau_1)\hat{q}_E(\tau_2) \rangle_{c,\beta} \equiv \langle \hat{q}_E(\tau_1)\hat{q}_E(\tau_2) \rangle_\beta - \langle \hat{q}_E(\tau_1) \rangle_\beta \langle \hat{q}_E(\tau_2) \rangle_\beta \tag{2.67}$$

the term with $m = 0$ in the sum (2.65) is absent, and this leads to an exponential decaying function for large time differences,

$$\lim_{\beta \to \infty} \langle \hat{q}_E(\tau_1)\hat{q}_E(\tau_2) \rangle_{c,\beta} = \sum_{m \geq 1} e^{-(\tau_1-\tau_2)(E_m-E_0)} \left|\langle 0|\hat{q}|m \rangle\right|^2 . \tag{2.68}$$

For large time differences $\tau_1 - \tau_2$, the term with $m = 1$ dominates the sum and

$$\langle \hat{q}_E(\tau_1)\hat{q}_E(\tau_2) \rangle_{c,\beta \to \infty} \longrightarrow e^{-(E_1-E_0)(\tau_1-\tau_2)} \left|\langle 0|\hat{q}|1 \rangle\right|^2 , \quad \tau_1 - \tau_2 \to \infty . \tag{2.69}$$

It follows that we can read off the energy gap $E_1 - E_0$ as well as the transition probability $|\langle 0|q|1 \rangle|^2$ from the asymptotics of the connected two-point function.

To arrive at the path integral representation for the thermal two-point correlation function, we consider the matrix elements

$$\left\langle q'|\hat{K}(\beta)\hat{q}_E(\tau_1)\hat{q}_E(\tau_2)|q \right\rangle, \quad \text{with} \quad \hat{q}_E(\tau) = \hat{K}(-\tau)\hat{q}\hat{K}(\tau) . \tag{2.70}$$

Here $\hat{K}(\tau) = \exp(-\tau\hat{H})$ denotes the evolution operator for imaginary time with path integral representation given in (2.53). Now we insert twice the resolution of the identity and obtain

$$\langle \ldots \rangle = \int dv du \left\langle q'|\hat{K}(\beta - \tau_1)|v \right\rangle v \left\langle v|\hat{K}(\tau_1 - \tau_2)|u \right\rangle u \left\langle u|\hat{K}(\tau_2)|q \right\rangle .$$

Inserting the path integral representations for the three propagators, we find the path integral representation: Firstly, we sum over all paths from $q \to u$ in the time interval τ_2 and multiply the result with the coordinate u at time τ_2. Next we sum over all paths $u \to v$ in the time interval $\tau_1 - \tau_2$ and multiply with the coordinate v at time τ_1. The last step includes the summation over all paths $v \to q'$ in the time interval $\beta - \tau_1$. The integration over the intermediate positions u and v means that the summation extends over *all* paths $q \to q'$ and not only over paths going through u at time τ_2 and v at time τ_1. Besides $\exp(-S_E)$, the integrand includes the multiplicative factor $vu = q(\tau_1)q(\tau_2)$. Since the entire propagation time is β, we end up with

$$\left\langle q' | e^{-\beta \hat{H}} \hat{q}_E(\tau_1) \hat{q}_E(\tau_2) | q \right\rangle = C \int_{q(0)=q}^{q(\beta)=q'} \mathscr{D}q \, e^{-S_E[q]} q(\tau_1) q(\tau_2), \quad \tau_1 > \tau_2 .$$

(2.71)

The thermal expectation value is given by the trace. Thus we set $q = q'$, integrate over q, and divide the result by the partition function $Z(\beta)$. Integrating over q is then equivalent to summing over all periodic paths with period β,

$$\left\langle \hat{q}_E(\tau_1) \hat{q}_E(\tau_2) \right\rangle_\beta = \frac{1}{Z(\beta)} \oint \mathscr{D}q \, e^{-S_E[q]} q(\tau_1) q(\tau_2)$$

(2.72)

with partition function given in (2.56). In the derivation we assumed the time order $\tau_1 > \tau_2$ when applying the Trotter formula.

The path integral representation of higher time-ordered correlation functions is obtained in a similar fashion. They are all generated by the kernel

$$Z(\beta, j, q', q) = C \int_{q(0)=q}^{q(\beta)=q'} \mathscr{D}q \, e^{-S_E[q] + \int d\tau j(\tau) q(\tau)} ,$$

(2.73)

in which one integrates over all paths from q to q', or by the partition function in presence of an external source,

$$Z(\beta, j) = \int dq \, Z(\beta, j, q, q) = C \oint_{q(0)=q(\beta)} \mathscr{D}q \, e^{-S_E[q] + \int d\tau j(\tau) q(\tau)} .$$

(2.74)

The functional (2.73) generates matrix elements as in (2.71) but with an arbitrary number of insertions of position operators and the functional $Z(\beta, j)$ all time-ordered thermal correlation functions. For example, the thermal two-point function follows by differentiating (2.74) twice:

$$\langle T \hat{q}_E(\tau_1) \hat{q}_E(\tau_2) \rangle_\beta = \frac{1}{Z(\beta, 0)} \frac{\delta^2}{\delta j(\tau_1) \delta j(\tau_2)} Z(\beta, j) \Big|_{j=0} ,$$

(2.75)

wherein T indicates the time ordering. Since the right-hand side is symmetric in its arguments τ_1, τ_2 and both sides are identical for $\tau_1 > \tau_2$, we must include the time ordering on the left-hand side. The ordering also results from a repeated calculation for $\tau_2 > \tau_1$.

The *connected correlation functions* are generated by the logarithm of the partition function, called *Schwinger functional*

$$W(\beta, j) \equiv \log Z(\beta, j) ,$$

(2.76)

by repeated differentiations with respect to the external source,

$$\langle T \hat{q}_E(\tau_1)\hat{q}_E(\tau_2)\cdots\hat{q}_E(\tau_n)\rangle_{c,\beta} = \frac{\delta^n}{\delta j(\tau_1)\cdots\delta j(\tau_n)} W(\beta, j)|_{j=0} . \qquad (2.77)$$

If we consider conservative systems and a time-independent source j, then the Schwinger functional is proportional to the free energy in the presence of the source.

2.6   The Harmonic Oscillator

We wish to study the path integral for the Euclidean oscillator with discretized time. The oscillator is one of the few systems for which the path integral can be calculated explicitly. For more such system, the reader may consult the text [19]. But the results for the oscillator are particularly instructive with regard to lattice field theories considered later in this book. So let us discretize the Euclidean time interval $[0, \tau]$ with n sampling points separated by a lattice constant $\varepsilon = \tau/n$. For the Lagrangian

$$L = \frac{m}{2}\dot{q}^2 + \mu q^2 \qquad (2.78)$$

the discretized path integral over periodic paths reads

$$Z = \int dq_1 \cdots dq_n \left(\frac{m}{2\pi\varepsilon}\right)^{n/2} \exp\left\{-\varepsilon\sum_{j=0}^{n-1}\left(\frac{m}{2}\left(\frac{q_{j+1}-q_j}{\varepsilon}\right)^2 + \mu q_j^2\right)\right\}$$

$$= \left(\frac{m}{2\pi\varepsilon}\right)^{n/2}\int dq_1 \cdots dq_n \exp\left(-\frac{1}{2}(q, Aq)\right) , \qquad (2.79)$$

where we assumed $q_0 = q_n$ and introduced the symmetric matrix

$$A = \frac{m}{\varepsilon}\begin{pmatrix} \alpha & -1 & 0 & \cdots & 0 & -1 \\ -1 & \alpha & -1 & \cdots & 0 & 0 \\ & & \ddots & & & \\ & & & \ddots & & \\ 0 & 0 & \cdots & -1 & \alpha & -1 \\ -1 & 0 & \cdots & 0 & -1 & \alpha \end{pmatrix} , \quad \alpha = 2\left(1 + \frac{\mu}{m}\varepsilon^2\right) . \qquad (2.80)$$

This is a *Toeplitz matrix* in which each descending diagonal from left to right is constant. This property results from the invariance of the action under lattice translations. For the explicit calculation of Z, we consider the *generating function*

$$
\begin{aligned}
Z[j] &= \left(\frac{m}{2\pi\varepsilon}\right)^{n/2} \int \mathrm{d}^n q \, \exp\left\{-\frac{1}{2}(q, \mathbf{A}q) + (j, q)\right\} \\
&= \frac{(m/\varepsilon)^{n/2}}{\sqrt{\det \mathbf{A}}} \exp\left\{\frac{1}{2}\left(j, \mathbf{A}^{-1}j\right)\right\} .
\end{aligned}
\tag{2.81}
$$

Here we applied the known result for Gaussian integrals. The n eigenvalues of $\mathbf{A}$ are

$$
\lambda_k = \frac{m}{\varepsilon}\left(\alpha - 2\cos\frac{2\pi}{n}k\right) = \frac{2}{\varepsilon}\left(\mu\varepsilon^2 + 2m\sin^2\frac{\pi k}{n}\right), \qquad k = 1, \ldots, n
\tag{2.82}
$$

and the corresponding orthonormal eigenvectors have the form

$$
\psi(k) = \frac{1}{\sqrt{n}}\left(z^k, z^{2k}, \ldots, z^{nk}\right)^T \quad \text{with} \quad z = \mathrm{e}^{2\pi\mathrm{i}/n} .
\tag{2.83}
$$

With the spectral resolution for the inverse matrix $\mathbf{A}^{-1} = \sum_k \lambda_k^{-1}\psi^\dagger(k)\psi(k)$, we obtain

$$
\left(\mathbf{A}^{-1}\right)_{pq} = \frac{\varepsilon}{2n}\sum_{k=1}^{n}\frac{\mathrm{e}^{2\pi\mathrm{i}k(p-q)/n}}{\mu\varepsilon^2 + 2m\sin^2\frac{\pi k}{n}} .
\tag{2.84}
$$

Note that the connected correlation function

$$
\langle q_{i_1}\cdots q_{i_m}\rangle = \frac{\partial^m}{\partial j_{i_1}\cdots\partial j_{i_m}}\log Z[j]\Big|_{j=0}
\tag{2.85}
$$

of the harmonic oscillator vanishes for $m > 2$. This means that all correlation functions are given in terms of the two-point function

$$
\langle q_i q_j\rangle_c = \langle q_i q_j\rangle = \frac{\partial^2}{\partial j_i\partial j_j}\left(j, \mathbf{A}^{-1}j\right) = \left(\mathbf{A}^{-1}\right)_{ij} .
\tag{2.86}
$$

As a consequence of time-translation invariance, the expectation value

$$
\langle q_i^2\rangle = \frac{\varepsilon}{2n}\sum_{k=1}^{n}\frac{1}{\mu\varepsilon^2 + 2m\sin^2\frac{\pi k}{n}} .
\tag{2.87}
$$

is independent of i. This and similar expectation values, together with the virial theorem, yield the ground state energies of Hamiltonians discretized on finite lattices. More details and numerical results are found in Chap. 4.

2.7   Problems

2.1 (Gaussian Integral) Show that

$$\int dz_1 d\bar{z}_1 \ldots dz_n d\bar{z}_n \, \exp\left(-\sum_{ij} \bar{z}_i A_{ij} z_j\right) = \pi^n (\det A)^{-1}$$

with A being a positive Hermitian $n \times n$ matrix and z_i complex integration variables.

2.2 (Harmonic Oscillator) In (2.43) we quoted the result for the kernel $K_\omega(\tau, q', q)$ of the d-dimensional harmonic oscillator with Hamiltonian

$$\hat{H} = \frac{1}{2m} \hat{p}^2 + \frac{m\omega^2}{2} \hat{q}^2$$

at imaginary time τ. Derive this formula.

Hint: Express the kernel in terms of the eigenfunctions of $\hat{H}$, which for $\hbar = m = \omega = 1$ are given by

$$\exp\left(-\xi^2 - \eta^2\right) \sum_{n=0}^{\infty} \frac{\alpha^n}{2^n n!} H_n(\xi) H_n(\eta) = \frac{1}{\sqrt{1 - \alpha^2}} \exp\left(\frac{-(\xi^2 + \eta^2 - 2\xi\eta\alpha)}{1 - \alpha^2}\right).$$

The functions H_n denote the Hermite polynomials.

Comment: This result also follows from the direct evaluation of the path integral.

2.3 (Free Particle on a Circle) A free particle moves on an interval and obeys periodic boundary conditions. Compute the time evolution kernel $K(t_b - t_a, q_b, q_a) = \langle q_b, t_b | q_a, t_a \rangle$. Use the familiar formula for the kernel of the free particle (2.26) and enforce the periodic boundary conditions by a suitable sum over the evolution kernel for the particle on $\mathbb{R}$.

2.4 (Connected and Unconnected Correlation Function) The unconnected thermal correlation functions are given by

$$\langle T \, \hat{q}_E(\tau_1) \ldots \hat{q}_E(\tau_n) \rangle_\beta = \frac{1}{Z(\beta)} \frac{\delta^n}{\delta j(\tau_1) \ldots \delta j(\tau_n)} Z(\beta, j) \Big|_{j=0}$$

with generating functional

$$Z(\beta, j) = \oint \mathcal{D}q \, \exp\left(-S_E[q] + \int_0^\beta j(\tau) q(\tau)\right),$$

wherein one integrates overall β-periodic paths. Assume that the Euclidean Lagrangian density

$$\mathcal{L}_E(q, \dot{q}) = \frac{1}{2}\dot{q}^2 + V(q)$$

contains an even potential, i.e., $V(-q) = V(q)$.

(a) Show that $\langle \hat{q}_E(\tau) \rangle_\beta = 0$.
(b) Express the unconnected four-point function $\langle T\hat{q}_E(\tau_1)\ldots\hat{q}_E(\tau_4)\rangle_\beta$ via connected correlation functions.

2.5 (Semi-classical Expansion of the Partition Function) In Sect. 2.5 we discussed the path integral representation of the thermal partition function, given by

$$Z(\beta) = C \int dq \int_{q(0)=q}^{q(\hbar\beta)=q} \mathcal{D}q \; e^{-S_E[q]/\hbar} \; .$$

We rescale the imaginary time and the amplitude according to

$$\tau \longrightarrow \hbar\tau \quad \text{and} \quad q(.) \longrightarrow \hbar q(.) \; .$$

After this rescaling, the "time interval" is of length β instead of $\hbar\beta$ and

$$Z(\beta) = C \int dq \int_{q(0)=q/\hbar}^{q(\beta)=q/\hbar} \mathcal{D}q \; \exp\left\{ -\int_0^\beta \left(\frac{1}{2}m\dot{q}^2 + V(\hbar q(.)) \right) d\tau \right\} \; .$$

For a moving particle, the kinetic energy dominates the potential energy for small $\hbar$. Thus we decompose each path into its constant part and the fluctuations about the constant part: $q(.) = q/\hbar + \xi(.)$. Show that

$$Z(\beta) = \frac{C}{\hbar} \int dq \int_{\xi(0)=0}^{\xi(\beta)=0} \mathcal{D}\xi \; \exp\left\{ -\int_0^\beta \left(\frac{1}{2}m\dot{\xi}^2 + V(q + \hbar\xi) \right) d\tau \right\} \; .$$

Determine the constant C by considering the limiting case $V = 0$ with the well-known result $Z(\beta, q, q) = (m/2\pi\beta\hbar^2)^{1/2}$. Then expand the integrand in powers of $\hbar$ and prove the intermediate result

$$Z = \frac{C}{\hbar} \int dq \; e^{-\beta V(q)} \int_{\xi(0)=0}^{\xi(\beta)=0} \mathcal{D}\xi \; e^{-\frac{1}{2}m\int d\tau\dot{\xi}^2}$$

$$\times \left\{ 1 - \hbar V'(q) \int \xi(\tau) - \frac{1}{2}\hbar^2 \left(V''(q) \int \xi^2(\tau) - V'^2(q) \int \xi(\tau) \int \xi(s) \right) \right.$$

$$\left. + \cdots \right\} \; .$$

Conditional expectation values as

$$\langle \xi(\tau_1)\xi(\tau_2)\rangle = \langle \xi(\tau_2)\xi(\tau_1)\rangle = C \int_{\xi(0)=0}^{\xi(\beta)=0} \mathscr{D}\xi \; e^{-\frac{1}{2}m\int d\tau\, \dot{\xi}^2} \xi(\tau_1)\xi(\tau_2)$$

are computed by differentiating the generating functional

$$C \int_{\xi(0)=0}^{\xi(\beta)=0} \mathscr{D}\xi \; e^{-\frac{1}{2}m\int d\tau\, \dot{\xi}^2 + \int d\tau\, j\xi}$$

$$= \sqrt{\frac{m}{2\pi\beta}} \exp\left(\frac{1}{m\beta} \int_0^\beta d\tau \int_0^\tau d\tau'(\beta-\tau)\tau' j(\tau) j(\tau') \right).$$

Prove this formula for the generating functional and compute the leading and sub-leading contributions in the semi-classical expansion.

2.6 (High-Temperature Expansion of the Partition Function) Analyze the temperature dependence of the partition function (set $\hbar = 1$). Repeat the calculation in Problem 2.5 but this time with the rescalings

$$\tau \longrightarrow \beta\tau \quad \text{and} \quad \xi \longrightarrow \sqrt{\beta}\xi \,,$$

and show that

$$Z(\beta) = \frac{C}{\sqrt{\beta}} \int dq \int_{\xi(0)=0}^{\xi(1)=0} \mathscr{D}\xi \exp\left\{ -\int_0^1 \left(\frac{m}{2}\dot{\xi}^2 + \beta V\left(q + \sqrt{\beta}\xi\right) \right) d\tau \right\}.$$

Expand $Z(\beta)$ in powers of the inverse temperature and use the generating functional in Problem 2.5 (with $\beta = 1$) to compute the correlation functions. The remaining integrals over correlation functions are easily calculated. Determine the contributions of order $T^{1/2}$, $T^{-1/2}$, and $T^{-3/2}$ in the high-temperature expansion of $Z(\beta)$.

References

1. P.A.M. Dirac, The Lagrangian in quantum mechanics. Phys. Z. Sowjetunion **3**, 64 (1933)
2. P.A.M. Dirac, *The Principles of Quantum Mechanics* (Oxford University Press, London, 1947)
3. N. Wiener, Differential space. J. Math. Phys. Sci. **2**, 132 (1923)
4. R. Feynman, Spacetime approach to non-relativistic quantum mechanics. Rev. Mod. Phys. **20**, 267 (1948)
5. R. Feynman, A. Hibbs, *Quantum Mechanics and Path Integrals* (Dover, New York, 2010)
6. M. Kac, Random walk and the theory of Brownian motion. Am. Math. Mon. **54**, 369 (1947)
7. I.M. Gel'fand, A.M. Yaglom, Integration in functional spaces and its applications in quantum physics. J. Math. Phys. **1**, 48 (1960)
8. L.S. Schulman, *Techniques and Applications of Path Integration* (Dover, New York, 2005)
9. E. Nelson, Feynman integrals and the Schrödinger equation. J. Math. Phys. **5**, 332 (1964)
10. S.G. Brush, Functional integrals and statistical physics. Rev. Mod. Phys. **33**, 79 (1961)

11. J. Zinn-Justin, *Path Integrals in Quantum Mechanics* (Oxford University Press, London, 2010)
12. J.R. Klauder, *A Modern Approach to Functional Integration* (Birkhäuser, Basel, 2011)
13. H. Kleinert, *Path Integral, in Quantum Mechanics, Statistics, Polymer Physics and Financial Markets* (World Scientific, Singapore, 2009)
14. U. Mosel, *Path Integrals in Field Theory: An Introduction* (Springer, Berlin, 2013)
15. W. Dittrich, M. Reuter, *Classical and Quantum Dynamics: From Classical Paths to Path Integrals*, 6th edn. (Springer, Berlin, 2020)
16. P.R. Chernoff, Note on product formulas for operator semigroups. J. Funct. Anal. **2**, 238 (1968)
17. P. Del Moral, *Feynman-Kac Formulae: Genealogical and Interacting Particle Systems with Applications* (Springer, Berlin, 2004)
18. J. Glimm, A. Jaffe, *Quantum Physics: A Functional Integral Point of View* (Springer, Berlin, 1981)
19. C. Grosche, F. Steiner, *Handbook of Feynman Path Integrals*. Springer Tracts in Modern Physics, vol. 145 (Springer, Berlin, 2013)

High-Dimensional Integrals **3**

Unfortunately, path integrals can be evaluated explicitly only for very simple systems like the free particle, harmonic oscillator, or topological field theories. More complicated systems are analyzed via perturbation theory (e.g., semi-classical expansion, perturbative expansion in powers of the interaction strength, strong-coupling expansion, high- and low-temperature expansions) or by numerical methods. In Chap. 2 we have demonstrated that path integrals for imaginary time or systems in thermal equilibrium may be approximated by finite-dimensional integrals. We discretized time as $s \in \{0, \varepsilon, \ldots, n\varepsilon = \tau\}$ and approximated the Euclidean action by a Riemann sum. The latter depends on the values

$$\boldsymbol{q} = \{q_0, q_1, \ldots, q_n\} = \{q(0), q(\varepsilon), \ldots, q(n\varepsilon)\}$$

of the path at the lattice points $s_k = k\varepsilon$. In this lattice approximation, every expectation value is given by a finite-dimensional integral,

$$\langle O \rangle = \frac{\int \mathscr{D}\boldsymbol{q}\, O(\boldsymbol{q})\, \mathrm{e}^{-S_{\mathrm{E}}(\boldsymbol{q})}}{\int \mathscr{D}\boldsymbol{q}\, \mathrm{e}^{-S_{\mathrm{E}}(\boldsymbol{q})}} \quad \text{with} \quad \int \mathscr{D}\boldsymbol{q} = \int_{-\infty}^{\infty} \prod_{1}^{n} \mathrm{d}q_j \,. \tag{3.1}$$

The function $S_{\mathrm{E}}(\boldsymbol{q}) = S_{\mathrm{E}}(q_1, \ldots, q_n)$ in the exponents is the Euclidean lattice action. For a particle on a line, it is given in Eq. (2.52). In this chapter we shall only consider Euclidean path integrals and thus skip the index E.

3.1 Numerical Algorithms

We are confronted with high-dimensional integrals in quantum statistics, solid-state physics, Euclidean quantum field theory, high-energy physics, and numerous other branches in natural sciences or even the financial market. For example, consider the expectation value of interest derivatives, which can be written as a high-dimensional

© The Author(s), under exclusive license to Springer Nature Switzerland AG 2021
A. Wipf, *Statistical Approach to Quantum Field Theory*, Lecture Notes
in Physics 992, https://doi.org/10.1007/978-3-030-83263-6_3

integral. Assuming a duration of 30×12 months and a separate interest rate each month, one is confronted with a 360-dimensional integral. Thus we are in need of efficient algorithms to compute these integrals with a controllable error. Richard Bellman coined the phrase *the curse of dimensionality* to describe the rapid growth in the difficulty as the number of integration variables increases [1]. Application of the concepts of quantum mechanics—including path integrals—to the modeling of interest rates and the theory of options are presented in the text [2].

3.1.1   Newton–Cotes Integration Method

We distinguish between two classes of numerical algorithms, depending on whether we evaluate the integrand at equidistant sampling points (Newton–Cotes integration method) or at carefully chosen, but not equidistant sampling points (Gaussian integration method). For particular integrands the Gaussian method may be much more efficient. For example, the maximum degree of exactness is obtained for the *Gauss–Legendre formula*, the sampling points of which are given by zeros of Legendre polynomials. An exhaustive representation of the numerical integration methods is found in [3]. Other textbooks are [4, 5].

Numerical algorithms are based on Riemann's definition of integrals. To see whether a given function $f : [a, b] \rightarrow \mathbb{R}$ is Riemann integrable, we choose a partition of the interval

$$\gamma : \quad a = x_0 < x_1 < x_2 < \cdots < x_{n-2} < x_{n-1} < x_n = b \tag{3.2}$$

and consider the associated lower and upper Riemann sum

$$L(f, \gamma) = \sum_{i=0}^{n-1} (x_{i+1} - x_i) \cdot \inf \{f(x) | x_i \le x \le x_{i+1}\}$$

$$U(f, \gamma) = \sum_{i=0}^{n-1} (x_{i+1} - x_i) \cdot \sup \{f(x) | x_i \le x \le x_{i+1}\} \; ,$$

with $U(f, \gamma) \ge L(f, \gamma)$. The function f is called Riemann integrable, if

$$\sup_{\gamma} L(f, \gamma) = \inf_{\gamma} U(f, \gamma)$$

and we denote by

$$\int_a^b f(x)\mathrm{d}x \equiv \sup_{\gamma} L(f, \gamma) \, . \tag{3.3}$$

the *Riemann integral* of f. This definition is easily extended to multidimensional integrals and serves as point of departure for numerical algorithms.

Many algorithms are based on the approximation of arbitrary smooth functions by *polynomial interpolations*. We remind the reader that there is a unique polynomial P_m of degree $\leq m$, which assumes the values $f_i = f(x_i)$ of a given function f at $(m + 1)$ sampling points $x_0, x_1, \ldots, x_{m-1}, x_m$. The construction of this interpolating polynomial makes use of the $m + 1$ *Lagrange polynomials* of degree m:

$$L_p^{(m)}(x) = \prod_{\substack{i=0 \\ i \neq p}}^{m} \frac{x - x_i}{x_p - x_i}, \qquad p = 0, \ldots, m \quad \text{with} \quad L_p^{(m)}(x_q) = \delta_{pq} . \quad (3.4)$$

The interpolating polynomial of degree m is given by

$$P_m(x) = \sum_{p=0}^{m} f(x_p) L_p^{(m)}(x) . \quad (3.5)$$

The error of the polynomial approximation is bounded by the following result:

Theorem 3.1 *Let f be a $(m + 1)$-times continuous differentiable function and P_m the interpolating polynomial of degree $\leq m$ with sampling points $x_0, \ldots, x_m \in \Delta$. Then there exists for every $x \in \Delta$ a point ξ_x (which lies in the smallest interval containing all sampling points) such that*

$$f(x) - P_m(x) = \frac{f^{(m+1)}(\xi_x)}{(m + 1)!} L^{(m)}(x), \qquad L^{(m)}(x) = \prod_{i=0}^{m}(x - x_i) . \quad (3.6)$$

This theorem leads to the following representation of the integral from the smallest to the largest sampling point:

$$\int_{x_0}^{x_m} dx\, f(x) = \sum_{p=0}^{m} f(x_p) \underbrace{\int dx\, L_p^{(m)}(x)}_{\gamma_p^{(m)}} + \int dx\, \frac{f^{(m+1)}(\xi_x)}{(m + 1)!} L^{(m)}(x) . \quad (3.7)$$

We call $\gamma_p^{(m)}$ the *weights* and x_p the *nodes* of the integration formula. For equidistant nodes at points

$$x_0, \quad x_1 = x_0 + \varepsilon, \quad x_2 = x_0 + 2\varepsilon, \quad \ldots, \quad x_m = x_0 + m\varepsilon \quad (3.8)$$

we find, after setting $x = x_0 + \varepsilon t$ with $t \in [0, m]$ in (3.7), the weights

$$\gamma_p^{(m)} = \varepsilon \int_0^m dt \prod_{\substack{i=0 \\ i \neq p}}^{m} \frac{t - i}{p - i} \equiv \varepsilon w_p^{(m)} = \varepsilon w_{m-p}^{(m)}, \qquad p = 0, 1, \ldots, m . \quad (3.9)$$

Table 3.1 Weights for the Newton–Cotes formulas

m	Name	$p =$	0	1	2	3	4	5	6
0	Rectangle rule	$w_p^{(0)} =$	1						
1	Trapezoidal rule	$2 \times w_p^{(1)} =$	1	1					
2	Simpson's rule	$3 \times w_p^{(2)} =$	1	4	1				
3	3/8-rule	$8 \times w_p^{(3)} =$	3	9	9	3			
4	Milne's rule	$45 \times w_p^{(4)} =$	14	64	24	64	14		
5	6-point rule	$288 \times w_p^{(5)} =$	95	375	250	250	288	95	
6	Weddle's rule	$140 \times w_p^{(6)} =$	41	216	27	272	27	216	41

Applying the formula (3.7) to constant functions yields the sum rule $\sum_p \gamma_p^{(m)} = m\varepsilon$ or equivalently

$$w_0^{(m)} + w_1^{(m)} + \cdots + w_m^{(m)} = m \; . \tag{3.10}$$

Hence the Newton–Cotes formulas can be written as

$$\int_{x_0}^{x_m} \mathrm{d}x f(x) \approx \sum_{p=0}^{m} \varepsilon \, w_p^{(m)} f(x_0 + \varepsilon p) \; . \tag{3.11}$$

The corresponding weights $w_p^{(m)}$ for $m = 0, \ldots, 6$ are listed in Table 3.1.

The integration step size ε can be chosen sufficiently small in order to ensure that the quadrature error is less than a prescribed tolerance. We illustrate how one can estimate the error for the Simpson formula. Thus we consider the difference between the integral of f from $-\varepsilon$ to ε and the corresponding approximation (3.11) with $m = 2$,

$$E_2(\varepsilon) = \int_{-\varepsilon}^{\varepsilon} \mathrm{d}x \, f(x) - \frac{\varepsilon}{3} \left(f(-\varepsilon) + 4f(0) + f(\varepsilon) \right) \; .$$

Differentiating the error $E_2(\varepsilon)$ three times with respect to ε leads to

$$E_2'''(\varepsilon) = -\frac{\varepsilon}{3} \left(-f'''(-\varepsilon) + f'''(\varepsilon) \right) \; .$$

The absolute value of this expression can be bounded as follows:

$$\left| E_2'''(\varepsilon) \right| = \frac{\varepsilon}{3} \left| f'''(\varepsilon) - f'''(-\varepsilon) \right| \le \frac{2\varepsilon}{3} M_3 \quad \text{with} \quad M_3 = \sup_{t \in [-\varepsilon, \varepsilon]} \left| f'''(t) \right| \; .$$

Integrating three times provides the error estimation

$$|E_2(\varepsilon)| \leq M_3 \cdot \frac{\varepsilon^4}{36} \ . \tag{3.12}$$

If the function f is at least four times differentiable, then we may apply the mean value theorem to E_2''' and obtain

$$E_2'''(\varepsilon) = \frac{2\varepsilon}{3}\varepsilon \cdot f^{(4)}(\xi) \ .$$

This results in an improved error bound,

$$|E_2(\varepsilon)| \leq M_4 \cdot \frac{\varepsilon^5}{90} \quad \text{with} \quad M_4 = \sup_{t \in [-\varepsilon, \varepsilon]} \left| f^{(4)}(t) \right| \ . \tag{3.13}$$

Hence with *Simpson's rule* (Kepler's rule for the calculation of wine casks), even cubic polynomials are integrated in an exact manner. Analogous error bounds for approximations based on interpolating polynomials of degree $m \leq 6$ with equidistant nodes are listed in Table 3.2. They bound the errors for integrals from the smallest to the largest sampling point and contain as factors $M_m = \sup_{[x_0, x_m]} |f^{(m)}|$.

With increasing degree m, the coefficients of the Newton–Cotes formulas grow rapidly and have alternating signs. This leads to integration formulas containing differences of large numbers. Largely for this reason higher-order Newton–Cotes formulas are rarely used in practice. In passing we also note that any method based on polynomial interpolation may lead to wrong results in case the function is not sufficiently often differentiable.

Composite Integration Formulas
By partitioning the integration interval into smaller subintervals of equal length, we arrive at the *composite* rectangle, trapezoidal, and Simpson rule or one of the higher-order composite integration formulas. Thereby the number of intervals should be a multiple of m. For the Simpson rule, the situation is depicted in Fig. 3.1. Let us study the composite Simpson rule in more detail. First we partition the interval $[a, b]$ into $2n$ subintervals of length ε such that $b - a = 2n\varepsilon$. The $2n + 1$ sampling points are

Table 3.2 Error estimation of the integration over the interval $[x_0, x_m]$ with $x_m = x_0 + m\varepsilon$

m	Name	$E_m(\varepsilon)$	m	Name	$E_m(\varepsilon)$
0	Rectangle rule	$\varepsilon^2 M_1/2$	4	Milne's rule	$8\varepsilon^7 M_6/945$
1	Trapezoidal rule	$\varepsilon^3 M_2/12$	5	6-point rule	$275\varepsilon^7 M_6/12096$
2	Simpson's rule	$\varepsilon^5 M_4/90$	6	Weddle's rule	$9\varepsilon^9 M_8/1400$
3	3/8-rule	$3\varepsilon^5 M_4/80$			

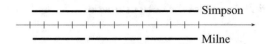

Fig. 3.1 In the composite formulas, the interval is divided into subintervals or length ε. We cluster m neighboring subintervals and apply the previous integration rules to each cluster

$x_j = a + \varepsilon j$ with $j = 0, 1, \ldots, 2n$. The integral is approximated by

$$S_2(f) \approx \frac{\varepsilon}{3}\Big(\{f(x_0) + 4f(x_1) + f(x_2)\} + \{f(x_2) + 4f(x_3) + f(x_4)\} + \ldots$$

$$\cdots + \{f(x_{2n-2} + 4f(x_{2n-1}) + f(x_{2n})\}\Big) .$$

The resulting formula is called *composite Simpson quadrature formula* and reads

$$S_2(f) = \frac{\varepsilon}{3}\left(f(x_0) + 4\sum_{j=0}^{n-1} f(x_{2j+1}) + 2\sum_{j=1}^{n-1} f(x_{2j}) + f(x_{2n}) \right) . \tag{3.14}$$

We can bound the error of the composite Simpson formula as follows:

$$\left| \int_a^b f(x)\mathrm{d}x - S_2(f) \right| \le \frac{1}{90}\varepsilon^5 \cdot n \underbrace{\sup_{t\in[a,b]} \left| f^{(4)}(t) \right|}_{M_4} = \frac{b-a}{180}\varepsilon^4 M_4 . \tag{3.15}$$

Similar bounds can be derived for other composite quadrature formulas. For the partition of $[a, b]$ into $m \times n$ subintervals of length $\varepsilon = (b-a)/mn$, we find

$$\left| \int_a^b f(x)\mathrm{d}x - S_m(f) \right| \le \frac{b-a}{m\varepsilon} E_m(\varepsilon) , \tag{3.16}$$

where the numbers $E_m(\varepsilon)$ are listed in Table 3.2. But now M_m in this table denotes the supremum of $|f^{(m)}|$ on the integration interval. Let us calculate and compare four approximations to the integral of $\exp(x)$ from 0 to 1: with the rectangle rule, the trapezoidal rule, the Simpson rule, and the Monte Carlo method. The latter will be explained in Sect. 3.2. For the numerical approximations, we applied the following composite quadrature formulas:

$$\text{rectangle rule:} \quad \sum_{i=0,1,2}^{n-1} \varepsilon f(x_i)$$

$$\text{trapezoidal rule:} \quad \sum_{i=0,1,2}^{n-1} \frac{\varepsilon}{2}\big(f(x_i) + f(x_{i+1})\big)$$

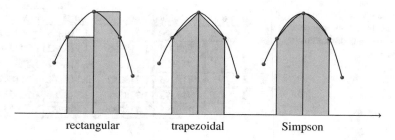

Fig. 3.2 Numerical integration via the rectangle rule, the trapezoidal rule, and the Simpson's rule

Simpson's rule:
$$\sum_{i=0,2,4}^{n-2} \frac{\varepsilon}{3}\bigl(f(x_i) + 4f(x_{i+1}) + f(x_{i+2})\bigr) . \qquad (3.17)$$

Recall that in the last formula n must be even. The interpolations used in these approximations are sketched in Fig. 3.2. With the program `1dintegral.c` on p. 48, one can estimate the definite integral

$$\int_0^1 \mathrm{d}x\, \mathrm{e}^x \quad \text{with} \quad \varepsilon \in \bigl\{10^{-n} \mid n = 1, 2, \ldots, 6\bigr\} \qquad (3.18)$$

with the help of Simpson's rule. With a slight modification of the code, we are able to integrate with the other quadrature formulas as well. The following table compares the results obtained with the piecewise constant, linear, or quadratic interpolations. For Simpson's rule we observe a rapid convergence to the exact value 1.7182818.

3.2   Monte Carlo Integration

The Monte Carlo method was presumably invented by Stanislaw Ulam. He developed this method in 1946 while he was thinking about the probability of profit for solitaire:

> The first thoughts and attempts I made to practice [the Monte Carlo Method] were suggested by a question which occurred to me in 1946 as I was convalescing from an illness and playing solitaires. The question was what are the chances that a Canfield solitaire laid out with 52 cards will come out successfully? After spending a lot of time trying to estimate them by pure combinatorial calculations, I wondered whether a more practical method than "abstract thinking" might not be to lay it out say one hundred times and simply observe and count the number of successful plays. . . .

Table 3.3 Convergence of different integration methods. The integer n is defined in (3.18)

n, log M	Rectangle	Trapezoidal	Simpson	Monte Carlo
1	1.633799	1.719713	1.718283	1.853195
2	1.709705	1.718296	1.718282	1.793378
3	1.717423	1.718282	1.718282	1.720990
4	1.718196	1.718282	1.718282	1.711849
5	1.718273	1.718282	1.718282	1.719329
6	1.718281	1.718282	1.718282	1.718257

A few years later, the method was applied to the problem of neutron diffusion which could not be solved analytically [6]. Of utmost importance in physics is the application of the Monte Carlo method to estimate high-dimensional integrals.

A basic algorithm to estimate the integral of a function $f : G \rightarrow \mathbb{R}$ could be as follows:

- Generate M uniformly distributed points $\{x_1, \ldots, x_M\}$ in G.
- Compute the function value $f(x_i)$ for every point x_i, $i = 1, \ldots, M$.
- Compute the mean value

$$I(M) = \frac{\text{Vol}(G)}{M} \sum_{i=1}^{M} f(x_i) . \tag{3.19}$$

For any Riemann integrable function, the mean $I(M)$ converges for large M to the integral. For the exponential integral in (3.18), we calculated the mean for $M = 10, 100, \ldots$, and the results are contained in the last column of Table 3.3. We used the simple Monte Carlo program listed on p. 48. In line 34 of the code, the random number generator is called to generate a random number in $[0, 1]$.

Figure 3.3 compares the convergence behavior of the rectangle and Simpson quadrature formulas and of the simple Monte Carlo integration. Simpson's integration rule, applied to the exponential function, agrees with the exact result in 1 ppm already for a partitioning of $[0, 1]$ in ten subintervals. When we consider the problem of numerical integration over n-dimensional domains,

$$I = \int_G dq_1 \ldots dq_n f(q_1, \ldots, q_n) \equiv \int_G d^n q \ f(\boldsymbol{q}) \tag{3.20}$$

then an obvious strategy is to apply a rule such as Simpson's rule in each dimension, creating what is called a product rule. But any product rule is prohibitively time-consuming when n is large. As an example, consider fixed integration limits 0 and 1 as well as a fixed distance ε between the sampling points in a given direction. Then the total number of sampling points is ε^{-n}, and the computing time is proportional to this number. To estimate this time, we choose a coarse partition of the interval $[0, 1]$ with $\varepsilon = 0.1$ for which the number of sampling points is $\approx 10^n$. Calculating the contribution of one node takes approximately 10^{-7} s of

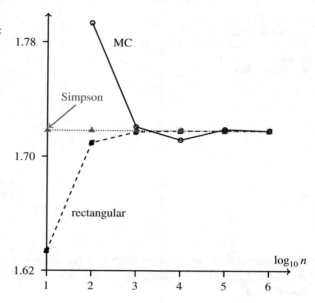

Fig. 3.3 Convergence of different integration methods: rectangle rule, trapezoidal rule, and Monte Carlo integration

CPU time such that the numerical integration of a 12-dimensional integral takes approximately a whole day. Nevertheless, there are other ways of tackling high-dimensional integrals, as demonstrated by Traub and Paskov when they treated a mathematical finance problem from Wall Street as an integration problem over the 360-dimensional unit cube [7]. They used what nowadays is called quasi-Monte Carlo method, a deterministic algorithm which is widely used in the financial sector to estimate financial derivatives. For reviews on this potentially interesting development, see [8].

3.2.1 Hit-or-Miss Monte Carlo Method and Binomial Distribution

We wish to estimate the value of a definite integral $I = \int d^n x \, f(x)$ with stochastic means. First we transform both the coordinates and function such we are dealing with an integration problem of a function $0 \le f \le 1$ over the n-dimensional unit cube. We are not touching the problem of how best to transform a given problem to the unit cube. Let us first consider a one-dimensional integral of a function on the unit interval. The integral is equal to the area below the graph of f; see Fig. 3.4.

How can we calculate this area with probabilistic means? From a random number generator which generates uniformly distributed numbers on the unit interval, we take two random numbers x and y. Now we shoot on the unit square in Fig. 3.4. A hit of the shaded area is then identified with $y \le f(x)$. The *probability* for a hit is

$$p = \frac{\text{number of hits}}{\text{number of trials}} = \frac{\text{grey area}}{\text{total area}} = \frac{I}{1} = I . \tag{3.21}$$

Fig. 3.4 The gray area underneath the graph of f is proportional to the relative frequency of hits when shooting on the unit square

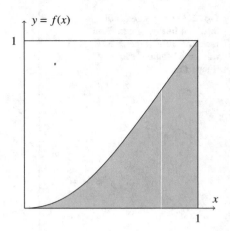

Let us repeat shooting M times such that the number of hits is in $\{0, \ldots, M\}$. For statistically independent shots, the probability for k hits is given by the *binomial distribution*

$$P(M, k) = \binom{M}{k} p^k (1 - p)^{M-k} \quad \text{with} \quad \sum_{k=0}^{M} P(M, k) = (p + 1 - p)^M = 1 \; .$$

(3.22)

The binomial coefficient in this distribution function counts the number of ways of picking k hits from M shots. The distribution (3.22) describes a bell-shaped curve, localized at pM. The *generating function* for the moments of the binomial distribution can be computed easily:

$$Z(t) = \left\langle e^{tk} \right\rangle = \sum_{k=0}^{M} e^{kt} P(M, k) = \left(e^t \, p + (1 - p) \right)^M \; .$$

(3.23)

Being the sum of all probabilities $Z(0) = \langle 1 \rangle = 1$. The expectation values of arbitrary powers of k are obtained by differentiating the generating function with respect to t.

For example, for a very large number of shots, the *relative frequency of hits* is

$$\left\langle \frac{k}{M} \right\rangle = \frac{1}{M} \sum_{k=0}^{M} k \, P(M, k) = \frac{1}{M} \frac{\mathrm{d}Z}{\mathrm{d}t} \bigg|_{t=0} = p \; ,$$

(3.24)

Table 3.4 Convergence behavior of different methods. M is defined as stated in (3.19)

$\log_{10} M$	p	$I - p$	σ	p_{impr}	$I - p_{\text{impr}}$	σ_{impr}
1	0.500000	−0.123630	0.158114	0.333333	+0.043037	0.000000
2	0.330000	+0.046370	0.047021	0.363333	+0.013037	0.017059
3	0.399000	−0.022630	0.015485	0.377333	−0.000963	0.006486
4	0.378900	−0.002530	0.004851	0.376833	−0.000463	0.002040
5	0.376570	−0.000200	0.001532	0.377693	−0.001323	0.000651
6	0.374857	+0.001513	0.000484	0.376305	+0.000065	0.000203
7	0.376273	+0.000097	0.000153	0.376303	+0.000067	0.000064

as expected. The relative frequency has mean square

$$\left\langle \frac{k^2}{M^2} \right\rangle = \frac{1}{M^2} \frac{d^2 Z}{dt^2}\bigg|_{t=0} = \frac{p}{M} + \left(1 - \frac{1}{M}\right) p^2 . \tag{3.25}$$

More interesting is the variance of the relative frequency,

$$\sigma^2 = \frac{1}{M^2} \left\langle \left(k - \langle k \rangle\right)^2 \right\rangle = \frac{1}{M^2} \frac{d^2 \log Z}{dt^2}\bigg|_{t=0} = \frac{p(1-p)}{M} . \tag{3.26}$$

Note that the standard deviation σ decreases only slowly with the number of trials, $\sigma \sim M^{-1/2}$, and this slow decrease is typical for probabilistic methods.

Numerical Experiment
An estimation of the probability p is given by the relative frequency of hits in M trials. Table 3.4 contains estimations p of the definite integral

$$I = \int_0^1 dx\, f(x) \approx 0.376370, \qquad f(x) = \frac{x^2 e^x}{1 - x + x e^x} , \tag{3.27}$$

for an increasing number of trials M and of the corresponding spreads σ around the estimated values. The numbers in columns 2, 3, and 4 were calculated with the program `hitandmissarea.c`, listed on p. 49.

The simple hit-or-miss method can be improved with little effort. Observe that the variance decreases when p approaches 0 or 1. As a warning we mention that at the same time the relative error may increase. Now we choose an analytically integrable function $g \leq f$, which approximates f rather well. Thus the first integral in

$$I = \underbrace{\int (f(x) - g(x))\, dx}_{\text{small}} + \underbrace{\int g(x) dx}_{\text{known}} \tag{3.28}$$

is small and the integrand lies between 0 and 1. It is estimated by means of the hit-or-miss method with reduced variance. For the function f in (3.27), we may choose

$$g(x) = x^2 \quad \text{with} \quad \int g(x)\, dx = 1/3 .$$

With this reweighing technique, we obtain improved estimates and smaller variances for integrals of interest. The last three columns in Table 3.4 contain the estimates and standard deviations for the improved method for various values of trials M. These values were computed with the program `hitandmissarea.c` on p. 49.

3.2.2   Sum of Random Numbers and Gaussian Distribution

The program `gaussdistr.c` on p. 50 generates the sum $s = x_1 + x_2 + \cdots + x_n$ of n equally distributed and independent random numbers on the unit interval. The sum itself is a random number with values in $[0, n]$ and generating function

$$Z(t) = \langle e^{ts} \rangle = \int_{I^n} d^n x\, e^{t(x_1 + \cdots + x_n)} = \left(\int_0^1 dx\, e^{tx} \right)^n = t^{-n} \left(e^t - 1 \right)^n . \quad (3.29)$$

We find the expected result for the mean value of s

$$m = \langle s \rangle = \frac{dZ}{dt}\Big|_{t=0} = \int_{I^n} d^n x\, (x_1 + \cdots + x_n) = \frac{n}{2} . \quad (3.30)$$

Similarly, we obtain for the square of the statistical spread

$$\frac{d^2 \log Z}{dt^2}\Big|_{t=0} = \sigma^2 = \langle s^2 \rangle - m^2 = \frac{n}{12} . \quad (3.31)$$

According to the law of large numbers,[1] we obtain the Gaussian distribution

$$P_s = \frac{1}{\sqrt{2\pi}\,\sigma}\, e^{-(s-m)^2/2\sigma^2} . \quad (3.32)$$

The program calculates the distribution of s for $n = 10, 50, 100$. For each n a histogram is generated from one million trials. Rescaling s with n leads to a maximum of the distribution at $s/n = 1/2$ and a variance of $1/12n$. Figure 3.5 compares the distributions obtained by Monte Carlo simulation and the Gaussian distributions (3.32).

---

[1] For a discussion and proof of this law, see p. 46.

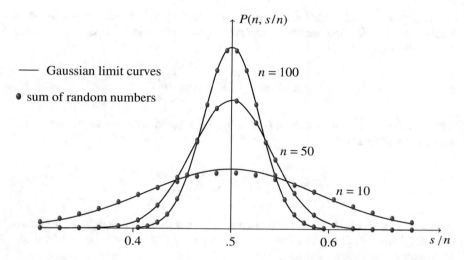

Fig. 3.5 The sum s of n random numbers forms a Gaussian distribution for large n. The limit distribution of the rescaled random variable s/n has the mean value $1/2$. Furthermore the square of the statistical spread is $1/12n$

The inequality (3.57) describes the probability that the random number s/n deviates more than δ from its mean value. In the present case, it reads

$$P\left[\left|\frac{s}{n} - \frac{1}{2}\right| \geq \delta\right] \leq \frac{1}{12n\delta^2} \, . \tag{3.33}$$

3.3 Importance Sampling

Numerical algorithms approximate integrals by finite sums,

$$\int d^n q \, f(\boldsymbol{q}) \sim \sum_{\mu=1}^{M} f(\boldsymbol{q}_\mu) \, \Delta \boldsymbol{q}_\mu \, .$$

In case of high-dimensional integrals and slowly varying functions f, it may be advantageous to choose the sampling points $\boldsymbol{q}_\mu$ randomly. But in physical problems, we are often confronted with integrands that vary by several orders of magnitude for different points in which case we may waste computing time at sampling points with small integrands. The idea of *importance sampling*, according to which one preferably samples points with large integrands, is incorporated in many stochastic algorithms for calculating high-dimensional integrals. The best-known examples are based on the Metropolis algorithm. Since the sampling points lie primarily in regions with high values of f, the variance of the estimate of the integral decreases.

To implement this idea, we choose a function $g(\boldsymbol{q})$ for which the integral is known analytically and which represents a good approximation of the function $f(\boldsymbol{q})$. We write

$$\int_0^1 f(\boldsymbol{q})\, \mathrm{d}^n q = \int_0^1 \mathrm{d}^n q\, g(\boldsymbol{q})\, \frac{f(\boldsymbol{q})}{g(\boldsymbol{q})} \ . \tag{3.34}$$

Through the generation of random points $\boldsymbol{q}_\mu$ distributed according to $g(\boldsymbol{q})\mathrm{d}^n q$, we come to the estimation

$$\int f(\boldsymbol{q})\mathrm{d}^n q \approx \bar{f} = \frac{1}{M}\sum_{\mu=1}^M \frac{f(\boldsymbol{q}_\mu)}{g(\boldsymbol{q}_\mu)} \ , \tag{3.35}$$

whereupon we performed M "measurements." Note that the reweighed summands vary only little. But keep in mind that the integral of $g(\boldsymbol{q})$ must be known in order to extract g-distributed random numbers.

Now let us consider expectation values in lattice field theories,

$$\langle O \rangle \approx \frac{1}{Z}\int \mathscr{D}\boldsymbol{q}\, O(\boldsymbol{q})\mathrm{e}^{-S(\boldsymbol{q})}, \quad \mathscr{D}\boldsymbol{q} = \mathrm{d}q_1 \cdots \mathrm{d}q_n, \quad Z = \int \mathscr{D}\boldsymbol{q}\, \mathrm{e}^{-S(\boldsymbol{q})} \ , \tag{3.36}$$

for which we would like to choose the Boltzmann distribution

$$P(\boldsymbol{q}) = \frac{1}{Z}\,\mathrm{e}^{-S(\boldsymbol{q})} \tag{3.37}$$

as function g. Then we only need to average over the in many cases slowly varying (in contrast to P) observables $O(\boldsymbol{q})$:

$$\langle O \rangle \approx \bar{O} = \frac{1}{M}\sum_{\mu=1}^M O(\boldsymbol{q}_\mu) \ . \tag{3.38}$$

Here M denotes the number of generated points $\boldsymbol{q}_\mu$ in configuration space. Thus the Monte Carlo estimation $\bar{O}$ of the mean value of O represents an arithmetic average.

We are confronted with the following algorithmic problem: The n-dimensional integrals

$$\langle O \rangle = \int \mathscr{D}\boldsymbol{q}\, O(\boldsymbol{q})P(\boldsymbol{q}), \quad \int \mathscr{D}\boldsymbol{q}\, P(\boldsymbol{q}) = 1 \ , \tag{3.39}$$

with fixed probability density P should be approximated for varying observables O. We therefore need algorithms which generate P-distributed points $\boldsymbol{q}_1, \boldsymbol{q}_2, \ldots$ in the n-dimensional domain of integration. The following *Metropolis algorithm* [6] generates a series of such points:

1. Start with $\mu = 0$ and an arbitrary initial point q_μ.
2. Choose a point q' at random and in addition a random number $r \in [0, 1]$.
3. If $P(q')/P(q_\mu) > r$, set $q_{\mu+1} = q'$. Otherwise set $q_{\mu+1} = q_\mu$.
4. Increase μ by one and repeat steps 2, 3, and 4.

The resulting points $\{q_\mu\}$ are distributed according to $P(q)$ and

$$\bar{O} = \frac{1}{M} \sum_{\mu=1}^{M} O(q_\mu) \tag{3.40}$$

represents an estimate of $\langle O \rangle$ which converges to $\langle O \rangle$ for large values of M. In lattice field theories, the points q_μ are called *lattice configurations*.

The program `samplingarea.c` on p. 51 estimates the definite integral

$$I = 128 \cdot \frac{\int_0^1 dxdydz \, x^3 y^2 z \exp\left(-x^2 - y^2 - z^2\right)}{\int_0^1 dxdydz \, \exp\left(-x^2 - y^2 - z^2\right)} \tag{3.41}$$

$$= 128 \cdot \langle x^3 y^2 z \rangle = 2.4313142\ldots$$

via the Metropolis algorithm with the exponential function, normalized by the integral in the denominator, as distribution P. We observe a slow convergence to the exact result with errors of the order $1/\sqrt{M}$. Table 3.5 contains the computed estimates for various M. The last entry results from 10^6 Monte Carlo iterations and has an error of 0.00555.

Table 3.5 Monte-Carlo estimates for the three-dimensional integral $I = 2.4313142\ldots$ in (3.41) via the Metropolis algorithm. $\bar{I}$ denotes the Monte-Carlo-estimate (MC-estimate), M the number of iterations, and E the error

M	5000	10,000	15,000	20,000	25,000	30,000	35,000
$\bar{I}$	2.33113	2.31536	2.33432	2.38934	2.3568	2.34805	2.35253
E	0.10018	0.11595	0.09699	0.04197	0.07449	0.08327	0.07878
M	40,000	45,000	50,000	55,000	60,000	65,000	70,000
$\bar{I}$	2.34528	2.34193	2.35193	2.35089	2.35659	2.35952	2.36130
E	0.08603	0.08939	0.07938	0.08043	0.07473	0.07179	0.07001
M	75,000	80,000	85,000	90,000	95,000	100,000	1,000,000
$\bar{I}$	2.36969	2.37196	2.36937	2.38248	2.38742	2.38448	2.43686
E	0.06162	0.05935	0.06194	0.04884	0.04390	0.04683	−0.00555

3.4   Some Basic Facts in Probability Theory

Knowing the basic facts about probability theory is of course useful in any statistical approach to physical problems. Thus it is worth summarizing the main concepts of probability and statistics. The axiomatic approach to this theory was developed by Kolmogorow in the 1930s. See [9] for an introduction to probability and [10] for an introduction to measure theory.

The set of elementary events, the *sample space*, is denoted by Ω. A general event is a subset of Ω, and the set of events is the *event space* Σ. The measure $P : \Sigma \to [0, 1]$ is a *probability measure* in case $P(\Omega) = 1$. The triple (Ω, Σ, P) consisting of a sample space, an event space, and a probability measure form a probability space.

The probability measure should fulfill the following axioms:

- The probability $P(A)$ of an event $A \subset \Omega$ is in $[0, 1]$.
- The probability of some elementary event occurring in Ω is $P(\Omega) = 1$.
- The probability of the union of countably many disjoint events is equal to the sum of the probabilities of the individual events (σ additivity):

$$P(A_1 \cup A_2 \cup \cdots) = \sum P(A_i) \quad \text{if} \quad A_i \cap A_j = \emptyset .$$

From Kolmogorov axioms one concludes

$$P(\Omega \backslash A) = 1 - P(A) \quad \text{and} \quad P(A_1 \cup A_2) = P(A_1) + P(A_2) - P(A_1 \cap A_2) .$$

The incidental result $X : \Omega \to \mathbb{R}$ is called *random variable* with expectation value (mean value)

$$\langle X \rangle = \sum_{w \in \Omega} X(w) \, P[w] = \sum_{x \in X(\Omega)} x \sum_{\substack{w \in \Omega \\ X(w)=x}} P[w] = \sum_{x \in X(\Omega)} x \, P_X(x) , \quad (3.42)$$

wherein $P[w]$ denotes the probability for the event w and

$$P_X(x) = P\left(X^{-1}(x)\right) \tag{3.43}$$

represents the probability for the random variable X to assume the value x. The mean value of repeated measurements of X is simply $\langle X \rangle$ and the expression

$$P_X(\Delta) = \sum_{x \in \Delta} P_X(x) . \tag{3.44}$$

is interpreted as probability for finding an event for which X has a value in Δ. If X is a real-valued random variable, then P_X is the probability density, and we have

$$P_X(\Delta) = \int_\Delta P_X(x)\,dx \ . \tag{3.45}$$

For a continuous function f, the expectation value of the random variable $f(X)$ is

$$\langle f(X) \rangle = \sum_{w \in \Omega} f(X(w)) = \sum_{x \in X(\Omega)} f(x) P_X(x) \ . \tag{3.46}$$

The average value of any linear combination of random variables is the corresponding linear combination of individual averages,

$$\langle X \rangle = \langle X_1 \rangle + \cdots + \langle X_N \rangle \tag{3.47}$$

The random variables are called *independent* if the probability of the events w with $X_i(w) = x_i$ factorizes for arbitrary $x_1, \ldots, x_N$:

$$P(\{w\,|X_1(w) = x_1, \ldots, X_N(x) = x_N\}) = P_{X_1}(x_1) \cdots P_{X_N}(x_N) \ . \tag{3.48}$$

The generating function of independent random variables factorizes,

$$\left\langle e^{i(t_1 X_1 + \cdots + t_N X_N)} \right\rangle = \sum_w e^{i(t_1 x_1 + \cdots + t_N x_N)} P(\{w|X_1(w) = x_1, \ldots, X_N(x) = x_N\})$$

$$= \sum_{x_1, \ldots, x_N} \prod_{k=1}^{N} e^{it_k x_k} P_{X_k}(x_k) = \left\langle e^{it_1 X_1} \right\rangle \cdots \left\langle e^{it_N X_N} \right\rangle \ . \tag{3.49}$$

For $t = t_1 = \cdots = t_N$, we obtain the following useful relation for the generating function of the connected correlations of independent random variables,

$$\log \left\langle e^{it(X_1 + \cdots + X_N)} \right\rangle = \sum_{i=1}^{N} \log \left\langle e^{it X_i} \right\rangle \ . \tag{3.50}$$

Differentiating twice with respect to t at $t = 0$, we conclude that the variance of the sum of independent random variables X_i is equal the sum of the variance of X_i,

$$\text{Var}[X] = \sum_{i=1}^{N} \text{Var}[X_i], \quad \text{Var}[X] \equiv \langle X^2 \rangle - \langle X \rangle^2 = \left\langle (\Delta X)^2 \right\rangle \ , \tag{3.51}$$

where ΔX denotes the random variable $X - \langle X \rangle$.

Theorem 3.2 (Markov's Theorem) *Let X be a random variable which assumes non-negative values only. Then for all $t \in \mathbb{R}^+$*

$$P[X \geq t] \leq \frac{1}{t} \langle X \rangle \ . \tag{3.52}$$

Proof The proof of Markov's theorem is quite simple:

$$\langle X \rangle = \sum_{x \geq 0} x \cdot P_X(x) \geq \sum_{x \geq t} x \cdot P_X(x) \geq \sum_{x \geq t} t \cdot P_X(x)$$

$$= t \sum_{x \geq t} P_X(x) = t \cdot P[X \geq t] \ .$$

This result leads to the useful Chebyshev inequality for the average deviation of a real-valued random number from its average value.

Theorem 3.3 (Chebyshev's Theorem) *Let X be a random variable and $t \in \mathbb{R}^+$. Then*

$$P\left[|\Delta X| \geq t\right] \leq \frac{1}{t^2} \mathrm{Var}[X] \ . \tag{3.53}$$

Proof

$$P\left[|\Delta X| \geq t\right] = P\left[(\Delta X)^2 \geq t^2\right] \overset{(3.52)}{\leq} \frac{\mathrm{Var}[X]}{t^2} \ .$$

Another very important theorem we have seen in action in Sect. 3.2.2 is the

Theorem 3.4 (Law of Large Numbers) *Given a random variable X and arbitrary but fixed numbers $\varepsilon, \delta > 0$. Define*

$$K := \frac{\mathrm{Var}[X]}{\varepsilon \cdot \delta^2} = const. \ , \tag{3.54}$$

let $X_1, \ldots, X_N$ be independent random variables with the same distribution as X and define $Z := (X_1 + \cdots + X_N)/N$. Then the following bound holds:

$$P\left[|\Delta Z| \geq \delta\right] \leq \varepsilon \ . \tag{3.55}$$

Proof Linearity of expectation values and (3.51) imply

$$\langle Z \rangle = \frac{1}{N} \sum \langle X_i \rangle = \langle X \rangle \quad \text{and} \quad \mathrm{Var}[Z] = \frac{1}{N^2} \sum \mathrm{Var}[X_i] = \frac{\mathrm{Var}[X]}{N} \ .$$

The latter inequality is interesting in its own. It is the well-known *square root law* for the relative fluctuations

$$\frac{\sqrt{\text{Var}[Z]}}{\langle Z \rangle} \leq \frac{1}{N} \frac{\sqrt{\text{Var}[X]}}{\langle X \rangle} \tag{3.56}$$

and it means that the scale of the relative fluctuations of Z is of order $O(N^{-1/2})$. It implies that for a large number of random variables X_i (trials), we may neglect the fluctuations. In other words, the statistical errors decrease with the number of trials. Combining Chebyshev's inequality with the result (3.56) leads to the bound

$$P\left[|\Delta Z| \geq \delta\right] \leq \frac{\text{Var}[Z]}{\delta^2} = \frac{\text{Var}[X]}{N \cdot \delta^2} \leq \varepsilon . \tag{3.57}$$

Now let $X_1, \ldots, X_N$ be a series of independent and identically distributed random variables with vanishing mean and covariance matrix $\langle X_i X_j \rangle = \delta_{ij}\sigma^2$. Furthermore we define the random variable

$$Y_N := \frac{1}{\sqrt{N}} \sum_{i=1}^{N} X_i . \tag{3.58}$$

The generating function for Y_N is given by

$$\left\langle e^{itY_N} \right\rangle = \prod_{i=1}^{N} \left\langle \exp\left(i\frac{t}{\sqrt{N}} X_i \right) \right\rangle$$

$$= \left\langle 1 + i\frac{t}{\sqrt{N}} X_1 - \frac{1}{2}\frac{t^2}{N} X_1^2 + \frac{1}{O(N^{3/2})} \right\rangle^N \xrightarrow{N \to \infty} \exp\left(-\frac{1}{2} t^2 \sigma^2 \right) .$$

On the other hand, the generating function of a Gaussian random variable with mean m and variance σ^2 is given by

$$\frac{1}{\sqrt{2\pi}\,\sigma} \int dx\, e^{-(x-m)^2/2\sigma^2}\, e^{itx} = \exp\left(imt - \frac{1}{2} t^2 \sigma^2 \right) .$$

It follows at once that for large N the random variables Y_N are Gaussian with mean 0 and variance σ^2.

3.5 Programs for This Chapter

The C-programs

- `1dintegral.c`
- `hitandmissarea.c`

- `gaussdistr.c`
- `samplingarea.c`

were mentioned and used in this chapter. They are listed and explained in this section. The first program `1dintegral.c` in Listing 3.1 calculates the definite integral of a function on the unit interval with Simpson's rule for lattice constants

$$\varepsilon \in \left\{10^{-n} \,|\, n = 1, 2, \dots, 6\right\}$$

and with the Monte Carlo (MC) method. The function is defined in line 10 of the listing.

Listing 3.1 One-dimensional integrals

```c
1  /*program 1dintegral.c
2  /*numerical integration of f(x) from alpha to beta
3  /*with Simpson algorithm and MC method*/
4  #include <stdio.h>
5  #include <stdlib.h>
6  #include <math.h>
7  #include <time.h>
8
9  /*function to be integrated*/
10 double f(double x)
11 {return exp(x);}
12 /*random number between 0 and 1*/
13 double randa(void)
14 {return (double)rand()/((double)RAND_MAX);}
15 int main(void)
16 {
17    double eps,sum,int_simpson,int_monte_carlo,x0,x1,x2;
18    double alpha=0,beta=1; /*limits of integration*/
19    long i,N;
20    srand ( time(NULL) );
21    printf("log_10(N)\t int_simpson\t int_monte_carlo\n");
22 for (N=10;N<1000001;N*=10)
23    {
24    eps=(beta-alpha)/N;
25 /*simpson rule*/
26    sum=0;
27    for (i=0;i<N-1;i=i+2)
28       {x0=alpha+eps*i;x1=x0+eps;x2=x1+eps;
29       sum=sum+(f(x0)+4.0*f(x1)+f(x2))/3.0;}
30    int_simpson=sum*eps;
31 /*MC method*/
32    sum=0;
33    for (i=0;i<N;i++)
34       {x0=randa();sum=sum+f(x0);}
35    int_monte_carlo=eps*sum;
36    printf("%8.0f \t%f \t%f \n",
37    log10(N),int_simpson,int_monte_carlo);
```

```
38  |        }
39  |    return 0;
40  | }
```

The C-program in Listing 3.2 computes the area below a given function with the help of the hit-or-miss MC method in a rather simple way. The function is defined in line 9 of the listing.

The program in Listing 3.3 computes the distribution of the sum s of 10, 50, and 100 random numbers $r \in [0, 1]$. Each time we perform one million trials. The values of the stochastic variable s form a histogram, which is saved in the array *mean[100]*. Furthermore we rescale the random variable s with $2m$ such that the maximum of the distribution lies at $s/(2m) = 1/2$.

Listing 3.2 hit_or_miss

```
1   | /*program hitormissarea.c
2   | /*integration of f(x) with hit-or miss method*/
3   | #include <stdio.h>
4   | #include <stdlib.h>
5   | #include <math.h>
6   | #include <time.h>
7   | #define M 10000001 /*number of attempts*/
8   | double f(double x) /*function to be integrated*/
9   | {return x*x*exp(x)/(1-x+x*exp(x));}
10  | double g(double x) /*function for improved method*/
11  | {return x*x*exp(x)/(1-x+x*exp(x))-x*x;}
12  | int main(void)
13  | {
14  | double sum1,sum2,I1,I2,sig1,sig2,x,y;
15  |     long n,m;
16  |     srand48(time(NULL));
17  |     printf("attemps\t\t p\t\t integ-p\t sigma_1\t");
18  |     printf("p_imp\t\t integ-p_imp\t sigma_2\n");
19  |     for (m=10;m<M;m*=10)
20  |        {sum1=0;sum2=0;
21  |         for (n=1;n<m+1;n++)
22  |            {x=drand48();y=drand48();
23  |             if (y<f(x)) sum1=sum1+1;
24  |             if (y<g(x)) sum2=sum2+1;
25  |            };
26  |         I1=sum1/m;I2=sum2/m;
27  |         sig1=sqrt(I1*(1-I1)/m);sig2=sqrt(I2*(1-I2)/m);
28  |         n=(int)log10(m);
29  |         printf("%8ld\t%8.5f\t%8.5f\t%8.5f\t%8.5f\t%8.5f
30  |         \t%8.5f\n",m,I1,0.376370-I1,sig1,1/3.0+I2,0.043037-I2,
31  |         sig2);
32  |        };
33  |     return 0;
34  | }
```

Listing 3.3 Gauss distribution

```
1   /*Programm gaussdistr.c
2   /*sum of random numbers from the interval [0,1]*/
3   #include <stdio.h>
4   #include <stdlib.h>
5   #include <math.h>
6   #include <time.h>
7   #define PI 3.1415926
8   #define numbersadded 50 /*number of added random numbers*/
9   #define M 1000000 /*number of MC iterations*/
10  int main(void)
11  { double sum,aux=6.0*numbersadded,mean[100];
12      double dM=100.0/(double)M; /*scaling factor 100*/
13      long i,j,sumi;
14      /*initial values*/
15      for (i=0;i<numbersadded;i++) mean[i]=0;
16        sum=0;srand48(time(NULL));
17      /*repeat experiment M times*/
18      for (i=0;i<M;i++)
19        {sum=0;
20          /*sum of random numbers in each experiment*/
21          for (j=0;j<numbersadded;j++)
22              sum=sum+drand48();
23          /*100 bins for histogram*/
24          sumi=(int)(0.5+100*sum/numbersadded);
25          ++mean[sumi];
26        };
27      printf("maximum at bin = 49.5\n");
28      printf("bin\t estimate\t Gaussian dist\n");
29      for (i=30;i<71;i=i+2)
30              {sum=i-50;
31              printf("%li\t%8.5f\t%8.5f\n",
32              i,mean[i]*dM,sqrt(aux/PI)*exp(-aux*sum*sum*dM));
33              };
34      return 0;
35  }
```

The program in Listing 3.4 applies the technique of *importance sampling*. There we only consider points associated with *large* integrands. Thus the variance of a single estimate is reduced. The program `samplingarea.c` computes the proper integral

$$128 \cdot \frac{\int_0^1 \mathrm{d}x\mathrm{d}y\mathrm{d}z\, x^3 y^2 z \exp\left(-x^2 - y^2 - z^2\right)}{\int_0^1 \mathrm{d}x\mathrm{d}y\mathrm{d}z\, \exp\left(-x^2 - y^2 - z^2\right)} \approx 2.4313142$$

with the help of the Metropolis algorithm.

Listing 3.4 Sampling area

```c
/*pogram samplingarea.c
/*three-dimensional integral with importance sampling.*/
#include <stdio.h>
#include <stdlib.h>
#include <math.h>
#include <time.h>
#define M 100000 /*number of measured MC-iterationen*/
#define MA 1000 /*every MA'th configuration is measured*/
/*distribution*/
double P(double *x)
{return exp(-x[0]*x[0]-x[1]*x[1]-x[2]*x[2]);}
/*function to be integrated*/
double f(double *x)
{return 128.0*x[0]*x[0]*x[0]*x[1]*x[1]*x[2];}
int main(void)
{ double integral,sum,x[3],y[3];
    long i,j;
    srand48(time(NULL));
    sum=0; x[0]=drand48();x[1]=drand48();x[2]=drand48();
    for (i=1;i<M+1;i++)
      { for(j=0;j<MA;j++)
            { y[0]=drand48(); y[1]=drand48(); y[2]=drand48();
              if (P(y)>P(x)*drand48() )
                  { x[0]=y[0];x[1]=y[1];x[2]=y[2];};
            };
        sum=sum+f(x);integral=sum/i;
        if (i%5000==0)
        printf("i = %ld\t integral = %.5f\t error = %8.5f\n",
          i,integral,2.4313142-integral);
        };
    return 0;
}
```

3.6 Problems

3.1 (Numerical Calculation of Integrals) Calculate the exponential integral

$$\int_0^1 dx \ e^x$$

with the help of Simpson's rule. Compare your result with the exact one.

3.2 (Error of Simpson Formula) For the function

$$f(x) = e^x \cos(x) \quad \text{in} \quad [0, \pi]$$

compute the minimum number of intervals such that the error of the composite Simpson formula is less than 10^{-4}.

3.3 (Volume of n-Dimensional Unit Ball) Write a program to find the volume of an n-dimensional ball using the MC technique. The volume will be the number of points with the sum,

$$\sum_{j=1}^{n} x_j^2 < 1$$

divided by the total number of points in the region from which the x_i are selected. A convenient region is the n-cube defined by

$$-1 < x_j < 1, \quad j = 1, \ldots, n$$

having volume 2^n.

3.4 (Particle Diffusion) A particle starts at the origin in two dimensions and after each time interval jumps with equal probability $1/4$ to one of the neighboring points. After the first jump, the particle is at one of the points $(\pm 1, 0)$ or $(0, \pm 1)$. Write a program to jump a large number of times, and print out the distance r of the particle from its starting point after all jumps are completed. Theory predicts that the expected value of r is $K \times \sqrt{n}$ after n jumps. Start the program sufficiently many times so that you get an estimate for the constant K.

References

1. R. Bellman, *Dynamic Programming*. Princeton Landmarks in Mathematics (Princeton University Press, Princeton, 2010)
2. B. Baaquie, *Quantum Finance: Path Integrals and Hamiltonians for Options and Interest Rates* (Cambridge University Press, Cambridge, 2007)
3. E. Hairer, C. Lubich, *Geometric Numerical Integration* (Springer, Heidelberg, 2010)
4. W.H. Press, S.A. Teukolsky, W.T. Vetterling, B.P. Flannery, *Numerical Recipes: The Art of Scientific Computing*, 3rd edn. (Cambridge University Press, Cambridge, 2007)
5. A. Quarteroni, F. Saleri, *Scientific Computing with Matlab and Octave* (Springer, Heidelberg, 2014)
6. N. Metropolis, S. Ulam, The Monte Carlo method. J. Am. Stat. Assoc. **44**, 335 (1949)
7. S.H. Paskov, J.F. Traub, Faster evaluation of financial derivatives. J. Portf. Manag. **22**, 113 (1995)
8. F.Y. Kuo, I.H. Sloan, Lifting the curse of dimensionality. Not. the Am. Math. Soc. **52**, 1320 (2005)
9. J. Pitman, *Probability*. Springer Texts in Statistics (Springer, Berlin, 1993)
10. P.R. Halmos, *Measure Theory*. Graduate Texts in Mathematics (Springer, Berlin, 2014)

Monte Carlo Simulations in Quantum Mechanics

4

This chapter provides an introduction to particular Markov processes which obey the detailed balance condition. We explain the Metropolis algorithm—still the workhorse in many simulations—the heat bath algorithm, and the hybrid Monte Carlo algorithm. We will apply these algorithms to simulate the anharmonic oscillator. Later in this book, we shall use these algorithms to analyze non-perturbative aspects of spin systems and quantum field theories.

4.1 Markov Chains

We begin our discussion with a particular realization of the method of "importance sampling" as introduced in Chap. 3. For simplicity we first consider a system with a discrete set of configurations $\{\omega\}$. With an appropriate Markov chain, we generate configurations distributed according to a prescribed probability distribution $P(\omega)$. Later on P will be an equilibrium distribution of a statistical system. A Markov chain shows one important property: it is a discrete-time process for which the future behavior, given the past and the present, only depends on the present and not on the past. Hence the system has a short-term memory. A Markov chain is characterized by transition probabilities $W(\omega, \omega') = W(\omega \to \omega')$ which are interpreted as probabilities for a given configuration ω to make a transition to ω' in one time-step. In passing we note that a Markov process is a continuous-time version of a Markov chain.

Stochastic Matrices and Stochastic Vectors
We consider $W(\omega, \omega')$ as matrix elements of a *stochastic matrix* W. If the number of configurations is not finite, then W is a linear operator. Clearly, the matrix elements

© The Author(s), under exclusive license to Springer Nature Switzerland AG 2021
A. Wipf, *Statistical Approach to Quantum Field Theory*, Lecture Notes
in Physics 992, https://doi.org/10.1007/978-3-030-83263-6_4

of a *stochastic matrix* must be positive and normalized:

$$W(\omega, \omega') \geq 0 \quad \text{and} \quad \sum_{\omega'} W(\omega, \omega') = 1 . \tag{4.1}$$

The probability for jumping from ω to ω' during two time-steps is given by the sum of probabilities of all realizations of this two-step process. This means that the probability for the transition $\omega \to \omega'$ after two time-steps is given by

$$W^{(2)}(\omega, \omega') = \sum_{\omega_1} W(\omega, \omega_1) W(\omega_1, \omega') , \tag{4.2}$$

where one sums over all intermediate configurations which could have been visited after one time-step. Similarly, we obtain the probabilities

$$W^{(n)}(\omega, \omega') = \sum_{\omega_1 \cdots \omega_{n-1}} W(\omega, \omega_1) W(\omega_1, \omega_2) \cdots W(\omega_{n-1}, \omega') \tag{4.3}$$

for the transitions from ω to ω' after n steps and conclude that the long-time behavior of the chain is determined by high powers of the stochastic matrix.

A *stochastic vector* p has non-negative entries p_ω, which add up to 1:

$$\sum_\omega p_\omega = 1, \qquad p_\omega \geq 0 . \tag{4.4}$$

p_ω represents the probability of finding the system in configurations ω. In statistical mechanics p is identified with a (mixed) state. In order not to confuse ω with p_ω, we call ω a configuration and not a state as it is called in textbooks on Markov chains. A stochastic matrix transforms stochastic vectors into stochastic vectors:

$$\sum_{\omega'} (pW)(\omega') = \sum_{\omega\omega'} p_\omega W(\omega, \omega') = \sum_\omega p_\omega = 1 .$$

Let us consider the following stochastic matrix of a system with two configurations,

$$W = \begin{pmatrix} a & 1 - a \\ 0 & 1 \end{pmatrix} \qquad 0 \leq a \leq 1 . \tag{4.5}$$

Its eigenvalues are $\{1, a\}$ and its powers converge exponentially fast to a stochastic matrix with identical rows,

$$W^n = \begin{pmatrix} a^n & 1 - a^n \\ 0 & 1 \end{pmatrix} \overset{n \to \infty}{\longrightarrow} \begin{pmatrix} 0 & 1 \\ 0 & 1 \end{pmatrix} .$$

Later we will prove that for most stochastic matrices, the powers W^n converges to a stochastic matrix with *identical rows*. A counterexample is given by

$$W = \begin{pmatrix} a & \frac{1}{2}(1-a) & \frac{1}{2}(1-a) \\ 0 & 0 & 1 \\ 0 & 1 & 0 \end{pmatrix} \quad \text{with} \quad 0 \le a < 1. \tag{4.6}$$

The even powers are

$$W^n = \begin{pmatrix} a^n & \frac{1}{2}(1-a^n) & \frac{1}{2}(1-a^n) \\ 0 & 1 & 0 \\ 0 & 0 & 1 \end{pmatrix} \longrightarrow \begin{pmatrix} 0 & \frac{1}{2} & \frac{1}{2} \\ 0 & 1 & 0 \\ 0 & 0 & 1 \end{pmatrix}, \quad n \quad \text{even}$$

and the odd powers read

$$W^n = \begin{pmatrix} a^n & \frac{1}{2}(1-a^n) & \frac{1}{2}(1-a^n) \\ 0 & 0 & 1 \\ 0 & 1 & 0 \end{pmatrix} \longrightarrow \begin{pmatrix} 0 & \frac{1}{2} & \frac{1}{2} \\ 0 & 0 & 1 \\ 0 & 1 & 0 \end{pmatrix}, \quad n \quad \text{odd}.$$

Any stochastic vector p is mapped into

$$p W^{2n} \overset{n \to \infty}{\longrightarrow} \left(0, p_2 + \frac{p_1}{2}, p_3 + \frac{p_1}{2}\right)$$

$$p W^{2n+1} \overset{n \to \infty}{\longrightarrow} \left(0, p_3 + \frac{p_1}{2}, p_2 + \frac{p_1}{2}\right).$$

For a generic p, the series pW^n approaches exponentially fast a periodic orbit with period 2. This lack of convergence to a stochastic vector is only possible since every column of W contains at least one zero. Note that W has the fixed point $(0, 0.5, 0.5)$ and that for exceptional stochastic vectors with $p_2 = p_3$ the series $W^n p$ converges to this fixed point.

4.1.1 Fixed Points of Markov Chains

Every stochastic matrix has the eigenvalue 1. The corresponding right eigenvector is given by $\sim (1, 1, \ldots, 1)^T$. But since W acts from the right, we are interested in the left eigenvector with eigenvalue 1. We follow [1] and consider the series

$$p_n = \frac{1}{n} \sum_{j=0}^{n-1} p W^j. \tag{4.7}$$

Since stochastic vectors form a compact set, the series has a convergent subsequence

$$\frac{1}{n_k} \sum_0^{n_k-1} p \, W^j \longrightarrow P \, .$$

We multiply this subsequence by W from the right,

$$\frac{1}{n_k} \sum_1^{n_k} p \, W^j \longrightarrow P W \, .$$

In the difference of the last two formulas, only two terms in the left-hand sides remain, and we obtain in the limit $n_k \to \infty$

$$\frac{1}{n_k} \left(p - p \, W^{n_k} \right) \longrightarrow P - P W \, .$$

Since the left-hand side converges to zero, this implies the eigenvalue equation

$$P W = P \, , \tag{4.8}$$

which means that every stochastic matrix W has at least one *fixed point* P, i.e., a left eigenvector with eigenvalue 1.

We now assume that W has at least one column with minimal element greater than a positive number δ. This means that *all configurations* can jump with non-vanishing probability to at least one configuration. Stochastic matrices with this property are called *attractive*, and for an attractive matrix, the series $W^n p$ converges for any stochastic vector p. Note that W in (4.6) is not attractive and this explains why the associated Markov chain does not converge. For the proof of convergence, we first note that for two real numbers p and p' we have

$$|p - p'| = p + p' - 2 \min(p, p') \, ,$$

and this relation implies

$$\| p - p' \| = 2 - 2 \sum_\omega \min(p_\omega, p'_\omega) \tag{4.9}$$

for two stochastic vectors. We used the ℓ_1-norm $\| p \| = \sum_\omega |p_\omega|$.

Next we wish to prove that an attractive W acts on vectors $\Delta = (\Delta_1, \Delta_2, \dots)$ with

$$\| \Delta \| \equiv \sum |\Delta_\omega| = 2 \quad \text{and} \quad \sum \Delta_\omega = 0 \tag{4.10}$$

in a contractive way. Let us assign the stochastic vector $\boldsymbol{e}_\omega$ to each ω which represents the probability distribution for finding ω with probability one. Thus all entries of $\boldsymbol{e}_\omega$ vanish with the exception of entry ω, which is 1. The stochastic vectors $\{\boldsymbol{e}_\omega\}$ form an orthonormal basis. In a first step, we prove that W is contractive for difference vectors $\boldsymbol{e}_\omega - \boldsymbol{e}_{\omega'}$. Thus we apply the identity (4.9) to the stochastic vectors $\boldsymbol{e}_\omega W$ and $\boldsymbol{e}_{\omega'} W$, i.e., to rows of W belonging to ω and ω'. In case of an attractive W, we find for $\omega \neq \omega'$

$$\|\boldsymbol{e}_\omega W - \boldsymbol{e}_{\omega'} W\| = 2 - 2 \sum_{\omega''} \min \left\{ W(\omega, \omega''), W(\omega', \omega'') \right\}$$

$$\leq 2 - 2\delta = (1 - \delta) \underbrace{\|\boldsymbol{e}_\omega - \boldsymbol{e}_{\omega'}\|}_{=2} \quad \text{with} \quad 0 < \delta < 1 . \quad (4.11)$$

This already proves that W is contractive on difference vectors $\boldsymbol{e}_\omega - \boldsymbol{e}_{\omega'}$. We used the inequality

$$\sum_{\omega''} \min_{\omega''} \left\{ W(\omega, \omega'') W(\omega', \omega'') \right\} \geq \min \left\{ W(\omega, \omega^*) W(\omega', \omega^*) \right\} \geq \delta ,$$

where ω^* belongs to the particular column of W with elements greater than or equal to δ. Now we shall prove the contraction property for all vectors $\boldsymbol{\Delta}$ in (4.10). Adding and subtracting the relations

$$\|\boldsymbol{\Delta}\| = \sum_{\omega:\Delta_\omega \geq 0} \Delta_\omega - \sum_{\omega:\Delta_\omega < 0} \Delta_\omega = 2 \quad \text{and} \quad \sum_{\omega:\Delta_\omega \geq 0} \Delta_\omega + \sum_{\omega:\Delta_\omega < 0} \Delta_\omega = 0 ,$$

we extract the values of the individual sums,

$$\sum_{\Delta_\omega \geq 0} \Delta_\omega = 1 \quad \text{and} \quad \sum_{\Delta_\omega < 0} \Delta_\omega = -1 . \quad (4.12)$$

To keep the notation simple, we denote in the following formulas the non-negative elements of $\boldsymbol{\Delta}$ by Δ_ω and the negative elements by $\Delta_{\omega'}$. Note that the index sets $\{\omega\}$ and $\{\omega'\}$ are disjunct. With (4.12) we obtain

$$\|\boldsymbol{\Delta}\| = 2 = -2 \sum \Delta_\omega \sum \Delta_{\omega'} = - \sum \Delta_\omega \Delta_{\omega'} \underbrace{\|\boldsymbol{e}_\omega - \boldsymbol{e}_{\omega'}\|}_{=2} , \quad (4.13)$$

where in the last step we took into account $\omega \neq \omega'$. Now we use the simple relations

$$\sum \Delta_\omega \boldsymbol{e}_\omega = - \sum \Delta_{\omega'} \sum \Delta_\omega \boldsymbol{e}_\omega$$

$$\sum \Delta_{\omega'} \boldsymbol{e}_{\omega'} = + \sum \Delta_\omega \sum \Delta_{\omega'} \boldsymbol{e}_{\omega'} ,$$

which follow from (4.12), to bound the norm of ΔW:

$$\|\Delta W\| = \left\| \sum \Delta_\omega e_\omega W + \sum \Delta_{\omega'} e_{\omega'} W \right\| = \left\| - \sum \Delta_{\omega'} \Delta_\omega (e_\omega - e_{\omega'}) W \right\|$$
$$\leq - \sum \Delta_\omega \Delta_{\omega'} \|(e_\omega - e_{\omega'}) W\| \leq - \sum \Delta_\omega \Delta_{\omega'} \|e_\omega - e_{\omega'}\| (1 - \delta) , \tag{4.14}$$

where in the last step we applied the inequality (4.11). A comparison with formula (4.13) proves the desired inequality

$$\|\Delta W\| \leq (1 - \delta) \|\Delta\| , \tag{4.15}$$

for vectors which satisfy the conditions (4.10). Since this inequality is linear in Δ, it also holds for vectors which do not fulfill the condition $\|\Delta\| = 2$. This shows that a stochastic matrix W is contractive on all vectors the elements of which add up to zero and especially on differences of any two stochastic vectors.

Iterating the inequality (4.15) yields

$$\|\Delta W^n\| \leq (1 - \delta)^n \|\Delta\| . \tag{4.16}$$

Let us apply this estimate to the difference vector $p - P$, where P is a fixed point of the Markov chain and p is an arbitrary stochastic vector. The elements of $p - P$ add up to zero such that the bound in (4.16) applies,

$$\|(p - P)W^n\| = \|p W^n - P\| \overset{n \to \infty}{\longrightarrow} 0$$

or, equivalently

$$p W^n \overset{n \to \infty}{\longrightarrow} P . \tag{4.17}$$

For the particular stochastic vectors e_ω, the left-hand side is the row of $\lim_{n \to \infty} W^n = W^{eq}$ belonging to ω, and we conclude that W^{eq} has identical rows,

$$W^{eq}(\omega, \omega') = \lim_{n \to \infty} W^n(\omega, \omega') = P_{\omega'} . \tag{4.18}$$

To summarize, for an attractive W, the situation is similar to that for the stochastic matrix in (4.5): the series $W^n p$ converges to a fixed point, and all elements in one single column of $W^{eq} = \lim_{n \to \infty} W^n$ are equal.

It is easy to show that for an attractive W the fixed point P is *unique*. Indeed, if there would exist a second fixed point P', then

$$P'_{\omega'} = \sum_\omega P'_\omega W(\omega, \omega') = \lim_{n\to\infty} \sum_\omega P'_\omega W^n(\omega, \omega') \overset{(4.18)}{=} \sum_\omega P'_\omega P_{\omega'} = P_{\omega'}$$

would hold, and this proves the uniqueness of the fixed point. With (4.18) we also conclude that for an attractive W, the stochastic matrix W^{eq} is unique.

The generalization of these results to continuous systems with infinitely many configurations should be self-evident. As an example, consider a mechanical system, the pure states of which correspond to the points $q \in \mathbb{R}^n$. We now have to deal with *probability densities* $p(q)$ instead of stochastic vectors. The sums over discrete indices ω turn into integrals over the continuous variables q. The conditions (4.1) now read

$$W(q, q') \geq 0 \quad \text{and} \quad \int \mathscr{D}q' \, W(q, q') = 1 \,, \tag{4.19}$$

and the fixed point condition (4.8) takes the form

$$P(q') = \int \mathscr{D}q \, P(q) \, W(q, q') \,. \tag{4.20}$$

In statistical physics and quantum field theory at finite temperature, the fixed point $P(q)$ corresponds to the canonical ensemble, and we look out for Markov chains with this equilibrium distribution as fixed point.

4.2   Detailed Balance

The *condition of detailed balance* is a simple and physically well-founded constraint on a Markov chain which implies the fixed point equation (4.8). Detailed balance means a balance between any two configurations: the equilibrium probability for ω, multiplied by the jump probability from ω to ω', is equal to the equilibrium probability of ω' multiplied by the jump probability from ω' to ω,

$$P_\omega W(\omega, \omega') = P_{\omega'} W(\omega', \omega) \,. \tag{4.21}$$

If in equilibrium the configuration ω is more likely occupied than the configuration ω', then the transition amplitude from ω to ω' is less than the amplitude for the reverse transition.

The condition of detailed balance guarantees that P is a fixed point of the chain,

$$\sum_\omega P_\omega W(\omega, \omega') = \sum_\omega P_{\omega'} W(\omega', \omega) = P_{\omega'} \,, \tag{4.22}$$

but it does not fix W uniquely. We may use the residual freedom to choose simple and efficient algorithms. In particular the fast *Metropolis* and *heat bath algorithms* are universally applicable and are often used. For statistical systems which can be dualized, there exist the more efficient *cluster algorithms*, which do not suffer from the problem of the so-called critical slowing down. These and other Monte Carlo algorithms will be introduced and applied in this book. More material can be found in the textbooks [2–5].

4.2.1 Acceptance Rate

The probabilities $W(\omega, \omega)$ of not jumping are not constrained by the condition of detailed balance. Thus we may change the probabilities $W(\omega, \omega')$ for the transition between different configurations without violating the sum rule (4.1) if we only readjust the unconstrained $W(\omega, \omega)$.

We factorize the transition probability into the product of a test probability and an acceptance rate

$$W(\omega, \omega') = T(\omega, \omega')\, A(\omega, \omega') , \qquad (4.23)$$

where $T(\omega, \omega')$ is the probability of testing the new configuration ω' with given initial configuration ω. If ω' is tested, then the quantity $0 \leq A(\omega, \omega') \leq 1$ corresponds to the probability that the transition to ω' is accepted. Note that the conditions

$$\frac{T(\omega, \omega')A(\omega, \omega')}{T(\omega', \omega)A(\omega', \omega)} = \frac{P_{\omega'}}{P_{\omega}} \qquad (4.24)$$

do not fix the ratios of acceptance rates. A good Monte Carlo algorithm requires the best possible choice for these rates. For too small rates, only a tiny fraction of jumps is accepted, and the system is stuck in its initial configuration. We waste valuable computing time without passing through the configuration space. Thus, in many cases, one sets the greater of two acceptance rates $A(\omega, \omega')$ and $A(\omega', \omega)$ equal to 1. The smaller rate is chosen such that the condition (4.24) is satisfied.

4.2.2 Metropolis–Hastings Algorithm

In the original Metropolis algorithm, the test probability (jump probability) is symmetric in ω and ω'. Most common is a normal distribution centered at ω the variance of which must be tuned to get a reasonable acceptance rate for new configurations. A generalization due to Hastings allows for asymmetric test distributions. Typically one takes a distribution $T(\omega, \omega')$ that is the same for all configurations ω' that can be *reached* from ω [6, 7]. The test probability of the remaining configurations is set to zero. Thus, if N is the number of accessible configurations, then

$$T(\omega, \omega') = \begin{cases} 1/N & \text{if } \omega \to \omega' \text{ is possible} \\ 0 & \text{otherwise}. \end{cases} \tag{4.25}$$

We choose the acceptance rate

$$A(\omega, \omega') = \min\left(\frac{P_{\omega'} T(\omega', \omega)}{P_\omega T(\omega, \omega')}, 1\right) \tag{4.26}$$

for which W in (4.23) fulfills the condition for detailed balance. In fact, the condition

$$P_\omega T(\omega, \omega') \times \min\left(\frac{P_{\omega'} T(\omega', \omega)}{P_\omega T(\omega, \omega')}, 1\right) = P_{\omega'} T(\omega', \omega) \times \min\left(\frac{P_\omega T(\omega, \omega')}{P_{\omega'} T(\omega', \omega)}, 1\right)$$

is fulfilled both for $P_{\omega'} T(\omega', \omega)$ larger and smaller $P_\omega T(\omega, \omega')$.

A good choice of the initial configuration ω may save computing time. For example, at high temperatures the degrees of freedom are uncorrelated, and we choose the variables at random in contrast to low temperatures where they are strongly correlated.

We now discuss a particular implementation of the algorithm for a one-dimensional quantum mechanical system discretized on time-lattice with n points. We choose an initial configuration $q = (q_1, \ldots, q_n)$. The first lattice-variable q_1 is altered or remains unchanged according to the following rules:

1. Suggest a provisional change of q_1 to a randomly chosen q_1'.
2. If the action decreases, that is, $\Delta S < 0$, then permanently replace q_1 by q_1'.
3. If the action increases, choose an uniformly distributed random number $r \in [0, 1]$. The suggestion q_1' is accepted if $\exp(-\Delta S) > r$. Otherwise the lattice-variable q_1 remains unaltered.
4. Proceed with the variables $q_2, q_3, \ldots$ in the same way till all variables have been tested.
5. If the last lattice point is reached, a "sweep through the lattice" or a *Monte Carlo iteration* is finished, and one starts again with the first lattice point.

A realistic simulation includes thousands of sweeps through the lattice in order to reduce statistical errors. Depending on the initial configuration, it may take some sweeps to generate configurations distributed according to the equilibrium distribution. In order to check whether the Markov chain is close to "equilibrium," one measures selected expectation values as function of the Monte Carlo time with a MC iteration as unit of time. After equilibrium is reached, only statistical fluctuations remain, and we measure expectation values according to (3.38).

Two-State System
Consider a system with two energy eigenstates

$$H \left| \ell \right\rangle = E_\ell \left| \ell \right\rangle \quad (\ell = 1, 2) \quad \text{with} \quad \Delta E = E_2 - E_1 > 0. \tag{4.27}$$

The transition $\left| 2 \right\rangle \rightarrow \left| 1 \right\rangle$ reduces the energy and therefore $W(2, 1) = 1$. On the other hand, the excitation probability $W(1, 2)$ is equal to the Boltzmann factor

$$B_{21} = e^{-\beta(E_2 - E_1)} < 1.$$

Since the elements in any row add up to 1, the stochastic matrix has the form

$$W = \begin{pmatrix} 1 - B_{21} & B_{21} \\ 1 & 0 \end{pmatrix}. \tag{4.28}$$

It is attractive, and its powers

$$W^n = \frac{1}{1 + B_{21}} \left\{ \begin{pmatrix} 1 & B_{21} \\ 1 & B_{21} \end{pmatrix} + (-B_{21})^n \begin{pmatrix} B_{21} & -B_{21} \\ -1 & 1 \end{pmatrix} \right\}$$

converge exponentially fast to the stochastic matrix

$$W^{\text{eq}} = \frac{1}{1 + B_{21}} \begin{pmatrix} 1 & B_{21} \\ 1 & B_{21} \end{pmatrix} = \frac{1}{Z} \begin{pmatrix} e^{-\beta E_1} & e^{-\beta E_2} \\ e^{-\beta E_1} & e^{-\beta E_2} \end{pmatrix}, \tag{4.29}$$

where $Z = \exp(-\beta E_1) + \exp(-\beta E_2)$ is the partition function of the two-state system. It follows that every initial probability distribution converges to the Boltzmann distribution

$$p \longrightarrow P = \frac{1}{Z} \left(e^{-\beta E_1}, e^{-\beta E_2} \right). \tag{4.30}$$

Three-State System
Let $\left| 1 \right\rangle, \left| 2 \right\rangle, \left| 3 \right\rangle$ be the energy eigenstates with energies $E_1 < E_2 < E_3$. The stochastic matrix takes the form

$$W = \frac{1}{2} \begin{pmatrix} 2 - B_{21} - B_{31} & B_{21} & B_{31} \\ 1 & 1 - B_{32} & B_{32} \\ 1 & 1 & 0 \end{pmatrix}, \quad B_{pq} = e^{-\beta(E_p - E_q)}, \tag{4.31}$$

and its powers converge to

$$W^{\text{eq}} = \frac{1}{Z} \begin{pmatrix} e^{-\beta E_1} & e^{-\beta E_2} & e^{-\beta E_3} \\ e^{-\beta E_1} & e^{-\beta E_2} & e^{-\beta E_3} \\ e^{-\beta E_1} & e^{-\beta E_2} & e^{-\beta E_3} \end{pmatrix}. \tag{4.32}$$

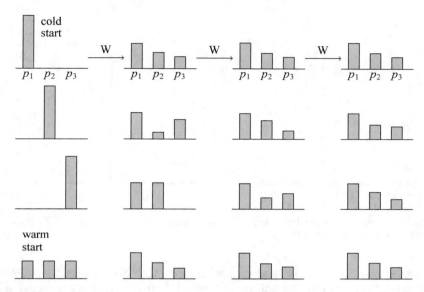

Fig. 4.1 Convergence to equilibrium for a system with three states, depending on the initial distribution

Thus every initial distribution converges to the Boltzmann distribution

$$P = \frac{1}{Z} \left(e^{-\beta E_1}, e^{-\beta E_2}, e^{-\beta E_3} \right) . \tag{4.33}$$

Figure 4.1 demonstrates the approach to equilibrium for different initial distributions. The energy differences are $\beta E_2 - \beta E_1 = 0.5$ and $\beta E_3 - \beta E_2 = 0.3$. The convergence to equilibrium is best for a "cold start" with initial state given by the ground state and a "warm start" with uniformly distributed initial probabilities. The convergence is worst when the system starts in the state with the highest energy.

4.2.3   Heat Bath Algorithm

For the heat bath algorithm, the transition probability $W(\omega, \omega')$ depends only on the final state ω' such that the condition of detailed balance (4.21) implies $W(\omega, \omega') \propto P_{\omega'}$. The normalization conditions for P and W lead to

$$W(\omega, \omega') = P_{\omega'} . \tag{4.34}$$

The algorithm is particularly useful when the equilibrium distribution P can be integrated or summed up easily. Let us first apply the heat bath algorithm to estimate one-dimensional integrals of the form $\langle O \rangle = \int O(x) P(x) \mathrm{d}x$ with fixed

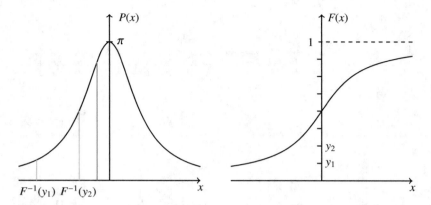

Fig. 4.2 A uniform distribution is mapped to a P-distribution by the inverse of F, where F denotes the anti-derivative of P

P and varying O. Thus we need random numbers distributed according to $P(x)$. To this end we first generate uniformly distributed random numbers y_i on the unit interval and consider the preimages $\{F^{-1}(y_i)\}$ of these numbers. Here F denotes the monotonically increasing anti-derivative of the probability density,

$$F(x) = \int_{-\infty}^{x} P(u)\, du \in [0, 1].$$

Because of the identity

$$y_2 - y_1 = \int_{F^{-1}(y_1)}^{F^{-1}(y_2)} P(u)\, du,$$

these preimages are distributed according to P. This is made clear in Fig. 4.2.

Box–Muller Method

This method has been introduced by George Box and Mervin Muller [8] and nicely illustrates how to extend the previous algorithm to higher dimensions. Uniformly distributed random points y in the unit square are mapped into normally distributed random numbers $x \in \mathbb{R}^2$ with mean $\bar{x}$ and variance σ. Thus we demand

$$d^2 y = \det\left(\frac{\partial y_i}{\partial x_j}\right) d^2 x = P(x)\, d^2 x, \qquad P(x) = \frac{1}{2\pi\sigma^2}\, e^{-(x-\bar{x})^2/2\sigma^2}. \qquad (4.35)$$

We introduce polar coordinates $x_1 - \bar{x}_1 = r\cos\varphi$ and $x_2 - \bar{x}_2 = r\sin\varphi$ and set $\varphi = 2\pi y_2$. Assuming that r only depends on y_1, we arrive at

$$d^2 y = \frac{dy_1}{dr} dr\, \frac{d\varphi}{2\pi} = \frac{1}{2\pi\sigma^2} e^{-r^2/2\sigma^2} r\, dr d\varphi \quad \text{or} \quad \frac{dy_1}{dr} = \frac{r}{\sigma^2} e^{-r^2/2\sigma^2}. \qquad (4.36)$$

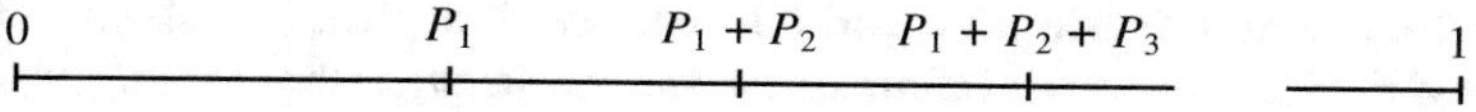

Fig. 4.3 The configurations are ordered such that $P_1 \geq P_2 \geq \cdots \geq P_n$

Since y_1 must take values in $[0, 1]$, we end up with $y_1 = 1 - e^{-r^2/2\sigma^2}$ or

$$r^2 = -2\sigma^2 \log(1 - y_1) .$$

Using the uniformly distributed random number $1 - y_1$ in place of y_1, we end up with

$$x_1 = \bar{x}_1 + \sigma\sqrt{-2\log y_1}\, \cos(2\pi y_2)$$
$$x_2 = \bar{x}_2 + \sigma\sqrt{-2\log y_1}\, \sin(2\pi y_2) . \qquad (4.37)$$

One needs to calculate the logarithm of the first random number and two trigonometric functions of the other random number and this takes some time. There exists other methods to generate normally distributed random numbers, for example, the polar method.

A Finite Number of Configurations
For systems with a finite number of configurations, the anti-derivative F is simply a step function. Let us order the n configuration $\omega = \{\omega_1, \omega_2, \ldots, \omega_n\}$ according to their probabilities, such that $P_1 \geq P_2 \geq \cdots \geq P_n$ as shown in Fig. 4.3. Then a possible implementation of the direct *heat bath algorithm* proceeds as follows:

1. Select n uniformly distributed random number $r \in [0, 1]$.
2. If $r < P_1$, choose the first configuration ω_1 and go to 1.
3. Otherwise, if $r < P_1 + P_2$, select the second configuration ω_2 and go to 1.
4. ...and so on.

The stochastic matrix corresponding to this algorithm is ideal,

$$W = \begin{pmatrix} P_1 & P_2 & \ldots & P_n \\ P_1 & P_2 & \ldots & P_n \\ \vdots & \vdots & \ldots & \vdots \\ P_1 & P_2 & \ldots & P_n \end{pmatrix} = W^{\mathrm{eq}} \Longrightarrow W^2 = W . \qquad (4.38)$$

This simple algorithm has an obvious disadvantage: it is only applicable to systems with a relatively small number of configurations and slows down when this number increases. Also note that for a Boltzmann equilibrium distribution, one needs to know the partition function in order to compute the probabilities.

So let us modify the simple algorithm such that it applies to systems with continuous variables. The modification uses a Metropolis algorithm for the conditional probabilities of the joint distribution $P(\boldsymbol{q})$. For example, $P(q_1|q_2, \dots, q_n)$ denotes the probability of q_1 for given $q_2, \dots, q_n$. If $\boldsymbol{q}$ denotes the configuration at a given time, then a configuration $\boldsymbol{q}'$ at the subsequent time is chosen according to

$$q_1' \sim P\left(q_1|\, q_2, q_3, \dots, q_n\right)$$
$$q_2' \sim P\left(q_2|\, q_1', q_3, \dots, q_n\right)$$
$$q_3' \sim P\left(q_3|\, q_1', q_2', \dots, q_n\right)$$

$$\vdots \qquad \vdots$$

$$q_n' \sim P\left(q_n|\, q_1', q_2', \dots, q_{n-1}'\right),$$

where $q_1' \sim P(q_1|\, q_2, q_3, \dots, q_n)$ means that the new q_1' has to be chosen according to the conditional probability $P(\dots)$. Later on we will discuss an implementation of this algorithm for spin systems.

4.3   The Anharmonic Oscillator

We return to one-dimensional quantum mechanical systems at imaginary time and discretized on a time-lattice. They are characterized by their Euclidean lattice action

$$S(\boldsymbol{q}) = \varepsilon \sum_{j=0}^{n-1} \left\{ \frac{m}{2} \frac{(q_{j+1} - q_j)^2}{\varepsilon^2} + V(q_j) \right\}. \tag{4.39}$$

In particular we shall consider the anharmonic oscillator with quartic potential

$$V(q) = \mu q^2 + \lambda q^4 \tag{4.40}$$

in more detail. The choice of the number n of lattice points and of the lattice constant ε is limited mainly by two aspects:

- ε should be sufficiently small to be near the continuum limit $\varepsilon \to 0$.
- The quantities of interest should fit into the interval $n\varepsilon$. For instance, the width of the ground state should be less than $n\varepsilon$.

If λ_0 is a typical length scale of the system at hand, then the quantities n and ε should satisfy constraints of the type

$$\varepsilon \lesssim \frac{\lambda_0}{10} \quad \text{and} \quad n\varepsilon \gtrsim 10\lambda_0. \tag{4.41}$$

Another problem concerns the size of *statistical fluctuations* in any Monte Carlo simulation. The relative standard deviation of a random variable O is

$$\Delta_O = \sqrt{\frac{\langle O^2 \rangle - \langle O \rangle^2}{\langle O \rangle^2}} \propto (\text{number of lattice points})^{-1/2} . \tag{4.42}$$

As an estimate for the expectation value $\langle O \rangle$, we take

$$\bar{O} = \frac{1}{M} \sum_{\mu=1}^{M} O(\boldsymbol{q}_\mu) \tag{4.43}$$

with Boltzmann-distributed configurations $\boldsymbol{q}_\mu$. Depending on the initial configuration, the Markov chain may need some "time" to equilibrate. In the simulations of the anharmonic oscillator presented below, equilibrium is reached after approximately 10–100 sweeps through the lattice. In addition, since configurations of successive sweeps are correlated, only every MA'th sweep is used to estimate expectation values. The number MA should be larger than the relevant *auto-correlation time*—the time over which the values $O(\boldsymbol{q}_\mu)$ are correlated. Different random variables may have vastly different auto-correlation times. As a general rule, they are large for spatially averaged quantities.

Hence we must generate $M \times MA$ configurations to obtain the Monte Carlo estimate (4.43). For the particular simulations of the anharmonic oscillator presented below, we take one in five configurations to estimate the correlation functions

$$\left\langle q_i^2 \right\rangle, \quad \left\langle q_i^4 \right\rangle \quad \text{and} \quad \langle q_i q_{i+m} \rangle . \tag{4.44}$$

Here we follow [9] and apply the well-known *virial theorem*

$$\frac{1}{2m} \left\langle 0 \left| \hat{p}^2 \right| 0 \right\rangle = \frac{1}{2} \left\langle 0 \left| \hat{q} V'(\hat{q}) \right| 0 \right\rangle \tag{4.45}$$

to relate the correlation functions (4.44) to the ground state energy of the oscillator. This theorem yields the following path integral representation for this energy,

$$E_0 = \left\langle 0 \left| \tfrac{1}{2} \hat{q} V'(\hat{q}) + V(\hat{q}) \right| 0 \right\rangle = \frac{1}{Z} \int \mathscr{D}q \, e^{-S[q]} \left(\frac{1}{2} q V'(q) + V(q) \right) . \tag{4.46}$$

Similarly, using (2.65) we can relate the energy of the first excited state to vacuum expectation values as follows,

$$E_1 = -\frac{1}{\Delta\tau} \lim_{\tau \to \infty} \log \frac{\left\langle 0 \left| \hat{q}_E(\tau + \Delta\tau) \, \hat{q}(0) \right| 0 \right\rangle}{\left\langle 0 \left| \hat{q}_E(\tau) \, \hat{q}_E(0) \right| 0 \right\rangle} + E_0 . \tag{4.47}$$

Finally, to extract information about the ground state wave function, we recall

$$K(\tau, q', q) = \sum_n e^{-\tau E_n} \psi_n(q') \psi_n(q) , \qquad (4.48)$$

where the ψ_n denote the normalized energy eigenfunctions in position space. Thus we may compute the probability density for finding the particle (in its ground state) at q by setting $q' = q$ and assuming large Euclidean times,

$$\lim_{\tau \to \infty} \frac{K(\tau, q, q)}{\int dq\, K(\tau, q, q)} = |\psi_0(q)|^2. \qquad (4.49)$$

The left-hand side is given by an imaginary-time path integral and hence is accessible to Monte Carlo simulations. In fact, the simulation program `anharmonic.c` on p. 77 counts for all MC configuration the number of coordinates that lie in each bin of a binning of coordinate space to calculate the left-hand side of (4.49).

4.3.1   Simulating the Anharmonic Oscillator

Since a CPU only knows numbers, we rescale the dimensionful constants and coordinates in the action (4.39) with powers of the dimensionful lattice constant ε to arrive at dimensionless lattice quantities $(m_L, \mu_L, \lambda_L, q_L)$:

$$q_L = q/\varepsilon, \quad m_L = \varepsilon m, \quad \mu_L = \varepsilon^3 \mu \quad \text{and} \quad \lambda_L = \varepsilon^5 \lambda . \qquad (4.50)$$

In terms of these dimensionless quantities, the lattice action takes the simple form

$$S(q) = \sum_{j=0}^{n-1} \left\{ \frac{m_L}{2} (q_{j+1} - q_j)_L^2 + \mu_L q_{j,L}^2 + \lambda_L q_{j,L}^4 \right\} . \qquad (4.51)$$

In a local Metropolis algorithm, we test a new configuration q' which differs from the old configuration q only on one lattice point, say lattice point j. Then the difference of the two actions is

$$S(q') - S(q) \approx (q'_j - q_j)_L \Big\{ -m_L (q_{j+1} + q_{j-1})_L$$
$$+ \left(q'_j + q_j \right)_L \Big\{ m_L + \mu_L + \lambda_L \left(q'^2_j + q^2_j \right)_L \Big\} \Big\}. \qquad (4.52)$$

Here the problem arises how to determine energies or lengths in physical units. They are only given in terms of the unknown lattice constant ε which does not even enter the lattice action $S(q)$. Thus one first calculates some observable (e.g., example an energy) which is then compared to the experimentally known value to set the scale.

Alternatively one may express all dimensionful quantities in units of a known and fixed unit of length ℓ,

$$\varepsilon = a\ell, \qquad q = \tilde{q}\ell, \qquad m = \tilde{m}/\ell, \qquad \mu = \tilde{\mu}/\ell^3 \quad \text{and} \quad \lambda = \tilde{\lambda}/\ell^5. \qquad (4.53)$$

In this system of units

$$S(q') - S(q) = \left(\tilde{q}'_j - \tilde{q}_j\right) \left\{ -\frac{\tilde{m}}{a}\left(\tilde{q}_{j+1} + \tilde{q}_{j-1}\right) \right.$$
$$\left. + \left(\tilde{q}'_j + \tilde{q}_j\right)\left\{\frac{\tilde{m}}{a} + a\tilde{\mu} + a\tilde{\lambda}\left(\tilde{q}'^2_j + \tilde{q}^2_j\right)\right\} \right\}. \qquad (4.54)$$

This formula is used in the header file `stdanho.h` on page 80 which is called by the program `anharmonic1.c` listed on p. 77. The parameters m, μ, λ, the number of lattice points (in the C-program n is renamed N), and the lattice constant a are all defined in another header file `constants.h` on p. 79. In order to have uncorrelated configurations, only one out of *MA* configuration is measured. In addition, since a Markov chain needs some time to reach equilibrium, we start measuring configurations only after *MA* × *ME* sweeps through the lattice.

To measure the square of $\psi_0(q)$ on the interval $[-INTERV, INTERV]$ with the help of (4.49), we divide the interval into *BIN* bins. With the parameter *DELTA*, we adjust the amplitude of a tentative coordinate change during a local update according to $q' = q + DELTA \times (1 - 2r)$, where r is a uniform random number on $[0, 1]$. The program `anharmonic1.c` generates the histogram of the probability distribution. In Fig. 4.4 we plotted the so obtained density $|\psi_0(q)|^2$, both for the harmonic and the anharmonic oscillator.

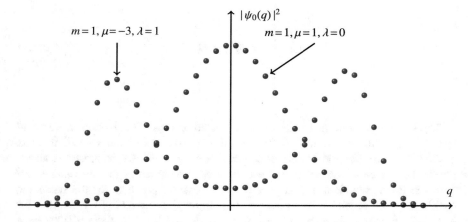

Fig. 4.4 Square of the ground state wave function for the harmonic and anharmonic oscillators as obtained with the Monte Carlo program `anharmonic1.c` listed on p. 77. The dimensionless parameters are given in the plot. More details are explained in the main text

Table 4.1 Ground state energies of oscillators with $(m, \mu) = (1, 0.5)$ and $(m, \mu, \lambda) = (1, -3, 1)$. The entries in the third column measure the violation of the Wick relation (4.55)

a	$E_0(1, 0.5)$	Wick	E_0(exact)	$E_0(1, -3, 1)$
1	0.4477	−0.0008	0.4473	−1.4624
1/2	0.4851	0.0010	0.4851	−1.1339
1/3	0.4928	0.0016	0.4932	−1.0177
1/4	0.4926	0.0014	0.4962	−0.9758
1/5	0.4970	0.0040	0.4976	−0.9466
1/6	0.4948	0.0006	0.4983	−0.9369
1/7	0.5016	0.0003	0.4988	−0.9173
1/8	0.4989	0.0067	0.4991	−0.9144
1/9	0.4992	0.0012	0.4993	−0.9160
1/10	0.4985	0.0009	0.4994	−0.9097

Fig. 4.5 Convergence of the MC results for the ground state energies of the harmonic oscillator (*top panel*) and anharmonic oscillator (*bottom panel*) with decreasing lattice constant a. In both cases m sets the scale

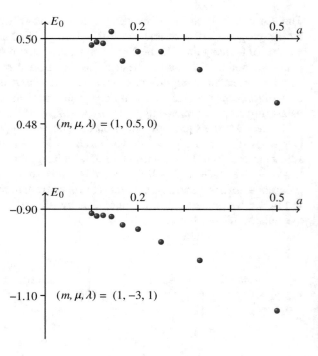

The similar program `anharmonic2.c` listed on page 78 calculates the ground state energy from the formula (4.46). We fix the size Na of the interval in units of ℓ and increase the number of lattice point N or equivalently decrease the lattice constant a. Table 4.1 shows how the ground state energies of the oscillators vary with the lattice constant a. The exact lattice values E_0(exact) for the harmonic oscillator in the second to last column are computed numerically. We see that for $a \lesssim 0.2$ the system is already close to its continuum limit. All simulations have been performed on the fixed interval $aN = 10$. How fast the ground state energies converge to their continuum values is plotted in Fig. 4.5.

An extrapolation of the lattice results to the continuum limit yield the following Monte Carlo estimates:

$$E_0(\text{harmonic oscillator}) \approx 0.50 \quad , \quad E_0(\text{anharmonic oscillator}) \approx -0.91.$$

Thus we reproduced the exact ground state energy of the harmonic oscillator in the continuum limit. Finally we note that for the harmonic oscillator

$$\left\langle 0 \left| \hat{q}^4 \right| 0 \right\rangle - 3 \left\langle 0 \left| \hat{q}^2 \right| 0 \right\rangle^2 = 0 . \tag{4.55}$$

The column labeled "Wick" in Table 4.1 contains estimates for the left-hand side of this identity. The deviations from zero are due to statistical errors in our simulations.

4.4 Hybrid Monte Carlo Algorithm

The powerful hybrid Monte Carlo algorithm has been developed by S. DUANE ET AL. [10]. A recommendable introduction can be found in [11], and in our presentation, we follow in part the reviews [12, 13]. The algorithm represents a combination of molecular dynamics (see [14]) and Metropolis algorithm. It aims at global updates of whole configurations with reasonable large acceptance rates, in order to minimize the time required to generate independent configurations.

Molecular dynamics (MD) simulations are frequently and successfully used to study (classical) many-body systems and are applied to problems in material science, astrophysics, and biophysics. In molecular dynamics simulations, one solves the Hamiltonian equations of motion

$$\dot{q}_i = \frac{\partial H}{\partial p_i} \quad , \quad \dot{p}_i = -\frac{\partial H}{\partial q_i} \tag{4.56}$$

numerically and makes use of the ergodic hypothesis, according to which the statistical ensemble averages are equal to time averages. From an initial configuration, represented by a point $(\boldsymbol{q}_0, \boldsymbol{p}_0)$ in phase space, one obtains a unique solution of Hamilton's equations of motion. Without any numerical errors, the energy is a constant of motion, and this simple observation will be relevant in what follows.

In Euclidean quantum mechanics discretized on an imaginary-time lattice Λ with n points $\{x\}$, we introduce an *extended phase space* with dimension $2n$. Each point in extended phase space consists of a broken-line path characterized by variables $\{q_x\}$ and their canonically conjugated momenta $\{p_x\}$. To construct a Markov process with *global updates* and high acceptance rate, we introduce the following auxiliary Hamiltonian in extended phase space:

$$H(\boldsymbol{q}, \boldsymbol{p}) = \frac{\boldsymbol{p}^2}{2} + S(\boldsymbol{q}), \quad \boldsymbol{p}^2 = \sum_{x \in \Lambda} p_x^2 , \tag{4.57}$$

where $q = (q_1, \ldots, q_n)$ and $p = (p_1, \ldots, p_n)$. This Hamiltonian generates a dynamics with respect to a fictitious time—it will be the time of the associated Markov process. Integrating the equations of motion over a certain "period of time" maps a point (q, p) in extended phase space to another point (q', p'). This mapping is used to suggest the image point as new configuration in a Markov chain in which the pair (q', p') is accepted with probability

$$A\left(q, p \rightarrow q', p'\right) = \min\left\{1, \exp\left(H(q, p) - H(q', p')\right)\right\} . \tag{4.58}$$

For an exact solution of the equations of motion (4.56), energy is conserved, and the acceptance probability would be one. But any numerical integration comes along with rounding errors such that H is only conserved up to discretization errors in which case the acceptance of the new configuration is not guaranteed. By tuning the period of integration, one can ensure that the acceptance rate stays close to 1. In almost all algorithms, the fictitious momenta are drawn from a Gaussian distribution,

$$P_G(p) = \mathcal{N}e^{-p^2/2} \tag{4.59}$$

Now we show that the corresponding Markov process obeys the condition of detailed balance if the integrator is time-reversible. Let us first calculate the transition probability from q to q'. Under time-reversal a point (q, p) in phase space is mapped to $(q, -p)$. Since the equations of motion and the integrator are both time-reversible, we conclude that the probability to move from an initial to a final point is invariant under time-reversal,

$$T\left(q, p \rightarrow q', p'\right) = T\left(q', -p' \rightarrow q, -p\right) . \tag{4.60}$$

With the known acceptance rate in (4.58) and the Gaussian distribution of momenta, we can calculate the transition probability $W(q, q')$ as follows:

$$W(q, q') = \int \mathscr{D}p\mathscr{D}p' \, P_G(p) \, T\left(q, p \rightarrow q', p'\right) A\left(q, p \rightarrow q', p'\right) , \tag{4.61}$$

where one averages and sums over initial and final momenta, respectively. Thus the left-hand side of the condition of detailed balance

$$e^{-S(q)} \, W(q, q') = e^{-S(q')} \, W(q', q) \tag{4.62}$$

can be written as

$$e^{-S(q)} \, W(q, q') = \mathcal{N} \int \mathscr{D}p\mathscr{D}p' e^{-H(q, p)} \, T\left(q, p \rightarrow q', p'\right) A\left(q, p \rightarrow q', p'\right) . \tag{4.63}$$

Similarly as in the proof of detailed balance for the Metropolis algorithm on p. 60, we can show that

$$e^{-H(q,p)} A\left(q, p \to q', p'\right) = e^{-H(q',p')} A\left(q', p' \to q, p\right) . \tag{4.64}$$

Inserting this relation into (4.63) and using that H and A are even functions of the momenta, we conclude

$$e^{-S(q)} W(q, q') = \mathcal{N} \int \mathscr{D}p \mathscr{D}p' e^{-H(q',p')} T\left(q, p \to q', p'\right) A\left(q', p' \to q, p\right)$$

$$= \mathcal{N} \int \mathscr{D}p \mathscr{D}p' e^{-H(q',p')} T\left(q', p' \to q, p\right) A\left(q', p' \to q, p\right)$$

$$= e^{-S(q')} W(q', q) .$$

The proof makes clear that the distribution of momenta cannot be arbitrary. We must impose that the product $P_G \exp(-S)$ is proportional to $\exp(-H)$ and for the Hamiltonian (4.57) this means that the distribution must be Gaussian. The described HMC scheme leaves room for improvement. In particular, one has considerable freedom in defining the auxiliary Hamiltonian that governs the molecular dynamics evolution. One method of speeding up the evolution is Fourier acceleration, in which the different Fourier modes are assigned different step sizes or masses. This techniques was introduced for Langevin simulations of field theories [15] and can be applied to HMC simulations as well [16]. Note that the proof of detailed balance also assumes that the molecular dynamics evolution defines a volume preserving symplectic map in phase space.

4.4.1 Implementing the HMC Algorithm

To fulfil the condition of detailed balance, we need a time-reversible and symplectic integrator to numerically solve the fictitious Hamiltonian dynamics. Using the naive forward difference operator $\dot{f}(\tau)h \approx f(\tau + h) - f(\tau)$ in the discretized equations of motion leads to the *time-irreversible* prescription

$$q(\tau + h) = q(\tau) + h p(\tau)$$
$$p(\tau + h) = p(\tau) + h F\left(q(\tau)\right) , \tag{4.65}$$

with force $F = -\nabla_q S$. This prescription must not be used in any HMC algorithm. In most simulation the time-reversible leapfrog integration is used instead. Here one first moves forward with the momenta a half step in fictitious time. Then one moves forward several time-steps alternately with coordinates q and momenta p. The last move of the momenta is again a half step in fictitious time.

One sweep amounts to an integration over a fictitious time-interval $\tau = ph$. We denote the position and momentum at fictitious time kh by $q_{(k)}$ and $p_{(k)}$. In particular $q_{(p)}$ and $p_{(p)}$ are the final position and momentum. From an initial configuration, we determine the final configuration as follows:

1. Begin with an initial lattice field $q_{(0)}$. Depending on the parameters (couplings, temperature), one chooses a cold or warm start.
2. Generate Gauss-distributed momenta $p_{(0)}$ with variance 1 and mean 0.
3. Move a half step forward with the momenta,

$$p_{(1/2)} = p_{(0)} + \frac{h}{2} F\left(q_{(0)}\right) . \tag{4.66}$$

4. Iterate the following two steps:

$$\begin{array}{lll} \text{(a)} & q_{(k)} = q_{(k-1)} + h\,p_{(k-1/2)}, & k = 1, 2, \ldots, p \\[2mm] \text{(b)} & p_{(k+1/2)} = p_{(k-1/2)} + hF\left(q_{(k)}\right), & k = 1, \ldots, p-1. \end{array} \tag{4.67}$$

5. Finally move a half step forward with the momenta,

$$p_{(p)} = p_{(p-1/2)} + \frac{h}{2} F\left(q_{(p)}\right) . \tag{4.68}$$

6. Accept the newly generated configuration $(q', p') = \left(q_{p}, p_{(p)}\right)$ with probability given in (4.58).
7. Start over again with the old or the new configuration at point 2.

The so obtained Markov chain converges to the Boltzmann distribution corresponding to the action S.

The leapfrog integrator used in the steps $3, 4,$ and 5 is time-reversible and defines a symplectic and therefore volume preserving map in phase space. This *second-order symplectic integrator* maps the initial point $q_{(0)}, p_{(0)}$ to the final point $q_{(p)}, p_{(p)}$. To test for a correct implementation of the algorithm, one observes that for any symplectic map $q, p \to q', p'$

$$\int \mathscr{D}q' \mathscr{D}p'\,\mathrm{e}^{-H(q',p')} = \int \mathscr{D}q\,\mathscr{D}p\,\mathrm{e}^{-H(q,p)-\Delta H(q,p)} , \tag{4.69}$$

where $\Delta H(q, p) = H(q', p') - H(q, p)$ denotes the increase in energy along the trajectory. With the help of the *Jensen inequality*, which expresses the convexity of the exponential function, one concludes that the inequality

$$1 = \left\langle \mathrm{e}^{-\Delta H} \right\rangle \geq \mathrm{e}^{-\langle \Delta H \rangle} \tag{4.70}$$

must hold in the simulations.

There are two parameters which must be adjusted to increase speed and efficiency of the algorithm. These are the step size h of the discretization of fictitious time and the interval $\tau = ph$ over which one follows a trajectory. In the leapfrog integration, one violates the conservation of energy due to discretization errors. According to [17]

$$\Delta H \propto h^3 + O(h^4). \tag{4.71}$$

For fixed integration time τ, the expectation value of ΔH shows the following dependency on the lattice volume V and the step size:

$$\langle \Delta H \rangle \propto V h^4. \tag{4.72}$$

Thus, for fixed volume, we can adjust the acceptance rate by tuning the step size h. Note that with increasing τ, we spend more time to generate new configurations, but at the same time these configurations are less correlated. As always one must find a good compromise to end up with an efficient algorithm. Also note that due to the unavoidable rounding errors, the leapfrog integration is not exactly time-reversible, and one should check that this does not screw up the results. For example, one can integrate first forward and then again backward to estimate the violation of time-reversibility. For more details, you may consult [13].

4.4.2 HMC Algorithm for Harmonic Oscillator

The explicit form of the HMC algorithm for the one-dimensional oscillator follows from the auxiliary Hamiltonian,

$$H = \frac{p^2}{2} + S(q), \qquad S(q) = \sum_x \left(\frac{1}{2}(\partial q)_x^2 + \frac{\omega^2}{2}q_x^2 + \frac{\lambda}{4}q_x^4 \right), \tag{4.73}$$

where ∂ is a discretization of the differential operator on the lattice with lattice constant one. The non-linear equations of motion read

$$\dot{q}_x = +\frac{\partial H}{\partial p_x} = p_x$$

$$\dot{p}_x = -\frac{\partial H}{\partial q_x} = (\partial^2 q)_x - \omega^2 q_x - \lambda q_x^3. \tag{4.74}$$

For the harmonic oscillator with $\lambda = 0$, we obtain a system of coupled linear differential equations

$$\dot{q} = p \quad \text{and} \quad \dot{p} = -Aq \tag{4.75}$$

with the positive matrix $A = -\partial^2 + \omega^2$. For the most naive discretization, the matrix ∂^2 is a tridiagonal Toeplitz matrix with -2 on the main diagonal and 1 on the upper and lower bidiagonals—it is just the matrix in (2.80) with particular parameters. The explicit solution of the HMC implementation on p. 74 reads

$$\begin{pmatrix} q_{(p)} \\ p_{(p)} \end{pmatrix} = M \begin{pmatrix} q_{(0)} \\ p_{0)} \end{pmatrix} \tag{4.76}$$

with *symplectic matrix*

$$M = \begin{pmatrix} P & Qh \\ AXQh & P \end{pmatrix}, \tag{4.77}$$

wherein

$$X = \frac{H}{4} - \mathbb{1} \quad \text{with} \quad H = Ah^2.$$

The commuting entries P, Q are matrix polynomials in H of degrees p and $p-1$, respectively. For example, for $p = 4$, we obtain

$$P = \frac{H^4}{2} - 4H^3 + 10H^2 - 8H + 1$$

$$Q = \qquad - H^3 + 6H^2 - 10H + 4.$$

These matrices satisfy the constraints

$$P^2 - AXQ^2 h^2 = \mathbb{1} \tag{4.78}$$

which express the fact that M is a symplectic matrix or that the map (4.76) in phase space is a *volume preserving* symplectic map—an integrator with this property is called symplectic.

For a detailed study of hybrid Monte Carlo algorithms for non-interacting models, the reader may consult [18]. This paper contains higher-order discretizations of the molecular dynamics equations of motion, results on the autocorrelations functions and on Metropolis acceptance probabilities as a function of the integration step size. For models with dynamical fermions, the use of higher-order symplectic integrators in combination with Fourier acceleration [16] often leads to improved algorithms, in particular for intermediate and strong coupling. For example, a fourth-order integrator introduced by FOREST and RUTH [19] and extended in [20] has been useful in simulations of supersymmetric models with bosonic and fermionic degrees of freedom [21].

For theories with local interactions, one may employ a local version of the HMC algorithm where single site (and link) variables are evolved in a HMC style [22]. In parameter regions where Metropolis and heat bath algorithms have low acceptance rates, the local HMC algorithm may be preferable [23].

4.5   Programs for Chap. 4

This chapter contains the C-programs

- anharmonic1.c
- anharmonic2.c
- constants.h
- stdanho.h

to simulate the anharmonic oscillator with the local Metropolis algorithm.

The program anhamonic1.c in Listing 4.1 estimates the probability density in position space $|\psi_0(q)|^2$ for the harmonic and anharmonic oscillators. The parameters of the model are stored in the header files constants.h and stdanho.h.

Listing 4.1   Anharmonic oscillator I

```
1    /*program anharmonic.c*/
2    /*Metropolis algorithm for the (an)harmonic oscillator*/
3    /*L=MASS/2 v^2+MU q^2+LAMBDA*q^4*/
4    /*calcules the square of the ground state wave function*/
5    #include <stdio.h>
6    #include <stdlib.h>
7    #include <math.h>
8    #include <time.h>
9    #include "constants.h"
10   /*defines: N,A,ME,MA,BIN,INTERV,MASS,MU,LAMBDA,DELTA*/
11   /*init: q[N],qneu,mass1,lambda1,muleff,reject,stretch,versch*/
12   #include "stdanho.h" /*function deltaS(qnew,qold,summenn)*/
13                        /*MA MC-iterations: mcsweep(*zgr,*q)*/
14   int main(void)
15   {
16     unsigned int i,j;
17     int *zgr,p,bin[BIN];
18     zgr=&reject; srand48(time(NULL));
19     /*initialization*/
20     for (i=0;i<N;i++)
21       q[i]=DELTA*(1-2*drand48());
22     for (i=0;i<BIN;i++)
23       bin[i]=0;
24     /*thermalisation*/
25     for (i=0;i<ME;i++)
26       mcsweep(zgr,q);
27     /*calculate binning*/
28     reject=0;
29     for (i=0;i<M;i++)
30       { mcsweep(zgr,q);binning(bin,q);};
31     /*output of probability density and rejection rate*/
```

```
32       printf("bin\t probability\n");
33       for (i=0;i<BIN;i++)
34          printf("%i\t %7.3f\n",i,20*bin[i]/(double)M);
35       printf("\nrejected configurations %.2f percent\n",
36             (float)100*reject/(N*M*MA));
37       return 0;
38    }
```

The program anharmonic2.c in Listing 4.2 estimates ground state energies E_0 of the harmonic and anharmonic oscillators with the Metropolis algorithm. Thereby the virial theorem (4.46) is used. The parameters are contained in header files called by the program.

Listing 4.2 Anharmonic oscillator II

```
1    /*program anharmonic2.c*/
2    /*Metropolis algorithm for anharmonic oscillator*/
3    /*L=MASS/2 v^2+MU q^2+LAMBDA*q^4 */
4    /*calculates ground state energy*/
5    #include <stdio.h>
6    #include <stdlib.h>
7    #include <math.h>
8    #include <time.h>
9    #include "constants.h"
10   #include "stdanho.h"
11   int main(void)
12   {
13     unsigned int i,j;
14     int *zgr,p; double moment2=0,moment4=0,wick;
15     zgr=&reject;
16     srand48(time(NULL));
17     /*initialize*/
18     for (i=0;i<N;i++)
19       q[i]=DELTA*(1-2*drand48());
20     /*thermalize*/
21     for (i=0;i<ME;i++)
22       mcsweep(zgr,q);
23     /*simulation and calculation of moments*/
24     reject=0;
25     for (i=0;i<M;i++)
26       {
27          mcsweep(zgr,q);
28          moment2=moment2+moments(2,q);
29          moment4=moment4+moments(4,q);
30       };
31     /*ground state energy, Wick-test, output*/
32     moment2=moment2/M;
33     moment4=moment4/M;
34     wick=3*moment2*moment2-moment4;
35     wick=3*moment2*moment2-moment4;
36     printf("2nd moment =\t\t %7.4f\n4th moment =\t\t %7.4f\n"
37       "ground state energy =\t %7.4f\nWick-test =\t\t %7.4f\n",
38         moment2,moment4,2*MU*moment2+3*LAMBDA*moment4,wick);
```

```
39        printf("\nrejected configurations %.2f\n",
40          (float)reject/(N*M*MA));
41      return 0;
42  }
```

Header Files

The following files are called by the programs `anharmonic.c`. In the header file `constants.h` in Listing 4.3, we defined the constants and variables called by the programs.

The header file `stanho.h` in Listing 4.4 contains four functions, which are called 1 by the main programs.

The first functions `delta_action` calculates the change of the action when the coordinate at a given lattice point is changed from x to y. The function expects the sum of coordinates on the neighboring lattice points, denoted by xs. The variables *muleff*, *lambdal*, and *massl* must be provided.

The next function `mcsweep` performs *MA* successive Monte Carlo sweeps through the lattice. $*q$ points to the array $q[N]$ and $*zgr$ to the variable *reject*, which counts how often a suggested change has been rejected. The constants N, *MA*, and *DELTA* must be provided.

Listing 4.3 Constants and variables

```
1   /*program constants.h*/
2   /*constants: N,A,ME,MA,BIN,INTERV*/
3   /*MASS,MU,LAMBDA,DELTA*/
4   /*initial values for q[N],qnew*/
5   /*massl,lambdal,muleff,reject,streck,versch*/
6   #define N 10      /*number of lattice points*/
7   #define A 1.0     /*lattice constant*/
8   #define M 1000000 /*number of iterations*/
9   #define ME 500    /*until equilibrium is reached*/
10  #define MA 5      /*every MA'th configuration is measured*/
11  #define BIN 40    /*number of  bins for wave function*/
12  #define INTERV 2 /*interval for binning [-INTERV,INTERV]*/
13  #define MASS 1.0
14  #define MU 1.0    /*coupling of q**2*/
15  #define LAMBDA 0.0 /*coupling of q**4*/
16  #define DELTA 0.5 /*change of variable = DELTA(1-2 random)*/
17  /*rescale to latticevariables*/
18  double massl=MASS/A;
19  double lambdal=A*LAMBDA;
20  double muleff=MASS/A+A*MU;
21  double qnew,q[N];
22  unsigned int reject=0;
23  double translate=(double)BIN/2.0;
24  double stretch=0.5*(double)BIN/(double)INTERV;
```

Listing 4.4 Functions for anharmonic oscillator

```
1   /*program stdanho.h*/
2   /*change of action*/
3   double delta_action(double y,double x,double xs)
4   { return (y-x)*((y+x)*(muleff+lambdal*(y*y+x*x))-massl*xs);};
5   /*MA sweeps*/
6   /*expects constants N,MA,DELTA*/
7   /*arguments: array q[N], pointer to reject*/
8   void mcsweep(int *zgr,double *q)
9   { int i,j; double qnew,dS;
10    for (i=0;i<MA;i++)
11      for (j=0;j<N;j++)
12        { qnew=q[j]+DELTA*(1-2*drand48());
13          dS=delta_action(qnew,q[j],q[(j+1)%N]+q[(j+N-1)%N]);
14          if (dS<0) q[j]=qnew;
15          else
16            if (exp(-dS)>drand48() ) q[j]=qnew;
17            else *zgr=*zgr+1;
18        };
19  }
20  /*binning of values in q[N]*/
21  void binning(int *bin,double *q)
22  { int i,p;
23    for (i=0;i<N;i++)
24      {p=(int)(q[i]*stretch+translate);
25       if ((0<=p)&&(p<BIN)) bin[p]++;};
26  }
27  /*calculation of moments*/
28  double moments(int n,double *q)
29  { int i; double sum=0;
30    for (i=0;i<N;i++)
31      sum=sum+pow(q[i],n);
32    return sum/N;
33  }
```

The third function `binning` bins the values of $q[N]$ between $-INTERV$ and $INTERV$ in the array $bin[BIN]$. The variables $q[N]$, $bin[BIN]$, $stretch$, $translate$, and BIN must be defined and initialized.

The last function *moments* calculates the sums $\frac{1}{N}\sum q_i^n$.

4.6   Problems

4.1 (Detailed Balance) A statistical system admits two configurations which in equilibrium are distributed with probabilities $P_\omega > 0$, $\omega = 1, 2$. Construct the most general stochastic matrix $W(\omega, \omega')$ which obeys the condition of detailed balance

$$P(\omega)W(\omega, \omega') = P(\omega')W(\omega', \omega) \quad \text{for} \quad \omega, \omega' \in \{1, 2\}.$$

What is the optimal choice for W such that W^n converges as quickly as possible to the limit matrix W^{eq}.

4.2 (Markov Process) Consider a system with three energy eigenstates the energies of which are ordered as $E_1 < E_2 < E_3$. The allowed transitions are $\omega \to \omega$ and $\omega \to (\omega + 1) \bmod 3$. This process cannot fulfill the condition of detailed balance. Show that there nevertheless is a stochastic matrix $W(\omega, \omega')$ with the Boltzmann distribution as fixed point.

4.3 (Precision of Integrators) Show that the numerical integration of an ordinary differential equation of type $df/dt = g(f, t)$ through a leapfrog integrator corresponds to a precision of 2nd order. This means that it should be demonstrated that the global error of the obtained solution is of order $\mathcal{O}(h^2)$, where h is the time-step.

4.4 (Verlet Algorithm) If we are not interested in the momenta, but just the trajectory of the particle, we can eliminate the momenta from the leapfrog algorithm and end up with the *Verlet algorithm*. Find this algorithm.

4.5 (Fourth-Order Symplectic Integrator) The simplest higher-order symplectic algorithm is that of Forest, Ruth, and Omelyan. If initially $q = q(\tau)$, $p = p(\tau)$, then, in the Forest-Ruth algorithm, the following steps:

$$q = q + \frac{\theta h}{2} p,$$
$$p = p + \theta h F(q),$$
$$q = q + (1 - \theta)\frac{h}{2} p,$$
$$p = p - (1 - 2\theta)h F(q),$$
$$q = q + (1 - \theta)\frac{h}{2} p,$$
$$p = p + \theta h F(q),$$
$$q = q + \frac{\theta h}{2} p.$$

with

$$\theta = \frac{1}{2 - 2^{1/3}} \approx 1.35120719195966$$

yield $q(\tau + h)$, $p(\tau + h)$. Prove that this integrator is symplectic.

Remark The algorithm requires three force evaluations per time-step. The middle step is larger in magnitude than h and goes "backwards in time." All higher-order integrators seem to have such steps "backwards in time." If one is willing to accept more than three force evaluations for a fourth-order integrator, one can avoid having a step greater in magnitude than h. The algorithm of Omelyan et al. [20] avoids large time-steps and hence is more accurate than the Forrest–Ruth algorithm. It reads

$$q = q + \xi h p,$$

$$p = p + (1 - 2\lambda)\frac{h}{2}F(q),$$

$$q = q + \xi h p,$$

$$p = p + \lambda h F(q),$$

$$q = q + (1 - 2(\chi + \xi)) h p,$$

$$p = p + \lambda h F(q),$$

$$q = q + \xi h p,$$

$$p = p + (1 - 2\lambda)\frac{h}{2}F(q),$$

$$q = q + \xi h p,$$

with parameters

$$\xi \approx 0.1786178958448091,$$

$$\lambda \approx -0.2123418310626054,$$

$$\chi \approx -0.06626458266981849.$$

References

1. K. Jacobs, *Markov-Prozesse mit endlich vielen Zuständen (Markov Processes with a Finite Number of States)*. Selecta Mathematica IV (Springer, Berlin, 1972)
2. M.E.J. Newman, G.G. Barkenna, *Monte Carlo Methods in Statistical Physics* (Clarendon, Oxford 1999)
3. K. Binder, D.W. Heermann, *Monte Carlo Simulation in Statistical Physics: An Introduction*. Graduate Texts in Physics (Springer, Berlin, 2019)
4. J.S. Liu, *Monte Carlo Strategies in Scientific Computing*. Springer Series in Statistics (Springer, Berlin, 2001)
5. A. Joseph, *Markov Chain Monte Carlo Methods in Quantum Field Theories: A Modern Primer*. Springer Briefs in Physics (Springer, Berlin, 2020)
6. N. Metropolis, A.W. Rosenbluth, M.N. Rosenbluth, A.H. Teller, Equations of state calculations by fast computing machines. J. Chem. Phys. **21**, 1087 (1953)
7. W.K. Hastings, Monte Carlo sampling methods using Markov chains and their applications. Biometrika **170**, 97 (1970)
8. G.E. Box, M.E. Muller, A note on the generation of random normal deviates. Ann. Math. Stat. **29**, 610 (1958)
9. M. Creutz, B.A. Freedman, A statistical approach to quantum mechanics. Ann. Phys. **132**, 427 (1981)

10. S. Duane, A.D. Kennedy, A.D. Pendleton, B.J. Roweth, Hybrid Monte Carlo. Phys. Lett. B **195**, 216 (1987)
11. I. Montvay, G. Münster, *Quantum Fields on a Lattice* (Cambridge University Press, Cambridge, 1997)
12. H.J. Rothe, *Lattice Gauge Theories: An Introduction* (World Scientific, Singapore, 2012)
13. C. Urbach, Untersuchung der Reversibilitätsverletzung beim HMC-Algorithmus. Thesis FU Berlin, 2002
14. M. Griebel, S. Knapek, G. Zumbusch, A. Caglar, *Numerical Simulation in Molecular Dynamics* (Springer, Berlin, 2010)
15. G.G. Batrouni, G.R. Katz, A.S. Kronfeld, G.P. Lepage, B. Svetitsky, K.G. Wilson, Langevin simulation of lattice field theories. Phys. Rev. D **32**, 2736 (1985)
16. D.H. Weingarten, D.N. Petcher, Monte Carlo integration for lattice gauge theories with fermions. Phys. Lett. B **99**, 333 (1981)
17. S. Gupta, A. Irback, F. Karch, B. Petersson, The acceptance probability in the hybrid Monte Carlo method. Phys. Lett. B **242**, 437 (1990)
18. A.D. Kennedy, B. Pendleton, Cost of the generalized hybrid Monte Carlo algorithm for free field theory. Nucl. Phys. B **607**, 456 (2001)
19. E. Forest, R.D. Ruth, Fourth-order symplectic integration. Physica D **43**, 105 (1990)
20. I.P. Omelyan, I.M Mryglod, R. Folk, Symplectic analytically integrable decomposition algorithms: classification, derivation, and application to molecular dynamics, quantum and celestial mechanics simulations. Comp. Phys. Comm. **151**, 272 (2003)
21. T. Kaestner, G. Bergner, S. Uhlmann, A. Wipf, C. Wozar, Two-dimensional Wess-Zumino models at intermediate couplings. Phys. Rev. D **78**, 095001 (2008)
22. P. Marenzoni, L. Pugnetti, P. Rossi, Measure of autocorrelation times of local hybrid Monte Carlo algorithm for lattice QCD. Phys. Lett. B **315**, 152 (1993)
23. B. Wellegehausen, A. Wipf, C. Wozar, Casimir scaling and string breaking in G2 gluodynamics. Phys. Rev. D **83**, 016001 (2011)

Scalar Fields at Zero and Finite Temperature

5

Scalar fields describe spinless particles and are often introduced and discussed in introductory textbooks to introduce novel concepts and techniques in quantum field theory [1–4]. Even more important than their educational value is their role in the electroweak theory, where a scalar field interacts with the fields of leptons, quarks, and gauge bosons. The scalar field is needed for the Higgs mechanism which is essential to explain the mass generation for fermions and electroweak gauge bosons.

Neglecting the interactions with the other fields, one is left with a ϕ^4 theory, the so-called Higgs sector of the Standard Model of particle physics. It is known that in *more than* four dimensions, the removal of a UV cutoff of a regularized ϕ^4-theory leads to a non-interacting system [5]. There are convincing arguments to believe that this applies to scalar field theories in four dimensions as well. In contrast, in less than four dimensions, we end up with an interacting theory. The triviality of the Higgs sector might very well be responsible for the electroweak gauge theory being only an effective theory below a cutoff Λ. This conclusion would not apply if there were a non-Gaussian fixed point (fixed points are discussed in Chap. 12). A non-Gaussian fixed point has been intensively searched after in lattice simulations—so far without success.

Scalar fields play a pivotal role in inflationary cosmological scenarios. They could be responsible for the anticipated exponential expansion of the early universe and could trigger various phase transitions, and, last but not least, their quantum fluctuations could contribute to the structures observed in the cosmic microwave background radiation [6, 7]

© The Author(s), under exclusive license to Springer Nature Switzerland AG 2021
A. Wipf, *Statistical Approach to Quantum Field Theory*, Lecture Notes
in Physics 992, https://doi.org/10.1007/978-3-030-83263-6_5

5.1   Quantization

This chapter is devoted to scalar field theories in d spacetime dimensions. In a given inertial frame, we identify an event with its coordinates $x = (x^\mu) = (ct, \boldsymbol{x})$ in $\mathbb{R}^d$. A real scalar field assigns to each spacetime point x a real number,

$$\phi : \mathbb{R}^d \longrightarrow \mathbb{R}, \quad x \longrightarrow \phi(x) . \tag{5.1}$$

It satisfies a covariant field equation which is the *Euler-Lagrange equation*

$$\frac{\delta S}{\delta \phi(x)} = 0 \Longrightarrow \partial_\mu \frac{\partial \mathscr{L}}{\partial(\partial_\mu \phi(x))} - \frac{\partial \mathscr{L}}{\partial \phi(x)} \tag{5.2}$$

of a Poincaré-invariant classical action

$$S[\phi] = \int \mathrm{d}^d x \, \mathscr{L}(\phi, \partial_\mu \phi) = \int \mathrm{d}^d x \, \left(\frac{1}{2} \partial_\mu \phi \partial^\mu \phi - V(\phi) \right) .$$

For a non-interacting field, $V(\phi) = \frac{1}{2}m^2\phi^2$, and the scalar field fulfills the linear *Klein–Gordon equation* $(\Box + m^2)\phi = 0$ with m representing the particle mass.

To quantize the scalar field, we apply the well-known quantization rules of quantum mechanics to a system with infinitely many degrees of freedom. The quantum mechanical results can be generalized according to the substitutions

$$q_i(t) \equiv q(t, i) \longrightarrow \phi(t, \boldsymbol{x}) = \phi(x) \quad \text{and} \quad \sum_i \longrightarrow \int \mathrm{d}^{d-1} x . \tag{5.3}$$

For example, the classical field at $\boldsymbol{x}$ becomes a position-dependent operator $\hat{\phi}(\boldsymbol{x})$— one operator at every point $\boldsymbol{x}$— the time dependence of which is determined by the Heisenberg equation.

With these substitution rules, we obtain the following (formal) functional integral representation for vacuum expectation values of time-ordered products of field operators:

$$\left\langle 0 | T\hat{\phi}(x_1) \cdots \hat{\phi}(x_n) | 0 \right\rangle = \frac{1}{Z} \int \mathscr{D}\phi \, \phi(x_1) \cdots \phi(x_n) \, \mathrm{e}^{\mathrm{i}S[\phi]/\hbar} . \tag{5.4}$$

The symbol $\int \mathscr{D}\phi$ means integration over all scalar fields ϕ. In particular, a one-dimensional field theory for $\phi \equiv q : \mathbb{R} \rightarrow \mathbb{R}$ describes a quantum mechanical system. The normalization factor Z in (5.4) represents the vacuum–vacuum amplitude

$$Z = \int \mathscr{D}\phi \, \mathrm{e}^{\mathrm{i}S[\phi]/\hbar} .$$

Similarly as in quantum mechanics at imaginary time, one introduces the Euclidean field operator

$$\hat{\phi}_E(x) \equiv \hat{\phi}_E(\tau, \boldsymbol{x}) = e^{\tau \hat{H}} \hat{\phi}(0, \boldsymbol{x}) e^{-\tau \hat{H}}, \quad x = (\tau, \boldsymbol{x}) = (-ix^0, \boldsymbol{x}) \tag{5.5}$$

and shows that vacuum expectation values of time-ordered products of such operators have the formal functional integral representations

$$\langle 0|T\hat{\phi}_E(x_1)\cdots\hat{\phi}_E(x_n)|0\rangle = \frac{1}{Z}\int \mathcal{D}\phi\, \phi(x_1)\cdots\phi(x_n)\, e^{-S_E[\phi]/\hbar}\,, \tag{5.6}$$

where S_E denotes the Euclidean action. These expectation values are the *Schwinger functions* $S_n(x_1, \ldots, x_n)$ and should have the following properties [8–10].

1. *Euclidean covariance:* The S_n are invariant (covariant for fields with spin) under translations and Euclidean "Lorentz transformations" $x_i \rightarrow Rx_i + a$, with rotation $R \in \mathrm{SO}(d)$.
2. *Reflection positivity:* Pick test functions $f_n(x_1, \ldots, x_n)$ with support in the "time-ordered" subsets $0 < \tau_1 < \cdots < \tau_n$. Choose one such f_n for each positive n. Given a point x, let $\bar{x}$ be the reflected point about the $\tau = 0$ hyperplane. Then,

$$\sum_{m,n \leq N} \int d^d x_1 \cdots d^d x_m \, d^d y_1 \cdots d^d y_n \, S_{m+n}(x_1, \ldots, x_m, y_1, \ldots, y_n)$$

$$\times f_m^*(\bar{x}_1, \ldots, \bar{x}_m) f_n(y_1, \ldots, y_n) \geq 0\,, \tag{5.7}$$

where $*$ represents complex conjugation. This property reflects the positivity of the Hilbert space in quantum theory.
3. *Permutation symmetry:* The S_n are symmetric functions of their arguments. This property replaces the locality property in Minkowski spacetime.
4. *Cluster property:* If there is a unique vacuum state, then the S_n cluster,

$$S_{m+n}(x_1, \ldots, x_m, y_1 + a, \ldots, y_n + a) \xrightarrow{|a|\to\infty} S_m(x_1, \ldots, x_m) S_n(y_1, \ldots, y_n)\,. \tag{5.8}$$

5. *Regularity:* There are different versions of regularity; see [8–10].

Slightly stronger axioms based on measure theory have been formulated in [11].

These Euclidean axioms are due to K. Osterwalder and R. Schrader, and they have significance for the analytically continued Minkowski space quantum fields: one can reconstruct a Minkowski space quantum field theory by assuming these axioms for the Schwinger functions. The vacuum expectation values of products of field operators in Minkowski space are called *Wightman functions* W_n. Thanks to the Euclidean axioms the Euclidean Schwinger functions S_n can be analytically

continued to the corresponding Wightman functions which possess all properties of
a relativistic Hilbert space theory.

5.2   Scalar Field Theory at Finite Temperature

Similarly as in quantum mechanics, one argues that thermal expectation values of
time-ordered products of Euclidean field operators have the following Euclidean
functional integral representations,

$$\left\langle T\hat{\phi}_E(x_1)\cdots\hat{\phi}_E(x_n)\right\rangle_\beta = \frac{1}{Z(\beta)}\operatorname{tr}e^{-\beta\hat{H}}T\hat{\phi}_E(x_1)\cdots\hat{\phi}_E(x_n)$$

$$= \frac{1}{Z(\beta)}\oint \mathscr{D}\phi\,\phi(x_1)\cdots\phi(x_n)\,e^{-S_E[\phi]/\hbar}\ , \quad (5.9)$$

where β denotes the inverse temperature and S_E the Euclidean action. We thereby
integrate over β-periodic fields $\phi(\tau+\beta,x)=\phi(\tau,x)$ as indicated by the circle on
the integration symbol [12–16]. In the zero-temperature limit $\beta\to\infty$, we integrate
overall fields and recover the vacuum expectation values or *Schwinger functions*.

The normalization factor in (5.9) is the *partition function*

$$Z(\beta) = e^{-\beta F} = \oint \mathscr{D}\phi\,e^{-S_E[\phi]/\hbar}\ , \quad (5.10)$$

the logarithm of which is proportional to the *free energy density* $f = F/V$, an
important quantity in quantum statistics. The thermal expectation values (5.9) are
generated by the partition function in presence of an external source,

$$Z[\beta,j] = \oint \mathscr{D}\phi\,e^{-S_E[\phi]+\int j(x)\phi(x)} \equiv e^{W[\beta,j]} \quad (5.11)$$

as follows:

$$\left\langle T\hat{\phi}_E(x_1)\ldots\hat{\phi}_E(x_n)\right\rangle_\beta = \frac{1}{Z[\beta,0]}\frac{\delta^n Z[\beta,j]}{\delta j(x_1)\cdots\delta j(x_n)}\bigg|_{j=0}. \quad (5.12)$$

In contrast, the *connected thermal correlation functions*

$$\left\langle T\hat{\phi}_E(x_1)\ldots\hat{\phi}_E(x_n)\right\rangle_{c,\beta} = \frac{\delta^n W[\beta,j]}{\delta j(x_1)\cdots\delta j(x_n)}\bigg|_{j=0} \quad (5.13)$$

are generated by the functional $W[\beta,j]$ defined in (5.11), which is proportional
to the free energy in presence of an external source. At zero temperatures it
becomes the *Schwinger functional* $W[j]$, which generates all the connected vacuum

expectation values

$$\left\langle 0 \left| T \hat{\phi}_E(x_1) \dots \hat{\phi}_E(x_n) \right| 0 \right\rangle_c = \left. \frac{\delta^n W[j]}{\delta j(x_1) \cdots \delta j(x_n)} \right|_{j=0} . \tag{5.14}$$

For non-interacting particles, the generating functionals at finite temperature can be calculated in closed forms.

5.2.1 Free Scalar Field

The Euclidean action of the free scalar field is quadratic,

$$S_{E,0}[\phi] = \frac{1}{2} \int d^d x \left(\nabla \phi \cdot \nabla \phi + m^2 \phi^2 \right) = \frac{1}{2} \int d^d x \, \phi \left(-\Delta + m^2 \right) \phi , \tag{5.15}$$

and the functional integral in (5.11) is Gaussian. Integrating over β-periodic fields yields the generating functional in closed form

$$Z[\beta, j] = \frac{\text{const}}{\det^{1/2}\left(-\Delta + m^2\right)} \exp\left(\frac{1}{2} \int d^d x d^d y \, j(x) \, S_\beta(x - y) j(y)\right) , \tag{5.16}$$

with the *thermal propagator* in position space

$$S_\beta(x - y) = \left\langle T \hat{\phi}_E(x) \hat{\phi}_E(y) \right\rangle_\beta = \left\langle x \left| \frac{1}{-\Delta + m^2} \right| y \right\rangle_\beta . \tag{5.17}$$

Its Fourier representation

$$S_\beta(x) = \frac{1}{\beta} \sum_n \int \frac{d^3 k}{(2\pi)^3} \frac{e^{-i\omega_n x^0 - i k x}}{\omega_n^2 + k^2 + m^2} , \quad \omega_n = \frac{2\pi}{\beta} n \tag{5.18}$$

contains a sum over n and an integral over the spatial momenta. The sum originates from the periodicity conditions for the fields at finite temperature. In the limit of very low temperatures, the *Matsubara frequencies* ω_n are dense, and the Riemann sum turns into a Riemann integral. Thus in the limit $T \to 0$, the function $S_\beta(x)$ approaches the Euclidean propagator

$$S(x) = \lim_{\beta \to \infty} S_\beta(x) = \int \frac{d^4 k}{(2\pi)^4} \frac{e^{-ikx}}{k^2 + m^2} . \tag{5.19}$$

The quadratic action $S_{E,0}[\phi]$ in (5.15) contains the linear operator

$$A = -\Delta + m^2 , \tag{5.20}$$

and the corresponding Gaussian functional integral for the partition function (5.10) yields the square root of $1/\det A$, see (5.16), such that the free energy reads

$$F(\beta) = -\frac{\ln Z(\beta)}{\beta} = \frac{1}{2\beta} \log \det A + \text{const} . \tag{5.21}$$

The differential operator A acts on β-periodic functions and the temperature dependence of its eigenvalues and determinant originates from these boundary conditions.

Determinants of differential operators—*functional determinants* — play a prominent role in theoretical and mathematical physics. They appear in a variety of investigations in quantum field theory, for example, in tunneling and semi-classical physics, self-consistent Hartree–Fock and Schwinger–Dyson equations, or lattice field theories with dynamical fermions to mention a few. Therefore we pause here and have a closer look at functional determinants.

Zeta Function Regularization
As a first simplification, we enclose the system in a finite box in order to find *discrete eigenvalues* $\{\lambda_n\}$ of A. We assume that the eigenvalues are positive—as happens for the particular operator in (5.20). Following [17, 18], we define the ζ-function of A,

$$\zeta_A(s) = \sum_n \lambda_n^{-s} . \tag{5.22}$$

For sufficiently large $\Re(s)$, these spectral sums converge and define analytic functions which can be analytically continued as meromorphic functions to the entire complex s-plane [19]. For example, the second-order differential operator

$$A = -\frac{\mathrm{d}^2}{\mathrm{d}\varphi^2} \quad \text{on the interval} \quad [0, \pi] \tag{5.23}$$

subject to Dirichlet boundary conditions has eigenvalues $1^2, 2^2, 3^2, \ldots$, and its zeta function is given by Riemann's celebrated zeta function,

$$\zeta_A(s) = \sum_{n=1}^{\infty} \frac{1}{n^{2s}} = \zeta_R(2s) . \tag{5.24}$$

The sum converges for $\Re(s) > 1/2$, and its analytic continuation is given by Riemann's zeta function.

The spectral sum for the operator in (5.20) converges for $\Re(s) > d/2$ and defines a meromorphic function on the complex s-plane; see Problem 5.1. More generally, with the help of Weyl's formula for the distribution of large eigenvalues [20], one can prove that for a second-order elliptic operator on a finite domain the spectral sum (5.22) converges for all s with $\Re(s) > d/2$. But for a finite-dimensional matrix A, the sum exists for all s, and with

$$\frac{\mathrm{d}\lambda_n^{-s}}{\mathrm{d}s} = -\lambda_n^{-s}\log(\lambda_n)$$

one arrives at the useful formula

$$-\left.\frac{\mathrm{d}\zeta_A(s)}{\mathrm{d}s}\right|_{s=0} = \sum_n \log\lambda_n = \mathrm{tr}\,\log A = \log\det A \;. \tag{5.25}$$

Now we *define* the determinant of an infinite-dimensional matrix, i.e., differential operator, by this formula. For any (elliptic) differential operator, the ζ-function is regular at $s = 0$ such that the left-hand side in this formula is well defined. The analytic continuation of the spectral sum to s-values where the sum diverges corresponds to a particular renormalization of functional determinants—the so-called zeta function regularization. We refer to [21] where the connection to other renormalizations is discussed.

Heat Kernel of a Differential Operator
Now we explore the intimate relation between the ζ-function and heat kernel of an elliptic differential operator A. They are related by a *Mellin transformation*:

$$\zeta_A(s) = \sum_n \frac{1}{\Gamma(s)} \int_0^\infty \mathrm{d}t\, t^{s-1} e^{-t\lambda_n} = \frac{1}{\Gamma(s)} \int_0^\infty \mathrm{d}t\, t^{s-1}\, \mathrm{tr}\left(e^{-tA}\right) \;. \tag{5.26}$$

If A were the Hamiltonian of a physical system, the trace of $K(t) = \exp(-tA)$ would be the partition function of that system at inverse temperature $\beta = t$. This explains why the integral kernel $K(t; x, y)$ of $K(t)$ is called a heat kernel. Clearly, the last trace in (5.26) is the integral of $K(t, x, x)$ over space such that

$$\zeta_A(s) = \frac{1}{\Gamma(s)} \int_0^\infty \mathrm{d}t\, t^{s-1} \int \mathrm{d}x\, K(t; x, x) \;. \tag{5.27}$$

In position space the heat kernel fulfills the differential equation

$$\frac{\partial}{\partial t} K(t; x, y) = -A_x K(t; x, y) \quad \text{with} \quad \lim_{t\to 0} K(t; x, y) = \delta(x - y) \;. \tag{5.28}$$

To calculate the determinant of A, one proceeds as follows:

1. Construct the unique solution of the initial value problem (5.28) for $x = y$.
2. Insert the solution $K(t, x, x)$ into the representation (5.27) of the ζ-function.
3. Calculate the determinant with the help of formula (5.25).

However, this approach meets several problems: For many operators, for example, wave operators in inhomogeneous background fields, one cannot calculate the kernel on the diagonal $K(t, x, x)$ in closed form. Furthermore, the s-integral in (5.27) only exists for sufficiently large $\Re(s)$. Thus we need the analytic continuation of the zeta function, which even for large $\Re(s)$ is not known analytically. Only for simple systems can one compute the heat kernel explicitly and construct the analytic continuation, for example, by a Poisson resummation. However, what is needed is not the zeta function itself but its derivative at the origin. Unlike ζ, this quantity is computable for many interesting systems. We refer to the book [19] for further facts about the zeta function method and the review [22, 23] for an introduction to the theory of heat kernels and their small-t expansions. In [24, 25] the zeta function is used to calculate Casimir energies, in [26] the regularization is applied to compute the fermionic path integral in the massless Schwinger model in all topological sectors (see Chap. 16), and in [27] it is used to find multi-instanton determinants in $four$-dimensional Yang–Mills theories.

Free Energy of Non-interacting Scalars

Let us return to the simple operator $A = -\Delta + m^2$ which defines the Euclidean action for free scalars. Since $-\Delta$ is the Schrödinger operator of a free particle, we may use the result (2.42) and obtain

$$K(t; x, x') = \frac{e^{-m^2 t}}{(4\pi t)^{d/2}} \sum_{n \in \mathbb{Z}} e^{-\{(\tau - \tau' + n\beta)^2 + (x - x')^2\}/4t}, \quad x = (\tau, x) . \quad (5.29)$$

The contribution with $n = 0$ corresponds to the heat kernel (Euclidean propagator) in $\mathbb{R}^d$. The summation over n enforces periodicity in imaginary time τ with period β and yields the heat kernel on the cylinder $[0, \beta] \times \mathbb{R}^{d-1}$. Inserting $K(t, x; x)$ into (5.27), we end up with the following integral representation of the ζ-function

$$\zeta_A(s) = \frac{\beta V}{(4\pi)^{d/2} \Gamma(s)} \int dt \, t^{s-1-d/2} e^{-m^2 t} \sum_{n=-\infty}^{\infty} e^{-n^2 \beta^2 / 4t} . \quad (5.30)$$

The spectrum of the operator A on the cylinder is not discrete. This fact expresses itself by a harmless volume divergence of the ζ-function. We get rid of this divergent factor through the transition to the free *energy density*. One finds the same energy density if one encloses the particles in a finite box and let the box size tend to infinity.

Now we can do the t-integral via the integral representation of the modified Bessel functions of second kind

$$\int_0^\infty dt \, t^a e^{-bt - c/t} = 2 \left(\frac{c}{b}\right)^{(a+1)/2} K_{a+1}\left(2\sqrt{bc}\right) , \quad (5.31)$$

and obtain

$$\zeta_A(s) = \frac{\beta V}{(4\pi)^{d/2}} \frac{m^{d-2s}}{\Gamma(s)} \left(\Gamma\left(s - \frac{d}{2}\right) + 4 \sum_1^\infty \left(\frac{nm\beta}{2}\right)^{s-d/2} K_{d/2-s}(nm\beta) \right).$$

(5.32)

The Gamma function $\Gamma(s)$ in the denominator has simple poles at $s = 0, -1, -2, \ldots$. Now we use the relations

$$\frac{\Gamma(s-2)}{\Gamma(s)} = \frac{1}{(s-1)(s-2)} \quad \text{and} \quad \frac{1}{\Gamma(s)} = s + O(s^2)$$

to find the explicit expression for the free energy density in four dimensions,

$$f(\beta) = -\frac{1}{2\beta V} \zeta_A'(0) = -\frac{m^4}{128\pi^2} \left(3 - 4\log\frac{m}{\mu} + 64 \sum_{n=1,2\ldots} \frac{K_2(nm\beta)}{(nm\beta)^2} \right).$$

(5.33)

Note that the density contains a scale parameter μ with the dimension of a mass. This parameter was introduced to rescale the operator A to a dimensionless operator A/μ^2 with dimensionless eigenvalues, zeta function, and determinant.

In order to obtain the free energy density for massless particles, we use $K_2(x) \sim 2/x^2$, so that

$$\lim_{m \to 0} f(\beta) = -\frac{T^4}{\pi^2} \zeta_R(4),$$

(5.34)

where *Riemann's zeta function* $\zeta_R(s)$ appears, which is analytic on $\mathbb{C}$, except for a simple pole at $s = 1$ with unit residue. Some specific values of ζ_R and its derivative are

$$\xi_R(0) = -\frac{1}{2}, \quad \xi_R(2) = \frac{\pi^2}{6}, \quad \xi_R(4) = \frac{\pi^4}{90} \quad \text{and} \quad \zeta_R'(0) = -\frac{1}{2}\log(2\pi).$$

(5.35)

Thus in the limit $m \to 0$, we obtain the following free energy density and internal energy density for *scalar particles*:

$$f(\beta) = -\frac{\pi^2 T^4}{90} \quad \text{and} \quad u(\beta) = \partial_\beta(\beta f) = \frac{\pi^2 T^4}{30}.$$

(5.36)

The black-body radiation of photons with two polarizations has twice the free energy density of massless scalars. Note that in the expression for massless scalars the scale parameter μ is absent.

High-Temperature Expansion

To calculate the high-temperature expansion of the free energy density for massive scalars in (5.33), we need the small-x expansions of the series

$$I(v, x) = \sum_{n=1}^{\infty} \frac{K_v(nx)}{n^v}, \qquad v \equiv \frac{d}{2}, \quad x \equiv m\beta . \tag{5.37}$$

These expansions have been calculated in [28]. In four spacetime dimensions, we need

$$I(2, x) = \frac{\pi^4}{45x^2} - \frac{\pi^2}{12} + \frac{\pi x}{6} + \frac{x^2}{16} \log \frac{x}{4\pi} - \frac{x^2}{32} \left(\frac{3}{2} - 2\gamma \right) + O(x^4) . \tag{5.38}$$

This leads to the following high-temperature expansion of the energy density

$$f(\beta) = -\frac{\pi^2 T^4}{90} + \frac{m^2 T^2}{24} - \frac{m^3 T}{12\pi} - \frac{\gamma m^4}{32\pi^2} + \frac{m^4}{32\pi^2} \log \left(\frac{4\pi T}{\mu} \right) + O \left(\frac{m^2}{T^2} \right) , \tag{5.39}$$

valid for $T \gg m$, with Euler's constant $\gamma \approx 0.5772$. The high-temperature expansion has been calculated with different regularizations in [12–14]. More details can be found in the textbooks [15, 16].

We may add a potential term to the action of the free field (5.15). This gives rise to the Euclidean action of a self-interacting scalar field

$$S_E = S_{E,0} + \int d^d x \, V(\phi) . \tag{5.40}$$

Expanding the functional integrals (5.9) or (5.10) in powers of the interaction potential V leads to a series expansion for thermal expectation values of time-ordered products of field operators. One can use the *zero-temperature* Feynman rules to calculate the terms in this expansion if one only uses the finite-temperature propagator and replaces each k^0-integral by the sum over Matsubara frequencies [12–16].

5.3   Schwinger Function and Effective Potential

The effective potential $u(\beta, \varphi)$ is useful in the study of phases and phase transitions in systems with an order parameter. It is the Legendre transform of the thermal Schwinger function $w(\beta, j)$, which is just the Schwinger functional $W[\beta, j]$ in (5.11) for a homogeneous source j, divided by the spacetime volume.[1] In other

---

[1] It should not be confused with the n-point Schwinger functions at zero temperature.

words, $u(\beta, \varphi)$ is the effective action density for a homogeneous field or the free energy density in the presence of an external source. One can consider alternatives to this "conventional" effective potential, e.g., the *constraint effective potential* [29], which will be introduced in Chap. 7.

The partition function in presence of a *homogeneous* external source j has the functional integral representation [12–14]

$$Z(\beta, j) \equiv e^{\beta V\, w(\beta, j)} = C \oint \mathscr{D}\phi \exp\left(-S_E[\phi] + j \int_0^\beta d^d x\, \phi(x)\right). \tag{5.41}$$

Again one integrates over β-periodic fields, i.e., fields on the cylinder $[0, \beta] \times V$ over the spatial region V. We denote the spatial region and its volume both by V. We can identify $w(\beta, j)$ with the negative free energy density of the system with shifted Hamiltonian $\hat{H}_j = \hat{H} - (j, \hat{\phi})$. At low temperatures, it converges to the negative ground state energy density $-E_0(j)/V$ of the system with shifted Hamiltonian.

From $w(\beta, j)$ we can extract the equilibrium value of the spatially averaged field in the presence of the source,

$$\frac{dw}{dj} = \frac{\oint \mathscr{D}\phi\, M\, e^{-S_E[\phi] + j \int \phi}}{\oint \mathscr{D}\phi\, e^{-S_E[\phi] + j \int \phi}} = \langle M \rangle_j, \qquad M = \frac{1}{\beta V} \int \phi(x). \tag{5.42}$$

A constant source is compatible with translational invariance of the system such that for periodic boundary conditions $\langle \phi(x) \rangle_j$ is independent of x. Hence we have

$$\langle \phi(x) \rangle_j = \frac{dw}{dj}. \tag{5.43}$$

One should keep in mind that expectation values depend on the external source. Also note that the formula (5.42) cannot be used at points where w is not differentiable.

The thermal Schwinger function is strictly convex, because its second derivative is equal to the expectation value of a positive quantity:

$$\frac{d^2 w}{dj^2} = \beta V \left((M - \langle M \rangle)^2 \right). \tag{5.44}$$

Now we define the finite-temperature *effective potential u* as the *Legendre transform* of the thermal Schwinger function,

$$u(\beta, \varphi) = (\mathscr{L}w)(\varphi) = \sup_j\, (j\varphi - w(\beta, j)). \tag{5.45}$$

The maximizing source j is conjugate to φ. The latter is an averaged macroscopic field, in contrast to the microscopic field ϕ entering the functional integral. If the minimum φ_0 of u is not degenerate, then it is equal to the expectation value of the field operator:

$$u(\beta, \varphi_0) \leq u(\beta, \varphi), \quad \forall \varphi \iff \varphi_0 = \left\langle \hat{\phi}(x) \right\rangle_{j=0} . \tag{5.46}$$

For a differentiable u, this follows from the second relation in (5.54).

Generalizations
Both in fundamental and applied physics, one is interested in how quantum systems react to a change of external conditions. A prominent example is the Casimir effect [30] which has been measured to great accuracy [31, 32] and which shows how the vacuum energy of a quantum field changes when one moves the walls of the enclosing cavity [33–35]. In a more general setting, one may ask the question how the vacuum structure of an Euclidean scalar field theory with Lagrangian density

$$\mathcal{L}(\phi(x)) = \int_\Omega d^d x \left\{ \frac{1}{2} \nabla \phi(x) \nabla \phi(x) + V(\phi(x)) \right\} \tag{5.47}$$

depends on the geometry of the quantization region Ω (which does not need be a cylinder as in thermal field theory) and on the boundary conditions imposed at the boundary $\partial \Omega$ of Ω [36,37]. In the *linear sigma models*, the field ϕ takes its values in a linear space and transforms non-trivially under a global inner symmetry group. In *non-linear sigma models*, the components of ϕ are coordinates of a target manifold, for example, a Lie group or a homogeneous space [38].

The "classical ground state" corresponds to a homogeneous field, which minimizes the potential V. For most systems it is not equal to the quantum mechanical expectation value $\langle \hat{\phi}(x) \rangle$ of the quantum field in Ω. In order to study the quantum corrections to the classical vacuum, one introduces a generalized Schwinger function w on the (Euclidean) spacetime region Ω according to[2]

$$Z(\Omega, j) = \int \mathcal{D}\phi \exp\left(-S_E[\phi] + j \int_\Omega \phi(x) \right) = e^{\Omega w(\Omega, j)} . \tag{5.48}$$

Analogously as in thermal field theory, one defines the effective potential $u(\Omega, \varphi)$ as Legendre transform of $w(\Omega, j)$.

5.3.1   The Legendre–Fenchel Transformation

The Legendre(-Fenchel) transformation (5.45) shows up in mechanics, thermodynamics, statistical mechanics, as well as quantum field theory. In this section we

---

[2] In non-linear sigma models, the coupling to the source may look differently.

shall collect some relevant properties of this transform. Related and more results are found in [39–41]. Let φ and j be elements of a convex set in $\mathbb{R}^d$.

Corollary 5.1 *The Legendre transform of a (for sufficiently large enough arguments) convex function is always convex.*

The proof is seen by considering the interpolating field between φ_1 and φ_2,

$$\varphi_\alpha = (1 - \alpha)\varphi_1 + \alpha\varphi_2, \quad 0 \le \alpha \le 1 . \tag{5.49}$$

Since the supremum of a sum is smaller or equal to the sum of the suprema of each summand, we have

$$u(\varphi_\alpha) = \sup_j \left\{ (1 - \alpha)(j, \varphi_1) + \alpha(j, \varphi_2) - \big((1 - \alpha) + \alpha\big)w(j) \right\}$$
$$\le (1 - \alpha) \sup_j \left\{ (j, \varphi_1) - w(j) \right\} + \alpha \sup_j \left\{ (j, \varphi_2) - w(j) \right\} \tag{5.50}$$
$$= (1 - \alpha)u(\varphi_1) + \alpha u(\varphi_2) ,$$

such that the graph of u lies below the straight line connecting the points $(\varphi_i, u(\varphi_i))$. This proves the convexity of u.

Corollary 5.2 *The Legendre transformation is involutive (squares to the identity) for convex functions.*

For every point $(j_0, w(j_0))$ on the graph of a convex w, we can find a touching hyperplane below the graph of w. Hence we can find a j_0 depending on φ_0 with

$$w(j_0) + (\varphi_0, j - j_0) \le w(j) \quad \text{for all} \quad j ,$$

or equivalently

$$(\varphi_0, j) - w(j) \le (\varphi_0, j_0) - w(j_0) \quad \text{for all} \quad j .$$

The supremum of the left-hand side with respect to j is the Legendre transform $\mathscr{L}w$ at the point φ_0. The resulting inequality can be written as

$$w(j_0) \le (\varphi_0, j_0) - (\mathscr{L}w)(\varphi_0) . \tag{5.51}$$

The right-hand side is the Legendre transform of $(\mathscr{L}w)(\varphi_0)$, and we conclude that $w(j_0)$ is *bounded from above* by $(\mathscr{L}^2 w)(j_0)$. On the other hand, from the very definition of the Legendre transformation, it is clear that

$$(\mathscr{L}w)(\varphi) \ge (\varphi, j) - w(j) \quad \text{for all} \quad \varphi \implies w(j) \ge (\varphi, j) - (\mathscr{L}w)(\varphi) . \tag{5.52}$$

Taking the supremum with respect to φ in the last inequality, we see that $w(j)$ is *bounded from below* by $(\mathscr{L}^2 w)(j)$. The two bounds imply that the Legendre transformation is involutive or in other words that $(\mathscr{L}^2 w)(j) = w(j)$.

Corollary 5.3 (Fenchel/Young Inequality) *For arbitrary φ and j, the inequality*

$$w(j) + u(\varphi) \geq (j, \varphi), \quad u = \mathscr{L}w \tag{5.53}$$

holds true. The inequality becomes an equality if φ and j are conjugate.

This inequality is an immediate consequence of inequality (5.52).

Corollary 5.4 *If a continuous w has a cusp, then $u = \mathscr{L}w$ has a plateau. In case of a one-component field, the width of the plateau is equal to the jump of the slope of w across the cusp. Inversely, a plateau is mapped into a cusp.*

This property follows from the graphical construction of the Legendre transformation as illustrated in Fig. 5.1. The Legendre transform $u(\varphi)$ is $-L(0)$, where $L(j)$ is the linear function with slope φ and tangent to $w(j)$. For a given φ and a convex and *differentiable* w, the conjugate source is defined by the requirement that $L(j)$ defines the hyperplane tangent to the graph of w at the point $(j, w(j))$.

Figure 5.2 illustrates a typical situation for a system which shows spontaneous symmetry breaking: The Schwinger function has a cusp for vanishing source and the effective potential u has a plateau.

Corollary 5.5 *The twofold Legendre transform of a for sufficiently large enough arguments convex function is the convex envelope of this function.*

This corollary results from the previously discussed properties of Legendre transformation and is illustrated in Fig. 5.2.

Fig. 5.1 Graphical construction of the Legendre transformation $u(\varphi)$ of a function $w(j)$. $L(j)$ is a linear function of j with slope φ touching the graph of w

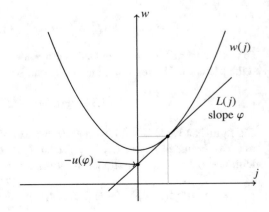

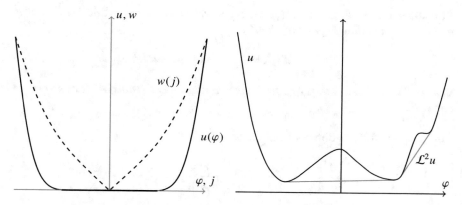

Fig. 5.2 Left panel: the Legendre transformation maps cusps to plateaus and plateaus to cusps. Right panel: the twofold application of a Legendre transformation produces the convex envelope

Corollary 5.6 *Let w and u be differentiable functions. Then the conjugate variables j and ϕ are related according to*

$$\varphi = w'(j) \quad and \quad j = u'(\varphi) . \tag{5.54}$$

Substituting (j, φ) by $(p, \dot{x})$ and (w, u) by (H, L), this is the well-known Legendre transformation in analytical mechanics. It describes the transition from the Hamiltonian to the Lagrangian formulation.

Defining the rescaled function of a function f according to

$$f_\alpha(x) = \alpha f\left(\frac{x}{\alpha^{1/2}}\right) \quad \alpha > 0 , \tag{5.55}$$

it is not difficult to show the

Corollary 5.7 *If $u = \mathscr{L}(w)$, then it follows $u_\alpha = \mathscr{L}(w_\alpha)$.*

The Legendre transform of the monomial with exponent $\alpha > 1$ is equal to the monomial with the dual exponent β,

$$w(j) = \frac{1}{\alpha} |j|^\alpha \iff u(\varphi) = \frac{1}{\beta} |\varphi|^\beta, \quad \text{with} \quad \frac{1}{\alpha} + \frac{1}{\beta} = 1 . \tag{5.56}$$

With increasing exponent β, the function u develops a plateau from -1 to 1. The exponent of the transformed function w approaches one such that it converges to the piecewise linear function $w(j) = |j|$.

Corollary 5.8 *If $w''(j)$ and $u''(j)$ are the matrices of the second derivatives of w and u, then the following identity holds:*

$$w''(j)\, u''(\varphi) = \mathbb{1}, \qquad (j, \varphi)\ \text{dual}. \tag{5.57}$$

Thus the second derivatives of a function and its Legendre transform are inverse to each other.

This property follows directly from the relations

$$\frac{\partial^2 w}{\partial j_r \partial j_s} = \frac{\partial \varphi_r}{\partial j_s} \quad \text{and} \quad \frac{\partial^2 u}{\partial \varphi_r \partial \varphi_s} = \frac{\partial j_r}{\partial \varphi_s} \ .$$

In passing we note how $\mathscr{L}$ acts on translated and inverted functions:

$$w(j) = w_1(j) + b \Rightarrow (\mathscr{L}w)(\varphi) = (\mathscr{L}w_1)(\varphi) - b$$

$$w(j) = w_1(j + k) \Rightarrow (\mathscr{L}w)(\varphi) = (\mathscr{L}w_1)(\varphi) - \varphi \cdot k \tag{5.58}$$

$$w(j) = w_1^{-1}(j) \Rightarrow (\mathscr{L}w)(\varphi) = -\varphi \cdot (\mathscr{L}w_1)\left(\frac{1}{\varphi}\right) \ .$$

5.4   Scalar Field on a Spacetime Lattice

In a lattice regularization of functional integrals, one first discretizes the Euclidean space $\mathbb{R}^d$ by an d-dimensional lattice Λ. Since the lattice constant a defines a minimal length, this regulates the quantum field theory at short distances. In a second step, one assumes that the lattice Λ is finite. Since the size of the system defines a maximal length, this regulates the quantum field theory at long distances. On a finite lattice, the formal functional integral (5.9) turns into a well-defined finite-dimensional integral which can be dealt with by the methods of statistical mechanics. To simplify matters we consider a *hypercubic lattice*. The lattice points $x \in \Lambda$ have coordinates

$$x^\mu = a\, n^\mu \quad \text{with} \quad n^\mu \in \mathbb{Z} \ , \tag{5.59}$$

and the extent of the lattice in direction μ is $L_\mu = aN_\mu$; see Fig. 5.3. Typically we impose periodic boundary conditions and thus identify the lattice points $x = (x^1, \ldots, x^\mu, \ldots x^d)$ and $x' = (x^1, \ldots, x^\mu + L_\mu, \ldots, x^d)$. With this identification the lattice becomes a discretized torus.

A scalar field on the lattice defines a map

$$\Lambda \ni x \longrightarrow \phi_x \in \mathscr{T} \tag{5.60}$$

Fig. 5.3 The lattice is characterized by the number of lattice points $N_1 \times N_2 \times \cdots \times N_d$ as well as the lattice constant a. A scalar field is defined on the lattice points

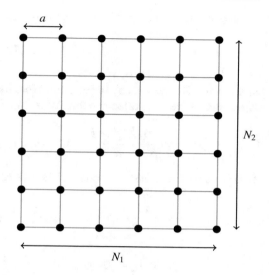

from Λ into the *target space* $\mathcal{T}$. For a real field, the target space is $\mathcal{T} = \mathbb{R}$, and in the standard model of the electroweak interaction $\mathcal{T} = \mathbb{C}^2$. Sigma models have a manifold as target space, and for Ising-like spin models, $\mathcal{T}$ is a finite group.

Boundary Conditions

To specify the lattice model, we must prescribe the *boundary conditions* for the scalar field. These conditions are classified as follows:

- *Periodic boundary conditions:* With these conditions the lattice is a discrete torus, and the lattice field theory is invariant under discrete translations and rotations.
- *Fixed boundary conditions:* Here we prescribe the field on the boundary $\phi|_{\partial\Lambda}$. Such boundary conditions are useful to describe entangled states in quantum field theory.
- *Open boundary conditions:* Here we switch off all interactions between sites on the lattice Λ with sites in the complement of Λ (viewed as subset of $\mathbb{Z}^d$). These boundary conditions are used in solid-state physics.
- *Antiperiodic boundary conditions:* They serve as a tool to inhibit unwanted long-range correlations or to study interfaces. This modification of the periodic boundary conditions is frequently used in lattice field theories.

In the remaining part of this chapter, we will consider non-interacting scalar fields subject to periodic boundary conditions.

Let us consider the free Klein–Gordon field with target space $\mathcal{T} = \mathbb{R}$. We need to approximate the continuum action (5.15) by an action for the lattice field. We thereby substitute the integral by a Riemann sum according to

$$\int d^d x \ldots \longrightarrow a^d \sum \ldots .$$ (5.61)

and replace differentials by differences. Note that we are free in the choice of the lattice derivative. One often chooses the *forward derivative* or *backward derivative*

$$(\partial_\mu \phi)(x) = \frac{\phi_{x+ae_\mu} - \phi_x}{a} \quad \text{or} \quad (\partial_\mu \phi)(x) = \frac{\phi_x - \phi_{x-ae_\mu}}{a} .$$ (5.62)

In both cases the discretized action of the free scalar field is given by

$$\begin{aligned} S &= \frac{a^{d-2}}{2} \sum_{\langle x,y \rangle} \left(\phi_x - \phi_y \right)^2 + \frac{m^2 a^d}{2} \sum_x \phi_x^2 \\ &= \frac{a^{d-2}}{2} \left(2d + (am)^2 \right) \sum_x \phi_x^2 - a^{d-2} \sum_{\langle x,y \rangle} \phi_x \phi_y , \end{aligned}$$ (5.63)

where the factor a^d results from the substitution (5.61) and the last sum includes all nearest neighbor pairs $\langle x, y \rangle$.

Similar as in quantum mechanics, we rescale dimensionful quantities with the appropriate power of the lattice constant in order to obtain dimensionless quantities. The dimensionless mass m_L and the dimensionless lattice field ϕ_L are

$$am = m_L \quad \text{and} \quad a^{(d-2)/2} \phi = \phi_L .$$ (5.64)

The rescaled distance between adjacent lattice points is one, and the rescaled lattice length in direction μ is $L_\mu = N_\mu$. The number of lattice points is given by $V = N_1 \cdots N_d$. In the following we suppress the index L.

The lattice action (5.63) defines a quadratic form of the field:

$$S = \frac{1}{2} \sum_{x,y \in \Lambda} \phi_x A_{xy} \phi_y, \quad A_{xy} = (m^2 + 2d) \delta_{xy} - \sum_{\mu=1}^d \left(\delta_{x,y+e_\mu} + \delta_{x,y-e_\mu} \right) .$$ (5.65)

The symmetric matrix (A_{xy}) is thereby positive for a positive m^2. For a linear target space, we regard the $\{\phi_x | x \in \Lambda\}$ as components of a vector. For a real field, this vector space is $\mathbb{R}^V$, equipped with the scalar product

$$(\phi, \chi) = \sum_{x \in \Lambda} \phi_x \chi_x .$$ (5.66)

The lattice action (5.65) can be rewritten as

$$S = \frac{1}{2}(\phi, A\phi) \quad \text{with} \quad A = (A_{xy}) , \tag{5.67}$$

and calculating the two-point function (or propagator) of the free Euclidean theory reduces to the computation of the simple Gaussian integral

$$\langle \phi_x \phi_y \rangle = \frac{1}{Z} \int \mathscr{D}\phi \, \phi_x \phi_y \, e^{-\frac{1}{2}(\phi, A\phi)}, \quad \mathscr{D}\phi = \prod_{x \in \Lambda} d\phi_x \tag{5.68}$$

with the partition function

$$Z = \int \mathscr{D}\phi \, e^{-\frac{1}{2}(\phi, A\phi)} = (2\pi)^{V/2} \det^{-1/2} A . \tag{5.69}$$

Such Gaussian integrals have been studied in Chap. 2 from which we take the result

$$\langle \phi_x \phi_y \rangle = A_{x,y}^{-1} \equiv G(x, y) \tag{5.70}$$

for the propagator of the free lattice field.

In order to calculate this propagator, we determine the eigenfunctions and eigenvalues of the symmetric *Toeplitz matrix* A. Thereby we assume that Λ has the same extent N in all directions and hence has $V = N^d$ sites. For periodic boundary conditions, A in (5.65) is circulant, and its V orthonormal eigenvectors ψ_p read

$$\psi_p(x) = \frac{1}{\sqrt{V}} \exp(ipx) \quad \text{with} \quad px = \sum_{\mu=1}^{d} p_\mu x^\mu . \tag{5.71}$$

The allowed lattice momenta lie on the *dual lattice* Λ^* with elements

$$p_\mu = \frac{2\pi}{N} n_\mu \in \Lambda^*, \qquad n_\mu \in \{1, 2, \ldots, N\} . \tag{5.72}$$

The corresponding V eigenvalues are

$$\lambda(p) = m^2 + 2d - 2 \sum_\mu \cos(p_\mu) = m^2 + \hat{p}^2, \quad \hat{p}_\mu = 2 \sin \frac{p_\mu}{2} . \tag{5.73}$$

With these eigenvalues and eigenvectors, we find the following spectral resolution of the propagator with V terms:

$$\langle \phi_x \phi_y \rangle = \sum_{p \in \Lambda^*} \frac{\psi_p(x)\psi_p^\dagger(y)}{\lambda(p)} = \frac{1}{V} \sum_{p \in \Lambda^*} \frac{e^{ip(x-y)}}{m^2 + \hat{p}^2} . \tag{5.74}$$

Fig. 5.4 The two-point function (5.75) of the free scalar field divided by the mass-dependent constant C_m as given in (5.76) together with its exponential fit $\exp(-mx)$ for different values of the propagator masses

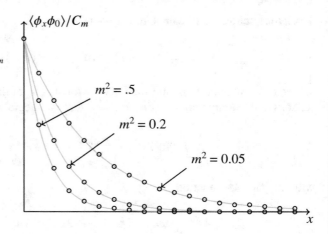

In case of different edge lengths $L_1, \ldots, L_d$, we have to substitute n_μ/L by n_μ/L_μ in the formula above. In the thermodynamic limit $N \to \infty$, the lattice momenta fill in the *Brillouin zone* $(0, 2\pi]^d$ and the Riemann sum

$$\langle \phi_x \phi_0 \rangle = \frac{1}{(2\pi)^d} \sum_{\{p_\mu\}} \Delta p_1 \cdots \Delta p_d \frac{e^{ipx}}{m^2 + \hat{p}^2}$$

where $\Delta p_\mu = 2\pi/N$ tends to an integral over the Brillouin zone,

$$\langle \phi_x \phi_0 \rangle \overset{N \to \infty}{\longrightarrow} \frac{1}{(2\pi)^d} \int_0^{2\pi} d^d p \, \frac{\cos(px)}{m^2 + \hat{p}^2}, \qquad \hat{p}_\mu = 2 \sin \frac{p_\mu}{2} . \tag{5.75}$$

The two-point function on $\Lambda = \mathbb{Z}^d$ is real and invariant under discrete translations and rotations which transform the lattice into itself. In $d = 1$, the value on the diagonal is

$$\langle \phi_0 \phi_0 \rangle \equiv C_m = \frac{1}{m\sqrt{m^2 + 4}}. \tag{5.76}$$

Figure 5.4 illustrates the normalized values of the two-point function for three different masses at the lattice points $x = 0, \ldots, 20$. The exponential fits are excellent.[3]

---

[3] However, for real x the integral (5.75) oscillates around the exponential fit.

5.5   Random Walk Representation of Green Function

The random walk representation is a reformulation of Euclidean field theory and was introduced by Kurt Symanzik in his studies of ϕ^4 models [42, 43]. Here we discuss the random walk representation for the two-point function (5.75) of the free scalar field on $\mathbb{Z}^d$ as "weighted sum over all paths on the lattice" from x to y. This result is useful both for estimates and approximations in lattice field theories.

Let us first consider the quantity

$$G(x) = e^{-\mu} \sum_{\text{paths } 0 \to x} e^{-\mu\ell} = e^{-\mu} \sum_{\ell=0}^{\infty} N_\ell(x) e^{-\mu\ell} \qquad (5.77)$$

on the infinite lattice $\mathbb{Z}^d$. The length ℓ of a path is measured in units of the lattice distance, and $N_\ell(x)$ is the number of paths of length ℓ from 0 to the lattice point x. The arbitrary parameter μ will be fixed later on. There exists a simple *generating function* for the N_ℓ:

$$\left(e^{ip_1} + e^{-ip_1} + \cdots + e^{ip_d} + e^{-ip_d} \right)^\ell = \sum_{x \in \mathbb{Z}^d} N_\ell(x) e^{i(p_1 x_1 + \cdots + p_d x_d)}, \qquad (5.78)$$

where $x_1 + \cdots + x_d = \ell$. This is seen when one computes the left-hand side explicitly. Thereby one obtains a sum over all possible combinations of ℓ coefficients $e^{\pm ip_\mu}$. Interpreting $e^{\pm ip_\mu}$ as a step towards the $\pm\mu$-direction, each of this terms corresponds to one special path of length ℓ on the lattice. Since $N_\ell(x)$ vanishes for all points x with a distance greater than ℓ from the origin, the sum over x terminates. Figure 5.5 indicates the number of paths of length 3 on a *two*-dimensional quadratic lattice.

Fig. 5.5 Number of possible paths of length 3 on the lattice. Each path starts at the origin, marked with a black bullet

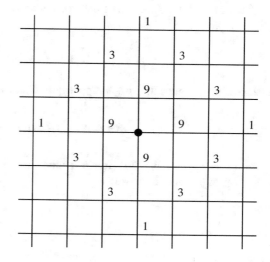

There are nine different paths connecting the origin (marked with a bullet) with an arbitrary nearest neighbor. In contrast, there is no path of length $\ell = 3$ that connects the origin with a next nearest neighbor. $P_3(x) = 0$ for points with a distance greater than 3. Thus the total number of paths with $\ell = 3$ is equal to $(2d)^3 = 64$.

Note that we get zero for the integration of the exponential function $\exp(ipx)$ over the Brillouin zone $p_\mu \in [0, 2\pi)$ as long as the exponent does not vanish. Thus we can extract the polynomials N_ℓ according to

$$N_\ell(x) = \frac{1}{(2\pi)^d} \int_0^{2\pi} d^d p \, e^{-ipx} \left(e^{ip_1} + e^{-ip_1} + \cdots + e^{ip_d} + e^{-ip_d} \right)^\ell . \tag{5.79}$$

Now we insert this result into (5.77). This yields the geometric series

$$
\begin{aligned}
G(x) &= \frac{e^{-\mu}}{(2\pi)^d} \int_0^{2\pi} d^d p \, e^{-ipx} \sum_\ell \left\{ 2e^{-\mu} \left(\cos p_1 + \cdots + \cos p_d \right) \right\}^\ell \\
&= \frac{e^{-\mu}}{(2\pi)^d} \int_0^{2\pi} d^d p \left(\frac{e^{-ipx}}{1 - 2e^{-\mu} \sum_\nu \cos p_\nu} \right) \tag{5.80} \\
&= \frac{1}{(2\pi)^d} \int_0^{2\pi} d^d p \left(\frac{e^{-ipx}}{e^\mu - 2d + 4 \sum_\nu \sin^2 \frac{p_\nu}{2}} \right) . \tag{5.81}
\end{aligned}
$$

We recognize the integral representation of the two-point function of the free Klein–Gordon field (5.75), provided

$$e^\mu - 2d = m^2 . \tag{5.82}$$

Inserting this relation into Eq. (5.77), we end up with the random walk representation for the Green function of the free scalar field on the lattice $\mathbb{Z}^d$ [11]

$$\langle \phi_x \phi_0 \rangle = \frac{1}{(m^2 + 2d)} \sum_{\text{paths } 0 \to x} \frac{1}{(m^2 + 2d)^\ell} . \tag{5.83}$$

This result implies the simple upper bound (see Problem 5.89)

$$\langle x | (-\Delta + m^2)^{-1} | y \rangle < \frac{1}{m^2} e^{-\mu |x - y|} \tag{5.84}$$

with $\mu = \log(1 + m^2/2d)$. It is a special case of the energy-entropy bounds as they appear in polymer expansions.

5.6   There Is No Leibniz Rule on the Lattice

When one proves the invariance of a continuum action under infinitesimal spacetime symmetries, then one employs the ubiquitous Leibniz rule. Unfortunately this rule does not hold on a lattice in accordance with the absence of infinitesimal translational and rotational symmetries and the absence of supersymmetry on a lattice [44]. Now we prove that there exists no lattice derivative which satisfies this rule.

The set of (complex) lattice functions $x \longrightarrow f_x \in \mathbb{C}$ form a linear space and any lattice derivative D is a linear operator on this space.

Lemma 5.1 *A linear operator on the set of lattice functions that fulfills Leibniz's rule*

$$D(f \cdot g) = (Df) \cdot g + f \cdot (Dg), \quad \forall\, f, g : \Lambda \to \mathbb{C}, \tag{5.85}$$

vanishes, i.e., $D = 0$. Here $f \cdot g$ is the pointwise product of functions, $(f \cdot g)_x = f_x g_x$.

Proof We write the Leibniz rule in component form in order to prove this lemma. For that purpose we identify a lattice function $f : x \to f_x$ with the V-component vector f_x and the linear operator D with a $V \times V$ matrix D_{xy}, i.e.,

$$(Df)_x = \sum_{y \in \Lambda} D_{xy} f_y \,. \tag{5.86}$$

Then Leibniz's rule may be written as

$$\sum_z D_{xz}\,(f_z g_z) = g_x \sum_z D_{xz} f_z + f_x \sum_z D_{xz} g_z \,. \tag{5.87}$$

For both functions in (5.87), we now choose the characteristic function of a *fixed* lattice site, say y. Then f_z and g_z vanish for all $z \neq y$ and in addition $f_y = g_y = 1$. Hence, Eq. (5.87) simplifies to $D_{xy} = 2\delta_{x,y} D_{xy}$. It immediately follows that $D_{xy} = 0$ for all sites x, y.

In case of periodic lattice functions, we will demand

$$\sum_{x \in \Lambda} (Df)_x = 0 \,, \tag{5.88}$$

following the corresponding formula for fields over $\mathbb{R}^d$. The backward and forward lattice derivatives satisfy this condition. Finally note that the Lemma (5.85) does not exclude the possibility that Leibniz's rule is satisfied for a particular subset of lattice functions.

5.7   Problems

5.1 (ζ-Function) Calculate the eigenvalues of $A = -\Delta + m^2$ in a finite box $[0, \beta] \times [0, L]^{d-1}$, subject to periodic boundary condition in imaginary time and all spatial directions. Show that the spectral sum defining the associated fuction $\zeta_A(s)$ converges for $s > d/2$.

5.2 (Black-Body Radiation for Massless Scalar Particles) Compute the free energy density as well as the internal energy density for massless scalar particles in two and three dimensions.

5.3 (High-Temperature Expansions) Calculate the high-temperature expansions for the free energy densities of massive scalars in two and three dimensions.

5.4 (Legendre Transformation) For which φ is the Legendre transform of $w(j) = e^j$ defined? Calculate the transform $u(\varphi)$.

5.5 (Two-Point Function in One Dimension) Calculate the Fourier integral (5.75) in one dimension to obtain the two-point function

$$\langle \phi_x \phi_0 \rangle = \frac{1}{m\sqrt{m^2 + 4}} \exp\left(-2x \log\left[\frac{m}{2} + \sqrt{1 + \frac{m^2}{4}} \right] \right)$$

and discuss the result for small m.
Hint: Assume a positive x and show that the integral over $[-\pi, \pi]$ can be replaced by the corresponding integral over the loop

$$-\pi \to \pi \to \pi + i\infty \to -\pi + i\infty \to -\pi .$$

Then convince yourself that the integrand has one pole in the region encircled by the loop.

5.6 (Bounds for the Free Propagator) Prove that for a positive m^2 the propagator (5.83) can be bounded as follows,

$$0 < \langle x|(-\Delta + m^2)^{-1}|y \rangle < \frac{1}{m^2}\, e^{-\mu|x-y|} \tag{5.89}$$

with a new mass parameter $\mu = \log(1 + m^2/2d)$.

5.7 (Generalization of Corollary 5.8) Let $W[j]$ be a twice differentiable *functional* and $\Gamma[\varphi]$ its Legendre transform,

$$\Gamma[\varphi] = \inf_{j(x)} \left(\int d^d x\, j(x)\varphi(x) - W[j] \right) . \tag{5.90}$$

The minimizing $j(x)$ is conjugated to the prescribed field $\varphi(x)$. Prove the following generalization of Corollary 5.8 in the section on the Legendre–Fenchel transformation:

$$\int \mathrm{d}^d z \, \frac{\delta^2 W}{\delta j(x)\delta j(z)} \, \frac{\delta^2 \Gamma}{\delta\varphi(z)\delta\varphi(y)} = \delta(x, y) \,, \tag{5.91}$$

whereby the second derivatives must be evaluated at conjugated fields $j \leftrightarrow \varphi$. In applications, $W[j]$ is the Schwinger functional and $\Gamma[\varphi]$ the effective action. This equation means that the second functional derivative of the effective action is the inverse of the connected two-point function.

References

1. M.E. Peskin, D.V. Schroeder, *An Introduction to Quantum Field Theory*, reprint edition (Taylor & Francis, Milton Park, 2019)
2. L. Brown, *Quantum Field Theory* (Cambridge University Press, Cambridge, 1994)
3. R. Haag, *Local Quantum Physics: Fields, Particles and Algebras* (Springer, Berlin, 2012)
4. M. Maggiore, *A Modern Introduction to Quantum Field Theory* (Oxford University Press, Oxford, 2004)
5. J. Fröhlich, On the triviality of $\lambda\phi_d^4$ theories and the approach to the critical point in $d \geq 4$ dimensions. Nucl. Phys. **B200**, 281 (1982)
6. A. Liddle, D. Lyth, *Cosmological Inflation and Large-Scale Structure* (Cambridge University Press, Cambridge, 2000); *Primordial Density Perturbation* (Cambridge University Press, Cambridge, 2009)
7. V. Mukhanov, *Physical Foundations of Cosmology* (Cambridge University Press, Cambridge, 2005)
8. K. Osterwalder, R. Schrader, Axioms for Euclidean Green's functions. Commun. Math. Phys. **31**, 83 (1973)
9. K. Osterwalder, R. Schrader, Axioms for Euclidean Green's functions II. Commun. Math. Phys. **42**, 281 (1975)
10. J. Fröhlich, Schwinger functions and their generating functionals. Helv. Phys. Acta **47**, 265 (1974)
11. J. Glimm, A. Jaffe, *Quantum Physics: A Functional Integral Point of View* (Springer, New York, 1987)
12. S. Weinberg, Gauge and global symmetries at high temperature. Phys. Rev. **D9**, 3357 (1974)
13. L. Dolan, R. Jackiw, Symmetry behavior at finite temperature. Phys. Rev. **D9**, 3320 (1974)
14. C.W. Bernard, Feynman rules for gauge theories at finite temperature. Phys. Rev. **D9**, 3312 (1974)
15. J.I. Kapusta, *Finite-Temperature Field Theory* (Cambridge University Press, Cambridge, 2011)
16. M. Le Bellac, *Thermal Field Theory* (Cambridge University Press, Cambridge, 2000)
17. J.S. Dowker, R. Critchley, Effective Lagrangian and energy-momentum tensor in de Sitter space. Phys. Rev. **D13**, 3224 (1976)
18. S.W. Hawking, Zeta-function regularization of path integrals in curved space. Commun. Math. Phys. **55**, 133 (1977)
19. E. Elizalde, S.D. Odintsov, A. Romeo, A.A. Bytsenko, S. Zerbini, *Zeta Regularization Techniques with Applications* (World Scientific, Singapore, 1994)
20. H. Weyl, Das asymptotische Verteilungsgesetz der Eigenschwingungen eines beliebig gestalteten elastischen Körpers. Rend. Circ. Mat. Palermo **39**, 1 (1915)

21. S. Blau, M. Visser, A. Wipf, Determinants, Dirac operators and one-loops physics. Int. J. Mod. Phys. **A4**, 1467 (1989)
22. P.B. Gilkey, *Invariance Theory: The Heat Equation and the Atiyah-Singer Index Theorem* (CRC Press, Boca Raton, 1994)
23. D.V. Vassilevich, Heat kernel expansion: user's manual. Phys. Rept. **388**, 279 (2003)
24. S. Blau, M. Visser, A. Wipf, Zeta functions and the Casimir energy. Nucl. Phys. **B310**, 163 (1988)
25. E. Elizalde, A. Romeo, Expressions for the zeta function regularized Casimir energy. J. Math. Phys. **30**, 1133 (1989)
26. I. Sachs, A. Wipf, Finite temperature Schwinger model. Helv. Phys. Acta **65**, 652 (1992)
27. E. Corrigan, P. Goddard, H. Osborn, S. Templeton, Zeta function regularization and multi - instanton determinants. Nucl. Phys. **B159** , 469 (1979)
28. H.W. Braden, Expansion for field theories on $S^1 \times \Sigma$. Phys. Rev. **D25**, 1028 (1982)
29. L. O'Raifeartaigh, A. Wipf, H. Yoneyama, The constraint effective potential. Nucl. Phys. **B271**, 653 (1986)
30. H. Casimir, On the attraction between two perfectly conducting plates. Proc. Kon. Nederland. Akad. Wetensch. **B51**, 793 (1948)
31. S.K. Lamoreaux, Demonstration of the Casimir force in the 0.6 to 6 μm range. Phys. Rev. Lett. **78**, 5 (1997)
32. G. Bressi, G. Carugno, R. Onofrio, G. Ruoso, Measurement of the Casimir force between parallel metallic surfaces. Phys. Rev. Lett. **88**, 041804 (201)
33. M. Bordag, U. Mohideen, V.M. Mostepanenko, New developments in the Casimir effect. Phys. Rep. **353**, 1 (2001)
34. K.A. Milton, *The Casimir Effect* (World Scientific, Singapore, 2001)
35. W.M.R. Simpson, U. Leonardt, *Forces of the Quantum Vacuum: An Introduction to Casimir Physics* (World Scientific, Singapore, 2015)
36. J.L. Cardy, I. Peschel, Finite size dependence of the free energy in two-dimensional critical systems. Nucl. Phys. **B300**, 377 (1988)
37. C. Wiesendanger, A. Wipf, Running coupling constants from finite size effects. Ann. Phys. **233**, 125 (1994)
38. J. Zinn-Justin, *Quantum Field Theory and Critical Phenomena* (Oxford University Press, Oxford, 2021)
39. R.T. Rockafellar, *Convex Analysis* (Princeton University Press, Princeton, 1997)
40. R. Balian, *From Microphysics to Macrophysics* vol. 1 (Springer, Berlin, 2006)
41. J.B. Hiriat-Urruty, J.E. Matinez-Legaz, New formulas for the Legendre-Fenchel transform. J. Math. Anal. Appl. **288**, 544 (2003)
42. K. Symanzik, A modified model of Euclidean quantum field theory. Courant Institute of Mathematical Sciences, IMM-NYU 327 (1964)
43. K. Symanzik, Euclidean quantum field theory, in *Local Quantum Field Theory*, ed. by R. Jost (Academic, New York, 1968)
44. G. Bergner, Complete supersymmetry on the lattice and a No-Go theorem. JHEP **01**, 024 (2010)

Classical Spin Models: An Introduction

6

One distinguishes between continuous and discrete spin models (lattice models) depending on whether the spins take their values in a continuous or discrete target space. The previously considered lattice scalar field theory with target space $\mathscr{T} = \mathbb{R}$ defines a *continuous spin model*. A typical representative of the class of *discrete spin models* is the ubiquitous *Ising model* and its generalizations. The target space of the Ising model is the finite group Z_2. It is a simple statistical model for *ferromagnetism* induced by "elementary spins" or "elementary magnets" sitting at the sites of a crystal lattice. The spins can be in one of two states, either spin up or spin down, and each spin interacts at most with its nearest neighbors. We will focus on a quantitative understanding of such systems with many or infinitely many degrees of freedom.

In this chapter *phase transitions* as observed in ferromagnets will be of particular interest. The two-dimensional Ising model is one of the simplest statistical models to show a phase transition. Below the *Curie temperature* , we observe a spontaneous magnetization where a majority of spins point in a given direction. Above T_c, the spontaneous magnetization vanishes. Iron, cobalt, and nickel with Curie temperatures of 1043, 1403, and 631 K are examples of ferromagnetic materials.

6.1 Simple Spin Models for (Anti)Ferromagnets

When modeling ferromagnets one often assumes that the spins of the atoms on the lattice sites can only attain discrete values. The Hamiltonian of such a spin model has the form

$$H = - \sum_{x,y \in \Lambda} J_{xy} s_x s_y - h \sum_{x \in \Lambda} s_x , \qquad (6.1)$$

where s_x denotes the spin associated with the atom at site x, h the magnetic field, and J_{xy} the interaction strength (exchange coupling) between the spins at sites x

© The Author(s), under exclusive license to Springer Nature Switzerland AG 2021
A. Wipf, *Statistical Approach to Quantum Field Theory*, Lecture Notes
in Physics 992, https://doi.org/10.1007/978-3-030-83263-6_6

and y. Often it is a good approximation to assume that J_{xy} is non-zero only if the sites x and y are nearest neighbors. In addition, for a translational invariant system, we expect that the exchange couplings are independent of the nearest neighbor pair x, y. In the Ising model, a spin s_x parallel to a fixed axis has value 1, and a spin anti-parallel to this axis has value -1. The interaction energy has the form

$$H = -J \sum_{\langle x,y \rangle} s_x s_y - h \sum_{x \in \Lambda} s_x, \quad s_x \in \{-1, 1\}, \tag{6.2}$$

where the first sum extends over all pairs of nearest neighbors. A ferromagnetic interaction $J > 0$ tends to align spins, and an anti-ferromagnetic interaction $J < 0$ tends to anti-align them.

6.1.1 Ising Model

The simple Ising model with energy function (6.2) has been intensively studied and is often referred to as "harmonic oscillator of statistical physics." It was introduced back in 1920 by Wilhelm Lenz, the thesis advisor of Ernst Ising [1], to model ferromagnetic systems. Five years later Ising solved the one-dimensional model, the *Ising chain* [2]. The thermodynamic potentials of the Ising chain can be calculated in closed form, for example, via its transfer matrix; see Sect. 8.1. Unfortunately this simple model does not show spontaneous magnetization at any finite temperature. With the C-program on p. 125, one can simulate the Ising chain and calculate its magnetization as function of the external magnetic field. In complete agreement with the analytic solutions for the infinitely long chain, one obtains a smooth curve for any positive temperature; see Fig. 6.1. This is expected for a system without spontaneous magnetization.

Fig. 6.1 Magnetization as a function of h for the one-dimensional Ising model

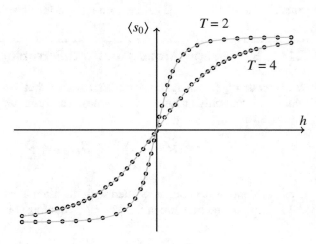

In 1936 Rudolf Peierls analyzed the *two-dimensional Ising model* and proved the existence of a low-temperature phase exhibiting spontaneous magnetization [3, 4]; see Sect. 10.1. For high temperatures there is no magnetization, and hence we expect a transition from an ordered low-temperature phase to a disordered high-temperature phase at a critical temperature $T_c > 0$. Without magnetic field this so-called Curie temperature was calculated by Kramers and Wannier in 1941 [5]. We shall discuss their duality transformation in Sect. 10.2. Three years later Lars Onsager found the exact solution by means of the transfer matrix method [6]. He obtained the following analytical formula for the free energy density, measured in units of $k_B T$, as function of $K = \beta J$ (for a lucid discussion, see [7–10])

$$-\beta f(T) = \log \cosh(2K) - 2K + \frac{2}{\pi} \int_0^{\pi/2} d\theta \, \log\left(1 + \sqrt{1 - \kappa^2 \sin^2\theta}\right) , \quad (6.3)$$

with $\kappa = 2 \tanh(2K)/\cosh(2K)$. The corresponding internal energy is

$$u(T) = J\frac{\partial}{\partial K}(\beta f(T)) = 2J - J \coth 2K \left(1 + (2\tanh^2 2K - 1)\frac{2K(\kappa)}{\pi}\right) , \quad (6.4)$$

where $K(\kappa)$ is the complete elliptic integral of the first kind, defined as

$$K(\kappa) = \int_0^{\pi/2} \frac{d\theta}{\sqrt{1 - \kappa^2 \sin^2\theta}} . \quad (6.5)$$

Differentiating the internal energy with respect to T yields the *specific heat*,

$$c = \frac{4K^2}{\pi \sinh^2 2K} \left\{ K(\kappa)\left(\sinh^2 2K + \frac{2}{\cosh^2 2K}\right) - E(\kappa)\cosh^2 2K - \frac{\pi}{2}\right\} \quad (6.6)$$

which contains the complete elliptic integrals of the second kind,

$$E(\kappa) = \int_0^{\pi/2} \sqrt{1 - \kappa^2 \sin^2\theta} \, d\theta .$$

The complete elliptic integral of the first kind has a singularity at $\kappa = 1$, and the phase transition occurs at this point. The temperature at which the phase transition occurs is then given by the condition

$$1 = 2\frac{\tanh(2K_c)}{\cosh(2K_c)} \quad \text{or} \quad 2K_c = \log\left(1 + \sqrt{2}\right) . \quad (6.7)$$

This yields the critical temperature $T_c = 2.269 J$. In Fig. 6.2 we plotted the internal energy density and the specific heat as functions of $K \propto 1/T$.

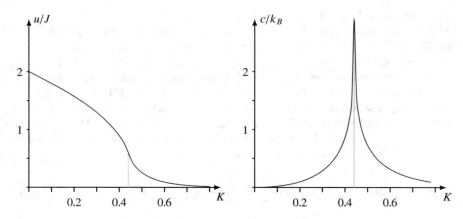

Fig. 6.2 The internal energy per site and the specific heat as a function of $K = \beta J$. The phase transition occurs at the critical value $K_c = 0.4407$

The Dutch physicist Hendrik Casimir was not well informed about what had happened in theoretical physics during the Second World War and asked Pauli about new developments. The latter answered:

> Not much interesting ... except Onsager's solution of the two-dimensional Ising model.

This comment should emphasize the importance of Onsager's solution in theoretical physics. After all, two-dimensional Ising-like models are the only non-trivial statistical systems that show a phase transition and can be solved analytically. Today we know several methods for solving these models. We will discuss some of them. The possibility to compare approximation schemes with the exact solution explains the importance of the Ising model in statistical physics.

So far, the *three-dimensional Ising model* could not be solved analytically, and one has to rely on approximations, for example, the mean field approximation, high-temperature or the low-temperature expansions, or numerical simulations to calculate thermodynamic potentials and (thermal) expectation values. The quality of the mean field approximation increases with the number of nearest neighbors, and the approximation yields the correct critical exponents in four or more dimensions.

6.2   Ising-Type Spin Systems

Besides the ubiquitous Ising model, there exist many other lattice spin models of interest with discrete or continuous target spaces. Examples with discrete target spaces are the standard and planar Potts models where the spin may assume q different values. Examples with continuous target spaces are non-linear O(N) models for which the spins take their values on the unit sphere in $\mathbb{R}^N$. Also lattice regularized Euclidean quantum field theories define particular spin models.

We use the same notation as in Chap. 5 and denote the coordinates of the sites in a hypercubic lattice $\Lambda \subset a\, \mathbb{Z}^d$ by x, where a is the lattice spacing. We impose periodic boundary conditions and thus identify the points x and $x + aN_\mu e_\mu$ (no sum), where e_μ denotes the unit vector in direction μ. At the end of the calculation, we often take the *thermodynamic limit* where the number of sites $N_1 \cdots N_d$ tends to infinity. After a suitable rescaling, we may assume that $a = 1$ such that the unit cell has unit volume and the lattice volume is $V = N_1 \cdots N_d$. To every site we assign a $\mathscr{T}$-valued spin variable s_x. Because of the periodic boundary conditions, we have

$$s_{x'} = s_x \quad \text{for} \quad x' = x + N_\mu e_\mu, \quad \mu = 1, \ldots, d \, . \tag{6.8}$$

Every point inside a hypercubic lattice has $2d$ nearest neighbors and $2d$ bonds towards these neighbors. A *configuration* $\omega = \{s_x \,|\, x \in \Lambda\}$ assigns a value to every spin

$$\omega : \Lambda \longrightarrow \mathscr{T} \times \mathscr{T} \times \cdots \times \mathscr{T} = \mathscr{T}^V, \quad V = |\Lambda| \, . \tag{6.9}$$

We denote the set of all configurations by Ω. If the target space has a finite number of elements, as happens for the Ising model with $|\mathscr{T}| = 2$, then there are $|\mathscr{T}|^V$ different configurations $\omega \in \Omega$.

6.2.1 Standard Potts Models

Cyril Domb suggested that his student Renfrey Potts study a class of generalized Ising models with $q > 2$ orientations for the spin. Nowadays these models are named after Potts who described the systems towards the end of his 1952 PhD thesis. His results for the Z_q-models and the standard q-state Potts models were published in [11]. Both lattice systems have q^V different configurations. In two dimensions the q-state Potts model is exactly solvable and shows a phase transition with order parameter. Both the order of the transition and the critical exponents depend on the number q.

In a lattice gas representation, every lattice point is occupied by one of q different atoms. We enumerate the different atoms by $\sigma \in \{1, 2, \ldots, q\}$. Identical nearest neighbor atoms have an interaction energy of $-J_p$, and different nearest neighbor atoms have no interaction energy. Thus the energy of a particular distribution of atoms, given by a configuration $\omega = \{\sigma_x \,|\, x \in \Lambda\}$, has the form

$$H_{\text{potts}}(\omega) = -J_p \sum_{\langle x,y \rangle} \delta(\sigma_x, \sigma_y) - h \sum_x \delta(\sigma_x, 1), \quad \sigma_x \in \{1, 2, \ldots, q\} \, , \tag{6.10}$$

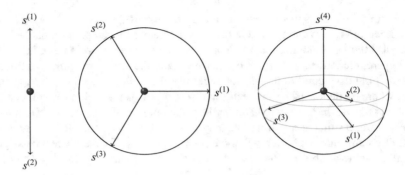

Fig. 6.3 The unit vectors of the vector Potts model with $q = 2, 3, 4$

where $\delta(\sigma_x, \sigma_y)$ is the Kronecker symbol and the first sum extends only over pairs of nearest neighbors. Note that we added an explicit symmetry-breaking term with magnetic field h. For $h = 0$ and $J_p > 0$, the energy is minimal if there is only one type of atoms on the lattice—clearly there are q such classical vacuum configurations.

We may rewrite the energy function (6.10) in the general form (6.2), with s_x being a unit vector pointing to one of q equally spaced points on the unit sphere in $\mathbb{R}^{q-1}$. The particular cases with $q = 2, 3, 4$ are illustrated in Fig. 6.3. So let us pick q unit vectors $\{s^{(1)}, \ldots, s^{(q)}\}$ which point from the origin to these q points on the sphere. Since $\sum_n s^{(n)} = 0$, the scalar product of any two of these unit vectors is

$$s^{(n)} \cdot s^{(m)} = \tilde{q}\, \delta_{nm} - \frac{1}{q-1} \quad \text{with} \quad \tilde{q} = \frac{q}{q-1} . \tag{6.11}$$

To prove the equivalence of the models (6.10) and (6.2), we map the variable σ_x at site x to the vector $s_x = s^{(\sigma_x)}$. This identification yields

$$s_x \cdot s_y = \tilde{q}\, \delta(\sigma_x, \sigma_y) - \frac{1}{q-1} . \tag{6.12}$$

Solving for the Kronecker symbol and inserting into (6.10) yields

$$H_{\text{potts}}(\omega) = -\tilde{J} \sum_{\langle x, y \rangle} s_x \cdot s_y - \tilde{h}\, s^{(1)} \cdot \sum_x s_x - C , \tag{6.13}$$

with $J_p = \tilde{q}\tilde{J}, h = \tilde{q}\tilde{h}$ and an additive constant C which depends on the volume and the parameters of the model. For positive J_p, all vectors are aligned in the classical ground states, and the systems show ferromagnetic behavior. For negative J_p, it is energetically favorable when two neighboring spins are different, and it is not easy to characterize all configurations with minimal energy, at least for $q \geq 3$.

The equivalence of the models (6.10) and (6.13) proves that the *two*-state Potts model is just the Ising model. In contrast, the *one*-state Potts model is very closely related to the bond percolation model. Several exact results are known for the Potts model in two dimensions. On a square lattice, a Kramers–Kronig type duality transformation exists and relates the partition functions at high and low temperatures and in particular fixes the critical temperature T_c [11, 12]. Baxter argued that for $q > 4$, the two-dimensional model shows a first-order transition, whereas for $q \leq 4$, the transition is of higher order [13]. The mean field approximation predicts a first-order phase transition for $q \geq 3$. The mean field results agree with all exact results when q is sufficiently large, and it is conjectured that the approximation is accurate to leading order in q in a large q expansion.

6.2.2 The Z_q Model (Planar Potts Model and Clock Model)

Now we introduce another generalization of the Ising model where one attaches a unit vector in a fixed plane to each lattice point. The unit vector points towards one of the corners of a *planar* equilateral q-gon, characterized by the angles

$$\theta_n = \frac{2\pi n}{q}, \quad n = 1, 2, \ldots, q . \tag{6.14}$$

Let $\theta_{xy} = \theta_x - \theta_y$ be the angle between the vectors at neighboring sites x, y and $J(\theta)$ a 2π-periodic function, and then the energy function for the generalized Z_q-model reads

$$H = -\sum_{\langle x, y\rangle} J(\theta_{xy}) . \tag{6.15}$$

Since the interaction only depends on the angle between vectors at neighboring sites, it is invariant under the global Z_q-transformations

$$\theta_x \longrightarrow \theta_x + \frac{2\pi n}{q} \quad \text{with} \quad n \in Z_q . \tag{6.16}$$

The special choice $J(\theta) = J_c \cos\theta$ leads to the *planar Potts model* or *clock model* with energy function

$$H_{\text{clock}} = -J_c \sum_{\langle x, y\rangle} \cos\left(\theta_x - \theta_y\right) . \tag{6.17}$$

Potts himself solved the two-dimensional model with *two*, *three*, and *four* states. The system with $q = 3$ states is equivalent to the standard Potts model with $J_p = 3J_c/2$,

$$H_{\text{clock}}(\theta) = H_{\text{Potts}}(\sigma) + \tfrac{1}{2} J_c, \quad \sigma_x \in \{1, 2, 3\} .$$

The planar Potts model with four states reduces to two standard Potts models with two states each. However, no relation between the standard and planar Potts models is known for $q > 4$.

6.2.3 The U(1) Model

We obtain the $U(1)$ model from the planar Potts model (clock model) in the limit of infinite q. In this limit the angle θ_n in (6.14) assumes any value between 0 and 2π, such that θ_x parametrizes the circle $S^1 \simeq U(1)$. The *two-dimensional U(1)-model* shows an unusual phase transition. Unusual because it is of infinite order, which means that the free energy is infinitely many times differentiable though at the critical point it is non-analytic. The transition is due to the existence of topologically stable vortex solutions and not related to any symmetry breaking. We may interpret the phase transition as the condensation of vortices and antivortices: at low temperature there are only bound pairs of vortices with opposite circulations, whereas above the critical temperature, we have a plasma of vortices and antivortices. In the disordered high-temperature phase, the correlation functions for the spin variables show exponential falloff, whereas in the "massless" low-temperature phase, they falloff polynomially. This type of transition is called *Kosterlitz–Thouless phase transition* [14–16].

6.2.4 Non-linear O(N) Models

In these models the spins take their values on a unit sphere in $\mathbb{R}^N$. For $N = 1, 2, 3$, we may interpret them as possible directions of elementary magnets generated by classical spins—these models come close to our classical conception of a ferromagnet. The energy of a configuration $\omega = \{s_x\}$ is given by

$$H_{O(N)}(\omega) = -J \sum_{\langle x, y \rangle} s_x \cdot s_y, \qquad s_x \in \mathbb{R}^N, \quad s_x \cdot s_x = 1 . \tag{6.18}$$

A ferromagnetic system is characterized by $J > 0$. The scalar product of two classical spins is invariant under a simultaneous rotation of the spins,

$$R s_x \cdot R s_y = s_x \cdot s_y \quad \text{with} \quad R \in O(N) .$$

Hence the energy function is invariant under global rotations of the spins:

$$H_{O(N)}(R\omega) = H_{O(N)}(\omega), \qquad R\omega = \{R s_x \,|\, x \in \Lambda\} . \tag{6.19}$$

The O(N) invariance justifies the name of these lattice systems. The model with $N = 1$ is the Ising model, the model with $N = 2$ is known as *XY model* (rotor model), and the model with $N = 3$ is referred to as *O(3)-model* or classical *Heisenberg model*.

The model with $O(4)$ is a toy model for the Higgs sector of the Standard Model of particle physics. Actually, the model with $N = 0$ describes self-avoiding walks [17].

6.2.5 Interacting Continuous Spins

So far we considered the energy functions of the form (6.1) with couplings J_{xy} and magnetic field h. Now we also admit an interaction of a spin with itself, in which case the energy function can be written as

$$H(\omega) = \frac{1}{2} \sum_{x,y} J_{xy} \left(s_x - s_y\right)^2 + \sum_x V(s_x) , \qquad (6.20)$$

For simplicity we assumed that the self-interaction V is the same at all lattice sites. If $V = 0$, then the energy function belongs to a *Gaussian model*. On the other hand, we may adjust the couplings J_{xy} in order to find the Euclidean action of an interacting scalar field on the lattice. Similarly, we may regard the energy (6.18) of the O(N) spin model as the Euclidean action for the non-linear O(N) sigma model on a lattice:

$$S_E(\omega) = \frac{1}{2g^2} \sum_{\langle x,y \rangle} \left(\phi_x - \phi_y\right)^2 , \quad \phi_x \in \mathbb{R}^N , \quad \phi_x \cdot \phi_x = 1 . \qquad (6.21)$$

The idea of identifying spin configurations as lattice fields of a discretized Euclidean field theory and energy functions as Euclidean actions according to

$$\omega = \{s_x | x \in \Lambda\} \longleftrightarrow \omega = \{\phi_x | x \in \Lambda\}$$
$$\beta H(\omega) \longleftrightarrow S_E(\omega)/\hbar \qquad (6.22)$$

has been very fruitful for the progress of non-perturbative quantum field theory as well as statistical physics.

6.3 Spin Systems in Thermal Equilibrium

In statistical physics we abandon the idea of solving the dynamics for each of the many microscopic degrees of freedom since this is not necessary to determine macroscopic variables such as pressure, temperature, or magnetization. Instead we describe the system by a so-called density matrix, which determines the probability of the occurrence of a configuration. In the following we are mainly interested in the *canonical ensemble*, where the energy $H(\omega)$ of a configuration ω determines the probability of its occurrence. In general the energy function depends on several parameters $\lambda = (\lambda_1, \ldots, \lambda_n)$. These parameters may characterize the coupling between two spins or the coupling between the spins and an external field. In the canonical ensemble, a configuration ω occurs with probability

$$\exp\left(-\beta H_\Lambda(\omega)\right), \quad \beta = \frac{1}{k_b T}, \tag{6.23}$$

where T denotes the absolute temperature and k_b the Boltzmann constant. The partition function is defined as

$$Z_\Lambda(\beta) = \sum_{\omega \in \Omega} \exp\left(-\beta H_\Lambda(\omega)\right), \tag{6.24}$$

where the sum extends over all configurations.

For a continuous target space, the sum turns into an integral over the lattice spins. For example, the partition function of the simple Gaussian model reads

$$Z_\Lambda(\beta) = \int \mathcal{D}\omega\, e^{-\beta H_\Lambda(\omega)}, \quad H_\Lambda(\omega) = \frac{1}{2} \sum_{x,y} J_{xy}\left(s_x - s_y\right)^2 \tag{6.25}$$

with the Lebesgue measure as single spin distributions,

$$\mathcal{D}\omega = \prod_{x \in \Lambda} ds_x, \quad s_x \in \mathbb{R}. \tag{6.26}$$

The partition function of the non-linear O(N) model with spins on the sphere S^{n-1} looks almost identical,

$$Z_\Lambda(\beta) = \int d\mu(\omega)\, e^{-\beta H_\Lambda(\omega)}, \quad H_\Lambda(\omega) = -J \sum_{\langle x,y \rangle} s_x \cdot s_y, \tag{6.27}$$

but the single spin distribution is the measure on the sphere induced from $\mathbb{R}^N$,

$$d\mu(\omega) = \prod_{x \in \Lambda} d\mu(s_x), \quad d\mu(s) = \delta\left(s^2 - 1\right) d^N s, \quad s_x \in \mathbb{R}^N. \tag{6.28}$$

In general, the partition function and thermodynamic potentials depend on the inverse temperature β, the parameters λ in the Hamilton function, and the lattice Λ.

Next we define expectation values of observables $O(\omega)$ in the canonical ensemble. For discrete spin systems, they are given by sums,

$$\langle O \rangle_\Lambda(\beta) = \frac{1}{Z_\Lambda(\beta)} \sum_\omega O(\omega)\, e^{-\beta H_\Lambda(\omega)} \tag{6.29}$$

and for continuous spin systems by high-dimensional integrals,

$$\langle O \rangle_\Lambda(\beta) = \frac{1}{Z_\Lambda(\beta)} \int d\mu(\omega)\, O(\omega)\, e^{-\beta H_\Lambda(\omega)} . \tag{6.30}$$

In the following we adopt the notation from the discrete models. The corresponding formulas for continuous models are obtained by the substitution $\sum_\omega \to \int d\mu(\omega)$.

The basic quantities of thermodynamics are derived from the partition function Z_Λ. For example, the *Helmholtz free energy* is proportional to the logarithm of Z_Λ,

$$F_\Lambda(\beta) = -\frac{1}{\beta} \log Z_\Lambda(\beta) . \tag{6.31}$$

Since this extensive quantity diverges in the thermodynamic limit, $V \to \infty$, one introduces the *free energy density*

$$f_\Lambda(\beta) = \frac{1}{V} F_\Lambda(\beta) \tag{6.32}$$

instead. For systems with short-range interactions, the densities f_Λ converge in the thermodynamic limit to the free energy density f of the infinite system,

$$f_\Lambda(\beta) \overset{V \to \infty}{\longrightarrow} f(\beta) . \tag{6.33}$$

Note that the energy of one single configuration is not accessible in an experiment. But we can measure the expectation value of the energy at equilibrium—the *internal energy*. It is given by

$$U_\Lambda(\beta) = \langle H \rangle_\Lambda(\beta) = -\frac{1}{Z_\Lambda(\beta)} \frac{\partial}{\partial \beta} \sum_w e^{-\beta H_\Lambda(\omega)} = -\frac{\partial}{\partial \beta} \log Z_\Lambda(\beta) . \tag{6.34}$$

Substituting the logarithm of the partition function by the free energy leads to the expression

$$U_\Lambda(\beta) = \frac{\partial}{\partial \beta} (\beta F_\Lambda(\beta)) = F_\Lambda(\beta) - T \frac{\partial}{\partial T} F_\Lambda(\beta) . \tag{6.35}$$

To find the macroscopic magnetization $m = \langle s_x \rangle$, we couple the spins to a homogeneous external magnetic field according to Eq. (6.1) and differentiate the corresponding h-dependent free energy density with respect to h. Making use of translational invariance on a lattice with periodic boundary conditions, we obtain

$$m := \langle M \rangle = \langle s_x \rangle = -\frac{\partial}{\partial h} f_\Lambda(\beta, h), \quad M = \frac{1}{V} \sum_x s_x . \tag{6.36}$$

We obtain further information about the system via its n-point correlation functions

$$G^{(n)}(x_1, \ldots, x_n) = \langle s_{x_1} \cdots s_{x_n} \rangle, \qquad x_1, \ldots, x_n \in \Lambda .$$ (6.37)

In particular, the two-point function

$$G^{(2)}(x, y) \equiv G(x, y) = \langle s_x s_y \rangle$$ (6.38)

measures the correlation of the spins at site x and at site y. If $G^{(2)}(x, y)$ is positive, these spins have the tendency to align. If this is the case for an arbitrary distance $|x - y|$ between the spins, then the system is *spontaneously magnetized*. If we knew all the correlation functions, then we could reconstruct the Gibbs state.

6.4   Variational Principles

There exists a variational characterization of the partition function and effective action which is useful in approximations, in particular the mean field approximation. The interested reader may also consult [19]. We use the notation for the continuous spin models to underline the connection to lattice field theories.

6.4.1   Gibbs State and Free Energy

Let P be a probability measure with density $p \geq 0$ over the configuration space Ω. Then we have

$$\mathrm{d}P(\omega) = p(\omega)\,\mathrm{d}\mu(\omega) \quad \text{with} \quad \int_\Omega \mathrm{d}P(\omega) = 1 .$$ (6.39)

The *Boltzmann-Gibbs–Shannon entropy* is defined as

$$S_\mathrm{B}(P) = -\int \mathrm{d}\mu(\omega)\, p(\omega) \log p(\omega) ,$$ (6.40)

where the fixed measure μ only depends on the geometries of the lattice and target space. The free energy has the following variational characterization:[1]

$$\beta F = \inf_P \left(\beta \int \mathrm{d}P(\omega)\, H(\omega) - S_\mathrm{B}(P) \right) ,$$ (6.41)

---

[1] There exists an alternative variational approach for the mean field approximation which is based on Jensen's inequality; see, for example, [18].

where the infimum is to be taken with respect to all possible probability measures on Ω, parametrized by density $p(\omega)$. We implement the constraint in (6.39) via the addition of the term $\lambda(\int dP(\omega) - 1)$ with Lagrange multiplier λ. Then the variation of the expression in brackets in (6.41) with respect to $p(\omega)$ leads to

$$0 = \int d\mu(\omega)\, \delta p(\omega)\, (\beta H(\omega) + \log p(\omega) + 1 + \lambda) \implies p(\omega) = C e^{-\beta H(\omega)} .$$

The constant C is fixed by the requirement that $pd\mu$ integrates to one. Hence the unique infimum with respect to all *probability measures* is just the Gibbs measure

$$dP_\beta(\omega) = \frac{1}{Z(\beta)} e^{-\beta H(\omega)} \, d\mu(\omega) \quad \text{with} \quad Z(\beta) = \int d\mu(\omega)\, e^{-\beta H(\omega)} . \tag{6.42}$$

By inserting this result into (6.41), we obtain the well-known expression

$$F(\beta) = -\frac{1}{\beta} \log Z(\beta) \tag{6.43}$$

for the free energy.

6.4.2 Fixed Average Field

The equivalent in statistical physics to the *effective action* in quantum field theory is the free energy functional with prescribed (inhomogeneous) average spin field m_x. Its variational characterization is given by (6.41) with additional constraints,

$$\beta F[m] = \inf_P \left(\beta \int dP(\omega)\, H(\omega) - S_B(P) \,\Big|\, \int dP(\omega)\, s_x = m_x \right) . \tag{6.44}$$

This means that we minimize with respect to probability densities with *prescribed average spins*. There is one constraint for every lattice site, and the constraints are given by the average spins m_x. Since the set of probability measures is convex, the resulting functional $F[m]$ is also convex.

Next we show that the functional F is the Legendre transform of the Schwinger functional W. Thereby we implement the constraints in (6.44) with the help of a Lagrange multiplier field j_x. First we minimize

$$\beta F[m] = \inf_P \left(\int dP(\omega)\, \{\beta H(\omega) - (j, s - m)\} - S_B(P) \right)$$

with respect to all probability measures, where $(j, s) = \sum_x j_x s_x$ denotes the ℓ_2 scalar product. The minimizing measure is given by

$$dP_j(\omega) = \frac{1}{Z[j]} e^{-\beta H(\omega) + (j,s)} d\mu(\omega) \quad \text{with} \quad Z[j] = \int d\mu(\omega) e^{-\beta H(\omega) + (j,s)} .$$

(6.45)

Inserting this result into (6.44) yields the functional,

$$\beta F[m] = (j, m) - W[j] \quad \text{with} \quad W[j] = \log Z[j] .$$ (6.46)

The Lagrangian multiplier field j must be chosen such that the constraints

$$m_x = \int dP_j(\omega) s_x = \frac{\delta W[j]}{\delta j_x}$$ (6.47)

are fulfilled on all sites. The formula (6.46) and (6.47) tell us that $F[m]$ is the Legendre transform of the Schwinger function, i.e.,

$$\beta F[m] = \sup_{j} \left((j, m) - W[j] \right) = (\mathscr{L}W)[m] .$$ (6.48)

If one is interested only in the average magnetization rather than in general correlation functions, then it suffices to evaluate the free energy functional $F[m]$ for a homogeneous average field $m_x = m$. Since for any translational invariant system the average of s_x is equal to the average of $M = \sum_x s_x / V$, we are lead to the following definition of the free energy density for a homogeneous field:

$$\beta f(m) = \frac{1}{V} \inf_{P} \left(\beta \int dP(\omega) H(\omega) - S_B(P) \,\Big|\, \int dP(\omega) M = m \right) .$$ (6.49)

This density is given by the Legendre transform of the Schwinger function,

$$\beta f(m) = (\mathscr{L}w)(m) \quad \text{with} \quad w(j) = \frac{1}{V} \log \int d\mu(\omega) e^{-\beta H(\omega) + j \sum s_x} .$$ (6.50)

At this point we emphasize the crucial difference between classical spin models and Euclidean lattice field theories. In spin models the temperature dependence originates from the temperature-dependent Boltzmann factor of the probability measure

$$dP_\beta(\omega) = \frac{1}{Z_\Lambda(\beta)} e^{-\beta H(\omega)} d\mu(\omega)$$ (6.51)

that is used to calculate expectation values. In an Euclidean lattice field theory with probability measure

$$dP_\beta(\omega) = \frac{1}{Z_\Lambda(\beta)} \, e^{-S(\omega)/\hbar} \, d\mu(\omega), \tag{6.52}$$

the temperature kT is replaced by the reduced Planck constant. Thus the transition from finite temperature spin models to lattice field theories is made by the substitutions

$$\beta H \longrightarrow S/\hbar, \quad \beta F[m] \longrightarrow \Gamma[\varphi]/\hbar, \quad \beta f(m) \longrightarrow u(\varphi)/\hbar \,. \tag{6.53}$$

For example, the effective action of a quantized scalar field is given by

$$\Gamma[\varphi] = \inf_P \left(\int dP(\omega) \, S(\omega) - \hbar S_B(P) \, \bigg| \int dP(\omega) \, \phi_x = \varphi_x \right), \tag{6.54}$$

where one minimizes with respect to probabilities on the space of lattice fields with prescribed average field φ_x. The convex functional Γ is the Legendre transform of the Schwinger functional,

$$\Gamma[\varphi] = (\mathscr{L}W)[\varphi] \quad \text{with} \quad W[j] = \log \int d\mu(\omega) \, e^{-S(\omega)/\hbar + (j,\phi)} \,. \tag{6.55}$$

For a homogeneous φ, the effective action density defines the effective potential

$$u(\varphi) = (\mathscr{L}w)(\varphi) \quad \text{with} \quad w(j) = \frac{1}{\beta V} \log \int d\mu(\omega) \, e^{-S(\omega) + j \sum \phi_x} \,. \tag{6.56}$$

In the classical limit, the effective action Γ converges to the classical action and the effective potential to the classical potential. Finite temperature effects are taken into account by the geometry of the underlying lattice: The lattice has extent $\beta \propto 1/T$ in the imaginary time direction, and all (bosonic) lattice fields are periodic with period β. The variational representations introduced above form an adequate starting point for the useful *mean field approximation* discussed in the Chap. 7.

6.5   Programs for Chap. 6

Here you find the C-programs

- `glgew1d.c`
- `constantsising.h`
- `stdmcising.h`

used in this chapter. The program `glgew1d.c` provides the basis for the simulation
of the one-dimensional Ising model with the energy function

$$H(\omega) = -J \sum_{x=1}^{N} s_x s_{x+1} - h \sum_{x=1}^{N} s_x, \quad s_x = \pm 1 \ . \tag{6.57}$$

The first 500 iterations are necessary to thermalize the system. Afterwards, only
every 20th configuration is evaluated in order to suppress correlations between
the configurations. The program should be self-explanatory. It reads the header
files `constantsising.h` containing the global variables and constants N, M,
MG, MA and J as well as `stdmcising.h`. The values for the magnetization at
temperature $T = 2.0$ are saved in the file `iisingT=2.0`. This file is stored in the
subdirectory `is1data` of the directory containing the program `glgew1d.c`.
glgew1d.c (needs `constantsising.h` and `stdmcising.h`):

```
1   /* program ising1d.c */
2   /* simulation of ferromagnetic 1d Ising-model */
3   /* calculates magnetization for different values */
4   /* of magnetic field. Saved in file  ./is1data/isingT */
5   #include <stdio.h>
6   #include <stdlib.h>
7   #include <math.h>
8   #include <string.h>
9   #include <time.h>
10  #include "constantsising.h"
11  #include "stdmcising.h"
12  int main(void)
13  {
14     srand48(time(NULL));
15     /* read temperature  */
16     puts("temperature (3 digits) = ");
17     scanf("%3s",temp);
18     beta=1/atof(temp);
19     strncat(ising1,temp,3);
20     a=4*beta*J;
21     fp=fopen(ising1,"w");
22     fprintf(fp,"# N = %i , T = %.3f\n",N,1/beta);
23     fprintf(fp,"# magnetization 1-d Ising\n");
24     /* initial configuration */
25     for (i=0;i<N;i++)
26       s[i]=-1; */ cold initial configuration */
27     /* if (rand()<1073741823) s[i]=1 else s[i]=-1 */
28     /* coming to equilibrium  */
29     h=-5.0;b=2*beta*h;
30     /* calculate Boltzmann weights */
31     boltzmann();
32     for (i=0;i<MG;i++) mcsweep(s);
33     /* simulation and calculation for h */
34     /* from -5 to 5 in in steps of 0.5 */
35     for (i=-10;i<11;i++) {
```

```
36        h=0.5*i;b=2*beta*h;
37        boltzmann();check();
38        ann=0;mean1=0;
39        for (j=0;j<M;j++){
40          mcsweep(s);
41          mean1=mean1+moments(1,s);
42        };
43        printf(\"accepted %.2f\n",(float)ann/(N*MA*M));
44        fprintf(fp,"%4.1f    %6.3f\n",h,2*mean1/M);
45        };\draw (100,16)node{$1$};
46      fclose(fp);
47      return 0;
48 }
```

constantsising.h defines constants and global variables:

```
1  /* file constantsising.h */
2  /* constants N,M,MG,MA,J */
3  /* variables s[N], ising1[], etc. */
4  #define N 128  /* number of lattice points */
5  #define M 10000 /* number of iterations */
6  #define MG 500 /* equilibrium */
7  #define MA 20 /* every MAth configuration is measured */
8  #define J 1.0
9  short nn,si,s[N],test[3][5];
10 unsigned int j,k;
11 double mean1;
12 float a,b,vorz,beta,boltz[3][5],h;
13 int i,ann=0;
14 FILE *fp;
15 char temp[20],ising1[]="./is1data/isingT=";
```

stdmcising.h contains functions called by the main program `glgew1d.c`. The constants, variables, and quantities

$$a = 4\beta J \quad \text{and} \quad b = 2\beta h ,$$

as declared in `constantsising.h` are used. The arrays *test* and *boltz* are needed for the Monte Carlo iterations. The first argument of these arrays is the value

```
1  /* header file stdmcising.h */
2  /* functions check and boltz: */
3  /* provides  arrays test and boltzmann.*/
4  /* mcsweep: MA sweeps over lattice */
5  /* moments: calculates average of spins */
6  void check(void)
7  {
8    if (b>0){
```

```
9         test[2][4]=1;test[2][2]=1;test[0][2]=0;test[0][4]=0;
10        if (b>a) {test[0][0]=0;test[2][0]=1;}
11        else {test[0][0]=1;test[2][0]=0;};
12      }
13      else{
14        test[2][0]=0;test[0][2]=1;test[2][2]=0;test[0][0]=1;
15        if(a+b>0) {test[2][4]=1;test[0][4]=0;}
16        else {test[2][4]=0;test[0][4]=1;};};
17    }
18    void boltzmann(void)
19    {
20     boltz[2][4]=exp(-a-b);boltz[2][2]=exp(-b);
21     boltz[2][0]=exp(a-b);boltz[0][4]=exp(a+b);
22     boltz[0][2]=exp(b);boltz[0][0]=exp(-a+b);
23    }
24    void mcsweep(short *s)
25    {
26      int p,q;
27      for (p=0;p<MA;p++)
28            for (q=0;q<N;q++){
29              nn=s[(q+1)%N]+s[(q+N-1)%N]+2;
30              si=s[q]+1;
31              if (test[si,nn]==0) {s[q]=-s[q];ann=ann+1;}
32              else
33                if (drand48()<boltz[si][nn]){
34                  s[q]=-s[q];ann=ann+1;};
35              };
36    }
37    /* calculation of moments */
38    double moments(short n,short *s)
39    {
40      int p,sum=0;
41      for (p=0;p<N;p++)
42     sum=sum+s[p];
43        /*sum=sum+pow(s[il],n);*/
44      return (double)sum/N;
45    }
```

of the spin s_x plus 1. The next argument is equal to the sum of the spins of the nearest neighbors plus 2. If the energy decreases during the change of s_x, we set $test = 0$. Otherwise we set $test = 1$. The Boltzmann weights are stored in the array $boltz$, in order to spare computing time for the calculation of the exponential function. The basic routine is mcsweep, where one finds the $MA \cdot MC$ iterations through the lattice. It is tested how often a change is accepted. We determined the magnetization as a function of h for different temperatures with this code. Figure 6.1 shows the results of the Monte Carlo simulations for $N = 128$. The agreement between the MC data for $N = 128$ and the exact solution for $N = \infty$ is remarkable.

6.6   Problems

6.1 (Two-Dimensional Ising Model: Part I) Determine the internal energy density and the magnetization via the summation over all configurations for a 2×2, 3×3 and 4×4 lattice. Assume thereby periodic boundary conditions and choose $\beta = 0$ to 1 in steps of 0.05. Assume that the external field h vanishes and set J in

$$H = -J \sum_{\langle xy \rangle} s_x s_y$$

equal to 1. Plot your results.
Calculate both $\langle m \rangle$ and $\langle |m| \rangle$. Is it really necessary to calculate $\langle m \rangle$?

6.2 (Two-Dimensional Ising Model: Part II) Adapt the program on p. 126 in order to simulate the *two-dimensional* Ising model via the Metropolis algorithm. Choose $\beta = 0.4406868$ and $h = 0$. Perform the simulations on 4×4, 8×8 and 32×32 lattices with 200 000 sweeps over the lattice each time. Determine

$$u = \frac{1}{V}\langle H \rangle, \quad \langle |m| \rangle \quad \text{and} \quad \langle m^2 \rangle .$$

Compare the result for the 4×4 lattice with the outcome of the analytical calculation in Problem 6.1.

6.3 (Minima of Energy Function) Find the configurations with minimal energy of the following spin models:

1. The Ising chain with first and second neighbor interactions

$$H = -J_1 \sum_x s_x s_{x+1} - J_2 \sum_x s_x s_{x+2}, \quad s_x \in \{-1, 1\}$$

   Consider both positive and negative values of the couplings J_1, J_2.
2. The one-dimensional clock model

$$H = -J_c \sum_x \cos \left\{ 2\pi (n_x - n_y + \Delta)/q \right\}$$

   for positive J_c and all values of Δ.
3. The anti-ferromagnetic Ising model on a triangular lattice,

$$H = J \sum_{\langle x,y \rangle} s_x s_y, \quad s_x \in \{-1, 1\}, \quad J > 0 .$$

References

1. W. Lenz, Beitrag zun Verständnis der magnetischen Erscheinungen in Festkörper. Z. Physik **21**, 613 (1920)
2. E. Ising, Beitrag zur Theorie des Ferromagnetismus. Z. Physik **31**, 253 (1925)
3. R. Peierls, Statistical theory of adsorption with interaction between the adsorbed atoms. Proc. Camb. Phil. Soc. **32**, 471 (1936)
4. R. Peierls, On Ising's model of ferromagnetism. Proc. Camb. Phil. Soc. **32**, 477 (1936)
5. H.A. Kramers, G.H. Wannier, Statistics of the two-dimensional ferromagnet. Part I. Phys. Rev. **60**, 252 (1941)
6. L. Onsager, Crystal statistics. I. A two-dimensional model with an order-disorder transition. Phys. Rev. **65**, 117 (1944)
7. R.P. Feynman, *Statistical Mechanics* (Westview Press, Boulder, 1998)
8. B.M. McCoy, T.T. Wu, *The Two-Dimensional Ising Model* (Harvard University Press, Harvard, 2014)
9. C. Thompson, *Mathematical Statistical Mechanics* (Princeton University Press, Princeton, 2015)
10. L.E. Reichl, *A Modern Course in Statistical Physics* (Wiley, New York, 2016)
11. R.B. Potts, Some generalized order-disorder transformations. Proc. Camb. Phil. Soc. **48**, 106 (1952)
12. L. Mittag, M. Stephen, Dual transformation in many-component Ising models. J. Math. Phys. **12**, 441 (1971)
13. R. Baxter, Potts model at the critical temperature. J. Phys. **C6**, L445 (1973)
14. J.M. Kosterlitz, D.J. Thouless, Ordering, metastability and phase transitions in two-dimensional systems. J. Phys. **C6**, 1181 (1973)
15. J.M. Kosterlitz, The critical properties of the two-dimensional XY-model. J. Phys. **C7**, 1046 (1974)
16. J.V. Jose, L.P. Kadanoff, S. Kirkpatrick, D.R. Nelson, Renormalization, vortices, and symmetry-breaking perturbations in 2-dimensional planar model. Phys. Rev. **B16**, 1217 (1977)
17. N. Madras, G. Slade, *The Self-Avoiding Walk* (Birkhäuser, Basel, 2012)
18. J.M. Yeomans, *Statistical Mechanics of Phase Transitions* (Clarendon, Oxford, 1992)
19. G. Roepstorff, *Path Integral Approach to Quantum Physics* (Springer, Berlin, 1996)

Mean Field Approximation

7

Since only a few lattice models can be solved explicitly, one is interested in efficient approximation schemes. A simple and universally applicable approximation is the mean field approximation (MFA) which yields qualitatively correct results for many lattice systems. When applied to ferromagnets, it is often called the *Curie–Weiss approximation* and when applied to lattice gases the *Bragg–Williams approximation*. The approximation appeared in Pierre Weiss' work back in 1907 [1], where, building on earlier results of Paul Langevin [2], he obtained a model which explains the Curie point below which ferromagnetism sets in. Often when one deals with systems with very many degrees of freedom, one uses the universally applicable MFA to gain information about the qualitative behavior of the system. In some cases the approximation even produces exact results, for example, for universal quantities.

In the MFA, one replaces the microscopic interaction of a spin with its neighboring spins by an approximate interaction of the spin with the averaged spin generated by all other spins. In the mean field ensemble, the spin variables are independently distributed, and under homogeneity assumptions on the interactions, they are even equally distributed. Hence the calculation of the free energy density or the order parameter, e.g., the magnetization of the system, reduces to a single spin problem. In this chapter we discuss the approximation for spin models and Euclidean lattice field theories in arbitrary dimensions. The MFA is discussed in many books on statistical mechanics and field theory. It may be useful to consult [3–6].

© The Author(s), under exclusive license to Springer Nature Switzerland AG 2021
A. Wipf, *Statistical Approach to Quantum Field Theory*, Lecture Notes
in Physics 992, https://doi.org/10.1007/978-3-030-83263-6_7

7.1   Approximation for General Lattice Models

Starting point is the variational characterization (6.41) of the free energy and of the free energy with fixed average field (6.44). In the MFA we only admit *product probability measures* on the configuration space Ω in the variational principle,

$$dP(\omega) = \prod_x d\nu_x(s_x), \quad d\nu_x(s) = d\mu(s)p_x(s) , \tag{7.1}$$

where $\nu_x(s)$ is the probability measure for the spin at site x and μ is the fixed and site-independent single spin distribution. Since we minimize the free energy functional only on a *subset* of all probability measures, the free energy in the mean field approximation $F_{mf}[m]$ bounds the exact free energy $F[m]$ from above,

$$F_{mf}[m] \geq F[m] . \tag{7.2}$$

Unlike the set of probability measures, the subset of product measures is not convex such that F_{mf} need not be convex. For a product probability measure, we have

$$dP(\omega) \log p(\omega) = \prod_x d\nu_x(s_x) \sum_y \log p_y(s_y), \quad \int d\nu_x(s) = 1 ,$$

such that the entropy of the total system is equal to the sum of single-site entropies,

$$S_B(P) = \sum_x s_B(p_x), \quad s_B(p_x) = - \int d\nu_x(s) \log p_x(s) . \tag{7.3}$$

In addition the constraint on the probability measure $P(\omega)$ in (6.44) turns into V independent constraints for the single-site measures ν_x,

$$\int d\nu_x(s) s = \int d\mu(s)p_x(s) s = m_x . \tag{7.4}$$

To proceed we must specify the energy functions to be considered. For general spin systems, the energy has the form

$$H(\omega) = - \sum_{x \neq y} J_{xy}s_x s_y + \sum_x Q_x (s_x) , \tag{7.5}$$

where the last term contains a possible coupling to an external field or a self-interaction of the spin variables. Due to the constraints (7.4), we find the average energy

$$\int dP(\omega)H(\omega) = - \sum_{x \neq y} J_{xy}m_x m_y + \sum_x \int d\nu_x(s) Q_x(s) . \tag{7.6}$$

It follows that for a product measure, the free energy functional takes the form

$$F_{mf}[m] = -\sum_{x \neq y} J_{xy} m_x m_y + \sum_x \alpha_x(m_x) , \tag{7.7}$$

where the function α_x only depends on the prescribed average field on site x and has the variational characterization

$$\alpha_x(m_x) = \inf_{p_x} \left(\int d\nu_x(s) \left\{ Q_x(s) + T \log p_x(s) \right\} \middle| \int d\nu_x(s) \, s = m_x \right) . \tag{7.8}$$

Thus for product measures, the difficult variational problem on the space of probability measures on Ω simplifies considerably to V variational problems on single sites. The minimizing probability density p_x in (7.8) is given by

$$p_x(s) = \frac{1}{z_x(j_x)} e^{-\beta Q_x(s) + j_x s}, \qquad z_x(j_x) = \int d\mu(s) \, e^{-\beta Q_x(s) + j_x s} , \tag{7.9}$$

whereby the multiplier j_x is determined by the *self-consistency equation*

$$m_x = \int d\mu(s) \, p_x(s) \, s = \frac{dw_x}{dj_x}, \qquad w_x(j_x) = \log z_x(j_x) . \tag{7.10}$$

This key equation is also called the *gap equation*. Inserting the result for the density into the expression for α_x finally yields

$$\beta \alpha_x(m_x) = j_x m_x - w_x(j_x) = (\mathscr{L} w_x)(m_x) . \tag{7.11}$$

To summarize: In the MFA the free energy functional $F_{mf}[m]$ for a prescribed average field is given by (7.7), whereby α_x is proportional to the Legendre transform of $w_x = \log z_x$ with one-site partition function z_x given in (7.9).

Let us turn to homogeneous spin systems with $Q_x = Q$ and assume that the average field is constant, in which case

$$\sum_{x \neq y} J_{xy} m_x m_y = \frac{V}{2} \tilde{J} m^2, \quad \text{with} \quad \tilde{J} = \frac{2}{V} \sum_{x \neq y} J_{xy} . \tag{7.12}$$

For lattice models with $J_{xy} = J$ for nearest neighbor pairs x, y and zero otherwise, the *effective coupling* is

$$\tilde{J} = qJ , \tag{7.13}$$

where q denotes the *coordination number*, i.e., the number of nearest neighbors of a given site. For a d-dimensional hypercubic lattice, $q = 2d$. Homogeneous systems have an extensive free energy, and it is advantageous to proceed with the intensive *free energy density*. In the MFA this density is given by

$$f_{\mathrm{mf}}(m) = -\frac{1}{2}\tilde{J}m^2 + T\left(\mathscr{L}w\right)(m) \; ; \tag{7.14}$$

and hence we remain with evaluating the Legendre transform of

$$w(j) = \log z(j), \quad \text{with} \quad z(j) = \int d\mu(s)\, e^{-\beta Q(s)+js} \; . \tag{7.15}$$

To calculate $\mathscr{L}w$, we must solve the gap equation. Note that in the MFA the dimension of the lattice enters only via the relations (7.12) and (7.13) between the microscopic couplings and the effective coupling $\tilde{J}$.

7.2   The Ising Model

Let us apply the general results to Ising models on hypercubic lattices with a single spin (Bernoulli) measure

$$d\mu(\omega) = \prod_x d\mu(s_x), \quad d\mu(s) = \frac{1}{2}\delta(s-1)ds + \frac{1}{2}\delta(s+1)ds \; . \tag{7.16}$$

For a vanishing magnetic field in (6.2), the interaction term Q in (7.15) is absent, and we obtain the simple one-site partition function $z(j) = \cosh(j)$. Note that in contrast to the discrete microscopic spin the average field in the free energy density

$$f_{\mathrm{mf}}(m) = -dJm^2 + T\left(\mathscr{L}w\right)(m) \tag{7.17}$$

is a continuous variable with values in the interval $[-1, 1]$. To calculate the Legendre transform of $w(j) = \log(\cosh j)$, we solve the gap equation $m = \tanh(j)$ for $j(m)$ and insert the result into

$$(\mathscr{L}w)(m) = mj(m) - \log\{\cosh j(m)\} \; . \tag{7.18}$$

Next we insert the Legendre transform into the defining equation (7.17) for the free energy density and obtain the simple expression

$$f_{\mathrm{mf}}(m) = -dJm^2 + \frac{1+m}{2\beta}\log(1+m) + \frac{1-m}{2\beta}\log(1-m) \; . \tag{7.19}$$

Figure 7.1 shows a plot of the density in units of $\tilde{J}$ in the regime near the critical temperature. We observe that the curvature at the origin changes sign at the temperature $T_{c,\mathrm{mf}} = 2dJ$. This defines the critical temperature in the MFA. *Below*

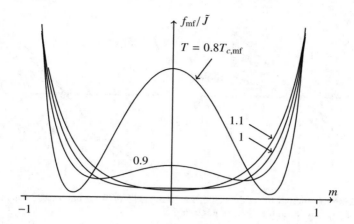

Fig. 7.1 Free energy density as a function of the magnetization m at different temperatures $T/T_{c,\mathrm{mf}}$

this temperature the free energy shows a local maximum at the origin as well as two global minima at $\pm m_0$. Above the critical temperature, there is one global minimum at the origin.

The variational characterization of the free energy functional $F[m]$ as discussed in Sect. 6.4 emphasizes that

$$P_{\mathrm{mf}}(m) = \frac{1}{Z_{\mathrm{mf}}} e^{-\beta V f_{\mathrm{mf}}(m)} \tag{7.20}$$

is to be interpreted as the probability distribution for finding the average field m in the MFA. Clearly, for large volumes this distribution shows distinct maxima at the minima of the free energy density.

A minimum m_0 of the density (7.19) solves the *gap equation*

$$2dJm_0 = \frac{1}{2\beta} \log \frac{1+m_0}{1-m_0} \implies m_0 = \tanh(2dJ\beta m_0) . \tag{7.21}$$

To simplify this self-consistency equation for the mean field, we define the dimensionless quantity $x := 2dJ\beta m_0$, which obeys

$$\frac{T}{2dJ} x = \tanh x . \tag{7.22}$$

The slope of the linear function on the left-hand side is greater than the slope of the tanh-function on the right-hand side for

$$T > T_{c,\mathrm{mf}} = 2dJ , \tag{7.23}$$

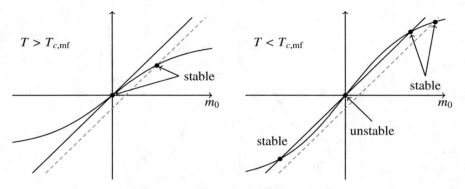

Fig. 7.2 Determination of the critical temperature in the MFA. The continuous lines through the origin belong to $h = 0$ and the dashed lines to $h \neq 0$

and then there exists only the solution $x = 0$. If the temperature is less than $T_{c,\mathrm{mf}}$, then the transcendental equation (7.22) admits three solutions: $x = 0$ and $x = \pm m_0$. To gain further insight, we couple the spins to an external field by adding the term $-h \sum s_x$ to the energy function. This amounts to adding the term $-hm$ to the free energy density f_{mf} such that now the minimizing magnetization obeys

$$m_0 = \tanh\left(2dJ\beta m_0 + \beta h\right) = \tanh\left(\frac{T_{c,\mathrm{mf}}}{T}m_0 + \beta h\right) . \tag{7.24}$$

Solutions of this *gap equation* are given by the intersection points in Fig. 7.2. For $h \neq 0$ the system is magnetized even for very high temperatures $T \gg T_{c,\mathrm{mf}}$.

7.2.1 An Alternative Derivation

Here we present a derivation of the MFA as it is contained in many textbooks on statistical mechanics; see for example, [7]. The main idea is the substitution of the microscopic interaction of a spin with its neighbors by an *averaged interaction* with all spins:

$$J_{xy} \longrightarrow \frac{1}{V}\sum_y J_{xy} = \frac{\tilde{J}}{V} . \tag{7.25}$$

For translationally invariant systems, the effective coupling strength $\tilde{J}$ does not depend on the site. In models with short-range interaction, we have $\tilde{J} = qJ$ with coordination number q. With the approximation (7.25), the energy function (6.1) becomes

$$H \to H_{\mathrm{mf}} = -\frac{V\tilde{J}}{2}m^2(s) - Vhm(s), \quad m(s) = \frac{1}{V}\sum s_x . \tag{7.26}$$

For the Ising model with $s_x = \pm 1$, the mean field $m(s)$ takes its values in

$$M = \{-1, -1+\delta, -1+2\delta, \ldots, 1-\delta, 1\}, \qquad \delta = \frac{2}{V}.$$

If the mean field $m \in M$ is fixed, then $V(1+m)/2$ spins point "upward" and $V(1-m)/2$ "downward," and the number of possible spin configurations is

$$d(m) = \frac{V}{[\frac{1}{2}V(1+m)]![\frac{1}{2}V(1-m)]!}. \tag{7.27}$$

With the help of Stirling's formula

$$\log(n!) = n(\log n - 1) + o(n)$$

we obtain the following approximation for the partition function

$$Z_{\mathrm{mf}} = \sum_{m \in M} d(m)e^{-\beta H_{\mathrm{mf}}(m)} = \sum_m e^{-\beta V f_{\mathrm{mf}}(m)}, \tag{7.28}$$

where $f_{\mathrm{mf}}(m)$ is the free energy density (7.19), up to corrections of order $o(V)/V$. In the thermodynamic limit, these corrections vanish, and at the same time m becomes a continuous field $m \in [-1, 1]$.

7.3 Critical Exponents α, β, γ, δ

The mean field approximation predicts that the Ising model shows a phase transition at some critical temperature $T_c > 0$. More sophisticated methods presented in the following chapters show that the mean field prediction is correct in two and more dimensions. A critical temperature separates the low-temperature ferromagnetic phase with non-zero magnetization from the high-temperature paramagnetic phase without magnetization. The free energy is no longer differentiable at T_c, and the resulting singularities are parametrized by universal *critical exponents*. In this section we calculate the most important critical exponents in the mean field approximation.

Let us introduce the *reduced temperature*, i.e., the relative deviation of the temperature from the critical temperature,

$$t = \frac{T_c - T}{T_c}. \tag{7.29}$$

The critical exponent λ associated with a macroscopic observable $g(t)$ is defined by

$$g(t) \sim |t|^\lambda. \tag{7.30}$$

Table 7.1 Most commonly used critical exponents for magnetic systems

Zero-field specific heat	α	$c_h \sim |t|^{-\alpha}$
Zero-field magnetization	β	$m \sim t^{\beta}$
Zero-field isoth. susceptibility	γ	$\chi_T \sim |t|^{-\gamma}$
Critical isotherm ($t = 0$)	δ	$h \sim |m|^{\delta}\text{sgn}(m)$
Correlation length	ν	$\xi \sim |t|^{-\nu}$
Pair correlation function at T_c	η	$G(x) \sim 1/r^{d-2+\eta}$

This equation only represents the asymptotic behavior of the function $g(t)$ as $t \to 0$. More generally one might expect

$$g(t) \sim A\,|t|^{\lambda}\left(1 + bt^{\lambda_1} + \ldots\right) \quad \text{with} \quad \lambda_1 > 0\,. \tag{7.31}$$

The definitions of the most commonly used critical exponents are listed in Table 7.1. The exponent δ characterizes the singular function $h = h(m)$ and the exponent η the anomalous behavior of the two-point function at the critical temperature. In compiling Table 7.1, we have made the unjustified assumption that the critical exponents associated with a given thermodynamic variable are the same as $T \to T_c$ from above and below. Early series expansions and numerical results suggested that this was the case, but it was only with the advent of the renormalization group that it was indeed proved to be so. A common notation was using a prime to distinguish the value of an exponent as $T \uparrow T_c$ from the value as $T \downarrow T_c$.

In what follows we obtain the mean field values of the critical exponents α, β, γ, and δ in the Ising model. They can be extracted from our previous results for the free energy and magnetization. Later we shall argue that the critical exponents are universal: two spin models in the same dimension for which the order parameter has the same symmetry should have identical critical exponents.

7.3.1 Susceptibility

Above the critical temperature, the magnetization $m_0(h)$ in Eq. (7.24) approaches zero for $h \to 0$, and the susceptibility

$$\chi = \left(\frac{\partial m_0}{\partial h}\right)\Big|_{h=0} = \sum_y \left(\langle s_x s_y \rangle - \langle s_x \rangle \langle s_y \rangle\right)\Big|_{h=0} \tag{7.32}$$

fulfills the *Curie–Weiss law*

$$\chi \overset{m_0(0)=0}{=} \beta\left(\chi T_c + 1\right) \quad \Longrightarrow \quad \chi = \frac{1}{T - T_c}\,. \tag{7.33}$$

In deriving the Curie–Weiss law, we made use of (7.21). We see that the susceptibility diverges at the critical point and obtain for $T \downarrow T_c$ or $T \uparrow T_c$ the scaling
law

$$\chi \sim |t|^{-\gamma}, \quad \gamma = 1 .\qquad (7.34)$$

Hence, in the MF approximation, the *critical exponent* of the susceptibility is $\gamma = 1$.
To keep the notation simple, we shall not always mark the mean values, fields, and
potentials by an index mf. For example, if it is clear from the context, we write T_c
for $T_{c,\mathrm{mf}}$, as we did in (7.33).

7.3.2 Magnetization as a Function of Temperature

Below the critical temperature and for $h > 0$, the magnetization $m_0(h)$ is given by
the largest solution to (7.24). In the limit $h \downarrow 0$, we find a spontaneous magnetization
$m_0(T)$ which approaches zero for temperatures $T \uparrow T_c$. This justifies a series
expansion of the tanh-function in (7.21) in powers of m_0 with the result

$$m_0 = \frac{T_c}{T} m_0 - \frac{1}{3} \left(\frac{T_c}{T} m_0 \right)^3 + \dots ,$$

where $T_c = 2dJ$. As expected, this equation has three solutions, given by

$$m_0 = 0 \quad \text{and} \quad m_0 \approx \pm \frac{T}{T_c} (3t)^{1/2} .\qquad (7.35)$$

The first solution belongs to an unordered paramagnetic state, and the two other
solutions describe ordered ferromagnetic low-temperature phases. The ordered
states minimize the free energy density at low temperature. Figure 7.3 illustrates

Fig. 7.3 The temperature
dependence of the
magnetization m in the mean
field approximation

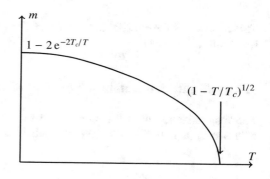

the temperature dependence of the spontaneous magnetization $m(T)$. It approaches zero as $T \uparrow T_c$ according to

$$m_0(T) = \frac{T}{T_c} (3t)^{1/2} .$$ (7.36)

Since $m_0 \neq 0$ corresponds to an ordered and $m_0 = 0$ to an unordered state, we call the spontaneous magnetization an *order parameter* of the system. We associate the critical exponent β with the order parameter of the transition,

$$m_0(T) \sim t^{\beta} .$$ (7.37)

Hence the MFA predicts a critical exponent[1] β of $1/2$.

7.3.3   Specific Heat

In the MFA the *internal energy density* without external field given by

$$u = \frac{\partial}{\partial \beta}(\beta f) = -d J m_0^2 ,$$ (7.38)

such that the specific heat depends on the magnetization as

$$c = \frac{\partial u}{\partial T} = -\frac{T_c}{2} \frac{\partial m_0^2}{\partial T} , \quad T_c = 2dJ .$$ (7.39)

Above T_c, the vanishing magnetization leads to a vanishing specific heat, and for T just below T_c, the formula (7.35) yields

$$c \approx -\frac{T_c}{2} \frac{\partial}{\partial T} \left(\frac{3T^2 t}{T_c^2} \right) \xrightarrow{T \nearrow T_c} \frac{3}{2} .$$

This means that the specific heat jumps at $T = T_c$ from the value $3k_B/2$ below T_c to the value 0 above T_c

7.3.4   Magnetization as a Function of the Magnetic Field

The magnetization as a function of the magnetic field h follows from the self-consistency equation (7.24). We expand the right-hand side of this equation at

---

[1] The exponent β should not be confused with the inverse temperature.

$T = T_c$ up to third order in h according to

$$m_0 = m_0 + \beta_c h - \frac{1}{3}(m_0 + \beta_c h)^3 + \ldots \quad \Rightarrow \quad \beta_c h = \frac{1}{3}(m_0 + \beta_c h)^3 + \ldots \quad (7.40)$$

The consistent assumption $\beta_c h \ll m_0$ leads to

$$m_0 \approx (3\beta_c h)^{1/3}, \quad (T = T_c) . \quad (7.41)$$

In general, the scaling law

$$m_0 \sim h^{1/\delta} \quad \text{at} \quad T = T_c \quad (7.42)$$

defines the critical exponent δ. Hence, our mean field studies predict $\delta = 3$.

7.3.5 Comparison with Exact and Numerical Results

This section is devoted to the comparison of the results of the mean field approximation with the exact ones near the phase transition. The critical temperature T_c in the MFA only depends on the number of nearest neighbors (the coordination number) q according to the relation

$$T_{c,\mathrm{mf}} = q J \quad (7.43)$$

This simple result points to a first deficiency of the MFA: it predicts a phase transition for the Ising chain at a $T_c > 0$ in conflict with the exact result. In Table 7.2 we compare the mean field predictions for the critical temperatures of Ising models on various two- and three-dimensional lattices with the known values. Note that with increasing coordination number q the MFA becomes more accurate. The MFA correctly predicts scaling laws for the magnetization and susceptibility as functions of the reduced temperature and for the magnetic field as a function of the magnetization in the vicinity of the critical point. But the mean field critical exponents $\beta = 1/2$, $\gamma = 1$ and $\delta = 3$ are independent of the space dimension and disagree with the exactly known exponents in two dimensions; see Table 7.3. In addition the MFA predicts a jump of the specific heat at T_c in contrast to the exact solution in two dimensions which shows a logarithmic singularity. Similar

Table 7.2 Critical temperatures depending on dimension d and number q of neighbors

Lattice	d	q	$T_{c,\mathrm{mf}}/T_c$
Square	2	4	1.763
Triangular	2	6	1.648
Simple cubic	3	6	1.330
Body-centered cubic	3	8	1.260
Face-centered cubic	3	12	1.225

Table 7.3 Critical exponents of the Ising model in two and three dimensions [8–10]

Quantity		$d = 2$ (exact)	$d = 3$ [8,9]	$d = 3$ [10]	Mean field
Zero-field specific heat	α	0 (log.)	0.110(1)	0.11008(1)	0 (jump)
Zero-field magnetization	β	1/8	0.3265(3)	0.326419(3)	1/2
Zero-field isoth. susceptibility	γ	7/4	1.2372(5)	1.237075(10)	1
Critical isotherm ($t = 0$)	δ	15	4.789(2)	4.78984(1)	3
Correlation length	ν	1	0.63002(10)	0.629971(4)	1/2
Pair correlation function at T_c	η	1/4	0.03627(10)	0.036298(2)	0

differences are seen in three dimensions; see Table 7.3. However, the critical exponents of the MFA are the correct ones in $d > 4$ dimensions.

We summarize the main results of the mean field approximation:

- The dimension of the lattice enters the MFA only via $T_{c,\mathrm{mf}} = 2dJ$.
- The order of the phase transition is correctly predicted for $d \geq 2$.
- The MF result for $T_{c,\mathrm{mf}}$ is greater than the exact T_c in $d \geq 2$.
- The critical MF exponents differ from the exact ones for $d < 4$.
- The MFA does not account for short-range interaction effects.
- The MFA may lead to long-range correlations.

7.4   Mean Field Approximation for Standard Potts Models

We minimize the free energy of the standard Potts model with the energy (6.10) on the set of product probabilities

$$P(\omega) = \prod_x p_x(\sigma_x), \quad \sigma_x \in \{1, \ldots, q\} . \tag{7.44}$$

Due to translational invariance, p_x does not depend on the site, and without magnetic field, we obtain the free energy density

$$f_{\mathrm{mf}}(\beta, p_n) = -dJ_p \sum_{n=1}^{q} p_n^2 + T \sum_{n=1}^{q} p_n \log p_n , \tag{7.45}$$

where p_n denotes the probability of $\sigma_x = n$. Clearly, in the symmetric phase, all probabilities are equal to $1/q$. We now minimize the free energy density with respect to the probability p_1 of σ_1 under the assumption of equal $p_2, \ldots, p_q$. We parametrize the corresponding probabilities as

$$p_1 = \frac{1}{q} + \frac{q-1}{q}x, \quad p_{n>1} = \frac{1}{q} - \frac{x}{q} \quad \text{with} \quad -\frac{1}{q-1} \leq x \leq 1 ,$$

where the value $x = 0$ characterizes the symmetric phase. Setting $q - 1 = q'$, the free energy density reads

$$f_{mf}(x) = -\frac{d J_p}{q}\left(1 + q'x^2\right) + \frac{1 + q'x}{\beta q} \log \frac{1 + q'x}{q} + \frac{q'(1 - x)}{\beta q} \log \frac{1 - x}{q} \tag{7.46}$$

and its curvature at the origin vanishes for $2d J_p = qT$. This defines the temperature

$$T'_c = \frac{2d J_p}{q} \tag{7.47}$$

below which the symmetric phase is unstable. But for Potts models with $q \geq 3$, the free energy density has a second minimum at $x_0 > 0$ for temperatures near T'_c. The second minimum turns into an absolute minimum below a critical temperature $T_c > T'_c$. When the temperature drops, the order parameter x jumps at the critical temperature

$$T_c = \frac{q - 2}{q - 1} \frac{d J_p}{\log(q - 1)} > T'_c \tag{7.48}$$

from 0 to $x_0 = (q - 2)/(q - 1)$. Figure 7.4 illustrates the change of the free energy density $\Delta f_{mf}(x) = f_{mf}(x) - f_{mf}(0)$ in units of $k_B T$ for Potts models with $q = 3, 4$ and temperatures close to T_c. With increasing q, we observe an increasing jump of the order parameter, and the first-order phase transition gets stronger. For $q = 2$, the jump vanishes and the phase transition is of second order.

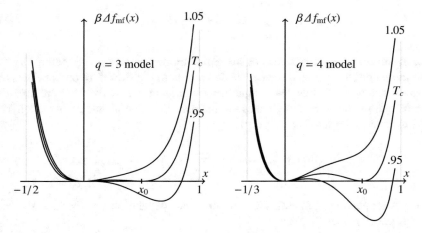

Fig. 7.4 The normalized free energy density for Potts models with $q = 3, 4$ in the MFA

For large q, the internal energy, free energy, and latent heat per site at the critical point have the following expansions:

$$e_{mf,c} \simeq -d J_p \left(1 - \frac{2}{q} \dots \right)$$

$$\beta_c f_{mf,c} \simeq -\log q \left(1 + \frac{1}{q} \dots \right) \tag{7.49}$$

$$\ell_{mf,c} \simeq d J_p \left(1 - \frac{3}{q} \dots \right).$$

To leading order in $1/q$, they agree with the exact values for the Potts models on a square lattice. This leads to the conjecture that the mean field treatment of the Potts model provides an accurate description of the transition in two or higher dimensions if the number of components q is large.

7.5   Mean Field Approximation for Z_q Models

In the MFA for the planar Potts models with energy functions (6.17), we minimize the free energy with respect to the product probabilities

$$P(\omega) = \prod_x p_x(\theta_x), \quad \theta_x \in \left\{ \theta_n = \frac{2\pi n}{q} \,\middle|\, n = 1, 2, \dots, q \right\}, \tag{7.50}$$

where we fix the average field

$$\sum_{n=1}^{q} p_x(\theta_n) \, e^{i\theta_n} = m_x . \tag{7.51}$$

The average spin m_x lies in the convex region in the complex plane spanned by the q extremal points $\exp(i\theta_n)$; see the shaded region in Fig. 7.5. If all probabilities are zero except $p_n = 1$, then the magnetization points towards the nth corner of the q-gon, and the corresponding states are *pure states* of the planar Potts model. The average energy for fixed m_x is given by

$$\sum_{\omega} p(\omega) H(\omega) = -\frac{J_p}{2} \sum_{\langle x,y \rangle} \left(m_x m_y^* + m_x^* m_y \right) . \tag{7.52}$$

For a homogeneous magnetization, p_x does not depend on the site, and the free energy density takes the form

$$f_{mf}(m) = \inf_{\{p_n\}} \left(-d J_p m m^* + T \sum_n p_n \log p_n \,\middle|\, \sum_n p_n e^{i\theta_n} = m \right), \tag{7.53}$$

Fig. 7.5 The possible values of the magnetization m in the complex plane for the planar Potts model (clock model) with five states

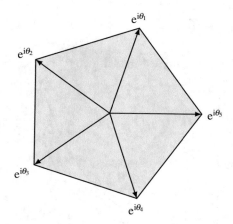

where we used the abbreviation $p(\theta_n) = p_n$. The solution of this simple variational problem with constraints leads to

$$f_{\mathrm{mf}}(m) = -dJ_p mm^* + T(\mathscr{L}w)(m), \quad e^{w(j)} = \sum_n \exp\left(je^{i\theta_n} + j^* e^{-i\theta_n}\right)$$

$$(7.54)$$

with the Legendre transform

$$(\mathscr{L}w)(m) = \sup_j \left\{ jm + j^* m^* - w(j) \right\}.$$

$$(7.55)$$

However, for the models with three and four states, we can minimize (7.53) directly. To do this we introduce the *moments* of the probability distribution

$$m_\ell = \sum_n p_n e^{in\theta_\ell} = m^*_{q-\ell}, \quad \text{with} \quad m_q = 1, \; m_1 \equiv m.$$

$$(7.56)$$

The individual probabilities can be reconstructed from the moments according to

$$p_n = \frac{1}{q}\left(1 + e^{-i\theta_n}m + e^{i\theta_n}m^*\right) + \frac{1}{q}\sum_{\ell=2}^{q-2} m_\ell\, e^{-i\ell\theta_n}.$$

$$(7.57)$$

The higher moments $m_2, \ldots, m_{q-2}$ as functions of the prescribed order parameter m follow from the requirement that the entropy contribution $\sum p_n \log p_n$ to (7.53) is minimal. For the *three*-state model, the last sum in (7.57) is absent, and m determines all $\{p_n\}$. This leads to the free energy density

$$f_{\mathrm{mf}}^{q=3} = -dJ_p mm^* - T\log 3 + \frac{T}{3}\sum_{n=1}^{3}\left(1 + 2\Re\left(e^{i\theta_n}m\right)\right)\log\left(1 + 2\Re\left(e^{i\theta_n}m\right)\right).$$

$$(7.58)$$

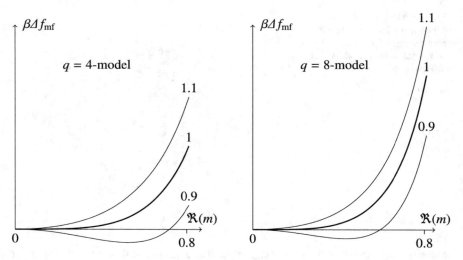

Fig. 7.6 The free energy density for the Z_q models (planar Potts models) with $q = 4$ and $q = 8$ in the vicinity of the critical temperature. The curves are labeled by T/T_c

The system shows a (weak) first-order phase transition at the critical temperature $T_c = 3d\,J_p/(4\log 2)$. For the *four*-state model, we rewrite the free energy density as sum of the densities of two Ising models with the coupling $J_p/2$ according to

$$f_{\mathrm{mf}}^{q=4} = f_{\mathrm{mf}}^{\mathrm{Ising}}\left(\frac{J_p}{2}, m_1 + m_2\right) + f_{\mathrm{mf}}^{\mathrm{Ising}}\left(\frac{J_p}{2}, m_1 - m_2\right), \tag{7.59}$$

where $f_{\mathrm{mf}}^{\mathrm{Ising}}$ is the free energy density of the Ising model as given in (7.19). It immediately follows that the *four*-state planar Potts model shows a second-order transition at $T_c = d\,J_p$. All planar Potts models with $q > 4$ show a second-order transition as well, and this fact is illustrated in Fig. 7.6. The critical temperature $T_{c,\mathrm{mf}} = d\,J_p$ is independent of q for $q \geq 4$, and with increasing q, the free energy density converges quickly to the density at $q \to \infty$.

7.6   Landau Theory and Ornstein–Zernike Extension

Lev Landau developed his celebrated theory of Fermi liquids in 1956 [11, 12]. Some years later experiments indicated that the prediction of his theory were satisfied, at least qualitatively, by liquid ^{3}He. The phenomenological theory is outlined in [13], and further results relying on the methods of quantum field theory are found in [14–16]. The Landau theory is an attempt to formulate a general theory of second-order phase transitions,[2] and it is based on very simple assumptions motivated by the MFA. The main assumption is that the free energy can be expanded as a power series in the order parameter m, where only those terms compatible with the symmetries of the system are admitted. Landau theory not only predicts a phase transition but also allows to reproduce the mean field exponents showing clearly how they depend on the symmetry of the order parameter. Landau theory is an extension of the MFA and thus has similar shortcomings as this approximation. In particular it fails to predict the correct critical exponents below four dimensions. Many details of the Landau theory of phase transitions can be found in the textbook [17].

Consider the free energy of a ferromagnet with real order parameter m. Without a magnetic field, the free energy is an even function of m such that in a series expansion only even powers of m occur,

$$f_L(m) = f_0 + a_2 m^2 + a_4 m^4 + \dots . \tag{7.60}$$

The coefficients in this expansion depend on the parameters of the system and in particular on the temperature. In case $a_4 > 0$, the series can be truncated after the term $O(m^4)$, because, as we shall see, subsequent terms cannot alter the critical behavior. The Landau free energy (7.60) is plotted as a function of m for decreasing values of the coefficient a_2 in Fig. 7.7. For positive a_2, the minimum of the free energy density is at $m_0 = 0$ corresponding to a paramagnetic phase. For negative a_2, the minimum is at finite values $\pm m_0$ corresponding to a ferromagnetic phase. The transition happens at $a_2 = 0$ which means that this value corresponds to the critical temperature. This suggests to write

$$a_2 = \tilde{a}_2 t \tag{7.61}$$

with a positive $\tilde{a}_2$ and the reduced temperature $t = (T - T_c)/T_c$. We take a positive coefficient a_4 in the quartic term such that the magnetization is bounded.

7.6.1   Critical Exponents in Landau Theory

The equilibrium *magnetization* corresponds to the minimum of the free energy and thus satisfies the equation

$$\frac{df_L}{dm}\bigg|_{m_0} = 2\tilde{a}_2 t m_0 + 4a_4 m_0^3 = 0 . \tag{7.62}$$

---

[2] Extensions of the theory are applicable to first-order phase transitions as well.

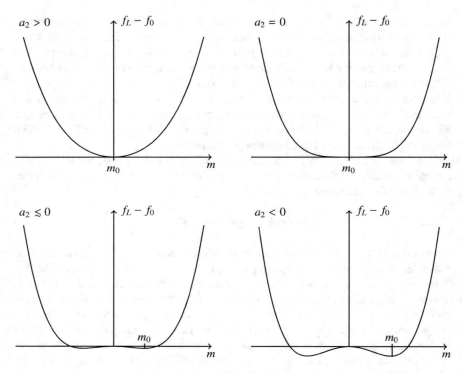

Fig. 7.7 Landau free energy as function of the magnetization for different couplings a_2

For negative t the minimum is at

$$m_0 = \pm \sqrt{\frac{\tilde{a}_2}{2a_4}} (-t)^{1/2} \theta(-t) \tag{7.63}$$

such that we recover the mean field critical exponent $\beta_{\mathrm{mf}} = 1/2$. The exponent α for the *specific heat* follows from differentiating the internal energy density u_{L} in

$$f_{\mathrm{L}}(m_0) = f_0 - \frac{\tilde{a}_2^2}{4a_4} t^2 \theta(-t), \quad u_{\mathrm{L}}(m_0) = u_0 + \frac{\tilde{a}_2^2}{4a_4} \left(\frac{T^2}{T_c^2} - 1 \right) \theta(-t)$$

with respect to temperature. We see that the specific heat jumps at the critical temperature,

$$c_{T \uparrow T_c} - c_{T \downarrow T_c} = \frac{1}{T_c} \frac{\tilde{a}_2^2}{2a_4} \bigg|_{T_c} \tag{7.64}$$

and we recover the mean field result $\alpha_{mf} = 0$. To find the exponents γ and δ, we add a magnetic field term $-hm$ to the Landau free energy,

$$f_L(m) = f_0 - hm + \tilde{a}_2 t\, m^2 + a_4 m^4 . \tag{7.65}$$

The minimizing equilibrium magnetization is determined by

$$2\tilde{a}_2 t\, m_0 + 4a_4 m_0^3 = h . \tag{7.66}$$

At the critical temperature, the first term vanishes, and the resulting relation $m_0^3 \sim h$ implies that the critical exponent δ attains the mean field value $\delta_{mf} = 3$. Finally, the variation of the equilibrium magnetization m_0 with the magnetic field at fixed temperature follows from (7.66). Inserting the zero-field value (7.63) yields the susceptibility

$$\chi = \left.\frac{\partial m_0}{\partial h}\right|_{h=0} = \frac{1}{3 - \text{sgn}(t)}\frac{1}{\tilde{a}_2 |t|} , \tag{7.67}$$

and we recover the mean field critical exponent $\gamma_{mf} = 1$.

The exponents extracted from the Landau free energy are the same as the mean field exponents of the Ising model. This is expected since the mean field free energy of the Ising model (7.19) has the reflection symmetry and hence must have the expansion (7.60). To see this more explicitly, we expand it in powers of m,

$$f_{mf}^{\text{Ising}}(J, m) = \frac{T_c}{2}t\, m^2 + \frac{T}{12}m^4 + \ldots, \quad T_c = qJ , \tag{7.68}$$

from which we can read off the coefficients $\tilde{a}_2$ and a_4. All reflection symmetric models with real order parameter have the Landau expansion (7.60) and thus have the same mean field critical exponents. However, for some models (7.60) is not the appropriate expansion. For example, the Z_q-symmetry of the Potts models with $q \geq 3$ is compatible with a m^3-term in the Landau free energy. This cubic term gives rise to a first-order phase transition.

7.6.2 Two-Point Correlation Function

To determine the mean field values of the exponents ν and η in Table 7.1, we need the asymptotic form of the two-point correlation function. For that purpose we invoke the Ornstein–Zernike extension [18] of Landau's theory. Inspired by the mean field result for the free energy with fixed inhomogeneous magnetization m_x in (7.7), we set

$$F_{rmL}[T, m] - F_L[T, 0] = g \int d^d x\, (\nabla m)_x^2 + \int d^d x\, f_L(m_x) \tag{7.69}$$

with an inhomogeneous magnetization. The first term on the right-hand side is the lowest-order term in a gradient expansion of the spin–spin interaction and takes into account the contribution from non-parallel spins. The second term is just the integral over the Landau free energy density for a constant order parameter. Here we are only interested in the long-distance behavior of the correlation function near a critical point. Near the critical point, the lattice spacing is small compared to the correlation length, and we may approximate lattice sums by integrals.

To calculate the correlation length and anomalous scaling dimension η, we need the connected two-point function of the microscopic spins,

$$G_c(x, y) = \langle s_x s_y \rangle - \langle s_x \rangle \langle s_y \rangle .$$ (7.70)

To relate this correlation function to the free energy, we recall that $\beta F[m]$ is the Legendre transform of the generating functional for the connected correlation functions,

$$\beta F[m] = (\mathscr{L} W)[m], \quad W[j] = \log \int d\mu(\omega) \, e^{-\beta H(\omega) + (j,s)} ,$$ (7.71)

cf. formula (6.46). Note that the second derivative of W is just the connected two-point function G_c. Now we make use of the result (5.91) on p. 109 to relate the second derivatives of W and F,

$$\beta \int d^d y \, G_c(x, y) \frac{\delta^2 F}{\delta m_y \delta m_z} = \delta(x, z) .$$ (7.72)

This result means that the second functional derivative of βF is the inverse of the connected two-point function,

$$G_c(x, y) = kT \left(\frac{\delta^2 F}{\delta m_x \delta m_y} \right)^{-1} .$$ (7.73)

The critical exponents ν and η are defined in the vicinity of a second-order transition point where the order parameter vanishes. Thus it suffices to take the Landau free energy to lowest order in the average field,

$$F_L[T, m] = g \int d^d x \, (\nabla m)_x^2 + \tilde{a}_2 t \int d^d x \, m_x^2 ,$$ (7.74)

in order to find the two-point function near T_c. The second derivative of F_L is

$$\frac{\delta^2 F_L}{\delta m_x \delta m_y} = -2g \Delta_{xy} + 2\tilde{a}_2 t \delta(x - y) ,$$ (7.75)

such that near the critical temperature the connected two-point function of the Landau theory takes the form

$$G_{L,c}(x, y) = \frac{kT}{2g} \langle x \left| \frac{1}{-\Delta + \tilde{a}_2 t/g} \right| y \rangle \qquad (7.76)$$

Up to the temperature-dependent factor, it is just the two-point function of the free scalar field with squared mass $m^2 = \tilde{a}_2 t/g$. It follows at once that for $t > 0$ and large separations $r = |x - y|$ in $d \geq 3$ dimensions

$$G_{L,c}(x, y) \sim \frac{e^{-mr}}{r^{d-2}}, \qquad T > T_c, \qquad (7.77)$$

We conclude that near the critical point the correlation length $\xi = 1/m$ diverges as $\xi \sim t^{-1/2}$ implying that the Landau theory predicts a critical exponent $\nu = 1/2$. On the other hand, at the critical point $t = 0$ such that for $d \geq 3$

$$G_{L,c}(x, y) = \frac{C_d}{r^{d-2}}, \qquad T = T_c. \qquad (7.78)$$

This means that the Landau theory predicts $\eta = 0$. All critical exponents of the Landau theory are collected in the last column of Table 7.3 on page 142.

7.7 Anti-ferromagnetic Systems

We return to general spin systems with energy functions

$$H(\omega) = -\sum_{x \neq y} J_{xy} s_x s_y + \sum_x Q_x(s_x), \qquad s_x \in \mathscr{T}. \qquad (7.79)$$

For ferromagnetic couplings $J_{xy} > 0$, the contribution $-J_{xy} s_x s_y$ to the energy is minimal for aligned spins and for anti-ferromagnetic couplings $J_{xy} < 0$ for anti-parallel spins. Hence, we call a system with negative couplings *anti-ferromagnet*. In this section we discuss such systems within the framework of the MFA. We thereby consider translation invariant systems with $Q_x = Q$ and $J_{xy} = J_{x-y}$. In general, the equilibrium state and average magnetization are not translation invariant in contrast to the energy function. Hence, in the MFA we dismiss the assumption of identical site probabilities p_x for the spins. Indeed, for anti-ferromagnets with nearest neighbor interactions, the lattice decomposes into two disjunct sublattices

$$\Lambda = \Lambda_e \cup \Lambda_o, \qquad \Lambda_e \cap \Lambda_o = \emptyset. \qquad (7.80)$$

The sites in the even sublattice Λ_e have an even $x_1 + \cdots + x_d$, and the sites in the odd sublattice Λ_o have an odd $x_1 + \cdots + x_d$. We now suppose that only the spins on the even (odd) sublattice are identically distributed:

$$p_x = p_e \quad \text{if} \quad x \in \Lambda_e \quad \text{and} \quad p_x = p_o \quad \text{if} \quad x \in \Lambda_o , \tag{7.81}$$

where the probabilities p_e and p_o on the sites of the even and odd sublattices satisfy the constraints

$$\int dv_{e,o}(s)\, s = \int d\mu(s) p_{e,o}(s)\, s = m_{e,o} . \tag{7.82}$$

As in (7.12) we introduce the mean coupling strengths

$$J_{ee} = \frac{2}{V_e} \sum_{x \neq y \in \Lambda_e} J_{x-y}, \quad J_{eo} = \frac{2}{V_o} \sum_{x \in \Lambda_e,\, y \in \Lambda_o} J_{x-y} \tag{7.83}$$

and similarly J_{oo} and J_{oe}. For a translationally invariant energy $J_{ee} = J_{oo}$ and $J_{eo} = J_{oe}$ and the internal energy density in the MFA takes the form

$$\frac{1}{V}\langle H \rangle = -\frac{J_{ee}}{4}\left(m_e^2 + m_o^2\right) - \frac{J_{eo}}{2} m_e m_o + \frac{1}{2}\int dv_e\, Q(s) + \frac{1}{2}\int dv_o\, Q(s) , \tag{7.84}$$

Now one proceeds as for ferromagnetic systems and determines the site probability densities which minimize the free energy subject to the constraints (7.82). With the help of two multipliers j_e and j_o, one finds

$$p_{e,o}(s) = \frac{1}{z(j_{e,o})} \exp\left(-\beta Q(s) + j_{e,o}s\right) , \tag{7.85}$$

where the multipliers are fixed by the constraints. Inserting p_e and p_o into the expression for the free energy density yields

$$f_{\mathrm{mf}}(m_e, m_o) = -\frac{J_{ee}}{4}\left(m_e^2 + m_o^2\right) - \frac{J_{eo}}{2} m_e m_o + \frac{T}{2}(\mathscr{L}w)(m_e) + \frac{T}{2}(\mathscr{L}w)(m_o) . \tag{7.86}$$

The last two terms contain the Legendre transform of the Schwinger function $w(j)$ on one lattice site as defined in (7.15).

For systems with only nearest neighbor interactions, $J_{ee} = 0$. If in addition the nearest neighbor couplings $J_{xy} = J$ are constant, then $J_{eo} = 2dJ$. In particular we find the following free energy density of the Ising model,

$$f_{\mathrm{mf}} = -dJ m_e m_o + \sum_{\alpha=o,e}\left(\frac{1+m_\alpha}{4\beta}\log(1+m_\alpha) + \frac{1-m_\alpha}{4\beta}\log(1-m_\alpha)\right) . \tag{7.87}$$

For a ferromagnetic coupling $J > 0$, the free energy is minimal for $m_e = m_o$, and we recover our previous result (7.19). To find the equilibrium magnetizations for anti-ferromagnetic couplings $J < 0$, we must solve the coupled gap equations

$$m_e = \tanh(2dJ\beta m_o) \quad \text{and} \quad m_o = \tanh(2dJ\beta m_e). \tag{7.88}$$

Besides the symmetric high-temperature phase with vanishing magnetizations on the two sublattices, there exists an anti-ferromagnetic low-temperature phase at temperatures below $T_c = 2d|J|$. In this phase the two magnetizations point in opposite directions, i.e., $m_e = -m_o$. The magnetization on the even sublattice is given by the same transcendental equation,

$$m_e = \tanh\left(2d|J|\beta m_e\right) \tag{7.89}$$

as we encountered for ferromagnetic systems. For more details on the MFA for anti-ferromagnetic systems, you may consult the textbook [19].

7.8 Mean Field Approximation for Lattice Field Theories

In a lattice field theory, the free energy for spin models with prescribed magnetization is replaced by the effective action; see (6.53). Hence we may use the results in Sect. 7.1 to obtain mean field approximations for lattice scalar field theories. As for the spin models, we only admit product measures

$$dP(\phi) = \prod_x dv_x(\phi_x) \qquad dv_x(\phi) = p_x(\phi)d\mu(\phi) , \tag{7.90}$$

in the variational principle for the effective action $\Gamma[\varphi]$ in (6.54). The site probability densities p_x fulfill the constraints

$$\int d\mu(\phi) p_x(\phi)\phi = \varphi_x \tag{7.91}$$

and should minimize the effective action (6.54). The lattice action of a scalar field has the form

$$S[\phi] = -\frac{1}{2}\sum_{x,y}\phi_x \Delta_{xy}\phi_y + \sum_x V(\phi_x) , \tag{7.92}$$

where Δ denotes the lattice Laplacian. Since we assume invariance under lattice translations, we have $\Delta_{xy} = \Delta_{x-y}$. We now decompose the derivative terms into a part fixed by the constraints $\langle \phi_x \rangle = \varphi_x$ and a remainder,

$$\sum_{x,y}\phi_x \Delta_{xy}\phi_y \longrightarrow \sum_{x\neq y}\varphi_x \Delta_{xy}\varphi_y + \Delta_0 \sum_x \phi_x^2 , \tag{7.93}$$

where $\Delta_0 \equiv \Delta_{xx}$. Except for the missing contribution $\Delta_0 \sum_x \varphi_x^2$, the first sum on the right-hand side is just the kinetic term of the mean field. Thus we add this contribution to the first sum and consequently subtract it from the second sum. Then we obtain for the action averaged with product measures

$$\int dP(\phi) S[\phi] = -\frac{1}{2} \sum_{x,y} \varphi_x \Delta_{xy} \varphi_y + \sum_x \int dv_x(\phi)\, V(\varphi_x, \phi) , \qquad (7.94)$$

wherein the shifted potential

$$V(\varphi, \phi) = \frac{\Delta_0}{2} \varphi^2 - \frac{\Delta_0}{2} \phi^2 + V(\phi) \qquad (7.95)$$

with $V(\phi, \phi) = V(\phi)$ occurs. To solve the variational problem, we proceed as we did for the spin models and end up with the following MFA for the effective action:

$$\Gamma_{\mathrm{mf}}[\varphi] = \frac{1}{2} \sum_x (\nabla \varphi_x)^2 + \sum_x u_{\mathrm{mf}}(\varphi_x) . \qquad (7.96)$$

Up to an additive term $\propto \varphi^2$, the effective potential $u_{\mathrm{mf}}(\varphi)$ is the Legendre transform of the convex $w(j) = \log z(j)$, where z is the Laplace transform of $-\Delta_0 \phi^2/2 + V(\phi)$:

$$u_{\mathrm{mf}}(\varphi) = \frac{\Delta_0}{2} \varphi^2 + (\mathscr{L}w)(\varphi) \quad \text{with} \quad e^{w(j)} = \int d\mu(\phi)\, e^{j\phi + \frac{1}{2}\Delta_0 \phi^2 - V(\phi)} . \qquad (7.97)$$

For a translation invariant theory with ferromagnetic couplings, we may choose a homogeneous average field $\varphi_x = \varphi$. The extensive effective action is then proportional to the lattice volume βV and defines the intensive *effective potential* u_{mf}:

$$\Gamma_{\mathrm{mf}}[\varphi] = \beta V\, u_{\mathrm{mf}}(\varphi), \quad \varphi_x = \varphi . \qquad (7.98)$$

One can show that u_{mf} represents the MFA of the *constraint effective potential*

$$e^{-\beta V u_c(\varphi)} = \int d\mu(\phi)\, \delta\left(\frac{1}{\beta V} \sum_x \phi_x - \varphi\right) e^{-S[\phi]} . \qquad (7.99)$$

introduced in [20]; see [21]. This result makes it clear that we must interpret

$$dP_{\mathrm{mf}}(\varphi) = \frac{1}{Z_{\mathrm{mf}}} e^{-\beta V u_{\mathrm{mf}}(\varphi)} d\mu(\varphi) \qquad (7.100)$$

as probability distribution for finding the mean constant field φ in the MFA. Since u_{mf} bounds the convex effective potential u from above, see (7.2), its convex hull

$$(\mathscr{L}^2 u_{\mathrm{mf}})(\varphi) \geq u(\varphi) \tag{7.101}$$

is a better approximation for $u(\varphi)$. This improvement is the well-known *Maxwell construction*. For a free theory with classical potential $V(\phi) = m^2\phi^2/2$, the effective potential (7.97) is (up to an additive constant) given by

$$w(j) = \frac{1}{2}\frac{j^2}{m_*^2} \implies (\mathscr{L}w)(\varphi) = \frac{1}{2}m_*^2\varphi^2, \quad m_*^2 = m^2 - \Delta_0, \tag{7.102}$$

such that the effective potential is simply the classical potential, $u_{\mathrm{mf}}(\varphi) = V(\varphi)$.

7.8.1 ϕ^4 and ϕ^6 Scalar Theories

An interacting ϕ^4-theory with classical potential

$$V(\phi) = \frac{m^2}{2}\phi^2 + \frac{\lambda}{4}\phi^4, \tag{7.103}$$

gives rise to the single site Schwinger function

$$w(j) = \log \int \mathrm{d}\phi \exp\left(j\phi - \frac{m_*^2}{2}\phi^2 - \frac{\lambda}{4}\phi^4 \right) \tag{7.104}$$

with effective mass m_* defined in (7.102). In order to localize the phase transition in the parameter space (m, λ), we need the second derivative of w at the origin

$$w''(0) = \frac{4z}{m_*^2}\frac{K_{3/4}(z) - K_{1/4}(z)}{K_{1/4}(z)}, \quad z := \frac{m_*^4}{8\lambda}, \tag{7.105}$$

where the modified Bessel functions of the second kind occur. For any Z_2-symmetric potential, the strictly convex and symmetric function w has its (unique) minimum at the origin such that $\varphi(j = 0) = 0$. According to (5.57) the curvature of $\mathscr{L}w$ at the origin is then equal to

$$(\mathscr{L}w)''(0) = \frac{1}{w''(0)}. \tag{7.106}$$

Thus the curvature of the potential u_{mf} in (7.97) changes sign at the origin for

$$-\Delta_0 = \frac{m_*^2}{4z}\frac{K_{1/4}(z)}{K_{3/4}(z) - K_{1/4}(z)}. \tag{7.107}$$

Fig. 7.8 The dependence of
the monotonically decreasing
function $F(z)$ in (7.108) as a
function of the variable
$z = m_*^4/8\lambda$

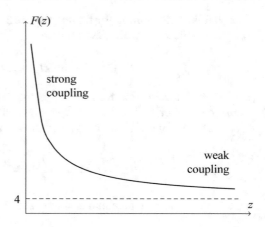

This gap equation determines a critical curve in the (m^2, λ)-plane, where the
symmetric phase with $\varphi = 0$ becomes unstable. For the simplest lattice Laplacian
based on the forward or backward derivative, we have $\Delta_0 = -2d$. Hence the
equation for the critical curve reads

$$\frac{4}{1 + m^2/2d} = \frac{1}{z} \frac{K_{1/4}(z)}{K_{3/4}(z) - K_{1/4}(z)} \equiv F(z) . \tag{7.108}$$

The monotonically decreasing function $F(z)$ approaches the value of 4 from above
for large arguments and is plotted in Fig. 7.8. It follows that Eq. (7.108) can only
be solved for negative m^2. Conversely, since $F(z)$ reaches infinity for $z \to 0$, there
always exists a solution $\lambda_c(m^2)$ for $-2d < m^2 < 0$.

Next we examine the weak-coupling regime for which z in (7.105) is large. For
$z \gg 1$, we insert the asymptotic expansions of the Bessel functions K_ν for large
arguments [22] into the gap equation (7.108) and find

$$-\frac{m^2}{2d} = \frac{3}{8z} - \frac{3}{8z^2} + \cdots = \frac{3\lambda}{m_*^4} \left(1 - \frac{8\lambda}{m_*^4} + \cdots\right) . \tag{7.109}$$

Neglecting terms of the order $O(\lambda^3)$ in this relation, we obtain the two solutions

$$\lambda_c(m) = \left(\frac{2d + m^2}{4}\right)^2 \left(1 \pm \sqrt{1 + \frac{16m^2}{3d}}\right) . \tag{7.110}$$

Since the "critical mass" vanishes for $\lambda = 0$, we only keep the solution with negative
sign. We have seen that m^2 must be negative for positive λ_c.

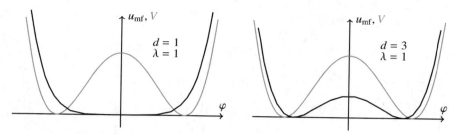

Fig. 7.9 The mean field effective potential of the interacting ϕ^4-theory with classical double-well potential $\lambda(\phi^2 - 1)^2$ in one and three dimensions. u_{mf} is the effective potential in the MFA (black) and V the classical potential (gray)

The classical double-well potential $V(\phi) = \lambda(\phi^2 - 1)^2$ together with its mean field effective potential u_{mf} in one and three dimensions are depicted in Fig. 7.9. Observe that with increasing dimension, the minima of u_{mf} approach the minima of the classical potential. An `octave` program which computes u_{mf} is contained in the Appendix to this chapter.

The Z_2-symmetric classical potential of the sixth order,

$$V(\phi) = \phi^6 - 3\phi^4 + \mu\phi^2 \, , \tag{7.111}$$

and its effective potential in $d = 3$ dimensions are plotted in Fig. 7.10 for four values of the mass parameter μ. In the MFA the system shows a weak first-order transition at $\mu = 2$. For $\mu < 2$, the Z_2-symmetry is spontaneously broken since $\varphi \neq 0$. However, for $\mu \geq 2$, the order parameter vanishes and the Z_2-symmetry is restored.

7.8.2   Non-linear O(N) Models

The Euclidean action of the non-linear O(N) sigma model in d dimensions is

$$S = -\frac{1}{2g^2} \int d^d x \, (\phi, \Delta\phi) \quad \text{with} \quad \phi(x) \in \mathbb{R}^N, \quad \phi(x) \cdot \phi(x) = 1 \, . \tag{7.112}$$

On the lattice we approximate the continuum action by

$$S = -\frac{1}{2g^2} \sum_{x,y} \phi_x \Delta_{xy} \phi_y \quad \text{with} \quad \phi_x \in \mathbb{R}^N, \quad \phi_x^2 = 1 \, , \tag{7.113}$$

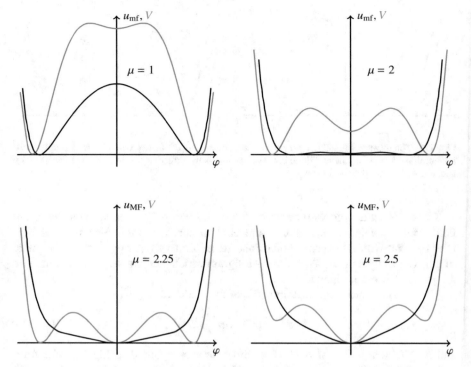

Fig. 7.10 The mean field effective potential u_{mf} and classical potential V of the ϕ^6-theory with potential (7.111) in $d = 3$ dimensions for different values of the mass parameter μ

where $\Delta_{xy} = \Delta_{x-y}$ is some translationally invariant lattice Laplacian. Product measures on the configuration space have the form (7.90) with single spin distribution

$$\mathrm{d}\mu(\phi) = \delta(\phi^2 - 1)\mathrm{d}^N\phi \qquad (7.114)$$

that enforces the constraint $\phi_x^2 = 1$ on every lattice site. The prescribed average field

$$\varphi_x = \int \mathrm{d}\nu_x(\phi)\,\phi \qquad (7.115)$$

takes its values inside or on the boundary of the unit ball in $\mathbb{R}^N$. For product measures the *averaged action* reads

$$\langle S \rangle = -\frac{1}{2g^2}\sum_{x,y}\varphi_x \Delta_{xy}\varphi_y - \frac{\Delta_0}{2g^2}\sum_x \left(1 - \varphi_x^2\right), \qquad (7.116)$$

where $\Delta_0 = \Delta_{xx}$ appeared previously. The constraints (7.115) are implemented by multiplier fields, and the solution of the associated variational problem for the site probability densities leads to the mean field effective action

$$\Gamma_{\mathrm{mf}}[\varphi] = -\frac{1}{2g^2} \sum_{x,y} \varphi_x \Delta_{xy} \varphi_y + \sum_x u_{\mathrm{mf}}(\varphi_x) . \tag{7.117}$$

The by now familiar effective potential

$$u_{\mathrm{mf}}(\varphi) = \frac{\Delta_0}{2g^2}(\varphi^2 - 1) + (\mathscr{L}w)(\varphi) \tag{7.118}$$

contains the Legendre transform of $w = \log z$ with one-site partition function

$$z(j) = \int \mathrm{d}^N \phi\, \delta(\phi^2 - 1)\, \mathrm{e}^{(j,\,\phi)}, \quad j \in \mathbb{R}^N . \tag{7.119}$$

To calculate the integral, we align the z-axis in ϕ-space with the direction defined by the source j and use polar coordinates in ϕ-space. Then the integral simplifies to

$$z(j) = \mathrm{Vol}\left(S^{N-2}\right) \int \mathrm{d}\theta\, \mathrm{e}^{|j|\cos\theta} (\sin\theta)^{N-2} . \tag{7.120}$$

With the integral representation for the modified Bessel function

$$I_\nu(z) = \frac{(z/2)^\nu}{\sqrt{\pi}\,\Gamma(\nu + 1/2)} \int_0^\pi \mathrm{d}\theta\, \mathrm{e}^{z\cos\theta} (\sin\theta)^{2\nu}, \tag{7.121}$$

the one-site partition function takes the form

$$z(j) = (2\pi)^{N/2} \frac{I_{N/2-1}(|j|)}{|j|^{N/2-1}} \xrightarrow{O(3)} z(j) = 4\pi \frac{\sinh(|j|)}{|j|} . \tag{7.122}$$

The site partition functions only depend on the modulus of j, in accordance with the underlying $O(N)$ symmetry. It immediately follows that u_{mf} in (7.118) only depends on the modulus of the mean field. To calculate its curvature at the origin,

$$u''_{\mathrm{mf}}(0) = \frac{\Delta_0}{g^2} + (\mathscr{L}w)''(0) ,$$

we use the small-j expansion $w(j) = \mathrm{const.} + j^2/2N + \ldots$ and conclude $(\mathscr{L}w)''(0) = N\mathbb{1}$, such that $u''_{\mathrm{mf}}(0)$ changes sign for the critical coupling

$$g^2 = g_c^2 = -\frac{\Delta_0}{N} . \tag{7.123}$$

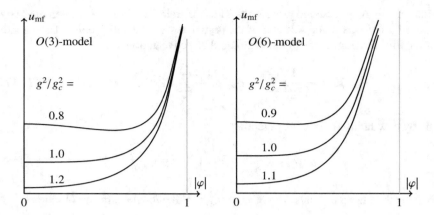

Fig. 7.11 Mean field effective potentials for the O(3) and O(6) models

The lattice Laplacian has diagonal elements $\Delta_0 = -2d$ such that $g_c^2 = 2d/N$. For weak couplings $g < g_c$, the MFA predicts a broken phase. Some typical effective potentials for the O(3) and O(6) models and $g \approx g_c$ are depicted in Fig. 7.11. The MFA predicts a *second-order* transition from an O(N)-symmetric strong coupling phase to a spontaneously broken weak-coupling phase. Note that in two dimensions this prediction is in conflict with the Mermin–Wagner theorem [23,24], according to which a continuous symmetry cannot be spontaneously broken in two dimensions.

7.9   Program for Chap. 7

This section contains the octave program

- mfscalar.m

which was used in this chapter. This program computes the effective potential u_{mf} in the mean field approximation with the classical potential $V(\phi) = \lambda(\phi^2 - 1)^2$. The plots on pp. 157 and 158 have been calculated with this program. The dimension of spacetime may be altered by changing d in the source code.

```
1   function mfscalar;
2   # calculation of effective potential for scalar field theory
3   # with potential V(phi)=lambda*(phi**2-1)**2 in the mean field
4   # approximation. Dimension and coupling lambda as input.
5   # Result is stored in file mfscalar.dat.
6   d=3; # dimension
7   lam=input("lambda ");
8   a=(d-2*lam);
9   closeplot;
10  Nx=501; eps=2/(Nx-1); # Nx must be odd number
```

```
11   x=linspace(-10,10,Nx);
12   x2=x.*x; x4=x2.*x2; eps=eps/3;
13   z=eps*exp(-a*x2-lam*x4-lam);
14   j=linspace(-20,20,80)'; N=length(j);
15   # for Simpson integration
16   for i=2:2:Nx-1;
17     z(i)=4*z(i);
18   endfor;
19   for i=3:2:Nx-2;
20     z(i)=2*z(i);
21   endfor;
22   int0=zeros(N,Nx);int1=int2=int0;
23   L=zeros(N,1);s0=s1=umf=umf1=L;
24   for i=1:N
25     int0(i,:)=z.*exp(j(i)*x);
26     s0(i)=sum(int0(i,:));
27     int1(i,:)=x.*int0(i,:);
28     s1(i)=sum(int1(i,:));
29   endfor;
30   # Schwinger function
31   w0=log(s0);
32   L=s1./s0;
33   # calculate, plot and store effective potential
34   umf=-d*L.*L+j.*L-w0;
35   # search for minimum of potential
36   [min1,nmin]=min(umf);
37   nmin=max(nmin,N+1-nmin);
38   umf(nmin)=umf(nmin)+.5; # mark minimum
39   data=[L,umf-min1]; # normalize potential
40   # classical potential
41   L2=L.*L;
42   V=lam*L2.*L2-2*lam*L2+lam;
43   [vmin1,vnmin]=min(V);
44   datav=[L,V-vmin1];
45   gplot [-1.5:1.5] data,datav;
46   mfscalar=fopen("mfscalar.dat","w","native");
47   for i=1:N
48   fprintf(mfscalar,"(%4.2f,%4.2f)",L(i),umf(i)-min1);
49   if (rem(i,5)==0) fprintf(mfscalar,"\n");
50   endif;
51   endfor;spotts17
52   fclose(mfscalar);
53   endfunction;
```

7.10   Problems

7.1 (MFA for the Z_3 Model) Determine the partition function of the Z_3 model with energy function

$$H = - \sum_{x,y \in \Lambda} J_{xy} \cos(\theta_x - \theta_y) \quad \text{with} \quad \theta_x \in \left\{ \frac{2\pi n}{3} \,\middle|\, n = 0, 1, 2 \right\}$$

in the mean field approximation. Introduce the order parameter

$$m = \frac{1}{V} \sum_{x \in \Lambda} \exp(i\theta_x)$$

and express the Hamiltonian as a function of m and $\bar{m}$. How many configurations do exist for a given m? Determine the function $f(m, \bar{m})$ occurring in

$$Z = \sum_m \exp(-\beta V f(m, \bar{m}))$$

and discuss the result.

Hint: Introduce a_0, a_1, a_2 with a_n representing the number of lattice points with $\theta = 2\pi n/3$. Express m as a function of the a_n. Are the a_n uniquely determined for fixed m and V? Use the a_n in order to determine the number of configurations.

7.2 (MFA Based on the Bogoliubov Inequality) An alternative way of deriving the mean field theory for a given microscopic system with energy H is to start with the *Bogoliubov inequality*

$$F \le \Phi \equiv F_0 + \langle H - H_0 \rangle_0 , \tag{7.124}$$

where $F = -kT \log Z$ is the free energy of the system of interest, H_0 is a trial energy function, F_0 is the corresponding free energy, and $\langle \ldots \rangle_0$ denotes the average taken in the ensemble defined by H_0.

- Prove the inequality (7.124). If you have any difficulties, you may consult the book of H.B. Callen [25].
- The mean field free energy is defined by minimizing Φ with respect to the variational parameters λ in the trial energy $H_0(\lambda)$,

$$F_{\text{mf}} = \min_{\lambda} \Phi(\lambda) .$$

The usual choice for H_0 is a free Hamiltonian as this allows for an explicit calculation of Φ. Consider the Ising model without magnetic field on a lattice with V sites and coordination number q. Choose as trial energy

$$H_0 = -\lambda \sum_x s_x \tag{7.125}$$

and compute Φ. Minimize the result with respect to the parameter λ. Calculate the corresponding F_{mf}. What is the meaning of the variational parameter λ?

7.3 (An Analytically Solvable Model) Consider a scalar field theory with

$$V(\phi) = -\log(1 + \phi^2) + \frac{1}{2}m^2\phi^2, \qquad m^2 > 0 .$$

For $m^2 < 2$, the origin is a local maximum. Plot the potential for various values of m. Calculate the one-site Schwinger function and discuss the resulting effective potential for small m^2.

7.4 (First-Order Transition in Landau Theory) Consider a Landau expansion of the free energy density of the form

$$f_L(m) = \frac{a}{2}m^2 + \frac{b}{4}m^4 + \frac{c}{6}m^6, \qquad c > 0 . \tag{7.126}$$

Show that there exists a line of critical transitions $a = 0$, $b > 0$ which joins a line of first-order transitions $b = -4(ac/3)^{1/2}$ at a tricritical point $a = b = 0$. Sketch $f_L(m)$ in each region of the (a, b) plane, on the transition lines, and at the tricritical point.

References

1. P. Weiss, L'Hypothèse du champs moleculairé et la propriete ferromagnéticque. J. Phys. **6**, 661 (1907)
2. P.J. Langevin, Magnétisme et théorie des électrons. J. Chim. Phys. **5**, 70 (1905)
3. L.E. Reichl, *Modern Course in Statistical Mechanics* (Wiley, New York, 2016)
4. D. Chandler, *Introduction to Modern Statistical Mechanics* (Oxford University Press, Oxford, 1987)
5. G. Parisi, *Statistical Field Theory* (Addison-Wesley, Reading, 1988)
6. M. Opper, D. Saad, *Advanced Mean Field Methods* (Bradford Books, Bradford, 2001)
7. K. Huang, *Introduction to Statistical Physics* (Taylor & Francis, Milton Park, 2009)
8. A. Pelissetto, E. Vicari, Critical phenomena and renormalization-group theory. Phys. Rep. **368**, 549 (2002)
9. M. Hasenbusch, A finite size scaling study of lattice models in the 3d Ising universality class. Phys. Rev. **B82**, 174433 (2010)

10. D. Simmons-Duffin, The Lightcone Bootstrap and the spectrum of the 3d Ising CFT. J. High Energy Phys. **3**, 086 (2017); S. Rychkov, 3D Ising model: a view from the Conformal Bootstrap island. Comptes Rendus Physique **21**, 185 (2020)
11. L.D. Landau, Theory of Fermi-liquids. Zh. Eksp. Teor. Fiz. **30**, 1058 (1956) [Sov. Phys. JETP **3** (1957) 920]
12. L.D. Landau, Oscillations in a Fermi-liquid. Zh. Eksp. Teor. Fiz. **32**, 59 (1957) [Sov. Phys. JETP **5** (1957) 101]
13. P. Nozières, *Theory of Interacting Fermi Systems*. Frontiers in Physics: Lecture Note and Reprint Series, vol. 19 (Benjamin-Cummings, Redwood City, 1964)
14. L. Landau, On the theory of Fermi liquid. Sov. Phys. JETP **35**, 97 (1958)
15. V. Galitskii, A. Migdal, Applications of quantum field theory to the many body problem. Sov. Phys. JETP **7**, 96 (1958)
16. A.A. Abrikosov, L.P. Gorkov, I.E. Dzyaloshinksi, *Methods of Quantum Field Theory in Statistical Mechanics* (Dover, New York, 1975)
17. J.C. Toledano, P. Toledano, *The Landau Theory of Phase Transitions* (World Scientific, Singapore, 1987)
18. L.S. Ornstein, F. Zernike, Accidental deviations of density and opalescence at the critical point of a single substance. Proc. Acad. Sci. Amst. **17**, 793 (1914)
19. J.S. Smart, *Effective Field Theories of Magnetism* (Saunders, Philadelphia, 1966)
20. L. O'Raifeartaigh, A. Wipf und H. Yoneyama, The constraint effective potential. Nucl. Phys. **B271**, 653 (1986)
21. Y. Fujimoto, A. Wipf, H. Yoneyama, Symmetry restoration of scalar models at finite temperature. Phys. Rev. **D38**, 2625 (1988)
22. M. Abramowitz, I.A. Stegun, *Handbook of Mathematical Functions with Formulas, Graphs, and Mathematical Tables* (Martino Fine Books, Eastford, 2014)
23. N.D. Mermin, H. Wagner, Absence of ferromagnetism or antiferromagnetism in one- or two-dimensional isotropic Heisenberg models. Phys. Rev. Lett. **17**, 1133 (1966)
24. S. Coleman, There are no Goldstone bosons in two dimensions. Commun. Math. Phys. **31**, 259 (1973)
25. H.B. Callen, *Thermodynamics and an Introduction to Thermostatistics* (Wiley, New York, 1985)

Transfer Matrices, Correlation Inequalities, and Roots of Partition Functions

8

In Chaps. 2 and 5, we quantized mechanical systems and classical field theories via the functional integral formalism. Through a Wick rotation, we arrived at a (formal) Euclidean functional integral. In a next step, the underlying Euclidean spacetime is replaced by a lattice, and this discretization leads to well-defined lattice field theories—these are particular spin models with continuous target spaces. To calculate thermal expectation values, one imposes (anti)periodic boundary conditions in the imaginary time direction for the lattice fields. By these steps one approximates quantum field theories in d spacetime dimensions by particular classical statistical systems in d dimensions.

In the first part of this chapter, we follow the opposite way. For particular spin models, we reconstruct a state space and a Hamiltonian, where the latter converges to the Hamiltonian of a relativistic quantum field theory. Thereby we shall make use of the *transfer-matrix method* which can be applied to general spin models with short-range interactions. To introduce the method, we first calculate thermodynamic potentials and correlation functions of simple one-dimensional spin models, so-called spin chains. Then we present the general formalism and apply it to scalar field theories on the lattice.

In the second part, we introduce and prove some useful correlation inequalities for general ferromagnetic systems. These inequalities tell us that certain expectation values or combinations of expectation values are non-negative. In particular the two Griffith–Kelly–Sherman (GKS) inequalities can be used to study the dependence of expectation values on the parameters of a system or to compare correlation functions of spin systems in different volumes or in different dimensions.

In the last part, we investigate the Lee–Yang zeros of partition functions, considered as functions of the complex magnetic field (or a complex chemical potential for lattice gases). For the target space $\mathscr{T} = \mathbb{R}$ and an even single-spin density, the partition function of a class of ferromagnetic systems can only vanish for imaginary "magnetic fields." This theorem of Lee and Yang holds for a system with ferromagnetic couplings provided it holds for zero interaction.

© The Author(s), under exclusive license to Springer Nature Switzerland AG 2021
A. Wipf, *Statistical Approach to Quantum Field Theory*, Lecture Notes
in Physics 992, https://doi.org/10.1007/978-3-030-83263-6_8

8.1   Transfer-Matrix Method for the Ising Chain

The transfer-matrix method is a powerful technique for solving problems in statistical physics. It has been used to find exact solutions and in particular the famous solution of the two-dimensional Ising model by Lars Onsager [1]. This section contains an introduction to transfer-matrix methods. For more information and applications to spin systems, you may consult the textbooks [2, 3]. We explain the method on the basis of the exactly solvable Ising chain on a one-dimensional lattice Λ with N sites [4]. We choose periodic boundary conditions for which s_x and s_{x+N} are identified and the energy function (6.2) takes the form

$$H_\Lambda(\omega) = -J \sum_{x=1}^{N} s_x s_{x+1} - h \sum_{x=1}^{N} s_x, \qquad s_x \in \{-1, 1\}. \qquad (8.1)$$

We will calculate the free energy, internal energy, magnetization, and two-point correlation function with the help of the transfer-matrix method.

8.1.1   Transfer Matrix

In a first step, we rewrite the partition function of the chain as follows:

$$Z_\Lambda(\beta) = \sum_\omega e^{-\beta H_\Lambda(\omega)} = \sum_{s_1,\dots,s_N} e^{K s_1 s_2 + \beta h (s_1+s_2)/2} \times e^{K s_2 s_3 + \beta h (s_2+s_3)/2} \times \cdots$$

$$= \sum_{s_1,\dots,s_N} T_{s_1 s_2} T_{s_2 s_3} \cdots T_{s_N s_1} = \operatorname{tr} \hat{T}^N, \qquad K = \beta J. \qquad (8.2)$$

Here we defined the two-dimensional *transfer matrix* with matrix elements

$$\langle s | \hat{T} | s' \rangle = e^{K ss' + \beta h (s+s')/2}, \qquad (8.3)$$

where s and s' take the values ± 1. It is real, is symmetric, and has positive entries,

$$\hat{T} = \begin{pmatrix} e^{K+\beta h} & e^{-K} \\ e^{-K} & e^{K-\beta h} \end{pmatrix}. \qquad (8.4)$$

To calculate the trace in (8.2), we diagonalize the transfer matrix with the help of a rotation matrix R,

$$\hat{T} = RDR^{-1}, \qquad R = \begin{pmatrix} \cos\gamma & -\sin\gamma \\ \sin\gamma & \cos\gamma \end{pmatrix}, \qquad D = \begin{pmatrix} \lambda_+ & 0 \\ 0 & \lambda_- \end{pmatrix}, \qquad (8.5)$$

where the diagonal matrix D contains the two positive eigenvalues

$$\lambda_\pm = e^K (\cosh \beta h \pm B) \quad \text{with} \quad B = \left(\sinh^2 \beta h + e^{-4K} \right)^{1/2} , \tag{8.6}$$

and $\lambda_+ > \lambda_-$. The angle γ in the rotation matrix is determined through

$$\sin 2\gamma = \frac{1}{B} e^{-2K} \quad , \quad \cos 2\gamma = \frac{1}{B} \sinh \beta h . \tag{8.7}$$

In terms of the eigenvalues, the trace in (8.2) is written as

$$Z_\Lambda(\beta, h) = \text{tr } \hat{T}^N = \lambda_+^N + \lambda_-^N = \lambda_+^N \left(1 + p^N \right), \quad p = \frac{\lambda_-}{\lambda_+} . \tag{8.8}$$

For all values of the magnetic field, the last ratio of eigenvalues is bounded, $p < 1$.

Thermodynamic Potentials
The result (8.8) means that the thermodynamic potentials are determined by the eigenvalues of the transfer matrix. For example, the *free energy density* is

$$f_\Lambda(\beta, h) = \frac{1}{N} F_\Lambda(\beta, h) = -\frac{1}{\beta} \log \lambda_+ - \frac{1}{\beta N} \log(1 + p^N) . \tag{8.9}$$

In the thermodynamic limit $N \to \infty$, the term p^N vanishes, and the free energy density is solely determined by the largest eigenvalue of $\hat{T}$,

$$f(\beta, h) = \lim_{N \to \infty} f_\Lambda(\beta, h) = -\frac{1}{\beta} \log \lambda_+ . \tag{8.10}$$

Similarly, the internal energy density

$$u_\Lambda(\beta, h) = -\frac{\partial}{\partial \beta} \left(\log \lambda_+ + \frac{1}{N} \log \left(1 + p^N \right) \right) \tag{8.11}$$

in the thermodynamic limit is determined by the largest eigenvalue according to

$$u(\beta, h) = \lim_{N \to \infty} u_\Lambda(\beta, h) = -\frac{\partial}{\partial \beta} \log \lambda_+ . \tag{8.12}$$

Without external field we simply have $u(\beta, 0) = -\tan K$.

Correlation Functions

The magnetization of the chain is calculated via the transfer matrix as follows

$$
m \equiv \langle s_1 \rangle = \frac{1}{Z} \sum_\omega e^{-\beta H(\omega)} s_1 = \frac{1}{Z} \sum_\omega s_1 T_{s_1 s_2} \cdots T_{s_N s_1} = \frac{1}{Z} \operatorname{tr} \left(\hat{s} \, \hat{T}^N \right) ,
$$

(8.13)

where $\hat{s} = \sigma_3$ denotes the third Pauli matrix. Using $\hat{T} = RDR^{-1}$ and that the trace is invariant under cyclic permutations of its arguments, we obtain

$$
m = \frac{1}{Z} \operatorname{tr} \left(R^{-1} \hat{s} \, R \, D^N \right), \quad \text{where} \quad R^{-1} \hat{s} \, R = \begin{pmatrix} \cos 2\gamma & -\sin 2\gamma \\ -\sin 2\gamma & -\cos 2\gamma \end{pmatrix} .
$$

(8.14)

Inserting the diagonal matrix D defined in (8.5) leads to

$$
m = \frac{1 - p^N}{1 + p^N} \cos 2\gamma \overset{N \to \infty}{\longrightarrow} \cosh(2\gamma) = \frac{\sinh \beta h}{\left(\sinh^2 \beta h + e^{-4K} \right)^{1/2}} .
$$

(8.15)

Of course, differentiating the free energy density (8.9) with respect to the external field yields the same magnetization.

For any positive temperature, the magnetization (8.15) vanishes when the field h is switched off—only at zero temperature a permanent magnetization remains. The Ising chain shares this property with many other one-dimensional lattice models with short-range interactions (and further mild assumptions), which show no phase transition at any positive temperature [5, 6]. This *van Hove theorem* does not completely exclude the possibility of phase transitions in one dimension; see, for example, [7]. For a further reading, I refer to the collection of reprints in [8].

For the two-point function, we proceed similarly as for the one-point function and obtain for $y \geq x$

$$
\langle s_x s_y \rangle = \frac{1}{Z} \sum_\omega e^{-\beta H(\omega)} s_x s_y
$$

$$
= \frac{1}{Z} \sum_{s_1, s_x, s_y} (T^{x-1})_{s_1 s_x} s_x (T^{y-x})_{s_x s_y} s_y (T^{N+1-y})_{s_y s_1}
$$

$$
= \frac{1}{Z} \operatorname{tr} \left(\hat{s} \, \hat{T}^{y-x} \hat{s} \, \hat{T}^{N-(y-x)} \right)
$$

$$
= \frac{1}{Z} \operatorname{tr} \left((R^{-1} \hat{s} \, R) \, D^{y-x} \, (R^{-1} \hat{s} \, R) \, D^{N-(y-x)} \right),
$$

with matrices D and $R^{-1}SR$ given in (8.5) and (8.14). A short calculation leads to

$$\langle s_x s_y \rangle = \cos^2 2\gamma + \frac{p^{y-x} + p^{N-(y-x)}}{1 + p^N} \sin^2 2\gamma, \quad y \geq x \,. \tag{8.16}$$

Translational invariance and $\langle s_x s_y \rangle = \langle s_y s_x \rangle$ imply that the two-point function only depends on the distance $|y - x|$ of the spins. In the thermodynamic limit, we find

$$\lim_{N \to \infty} \langle s_x s_y \rangle = \cos^2 2\gamma + e^{-|y-x|/\xi} \sin^2 2\gamma \,, \tag{8.17}$$

where we introduced the *correlation length*

$$\xi^{-1} = \log \frac{1}{p} \xrightarrow{h \to 0} - \log \tanh K, \quad K = \beta J \tag{8.18}$$

which is finite for all $T > 0$. The correlation function satisfies the *cluster property*

$$\langle s_x s_y \rangle - \underbrace{\langle s_x \rangle \langle s_y \rangle}_{=m^2} \xrightarrow{N \to \infty} \sin^2 2\gamma \; e^{-|y-x|/\xi} \xrightarrow{|y-x| \to \infty} 0 \,. \tag{8.19}$$

At first glance it seems surprising that the Ising chain shows no ordered phase at low temperature since the ordered states with all spins pointing in the same direction have lowest energy. However, they do not minimize the free energy $F = U - TS$. To see this let us consider particular configurations like $\uparrow\uparrow\uparrow\uparrow\uparrow\downarrow\downarrow\downarrow\downarrow$, where the spins are partially inverted. Due to the interface between the two sub-chains with aligned spins, the internal energy increases by $\Delta U = 4J$. But the N configurations with one interface contribute $k \log N$ to the entropy. We see that entropy wins over energy and that for large N the free energy is lowered by the interface.

8.1.2 The "Hamiltonian"

We now extract the two-dimensional Hermitian matrix $\hat{H}$ via the relation

$$\hat{T} = e^{-\hat{H}} \tag{8.20}$$

from the positive transfer matrix $\hat{T}$ such that

$$Z_\Lambda(\beta) = \text{tr}\, \hat{T}^N = \text{tr}\, e^{-N\hat{H}} \,. \tag{8.21}$$

The operator $\hat{H}$ should not be mistaken with the classical energy H of spin configurations. The result (8.21) suggests that $\hat{H}$ represents the discretized *Hamiltonian* of a quantum theory associated to the spin system. To actually calculate $\hat{H}$ for the Ising chain, we use the formula

$$\exp\left(\alpha \sum_i n_i \sigma_i\right) = \cosh \alpha + \sinh \alpha \sum_{i=1}^{3} n_i \sigma_i \, , \qquad (8.22)$$

where $\{\sigma_1, \sigma_2, \sigma_3\}$ are the Pauli matrices and $\boldsymbol{n}$ is a unit vector. To simplify matters we assume $h = 0$ such that the transfer matrix (8.4) takes the simple form

$$\hat{T} = e^K \mathbb{1} + e^{-K} \sigma_1 = \frac{e^K}{\cosh K^*} e^{K^* \sigma_1}, \quad K = \beta J \, . \qquad (8.23)$$

Here we introduced the so-called dual coupling K^* in virtue of the relation

$$\tanh K^* = e^{-2K} \quad \text{or} \quad \tanh K = e^{-2K^*} \, . \qquad (8.24)$$

From the last representation in (8.23), it is easy to extract the *Hamiltonian*

$$\hat{H} = -\log \hat{T} = -K + \log \cosh K^* - K^* \sigma_1 = E_0 + K^*(1 - \sigma_1) \qquad (8.25)$$

with ground state energy $E_0 = \log \cosh K^* - K - K^*$. Using the duality relation between K and K^*, we can write this energy as

$$E_0 = -\log(2 \cosh K) = \lim_{N \to \infty} \beta f(\beta, 0) \, . \qquad (8.26)$$

The only excited energy level lies $2K^*$ above the ground state and yields the inverse correlation length $2K^*$, in agreement with (8.18).

Two Dimensions
While it is a relatively simple problem to find the transfer matrix of the Ising chain, the corresponding problem for the two-dimensional model is highly non-trivial. The solution without external field is due to Onsager, who was able to find the analytical expression (6.3) for the free energy in the thermodynamic limit via the transfer matrix approach. This then gives an exact set of critical exponents; see the discussion on page 113 in Chap. 6. For a derivation, I refer to the textbook [9]. To date, the three-dimensional Ising model remains unsolved.

8.1.3 The Anti-Ferromagnetic Chain

For the anti-ferromagnetic Ising chain with negative $K = \beta J$, the eigenvalue λ_+ of the transfer matrix in (8.6) remains positive, whereas the eigenvalue λ_- becomes negative. Actually, without external field the larger eigenvalue does not change under $J \rightarrow -J$, whereas the smaller eigenvalue changes its sign,

$$\lambda_+ = 2\cosh(K) > 0 \quad \text{and} \quad \lambda_- = 2\sinh(K) < 0 . \tag{8.27}$$

Thus for $h = 0$, the anti-ferromagnetic and ferromagnetic chains share the same free energy density in the thermodynamic limit. However, since $p < 0$, the correlation function (8.16) becomes an oscillating function with different signs at adjacent lattice sites,

$$\lim_{N \to \infty} \langle s_x s_y \rangle_c = (\tanh K)^{|y-x|} = (-1)^{|y-x|} (\tanh |K|)^{|y-x|} . \tag{8.28}$$

Here we have taken into account that in the absence of the external field $\sin 2\gamma = 1$. The oscillatory behavior originates from the non-positivity of the transfer matrix. Since the exponential of a Hermitian matrix is a positive matrix, we cannot define a Hermitian Hamiltonian for the anti-ferromagnetic chain as we did in (8.20). However, we may define a Hamiltonian by using the positive matrix $\hat{T}^2$ according to

$$\hat{T}^2 = e^{-2\hat{H}} . \tag{8.29}$$

The Hermitian operator $\hat{H}$ agrees with the Hamiltonian of the ferromagnetic Ising chain and determines the long-range behavior of the anti-ferromagnetic system.

8.2   Potts Chain

A configuration $\omega = \{\sigma_1, \ldots, \sigma_N\}$ of the periodic q-state Potts chain has the energy

$$H_\Lambda(\omega) = -J \sum_{x=1}^{N} \delta(\sigma_x, \sigma_{x+1}) - 2h \sum_{x=1}^{N} \delta(\sigma_x, 1), \quad \sigma_x \in \{1, 2, \ldots, q\} . \tag{8.30}$$

For later convenience we rescaled the external field as compared to equation (6.10). The partition function can be written as

$$Z_\Lambda(\beta, J, h) = \sum_{\sigma_1, \ldots, \sigma_N} e^{K\delta(\sigma_1, \sigma_2) + 2\beta h \, \delta(\sigma_1, 1)} \times e^{K\delta(\sigma_2, \sigma_3) + 2\beta h \, \delta(\sigma_2, 1)} \times \cdots$$

$$= \sum_{\sigma_1, \ldots, \sigma_N} T_{\sigma_1 \sigma_2} T_{\sigma_2 \sigma_3} \cdots T_{\sigma_N \sigma_1} = \operatorname{tr} \hat{T}^N, \quad K = \beta J , \tag{8.31}$$

where we introduced the q-dimensional *non-symmetric* transfer matrix

$$(T_{\sigma\sigma'}) = \langle\sigma|\hat{T}|\sigma'\rangle = \begin{pmatrix} \zeta z & z & \cdots & z \\ 1 & \zeta & \cdots & 1 \\ \vdots & \vdots & \ddots & \vdots \\ 1 & 1 & \cdots & \zeta \end{pmatrix}, \qquad \zeta = e^K, \ z = e^{2\beta h}. \tag{8.32}$$

Following [10] we calculate the characteristic polynomial $\det(\hat{T} - \lambda\mathbb{1})$ explicitly. Its roots $\lambda_1, \ldots, \lambda_q$ are just the eigenvalues of the transfer matrix and determine the thermodynamic potentials.

Let us subtract the second row of $\hat{T} - \lambda\mathbb{1}$ from the third and each of the subsequent rows, followed by adding the third and all subsequent columns to the second column. The determinant of the resulting matrix

$$\begin{pmatrix} \zeta z - \lambda & z(q-1) & z & \cdots & z \\ 1 & \zeta + q' - \lambda & 1 & \cdots & 1 \\ 0 & 0 & \zeta - 1 - \lambda & \cdots & 0 \\ \vdots & & & \ddots & \\ 0 & & 0 & \cdots & \zeta - 1 - \lambda \end{pmatrix}$$

is equal to the characteristic polynomial of the transfer matrix. It factorizes into a polynomial of degree $q' = q - 2$ and a quadratic polynomial,

$$\det(\hat{T} - \lambda\mathbb{1}) = (\zeta - 1 - \lambda)^{q'} \left(\lambda^2 - (\zeta z + \zeta + q')\lambda + z(\zeta - 1)(\zeta + q - 1)\right),$$

The first factor has the root $\zeta - 1$ of order q' and the second factor the simple roots

$$\lambda_\pm = \frac{1}{2}\left((z+1)\zeta + q' \pm \sqrt{(\zeta z - \zeta - q')^2 + 4(q-1)z}\right)$$

$$= e^{K+\beta h}\cosh\beta h + \frac{q'}{2} \pm \sqrt{\left(e^{K+\beta h}\sinh\beta h - \tfrac{1}{2}q'\right)^2 + e^{2\beta h}(q-1)}. \tag{8.33}$$

Hence, we obtain the following explicit expression for the partition function

$$Z_\Lambda(\zeta, z) = \operatorname{tr}\hat{T}^N = \lambda_+^N + \lambda_-^N + q'(\zeta - 1)^N \tag{8.34}$$

which for $q = 2$ is proportional to that of the Ising chain. In the thermodynamic limit, only the contribution of the largest eigenvalue λ_+ remains and determines the free energy and internal energy density according to (8.10) and (8.12), respectively.

8.3 Perron–Frobenius Theorem

We have seen that the largest eigenvalue λ_+ of the transfer matrix determines the thermodynamic potentials in the thermodynamic limit. A theorem of Oskar Perron [11] and Georg Frobenius [12] asserts that a real square matrix with positive entries has a unique largest real eigenvalue and that there is a corresponding eigenvector with strictly positive components:

Theorem 8.1 (Perron-Frobenius) *Let $\hat{T}$ be a Hermitian matrix with positive matrix elements T_{ij}. Then the eigenvector with largest eigenvalue $\|\hat{T}\|$ is unique, and its components are all unequal to zero and may be chosen as positive numbers.*

Proof We choose the norm given by $\|\psi\|^2 = \sum \psi_i^* \psi_i$ for a vector in $\mathbb{C}^n$. The operator norm of a n-dimensional matrix $\hat{T}$ is defined as

$$\|\hat{T}\| = \sup_{\psi \neq 0} \frac{\|\hat{T}\psi\|}{\|\psi\|} . \tag{8.35}$$

Now let $\tilde{\Omega} = (\tilde{\Omega}_1, \ldots, \tilde{\Omega}_n)^T$ be the eigenvector of $\hat{T}$ with the largest eigenvalue, i.e.,

$$(\tilde{\Omega}, \hat{T}\tilde{\Omega}) = \|\hat{T}\|(\tilde{\Omega}, \tilde{\Omega}).$$

Clearly, the vector $\Omega := (|\tilde{\Omega}_1|, \ldots, |\tilde{\Omega}_n|)^T$ has the same norm as $\tilde{\Omega}$. Since the matrix elements of $\hat{T}$ are non-negative, we conclude that

$$(\Omega, \hat{T}\Omega) \geq (\tilde{\Omega}, \hat{T}\tilde{\Omega}) = \|\hat{T}\|(\tilde{\Omega}, \tilde{\Omega}) = \|\hat{T}\|(\Omega, \Omega) \tag{8.36}$$

holds. Using the *Cauchy–Schwarz inequality*, we have

$$(\Omega, \hat{T}\Omega) \leq \|\Omega\| \|\hat{T}\Omega\| \leq \|\hat{T}\| \|\Omega\|^2 . \tag{8.37}$$

The two inequalities (8.36) and (8.37) imply

$$(\Omega, \hat{T}\Omega) = \|\hat{T}\| (\Omega, \Omega)$$

such that also Ω is eigenvector with the same maximal eigenvalue $\|\hat{T}\|$ as $\tilde{\Omega}$. None of the components of this real eigenvector is zero as

$$0 < \sum_j T_{ij}\Omega_j = (\hat{T}\Omega)_i = \|\hat{T}\|\Omega_i \implies \Omega_i > 0$$

shows. Now we can prove that the eigenvectors $\tilde{\Omega}$ and Ω are *linearly dependent*. For that purpose we insert

$$\tilde{\Omega}_j = e^{i\varphi_j}\,\Omega_j$$

into $(\tilde{\Omega}, \hat{T}\tilde{\Omega}) = (\Omega, \hat{T}\Omega)$ and find

$$\sum \tilde{\Omega}_j^* T_{jk}\tilde{\Omega}_k = \sum \Omega_j T_{jk}\Omega_k e^{i(\varphi_k - \varphi_j)} = \sum \Omega_j T_{jk}\Omega_k \ .$$

Since $\Omega_j T_{jk}\Omega_k$ is positive, we conclude $\varphi_k = \varphi_j \equiv \varphi$ which means that the two eigenvectors Ω and $\tilde{\Omega}$ are linearly dependent, $\tilde{\Omega} = e^{i\varphi}\,\Omega$. Now it is not difficult to prove the Perron–Frobenius theorem: Let $\Omega^{(1)}$ and $\Omega^{(2)}$ be two linearly independent eigenvectors corresponding to the largest eigenvalue. According to the previous discussion, we may assume that all components of these vectors are positive numbers. Then there exists an $\alpha > 0$, such that the eigenvector corresponding to the same largest eigenvalue

$$\Omega^{(1)} - \alpha\Omega^{(2)}$$

has positive and negative components. But this is impossible as verified above.

8.4　The General Transfer-Matrix Method

A transfer matrix can be defined for spin systems on lattices of the form $\Lambda = \mathbb{Z} \times \mathscr{R}$. The coordinate $\tau \in \mathbb{Z}$ is sometimes called time coordinate in contrast to the coordinates x on the spatial lattice $\mathscr{R}$. A lattice site is characterized by its time coordinate and its spatial coordinates, $x = (\tau, x)$. We may view a spin configuration on Λ as the set of spin configurations on $\mathscr{R}$, labeled by τ,

$$\omega = \{s_x \,|\, x \in \Lambda\} = \{\omega_\tau \,|\, \tau \in \mathbb{Z}\}, \quad \text{with} \quad \omega_\tau = \{s_{\tau,x} \,|\, x \in \mathscr{R}\} \ . \tag{8.38}$$

For simplicity we will assume that Λ is a d-dimensional hypercubic lattice in which case $\mathscr{R}$ is a $d-1$-dimensional hypercubic lattice. As earlier on we denote the target space, sometimes called space of *local states*, by $\mathscr{T}$.

To construct the "Hilbert space" of states on the spatial lattice, we first introduce the one-site vector space $\mathscr{H}_x$ of complex-valued functions ψ on the target space,

$$\mathscr{H}_x = \{\psi \,|\, \psi : \mathscr{T} \to \mathbb{C}\} \ . \tag{8.39}$$

The local states characterize a basis of $\mathscr{H}_x$ consisting of characteristic functions,

$$|s\rangle = \psi_s \quad \text{with} \quad \psi_s(s') = \delta_{s,s'} \ . \tag{8.40}$$

We now define the scalar product according to

$$\langle s | s' \rangle = \delta_{s,s'} \,. \tag{8.41}$$

If $\mathscr{T}$ is infinite discrete, we demand ψ to be square summable, $\psi \in \ell_2$, and if $\mathscr{T}$ is continuous, then we demand ψ to be square integrable $\psi \in L_2(\mathscr{T})$, similar as in quantum mechanics. The one-site vector space $\mathscr{H}_x$ is $\mathbb{C}^2$ for the Ising model and $L_2(\mathbb{R})$ for a real scalar field. We now define a *state space*

$$\mathscr{H} = \bigotimes_{x \in \mathscr{R}} \mathscr{H}_x \,, \tag{8.42}$$

associated with the spatial lattice. This is the space of all (complex-valued) functions over the configurations on $\mathscr{R}$. A basis of $\mathscr{H}$ is given by the product states

$$| \omega \rangle = \bigotimes_{x \in \mathscr{R}} | s_x \rangle \,. \tag{8.43}$$

They are characteristic functions on configurations $\omega = \{ s_x \, | \, x \in \mathscr{R} \}$ at fixed time. We refer to this basis as the *configuration space basis*. The scalar product of two basis elements is defined as product of the scalar products of their factors in $\mathscr{H}_x$.

Along with the decomposition of the spacetime lattice $\Lambda = \mathbb{Z} \times \mathscr{R}$ goes a decomposition of the energy function. Assuming that the interaction is restricted to nearest neighbors, we obtain

$$H(\omega) = \sum_\tau H_0(\omega_{\tau+1}, \omega_\tau) + \sum_\tau U(\omega_\tau) \,. \tag{8.44}$$

The first term containing the interactions between spins on adjacent time slices does not change under an interchange of the configurations, i.e., $H_0(\omega', \omega) = H_0(\omega, \omega')$. We may write H_0 and U as

$$H_0 = \sum_x h_0(s_{\tau+1,x}, s_{\tau,x}) \quad \text{and} \quad U = \sum_{\langle x, y \rangle} u(s_{\tau,x}, s_{\tau,y}) \,. \tag{8.45}$$

We now introduce the *transfer matrix* as a linear operator on $\mathscr{H}$ through its matrix elements in the configuration space basis:

$$T_{\omega\omega'} = \langle \omega | \hat{T} | \omega' \rangle = \exp\left(-H_0(\omega, \omega') - \frac{1}{2} U(\omega) - \frac{1}{2} U(\omega') \right) \,. \tag{8.46}$$

This real and symmetric matrix with positive entries has real eigenvalues, and the Perron–Frobenius theorem ensures that the largest eigenvalue $\lambda_{\max}$ is positive and non-degenerate. If some eigenvalues of $\hat{T}$ are negative, such as for anti-ferromagnets, we choose the positive matrix $\hat{T}^2$ to extract a *lattice Hamiltonian* $\hat{H}$,

$$\hat{T}^2 = e^{-2\hat{H}}, \quad \hat{H} \text{ self-adjoint}. \tag{8.47}$$

The partition function on a periodic lattice with N time slices is given by

$$Z_\Lambda = \text{tr}\,\hat{T}^N = \sum_n \lambda_n^N \overset{N\to\infty}{\longrightarrow} \lambda_{\max}^N. \tag{8.48}$$

To discuss the continuum limit, we define a set of operators in the configuration space basis: the diagonal operators $\hat{s}_x$ through

$$\hat{s}_x|\boldsymbol{\omega}\rangle = s_x|\boldsymbol{\omega}\rangle \tag{8.49}$$

and the operators $\hat{\pi}_x$ which map basis vectors into different basis vectors,

$$\hat{\pi}_x|\boldsymbol{\omega}\rangle = |\boldsymbol{\omega}_\delta\rangle, \tag{8.50}$$

where $\boldsymbol{\omega}_\delta$ arises from $\boldsymbol{\omega}$ through a shift of s_x. The second operator corresponds to the quantum mechanical momentum operator and changes local states at x by a constant. The exact form of the $\hat{\pi}_x$ depends on the theory of interest. If s_x is real, then $\hat{\pi}_x$ corresponds to the addition of a constant, whereas if s_x is group valued, then (for a cyclic group) it is the multiplication with a generating group element. Every operator on $\mathcal{H}$ may be expressed in terms of the operators $\hat{s}_x$ and $\hat{\pi}_x$. For the simple Ising chain, the spatial lattice $\mathcal{R}$ is simply a point. If we choose the basis

$$|1\rangle = \begin{pmatrix} 1 \\ 0 \end{pmatrix} \quad \text{and} \quad |-1\rangle = \begin{pmatrix} 0 \\ 1 \end{pmatrix}, \tag{8.51}$$

then the matrices $\hat{s}$ and $\hat{\pi}$ are given by

$$\hat{s} = \sigma_3 \quad \text{and} \quad \hat{\pi} = \sigma_1. \tag{8.52}$$

8.5   Continuous Target Spaces

Now we consider lattice models with continuous target spaces $\mathcal{T}$ where the transfer "matrix" is actually an operator. In a suitable basis, the transfer matrix becomes a positive integral operator acting on real-valued functions on $\mathcal{T}$.

8.5.1 Euclidean Quantum Mechanics

For a particle on the line, we approximate the line by a one-dimensional lattice with N lattice points and lattice spacing ε. In terms of rescaled *dimensionless* variables, the Euclidean lattice action reads

$$S = \sum_{j=1}^{N} \left\{ \frac{m}{2}(q_{j+1} - q_j)^2 + V(q_j) \right\} . \tag{8.53}$$

As transfer "matrix" we choose the integral operator with symmetric kernel

$$\langle q|\hat{T}|q'\rangle = T_{qq'} = \sqrt{\frac{m}{2\pi}}\, e^{-\frac{1}{2}mq^2 - \frac{1}{2}V(q)}\, e^{mqq'}\, e^{-\frac{1}{2}mq'^2 - \frac{1}{2}V(q')} , \tag{8.54}$$

such that the partition function at inverse temperature $\beta = \varepsilon n$ is given by

$$Z_N = \operatorname{tr} \hat{T}^N = \int dq_1 \dots dq_N \, T_{q_1 q_2} T_{q_2 q_3} \times \dots \times T_{q_N q_1} . \tag{8.55}$$

For the harmonic oscillator with potential $V(q) = m\omega^2 q^2/2$, the kernel is Gaussian, and hence $\hat{T}$ has a Gaussian eigenfunction,

$$\int dq'\, T_{qq'}\, \psi_0(q') = \lambda_0 \psi_0(q), \quad \text{with} \quad \psi_0(q) \propto e^{-m\alpha q^2/2} . \tag{8.56}$$

The parameter α and the eigenvalue λ_0 are

$$\alpha = \omega\sqrt{1 + \omega^2/4} \quad \text{and} \quad \lambda_0 = \sqrt{1 - \alpha + \omega^2/2} < 1 . \tag{8.57}$$

Actually one can find all eigenfunctions and eigenvalues of the transfer matrix in closed form. The solution to Problem 8.3 tells us that

$$\hat{T}|\psi_n\rangle = \lambda_n |\psi_n\rangle \quad \text{with} \quad \lambda_n = \lambda_0^{2n+1} < \lambda_0, \quad n \in \mathbb{N}_0 , \tag{8.58}$$

where the $\psi_n(q)$ are calculated from the ground state wave function according to,

$$\psi_n(q) \propto \hat{a}^{\dagger n} \psi_0(q), \quad a^\dagger = \frac{\partial}{\partial q} - m\alpha q . \tag{8.59}$$

This means that the eigenvalues of the lattice Hamiltonian $\hat{H}$ are equidistant,

$$\operatorname{spectrum}(\hat{H}) = \left\{ -(2n+1)\log\lambda_0 \mid n \in \mathbb{N}_0 \right\} , \tag{8.60}$$

similarly as for the harmonic oscillator on $\mathbb{R}$.

Continuum Limit

The dimensionless lattice quantities in the previous formulas (now denoted by an index L, similarly as on page 68) are related to the dimensionful physical quantities as follows:

$$\omega_L = \varepsilon\omega, \quad m_L = \varepsilon m, \quad \hat{H}_L = \varepsilon\hat{H}, \quad q_L = \varepsilon^{-1}q \ . \tag{8.61}$$

For a small lattice spacing ε, the eigenvalues of the dimensionful Hamiltonian $\hat{H}$ have the expansions

$$-\frac{1}{\varepsilon}\log\lambda_n = \left(\frac{1}{2}+n\right)\omega + O(\varepsilon^2)$$

and in the continuum limit $\varepsilon \to 0$, we recover the energies of the harmonic oscillator on the line together with its ground state wave function $\exp(-m\omega q^2/2)$. In terms of physical parameters, the transfer kernel takes the form

$$T_{qq'} = \sqrt{\frac{m}{2\pi\varepsilon}} \exp\left\{-\frac{m}{2\varepsilon}(q-q')^2 - \frac{1}{2}V_\varepsilon(q) - \frac{1}{2}V_\varepsilon(q')\right\} \ ,$$

where V_ε is the potential one obtains by substituting q/ε for q_L and the dimensionful couplings for the dimensionless lattice couplings in the lattice potential. Finally, from the proof of the Feynman–Kac formula in section 2.3, it follows that

$$\lim_{\varepsilon\to 0} T_{qq'} = \lim_{\varepsilon\to 0} \langle q|e^{-\varepsilon\hat{H}}|q'\rangle \quad \text{with} \quad \hat{H} = -\frac{1}{2m}\frac{d^2}{dq^2} + V(q) \ . \tag{8.62}$$

Thus in the continuum limit, one recovers the well-known Hamiltonian, energies, and eigenfunctions of the harmonic oscillator on the line.

8.5.2 Real Scalar Field

Finally we apply the transfer matrix method to a non-interacting scalar field theory with lattice action

$$S = \frac{1}{2}\sum_{\langle x,y\rangle}(\phi_x - \phi_y)^2 + \frac{m^2}{2}\sum_x \phi_x^2 \ . \tag{8.63}$$

The local degrees of freedom are the real ϕ_x, and the one-site Hilbert space is $L_2(\mathbb{R})$. The full Hilbert space is the tensor product of these spaces over all points of the spatial lattice, i.e., $\mathcal{H} = \otimes_{x\in\mathcal{R}}L_2(\mathbb{R})$. In addition, we connect every field configuration ω on the spatial lattice to a basis vector $|\omega\rangle = |\{\phi_x | x \in \mathcal{R}\}\rangle$ in the configuration basis of $\mathscr{H}$. The symmetric transfer matrix can be expressed as

$$\langle \omega | \hat{T} | \omega' \rangle = e^{-F(\omega)} \exp\left(-\frac{1}{2} \sum_{x \in \mathscr{R}} (\phi_x - \phi'_x)^2 \right) e^{-F(\omega')} , \qquad (8.64)$$

where F only depends on the field values in a given time slice,

$$F(\omega) = \frac{1}{4} \sum_{\langle x,y \rangle} (\phi_x - \phi_y)^2 + \frac{m^2}{4} \sum_x \phi_x^2 . \qquad (8.65)$$

The diagonal operators are represented by the field operators and the displacement operators by the conjugated momentum operators

$$\hat{\phi}_x | \omega \rangle = \phi_x | \omega \rangle \quad , \quad \hat{\pi}_x = \frac{1}{i} \frac{\delta}{\delta \phi_x} . \qquad (8.66)$$

To rewrite the transfer matrix in terms of these operators, we recall the explicit heat kernel of the kinetic operator $K(\hat{\pi}) = \frac{1}{2} \sum_x \hat{\pi}_x^2$,

$$\langle \omega | e^{-K(\hat{\pi})} | \omega' \rangle = \frac{1}{(2\pi)^{|\mathscr{R}|/2}} \exp\left\{ -\frac{1}{2} \sum_x (\phi_x - \phi'_x)^2 \right\} , \qquad (8.67)$$

to obtain the following representation of the matrix elements of the transfer matrix:

$$\langle \omega | \hat{T} | \omega' \rangle = (2\pi)^{|\mathscr{R}|/2} e^{-F(\hat{\phi})} e^{-K(\hat{\pi})} e^{-F(\hat{\phi})} , \qquad (8.68)$$

where the *operator* $F(\hat{\phi})$ is the function (8.65) with the ϕ_x replaced by the operators $\hat{\phi}_x$. From the result (8.68), it is evident that $\hat{T}$ is not just the exponential function of a simple differential operator $\hat{H}$. Only in the continuum limit, we recover the familiar and simple Hamiltonian of the free scalar field.

8.6 Correlation Inequalities

Correlation inequalities are general inequalities between correlation functions of statistical systems. Many useful inequalities are known for Ising-type models; see the textbooks [6, 13]. With the help of these inequalities, one can compare correlation functions of lattice models with different couplings or in different dimensions. For example, knowing that the two-dimensional Ising model shows a spontaneous magnetization at low temperature, one can prove that the same holds true for the model in higher dimensions. We shall consider spin models with configuration space $\Omega = \mathbb{R}^V$ and even single-spin measures $d\mu_x$. Thus we examine *ferromagnetic* lattice systems with *real spin variables* and energy functions

$$H(\omega) = -\sum_{K \subset \Lambda} J_K s_K, \quad \text{where} \quad J_K \geq 0, \quad s_K = \prod_{x \in K} s_x . \tag{8.69}$$

The monomial s_K is the product of spin variables s_x on sites in K (with possible duplications in K). The expectation value of a function $O : \Omega \to \mathbb{R}$ is defined as

$$\langle O \rangle = \frac{1}{Z} \int_\Omega O(\omega) \, e^{-\beta H(\omega)} d\mu(\omega), \qquad d\mu(\omega) = \prod_{x \in \Lambda} d\mu_x(s_x) . \tag{8.70}$$

With a suitable choice of the single-spin measure $d\mu_x$, we can recover Ising-type systems or scalar field theories on a lattice. We now prove that the correlation of any monomial of the spins is non-negative:

Lemma 8.1 (First GKS Inequality) *For a ferromagnetic system with energy function (8.69) and even single-spin distribution, the inequality*

$$\langle s_A \rangle \geq 0, \quad s_A = \prod_{x \in A} s_x \tag{8.71}$$

holds for all $A \subset \Lambda$ (we allow duplications in A).

Proof To prove this inequality, we absorb the inverse temperature β in the couplings J_K and expand the Boltzmann factor according to

$$e^{-H(\omega)} = \sum_{n=0}^{\infty} \frac{1}{n!} \left(\sum_K J_K s_K \right)^n = \sum_{n_1,\dots,n_V} a_{n_1\dots n_V} s_1^{n_1} \cdots s_V^{n_V}$$

with positive coefficients $a_{n_1\dots n_V}$ and obtain

$$Z \cdot \langle s_A \rangle = \sum_{n_1,\dots,n_V} a_{n_1\dots n_V} \int s_1^{m_1} \cdots s_V^{m_V} \, d\mu_1(s_1) \cdots d\mu_V(s_V) ,$$

where m_x is the sum of n_x and the multiplicity of s_x in s_A. For even single-spin measures $d\mu_x$, the last integral vanishes if only one exponent m_x is odd, and it is non-negative if all exponents are even. This means that $\langle s_A \rangle$ is a sum of non-negative terms and hence is non-negative.

To proceed further we introduce the important concept of *replicas* where one considers several independent copies of the original system. Here we consider only two copies characterized by the configurations $(\omega, \omega') \in \Omega \times \Omega$. Since they should be independent, we assume that

$$d\mu(\omega, \omega') = d\mu(\omega) d\mu(\omega') \quad \text{and} \quad H(\omega, \omega') = H(\omega) + H(\omega') . \tag{8.72}$$

We also introduce the rotated configurations in $\Omega \times \Omega$,

$$\sigma = \frac{1}{\sqrt{2}}(\omega + \omega') \quad \text{and} \quad \sigma' = \frac{1}{\sqrt{2}}(\omega - \omega') . \tag{8.73}$$

The inverse transformation is

$$\omega = \frac{1}{\sqrt{2}}(\sigma + \sigma') \quad \text{and} \quad \omega' = \frac{1}{\sqrt{2}}(\sigma - \sigma') . \tag{8.74}$$

The spins of the rotated configuration σ are denoted by σ_x, and σ_A is defined similarly to s_A. A second interesting inequality is the content of

Lemma 8.2 (Ginibre's Inequality) *For a ferromagnetic system with energy function (8.69) and even single-spin distributions, the rotated spins obey the inequalities*

$$\langle \sigma_A \sigma_B' \rangle \geq 0 \tag{8.75}$$

for all subsets $A, B \subset \Lambda$.

Proof The negative energy

$$- H(\omega, \omega') = \sum_{K \subset \Lambda} J_K \left[\left(\frac{\sigma + \sigma'}{\sqrt{2}} \right)_K + \left(\frac{\sigma - \sigma'}{\sqrt{2}} \right)_K \right] \tag{8.76}$$

is a polynomial in σ and σ' with only *positive coefficients*. All terms with negative coefficients in the expansion of the second contribution on the right-hand side in (8.76) chancel against terms with positive coefficients in the expansion of the first contribution. Hence, we obtain an expansion similar to the one discussed in the context of the first GKS inequality. What is left is the proof of

$$I_{mn} = \int_{\mathbb{R}^2} \sigma^m \sigma'^n \, d\mu(s) d\mu(s') \geq 0 .$$

Clearly, the inequality holds for even exponents m and n. But for odd m or n, the integral vanishes since the measure is even both in σ and σ' as follows from the assumption that $d\mu(s)$ is even and from

$$(-\omega', -\omega) \longrightarrow (-\sigma, \sigma') \quad , \quad (\omega', \omega) \longrightarrow (\sigma, -\sigma') .$$

With the help of Ginibre's inequality, we can prove that the connected two-point correlation function of two monomials of the spins is non-negative:

Lemma 8.3 (Second GKS Inequality) *With the same assumption as in the previous lemma, we have for all $A, B \subset \Lambda$*

$$\langle s_A s_B \rangle - \langle s_A \rangle \langle s_B \rangle \geq 0 . \tag{8.77}$$

Proof We have

$$\langle s_A s_B \rangle - \langle s_A \rangle \langle s_B \rangle = \langle s_A (s_B - s'_B) \rangle$$

$$= \left\langle \left(\frac{\sigma + \sigma'}{\sqrt{2}} \right)_A \left[\left(\frac{\sigma + \sigma'}{\sqrt{2}} \right)_B - \left(\frac{\sigma - \sigma'}{\sqrt{2}} \right)_B \right] \right\rangle ,$$

where the product $\langle s_A \rangle \langle s_B \rangle$ turns into one expectation value in the doubled system. The expression between square brackets is a polynomial in σ, σ' with positive coefficients. By using Ginibre's inequality, we verify the inequality (8.77).

Now we specialize to ferromagnetic systems with pair interactions,

$$H(\omega) = - \sum_{x,y \in \Lambda} J_{xy} s_x s_y - \sum_x h_x s_x \qquad J_{xy} > 0 \tag{8.78}$$

for which the following inequality holds:

Lemma 8.4 (Percus' Inequality) *In the doubled ferromagnetic system, we have*

$$\langle \sigma'_A \rangle \geq 0, \qquad \forall A \subset \Lambda . \tag{8.79}$$

Proof The linear transformation (8.74) is a (improper) rotation such that

$$-H(\omega) - H(\omega') = \sum_{x,y \in \Lambda} J_{xy} \left(\sigma_x \sigma_y + \sigma'_x \sigma'_y \right) + \sqrt{2} \sum_x h_x \sigma_x .$$

Now we Taylor expand the exponential of $\sum J_{x,y} \sigma'_x \sigma'_y$ in the expectation value

$$\langle \sigma'_A \rangle = \frac{1}{Z^2} \int_{\Omega \times \Omega} d\mu(\omega) d\mu(\omega') \, \sigma'_A \, e^{-H(\omega) - H(\omega')}$$

and obtain a series with terms

$$\int_{\Omega \times \Omega} d\mu(\omega) \, d\mu(\omega') \, \sigma'^{n_1}_1 \cdots \sigma'^{n_V}_V \exp \left(\sum_{x,y} J_{xy} \sigma_x \sigma_y + \sqrt{2} \sum_x h_x \sigma_x \right)$$

and positive coefficients. Since the measure is even in σ', the last integral vanishes if one or more exponents n_x are odd. But for even n_x, the integral is non-negative, and this proves the inequality of Percus.

Apart from the inequalities discussed in this section, there are numerous other correlation inequalities, for example, inequalities for the expectation values of products of three or four spin functions. For a detailed discussion, I recommend [14] and the textbooks [6, 13, 15]. In passing we note that almost all inequalities apply to ferromagnetic systems only, the well-known exception being the FKG inequalities. For correlation inequalities for anti-ferromagnetic systems, you may consult [16].

Application of Correlation Inequalities

The second GKS inequality is of particular importance, since in the form

$$\langle s_A s_B \rangle - \langle s_A \rangle \langle s_B \rangle = \frac{\partial \langle s_A \rangle}{\partial J_B} \geq 0 \tag{8.80}$$

it proves the monotony of correlation functions as functions of the ferromagnetic couplings. The expectation value of any monomial of the spins $\langle s_A \rangle$ increases monotonically with

- An increasing external field
- An increasing ferromagnetic coupling
- A decreasing temperature

Furthermore, the inequality (8.80) allows for a comparison of different models in statistical mechanics. For example, if a ferromagnetic spin model with nearest neighbor interaction has an ordered phase, it remains in the ordered phase if additional *ferromagnetic long-range interactions* are "switched on" since both the magnetization and the critical temperature increase. Moreover, the inequality shows that the spontaneous magnetization decreases if we heat the system. The same happens if we increase the coordination number of a lattice system since we add additional ferromagnetic terms. For example, for fixed coupling and temperature, the magnetization in the three- and four-dimensional hypercubic Ising models are larger than in the two-dimensional model. Also, at fixed Ising coupling J, the critical temperature increases monotonically with the dimension of the system. This qualitative but rigorous prediction agrees with the mean field result (7.23). The same inequality shows that $\langle s_A \rangle$ increases monotonically with the lattice size. For the Ising model, these expectation values are bounded by 1 such that they converge in the thermodynamic limit. The Ginibre inequality was applied to prove the *existence of the thermodynamic limit* for the free energy density and spin correlations of the ferromagnetic classical XY-model [17].

8.7   Roots of the Partition Function

The zeros of the partition function $Z_\Lambda(\beta, h)$ are intimately related to possible phase transitions as was observed by Lee and Yang in their two pioneering papers [18].[1] As a sum of positive terms, the partition function is positive in the physical domain with real magnetic field. Thus, in a search of zeros, we must admit complex parameters and in particular a complex magnetic field. Following Lee and Yang, we study a *lattice gas* with energy function,

$$H = -\varepsilon \sum_{\langle x,y \rangle} n_x n_y - \mu \sum_x n_x, \quad n_x \in \{0, 1\} \tag{8.81}$$

or, what amounts to the same, an Ising model with magnetic field,

$$H(\omega) = -J \sum_{\langle x,y \rangle} s_x s_y - h \sum_x s_x, \quad s_x \in \{-1, 1\} . \tag{8.82}$$

The occupation numbers n_x, coupling, and chemical potential of the lattice gas are related to the spins s_x, coupling, and magnetic field of the Ising model by

$$n_x = (s_x + 1)/2, \qquad \varepsilon = 4J \quad \text{and} \quad \mu = 2h - 4dJ .$$

In the expansion of the Ising model partition function in powers of $\exp(\beta h)$,

$$Z_\Lambda(h) = \sum_{m=-V,-V+2,\dots}^{V} e^{\beta h m} A_m, \quad A_m = \sum_{\sum s_x = m} \exp\left(K \sum_{\langle x,y \rangle} s_x s_y \right) , \tag{8.83}$$

we set $m + V = 2n$ such that n takes the values $0, 1, \dots, V$ and obtain

$$Z_\Lambda(z) = z^{-V/2} \sum_{n=0}^{V} a_n z^n, \quad \text{where} \quad a_n = A_{2n-V}, \quad z = e^{2\beta h} . \tag{8.84}$$

The sum defines a polynomial of degree V in the *fugacity* z. Since all a_n are positive, we conclude that on a finite lattice the partition function is non-zero and analytic in a vicinity of the positive real semi-axis in the complex fugacity plane. This means that the free energy has no singularities for any real magnetic field and thus shows no phase transition in a finite volume when h varies.

However, in the thermodynamic limit $V \to \infty$, the partition function zeros may pinch the semi-axis $\mathbb{R}^+$. Phase transitions are localized at the intersection of the

---

[1] An elementary discussion of the subject can be found in [9].

set of zeros with the semi-axis. The intersection points determine the critical fields where phase transitions may occur. On the other hand, if there exists a region G in the complex z-plane which contains no zeros and encloses the positive real semi-axis, then the thermodynamic potentials are analytic for $V \to \infty$.

Because the coefficients of the polynomial in (8.84) are real and positive, all roots appear in complex-conjugated pairs in the complex fugacity plane, away from the real, positive semi-axis. There is a further property of the roots which follows from the symmetry $H(\omega, h) = H(-\omega, -h)$, where $-\omega$ is obtained from ω by a flip of all spins. This symmetry implies that the partition function is an even function of h and thus is invariant under $z \to 1/z$:

$$Z_\Lambda(z) = \sum_\omega e^{-\beta H(\omega, h)} = Z_\Lambda \left(\frac{1}{z} \right) . \tag{8.85}$$

It follows that the coefficients of the polynomial in (8.84) are not independent,

$$a_n = a_{V-n} \tag{8.86}$$

and that with z also $1/z$ is a root of the partition function. Thus, the non-real roots which are not on the unit circle in the fugacity plane are members of a quartet $\{z, \bar{z}, 1/z, 1/\bar{z}\}$ of roots. The non-real roots on the unit circle or the real roots which are not on the unit circle are members of a doublet $\{z, 1/z\}$ of roots. Only the root $z = -1$ is not related to another root.

In the high-temperature limit $K = \beta J = 0$ and the partition function,

$$\lim_{T \to \infty} Z_\Lambda(z) = z^{-V/2}(1 + z)^V \tag{8.87}$$

has the V-fold zero $z = -1$. On the other hand, in the low-temperature limit $K = \infty$ and the partition function,

$$\lim_{T \to 0} Z_\Lambda(z) = z^{-V/2} e^{dVK} \left(1 + z^V \right) \tag{8.88}$$

has its roots uniformly distributed on the unit circle,

$$z_n = e^{2\pi i(n-1/2)/V}, \quad n = 1, \ldots, V .$$

Below we shall argue that for all K between these two extreme values, the roots of the partition functions lie on the unit circle. This theorem of Lee and Yang applies to general ferromagnetic systems for which the couplings J_{xy} fall off with the distance rapidly enough in order to ensure the existence of the thermodynamic limit.

8.7.1   Lee–Yang Zeroes of Ising Chain

To find the Lee–Yang zeros for the one-dimensional Ising model, we observe that the partition function (8.8) is proportional to a polynomial of degree N in the fugacity $z = \exp(2\beta h)$. It vanishes if and only if

$$\lambda_+ = e^{in\pi/N}\lambda_-, \quad n = 1, 3, \ldots, 2N - 1 .$$

With the expressions (8.6) for the eigenvalues of the transfer matrix, we obtain

$$i \sin\left(\frac{n\pi}{2N}\right) \cosh(\beta h_n) = \sqrt{e^{-4K} + \sinh^2(\beta h_n)} \cos\left(\frac{n\pi}{2N}\right) ,$$

and after setting $\beta h_n = i\theta_n$, we end up with

$$\sin\frac{n\pi}{2N} \cos\theta_n = \sqrt{\sin^2\theta_n - e^{-4K}} \cos\left(\frac{n\pi}{2N}\right) .$$

Squaring this equation yields the following formula for the phases of the Lee–Yang zeros $z_n = \exp(2i\theta_n)$ in the complex fugacity plane:

$$\cos\theta_n = \sqrt{1 - e^{-4K}} \cos\left(\frac{n\pi}{2N}\right), \quad n = 1, 3, \ldots, 2N - 1 . \tag{8.89}$$

Since the square root takes its values in the interval $[0, 1]$, we obtain a *real solution* θ_n for every n such that all zeros z_n lie on the unit circle in the complex z–plane. In Fig. 8.1 we plotted the phases

$$2\theta_n = 2\arccos(\alpha \cos x_n) \quad \text{with} \quad \alpha^2 = 1 - e^{-4K}, \quad x_n = \frac{n\pi}{2N} , \tag{8.90}$$

Fig. 8.1 The phases $2\theta_n$ of the Lee–Yang zeros z_n for the Ising chain with 16 sites

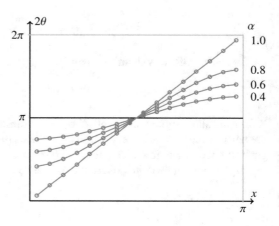

for different values of α. At zero temperature $\alpha = 1$ and the phases $2\theta_n = 2x_n$ are uniformly distributed within the interval $[0, 2\pi]$. At finite temperatures $\alpha < 1$ and the $2\theta_n$ take their values in the interval $[\Delta, 2\pi - \Delta]$ with $\Delta = 2\arccos\alpha > 0$. This estimate is independent of the lattice size. We conclude that the free energy density shows not singularities at finite temperatures and for any real magnetic field. Only for $T = 0$ and in the thermodynamic limit do the zeros of the partition function pinch the positive real axis. This points to a singularity of the free energy at $z = 1$ and thus to a phase transition at vanishing magnetic field and zero temperature. Figure 8.2 shows the distribution of zeros of $Z_A(z)$ on the unit circle in the fugacity plane for 30 sites and different values of the temperature-dependent parameter α. Figure 8.3 shows the zeros for $\alpha = 0.9$ for three different lattice sizes. In the thermodynamic limit $N \to \infty$, we observe an accumulation of the partition function zeros at the so-called edge singularity at

$$\Re z = \cos\left(2\arccos\alpha\right) \quad \text{and} \quad \Im z = \sin\left(2\arccos\alpha\right).$$

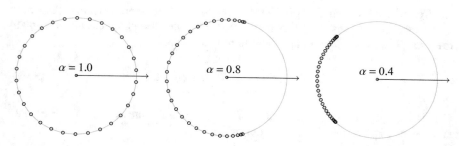

Fig. 8.2 The Lee–Yang zeros in the complex fugacity plane at different temperatures. At $T = 0$, the zeros are uniformly distributed on the unit circle. With increasing temperature, the zeros move away from the positive real axis

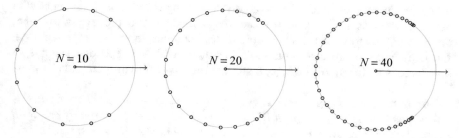

Fig. 8.3 The Lee–Yang zeros, located in the complex fugacity plane for different values of N and at a fixed positive temperature corresponding to $\alpha = 0.9$

8.7.2　General Ferromagnetic Systems

In their two classic papers, Lee and Yang showed that the zeros of the Ising model partition function lie on the unit circle and proposed a program to analyze phase transitions in terms of these zeros. Since their proposal, the Lee–Yang theorem has been used to prove the existence of infinite volume limits for spin models, to prove correlation inequalities and inequalities for critical exponents; see [13].

The Lee–Yang theorem states that the partition function

$$Z_\Lambda(\beta) = \int d\mu(\omega) \exp(-\beta H_\Lambda(\omega)), \quad d\mu(\omega) = \prod_{x\in\Lambda} d\mu(s_x) , \tag{8.91}$$

of ferromagnetic systems with energies

$$H_\Lambda = - \sum_{x\neq y\in\Lambda} J_{xy}s_x s_y - \sum_{x\in\Lambda} h_x s_x, \quad J_{xy} \geq 0 , \tag{8.92}$$

and a certain class of single-spin distributions $d\mu(s)$ is non-zero whenever $\Re h_x > 0$ on all sites. It was proved by Newman [19–21] for one-component systems with an even single-spin distribution $d\mu$ with the property that

$$\int e^{hs} d\mu(s) \neq 0 \quad \text{for} \quad \Re(h) \neq 0 . \tag{8.93}$$

For lattice systems with continuous spins, for example, lattice field theories, the single-spin distribution should also fall off rapidly,

$$\int e^{\alpha s^2} d\mu(s) < \infty, \quad \alpha < \infty , \tag{8.94}$$

for the partition function to exist. A rapidly decreasing and even measure on $\mathbb{R}$ with property (8.93) is said to have the *Lee–Yang property*. This property means that the Fourier transform of the single-spin distribution has its zeros on the real axis. The Lee–Yang theorem also applies to spin systems with site-dependent single-spin distributions in $d\mu(\omega) = \prod d\mu_x(s_x)$, provided all $d\mu_x$ fulfill the condition (8.93):

Theorem 8.2 (Lee–Yang Circle Theorem) *If all single-spin distributions $d\mu_x$ of a ferromagnetic spin system with $\mathscr{T} = \mathbb{R}$ satisfy the Lee–Yang property and if all h_x have a positive real part, then the partition function $Z_\Lambda[h]$ is non-zero. If the external field is constant, $h_x = h$, then all zeros of $Z_\Lambda(h)$ are imaginary.*

This theorem means that the Lee–Yang property holds for $J_{xy} \geq 0$ in case it holds for $J_{xy} = 0$. For the Ising model with identical single-spin distributions (7.16), the integral (8.93) is $\cosh(h)$ and the theorem applies: all zeros of the Ising model partition function as function of the fugacity lie on the unit circle, similarly as for the Ising chain. The single-spin distribution of a lattice ϕ^4 theory,

$$d\mu(\phi) = \exp(-\lambda\phi^4 - \mu\phi^2)\,d\phi, \quad \lambda > 0, \ \mu \in \mathbb{R}, \tag{8.95}$$

also satisfies the Lee–Yang property such that the circle theorem applies to the Schwinger functional of a lattice ϕ^4 theory.

Recent proofs of the circle theorem use multi-affine Lee–Yang polynomials together with the *Asano contraction*; see [22–24]. For scalar field theories, one can easily prove a weaker result [13,25] which is based on correlation inequalities:

Theorem 8.3 (Dunlop) *Consider a ferromagnetic system with energy function (8.92) and single-spin distribution*

$$d\mu(s) = e^{-P(s)}, \quad P(s) = \lambda s^4 + \mu s^2, \quad \lambda > 0, \ \mu \ \text{real}. \tag{8.96}$$

If all h_x are in the wedge $|\Im h_x| \leq \Re h_x$, then the partition function is non-zero,

$$|Z_\Lambda[h]| \geq Z_\Lambda[h = 0] > 0. \tag{8.97}$$

Proof We introduce two independent copies of the spin system to write

$$|Z[h]|^2 = \int d\mu(\omega)d\mu(\omega')\,e^{-\beta H(\omega) - \beta H^*(\omega')} \tag{8.98}$$

where H^* is the energy function (8.92) with the h_x replaced by their complex conjugate h_x^*. First we transform to the rotated frame with coordinates σ_x, σ'_x introduced in (8.73), and for these we use polar coordinates

$$\sigma_x = \varrho_x \cos\theta_x \quad , \quad \sigma'_x = \varrho_x \sin\theta_x.$$

In terms of polar coordinates, the single site measure of the doubled system contains the polynomial

$$P(s) + P(s') = \frac{\lambda\varrho^4}{4}(3 - \cos 4\theta) + \mu\varrho^2$$

such that the single site measure takes the form

$$e^{-P(s)-P(s')}dsds' = \varrho \exp\left(-\tfrac{3}{4}\lambda\varrho^4 - \mu\varrho^2\right)\exp\left(\tfrac{1}{4}\lambda\varrho^4 \cos 4\theta\right)\varrho\,d\varrho\,d\theta. \tag{8.99}$$

The second step in the proof is to rewrite the integrand in (8.98) in polar coordinates. With

$$\sum_{x \neq y} J_{xy}(s_x s_y + s_x s_y') = \sum_{x \neq y} J_{xy}\, \varrho_x \varrho_y \cos(\theta_x - \theta_y)$$

$$\sum_x (h_x s_x + h_x^* s_x) = \sum_x \varrho_x \left(\Delta_+ e^{i\theta_x} + \Delta_- e^{-i\theta_x}\right) ,$$

where we abbreviated

$$\sqrt{2}\, \Delta_\pm = \Re h_x \pm \Im h_x$$

we obtain for the exponent in (8.98)

$$-\beta H(\omega) - \beta H^*(\omega') = \sum_{x \neq y} J_{xy}\, \varrho_x \varrho_y \cos(\theta_x - \theta_y) + \sum_x \varrho_x \left(\Delta_+ e^{i\theta_x} + \Delta_- e^{-i\theta_x}\right) .$$

At this point we use the notion of a positive definite function $f(\boldsymbol{\theta}) = f(\theta_1, \ldots, \theta_V)$ of angular variables $\theta_x \in [0, 2\pi]$. Such a function is called *positive definite* if the coefficients $f_{n_1 \ldots n_V}$ in its Fourier series

$$f(\theta_1, \ldots, \theta_V) = \frac{1}{(2\pi)^{V/2}} \sum_{n_i \in \mathbb{Z}} f_{n_1 \ldots n_V} \exp\left(i \sum n_x \theta_x\right)$$

are all non-negative, $f_{n_1 \ldots n_V} \geq 0$. The set of positive definite functions is closed under exponentiation, multiplication, and multiplication with a positive real constant.

Clearly, for positive λ, the last exponent in (8.99) is a positive definite function such that the density of $d\mu(\omega)d\mu(\omega')$ is a positive definite function as well. In addition, the condition $|\Im h_x| \leq \Re h_x$ in the theorem implies that $\Delta_\pm$ are both non-negative, such that $-\beta H(\omega) - \beta H^*(\omega)$, and its exponentials are both positive definite functions. We conclude that under the assumption in the theorem the integrand of

$$|Z_\Lambda[h]|^2 = \int \prod_x \varrho_s d\varrho_x d\theta_x\, F(\varrho_1, \ldots, \varrho_V; \theta_1, \ldots, \theta_V)$$

is a positive definite function, the Fourier coefficients of which are monotonic functions of Δ_+ and Δ_-. It follows that the integral, which is proportional to the Fourier coefficient $F_{0 \ldots 0}$, decreases when we replace the larger of the two numbers Δ_+, Δ_- by the smaller one, which is $\Re h_x - |\Im h_x|$. But then $|Z|^2$ becomes the square of the partition function for the real magnetic field $x \to \Re h_x - |\Im h_x|$ such that

$$|Z_\Lambda[h]| \geq Z_\Lambda\left[\Re h - |\Im h|\right] \geq Z_\Lambda[h = 0] > 0 . \tag{8.100}$$

In addition, since $Z_\Lambda[h] = Z_\Lambda[-h]$, it follows that the partition function is also non-zero if all h_x are in the wedge $|\Im h_x| \le -\Re h_x$.

In many applications one considers homogeneous fields for which the zeros of the partition function must lie in the forward or backward "light cones" in the complex h-plane. Note that the theorem is not as strong as the circle theorem which nails down the zeros to the imaginary axis in h-space. But the proof given by Dunlop also extends to Z_2 gauge theories [26]. For a general lattice field theory with multidimensional target space (and some invariant single-spin distribution), no variant of the Lee–Yang theorem may exist. Only for simple systems like the $O(2)$ and $O(3)$ models could such theorems be established; see [19–21, 27].

A decade after the work of Lee and Yang, Fisher extended the study of the Ising partition function zeros to the complex temperature plane [28]. After these early results, there have been numerous studies of Lee–Yang and Fisher zeros in a variety of models.

8.8 Problems

8.1 (Transfer Matrices for Modified Ising Chains)

(a) Find the transfer matrix for a Ising chain in which the spins may take the three values $\{+1, 0, -1\}$.
(b) Find the transfer matrix for a Ising chain with first and second neighbor interactions,

$$H(\omega) = -J \sum_x s_x s_{x+1} - K \sum_x s_x s_{x+2} - h \sum_x s_x, \quad s_x = \pm 1 .$$

Hint: consider the transfer matrix of a pair to the neighboring pair. An analysis of this model which circumvents diagonalizing the resulting 4×4 transfer matrix is given in [29].

8.2 (Potts Chain Revisited) In Sect. 8.2 we calculated the free energy density for the Potts chain in the thermodynamic limit. Compute the magnetization $m(T, h)$ and the susceptibility $\chi(T, h)$. Compare the results with the corresponding results for the Ising chain.

8.3 (Transfer Matrix for Harmonic Oscillator) Consider the transfer kernel (8.54) for the oscillator on the lattice with harmonic oscillator potential $V(q) = m\omega^2 q^2/2$.

- Prove the useful identity

$$\lambda_0^2 \hat{a}^\dagger T(q, q') = \hat{a}' T(q, q') , \tag{8.101}$$

where $\hat{a}^\dagger$ acts on the argument q and $\hat{a}'$ on the argument q',

$$\hat{a}^\dagger = \frac{\mathrm{d}}{\mathrm{d}q} - m\alpha q, \quad \hat{a}' = -\frac{\mathrm{d}}{\mathrm{d}q'} - m\alpha q' .$$

The explicit expressions for α and λ_0 are found in Sect. 8.5.1.

- Use the result to show that if ψ is an eigenfunction of the transfer matrix $\hat{T}$ with eigenvalue λ

$$\int \mathrm{d}q' T(q, q') \, \psi(q') = \lambda \psi(q') ,$$

and that $\hat{a}^\dagger \psi$ is an eigenfunction as well with eigenvalue $\lambda_0^2 \lambda$.

- In Sect. 8.5.1 we calculated the eigenstate ψ_0 with largest eigenvalue λ_0 of the transfer matrix. Determine the other eigenvalues and eigenfunctions with purely algebraic means.

References

1. L. Onsager, Crystal statistics. I. A two-dimensional model with an order-disorder transition. Phys. Rev. **65**, 117 (1944)
2. E. Lieb, F.Y. Wu, Two-dimensional ferroelectric models, in *Phase Transitions and Critical Phenomena*, vol. 1, ed. by C. Domb, M.S. Green (Academic Press, London, 1972)
3. R.J. Baxter, *Exactly Solved Models in Statistical Mechanics* (Dover Books, New York, 1982/2008)
4. E. Ising, Beitrag zur Theorie des Ferromagnetismus. Zeitschrift f. Physik **31**, 253 (1925)
5. L. van Hove, Sur l'intégrale de configuration des systèmes des particules à une dimension. Physica **16**, 137 (1950)
6. D. Ruelle, *Statistical Mechanics: Rigorous Results* (World Scientific Publishing, Singapore, 1999)
7. T. Dauxois, M. Peyrand, Entropy-driven transition in a one-dimensional system. Phys. Rev. **E51**, 4027 (1995)
8. E. Lieb, D. Mattis, *Mathematical Physics in One Dimension* (Academic Press, New York, 1966)
9. K. Huang, *Statistical Mechanics* (Wiley, New York, 1987)
10. F.Y. Wu, Self-dual property of the Potts model in one dimension. arXiv:cond-mat/9805301
11. O. Perron, Zur Theorie der Matrizen. Mathematische Annalen **64**, 248 (1907)
12. G. Frobenius, Über Matrizen aus nicht negativen Elementen. Sitzungsber. Königl. Preuss. Akad. Wiss., 456 (1912)
13. B. Simon, *The P(ϕ)$_2$ Euclidean (Quantum) Field Theory* (Princeton Legacy Library, Princeton, 2016)
14. G.S. Sylvester, Inequalities for continuous-spin Ising ferromagnets. J. Stat. Phys. **15**, 327 (1975)
15. S. Friedli, Y. Velenik, *Statistical Mechanics of Lattice Systems* (Cambridge University Press, Cambridge, 2017)
16. S.B. Shlosman, Correlation inequalities for antiferromagnets. J. Stat. Phys. **22**, 59 (1976)
17. J. Ginibre, General formulation of Griffiths' inequalities. Commun. Math. Phys. **16**, 310 (1970)

18. C.N. Yang, T.D. Lee, Statistical theory of equations of state and phase transitions. I. Theory of Condensation. Phys. Rev. **87**, 404 (1952); Statistical theory of equations of state and phase transitions. II. Lattice gas and Ising model. Phys. Rev. **87**, 410 (1952)

19. C. Newman, Zeros of the partition function of generalized Ising systems. Commun. Pure Appl. Math. **27**, 143 (1974)

20. K.C. Lee, Zeros of the partition function for a continuum system at first-order transitions. Phys. Rev. **E53**, 6558 (1996)

21. W.T. Lu, F.Y. Wu, Partition function zeroes of a self-dual Ising model. Physica **A258**, 157 (1998)

22. T. Asano, Theorems on the partition functions of the Heisenberg ferromagnets. J. Phys. Soc. Jpn. **29**, 350 (1970)

23. D. Ruelle, Extension of the Lee-Yang circle theorem. Phys. Rev. Lett. **26**, 303 (1971)

24. E.H. Lieb, A.D. Sokal, A general Lee-Yang theorem for one-component and multicomponent ferromagnets. Commun. Math. Phys. **80**, 153 (1981)

25. F. Dunlop, Zeros of partition function via correlation inequalities. J. Stat. Phys. **17**, 215 (1977)

26. F. Dunlop, in *Colloquia Mathematica Sociatatis Janos Bolyai 27: Random Fields*, ed. by J. Fritz, J.L. Lebowitz, D. Szasz (North Holland, Amsterdam, 1981)

27. F. Dunlop, C. Newman, Multicomponent field theories and classical rotators. Commun. Math. Phys. **44**, 223 (1975)

28. M.E. Fisher, in *Lectures in Theoretical Physics*, vol. 7c, ed. by W.E. Brittin (University of Colorado Press, Boulder, 1965)

29. J. Stephenson, Two one-dimensional Ising models with disorder points. Can. J. Phys. **48**, 1724 (1979)

High-Temperature and Low-Temperature Expansions

9

Series expansions remain, in many cases, one of the most accurate ways of estimating critical exponents. Historically it was the results from series expansions that suggested universality at criticality. Two expansions will be considered in this chapter. In the *high-temperature series*, the Boltzmann factor is expanded in powers of the inverse temperature, and the sum over all configurations is taken term by term. In the Ising model, this leads to an expansion in powers of $\tanh(J/T) \ll 1$. In the *low-temperature expansion*, configurations are counted in order of their importance as the temperature is increased from zero. Starting from the ground state, the series is constructed by successively adding terms from $1, 2, 3, \ldots$ flipped spins. In the Ising model, this leads to an expansion in powers of $\exp(-2J/T) \ll 1$.

Each term in a series is represented by a graph on a lattice, and constructing the series amounts to counting the graphs belonging to a fixed order in the expansions. The expansions can be used to approximate the thermodynamic potentials and correlation functions at low and high temperatures. The hope is that the expansions are sufficiently well-behaved in order to extract the singular behavior from a limited number of lowest-order terms. Confidence in the method lies in the large body of circumstantial evidence available. Series expansions agree well with high-accuracy Monte Carlo simulations, renormalization group and conformal bootstrap methods, and results for exactly solvable models.

The high- and low-temperature expansions treated in this chapter are covered in numerous textbooks and papers. For a further reading beyond the introductory material presented in this chapter, you may consult the textbooks [1–5].

9.1 Ising Chain

To become acquainted with the high- and low-temperature expansions, we first consider these expansions for the Ising chain.

© The Author(s), under exclusive license to Springer Nature Switzerland AG 2021
A. Wipf, *Statistical Approach to Quantum Field Theory*, Lecture Notes in Physics 992, https://doi.org/10.1007/978-3-030-83263-6_9

Fig. 9.1 Spin configurations
of the Ising chain for different
energies

$\uparrow \uparrow \uparrow \uparrow \uparrow$ $-5J - 5h$　$\downarrow \downarrow \downarrow \downarrow \downarrow$ $-5J + 5h$

$\uparrow \uparrow \uparrow \uparrow \downarrow$ $-J - 3h$　$\downarrow \downarrow \downarrow \downarrow \uparrow$ $-J + 3h$

$\uparrow \uparrow \uparrow \downarrow \downarrow$ $-J - h$　$\downarrow \downarrow \downarrow \uparrow \uparrow$ $-J + h$

$\uparrow \downarrow \uparrow \downarrow \uparrow$ $3J - h$　$\downarrow \uparrow \downarrow \uparrow \downarrow$ $3J + h$

9.1.1　Low Temperature

The configuration with minimal energy is characterized by an alignment of *the
spins*—all spins point in the direction of the magnetic field. Configurations with
some spins flipped have higher energy and can only be populated for nonzero
temperature. But at low temperature, the number of misaligned spins is exponentially small. For warming up we consider the chain Λ with only five sites.
The 2^5 configurations form eight classes and each class is characterized by its
energy. Representatives of the classes are depicted in Fig. 9.1. The two classes with
aligned spins in the first row contain one element each, whereas all other classes
contain five elements. Assuming $h > 0$, the energy of the vacuum configuration is
$E_0 = -5J - 5h$, and we obtain the following *low-temperature expansion* of the
partition function for $e^{-\beta J} \ll 1$:

$$Z_\Lambda = e^{-\beta E_0} \left(1 + e^{-10\beta h} + 5e^{-\beta(4J+2h)} + 5e^{-\beta(4J+8h)}\right.$$

$$\left. + 5e^{-\beta(4J+4h)} + 5e^{-\beta(4J+6h)} + 5e^{-\beta(8J+4h)} + 5e^{-\beta(8J+6h)}\right).$$

The systematic low-temperature expansion based on graphs is discussed in Sect. 9.3
for the more interesting Ising models in higher dimensions.

9.1.2　High Temperature

For simplicity we consider the high-temperature expansion for the zero-field Ising
model. Thereby we expand the partition function on a hypercubic lattice Λ in d
dimensions in powers of the small parameter $v = \tanh K = \tanh J/T$,

$$Z_\Lambda = \sum_\omega \prod_{\langle x,y \rangle} e^{K s_x s_y} = (\cosh K)^P \sum_\omega \prod_{\langle x,y \rangle} (1 + v s_x s_y), \tag{9.1}$$

where $P = dV$ denotes the number of nearest-neighbor pairs and the last equation
holds because $s_x s_y \in \{-1, 1\}$.

The periodic *Ising chain* with only three sites has three nearest-neighbor pairs, and the product in (9.1) contains three factors:

$$\prod_{\langle x,y\rangle} (1 + v s_x s_y) = (1 + v s_1 s_2)(1 + v s_2 s_3)(1 + v s_3 s_1) \,.$$

The expansion of this expression in powers of the parameter v yields $2^P = 8$ terms:

$$Z_\Lambda = (\cosh K)^3 \sum_\omega \Big(1 + v(s_1 s_2 + s_2 s_3 + s_3 s_1)$$

$$+ v^2(s_1 s_2 s_2 s_3 + s_1 s_2 s_3 s_1 + s_2 s_3 s_3 s_1) + v^3(s_1 s_2 s_2 s_3 s_3 s_1)\Big) \,. \quad (9.2)$$

Next we assign to each term in this expansion a diagram on the periodic chain as follows: we connect two adjacent sites x and y by a line—in the following called bond or link—if the product of their spins occurs in the term under consideration. Those sites where at least one bond ends form the vertices of the associated diagram. We call the number of bonds attached to a vertex the *order of the vertex*.

Figure 9.2 shows the eight diagrams corresponding to the high-temperature expansion (9.2). Since the small parameter v enters via the combination $v s_x s_y$, we see that all diagrams of order n in v have n bonds. Because of the identity

$$\sum_{s_x = -1,1} s_x^n = \begin{cases} 2 & n \text{ even} \\ 0 & n \text{ odd} \,, \end{cases} \quad (9.3)$$

only diagrams for which all vertices have an even order contribute to Z_Λ. Such diagrams are called closed. On the chain with three sites only, the first and last diagrams in Fig. 9.2 are closed such that

$$Z_\Lambda = \cosh^3 K \left(8 + 8v^3\right) = 2^3 \left(\cosh^3 K + \sinh^3 K\right) \,.$$

More generally, the chain with N sites has only two closed diagrams: one diagram has no bond and the other has the maximal number of N bonds. It follows that only

Fig. 9.2 Diagrams associated with the high-temperature expansion of the Ising chain

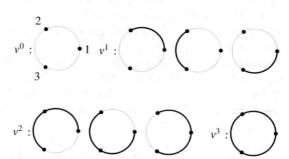

two diagrams contribute to the partition function:

$$Z_\Lambda = 2^N \left(\cosh^N K + \sinh^N K \right), \quad K = J/T . \tag{9.4}$$

The high-temperature expansion leads to the exact result of the partition function.

9.2   High-Temperature Expansions for Ising Models

The high-temperature expansion is an expansion in the small parameter $v = \tanh K$, where $K = J/T$. It corresponds to the strong-coupling expansion in quantum field theory.

9.2.1   General Results and Two-Dimensional Model

We consider the zero-field Ising model defined by the Hamiltonian (6.2) with $h = 0$. As starting point we choose the high-temperature expansion (9.1) in powers of v:

$$Z_\Lambda = (\cosh K)^P \sum_\omega \left(1 + v \sum_{\langle x,y \rangle} s_x s_y + v^2 \sum_{\langle x,y \rangle \neq \langle x',y' \rangle} s_x s_y s_{x'} s_{y'} + \dots \right), \tag{9.5}$$

To each spin product, we assign the diagram where nearest neighbors are connected by a line if their product occurs in the spin product. Figure 9.3 shows the diagram associated with the spin product $s_x s_y^3 s_z^2 s_u^2 s_v^2$. The diagram does not contribute to the partition function since it has the odd vertices x and y. Only closed diagrams contribute, such that on a *finite* lattice with P nearest-neighbor pairs,[1]

$$Z_\Lambda = (\cosh K)^P \, 2^V \sum_{\ell=0}^{P} z'_\ell \, v^\ell, \quad z'_0 = 1 . \tag{9.6}$$

where z'_ℓ counts the number of closed diagrams with ℓ bonds. Table 9.1 contains the closed diagrams for the *periodic* Ising model on a square lattice with V sites and $P = 2V$ nearest-neighbor pairs with eight or less bonds. The number in the third column counts the number of diagrams of the corresponding class. For example, the number $V(V-5)/2$ in the second to last row of this table is explained as follows: We may place the first plaquette somewhere on the lattice and we have V possibilities to do this. Then the remaining number of locations of the second plaquette subject to the constraint that none of the edges of the two plaquettes coincide is $V - 5$. Hence, there are $V(V - 5)$ possibilities to place the two plaquettes on the lattice. However,

---

[1] Coefficients in an expansion in v are marked with a stroke.

Fig. 9.3 Diagram as it occurs in the high-temperature expansion (9.5). The graph corresponds to the spin product $s_x s_y^3 s_z^2 s_u^2 s_v^2 = s_x s_y$

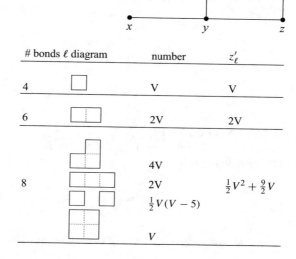

Table 9.1 Closed diagrams contributing to the high-temperature expansion of the partition function of the two-dimensional Ising model to order of v^8

# bonds ℓ	diagram	number	z'_ℓ
4		V	V
6		2V	2V
8		4V	
		2V	$\frac{1}{2}V^2 + \frac{9}{2}V$
		$\frac{1}{2}V(V-5)$	
		V	

since we obtain the same diagram under a permutation of the plaquettes, we finally end up with $V(V-5)/2$ different diagrams.

The corresponding expansion for the $2d$-Ising model partition function reads

$$Z_\Lambda = (\cosh K)^P \, 2^V \left(1 + Vv^4 + 2Vv^6 + \frac{V}{2}(V+9)v^8 + 2V(V+6)v^{10} + \ldots \right).$$

(9.7)

Only intensive quantities have a well-defined thermodynamic limit, and thus we turn to the free energy density, given by

$$-\beta f = \lim_{V \to \infty} \frac{1}{V} \log Z_\Lambda = \log\left(2\cosh^2 K\right) + \sum_{\ell \geq 4} f'_\ell v^\ell .$$

(9.8)

Inserting the power series for the partition function, one obtains the expansion coefficients f'_ℓ of the free energy density. Only connected diagrams contribute to the free energy density such that the coefficients f'_ℓ are volume-independent. For the Ising model on a square lattice,

$$-\beta f = \log\left(2\cosh^2 K\right) + v^4 + 2v^6 + \frac{9}{2}v^8 + 12v^{10} + \ldots .$$

(9.9)

Fig. 9.4 The high-temperature expansion of the two-point correlation function of the two-dimensional Ising model. The diagram corresponds to the term $s_x^2 s_y^4 s_{u'}^4 s_u^2 s_v^2 s_{u''}^2 s_{v''}^2$

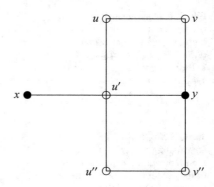

Correlation Functions

Above the critical temperature, the magnetization vanishes for zero field, and the susceptibility χ is given by

$$\chi = \frac{1}{V} \sum_{x,y} \langle s_x s_y \rangle = \sum_y \langle s_x s_y \rangle . \tag{9.10}$$

Again, we associate each term of the high-temperature expansion

$$\langle s_x s_y \rangle = \frac{\cosh^P K}{Z_\Lambda} \sum_\omega s_x s_y \left(1 + v \sum_{\langle uv \rangle} s_u s_v + v^2 \sum_{\langle uv \rangle \neq \langle u'v' \rangle} s_u s_v s_{u'} s_{v'} + \dots \right) \tag{9.11}$$

with a diagram. For example, consider the diagram in Fig. 9.4 which contributes $2^V v^8$ to the sum over all configurations in (9.11). More generally, a diagram with ℓ bonds contributes $2^V v^\ell$ if the vertices x and y are odd and all other vertices are even. Because of the relation (9.3), the other diagrams do not contribute. The factor $2^V \cosh^P K$ cancels against the same factor in Z_Λ. Thus we have the

Lemma 9.1 *The high-temperature expansion of the two-point function reads*

$$\langle s_x s_y \rangle = \frac{\sum_\ell g'_\ell v^\ell}{\sum_\ell z'_\ell v^\ell} , \quad v = \tanh K \ll 1 , \tag{9.12}$$

where g'_ℓ denotes the number of diagrams with ℓ bonds and only even vertices, with the exception of the odd vertices x and y.

This lemma immediately leads to the

Proposition 9.1 *The coefficient g'_ℓ vanishes if ℓ is less than the shortest distance $d(x, y)$ (on the lattice) between x and y.*

If the corollary was not true, then there would exist a diagram with less than ℓ bonds and odd vertices x and y, with all other vertices even. It follows that the sites x and y are not connected by a line of bonds or that the diagram consists of at least two disconnected sub-diagrams. At least one connected sub-diagram contains x but not y. According to the relation

$$\sum_{\text{vertices}} \text{vertex order} = 2 \cdot \text{number of bonds} \qquad (9.13)$$

the sum of vertex orders for every connected sub-diagram must be even. But for the sub-diagram containing x but not y and contributing to $\langle s_x s_y \rangle$, the sum of vertex orders is odd, since all vertices are even, with the exception of the odd vertex x. But this contradicts the selection rule (9.13) and thus proves Proposition 9.1.

The proposition tells us that for $T \gg J$ the *two*-point function falls off exponentially with the shortest distance $d(x, y)$ of x and y on the lattice:

$$\langle s_x s_y \rangle = O\left(v^{d(x,y)}\right) = O\left(e^{-d(x,y)/\xi}\right), \qquad \frac{1}{\xi} = \log\frac{1}{v} \gg 1, \qquad (9.14)$$

As expected, the correlation length ξ decreases with increasing temperature since the thermal fluctuations suppress spin-spin correlations. Figure 9.5 shows all diagrams that contribute to the numerator in (9.12) for fixed x and y to order v^6. In the two-point functions (9.12), the volume factor cancels in the ratio of formal power series. Together with the high-temperature expansion of the denominator given in (9.7), we obtain for next-to-nearest neighbors

$$\langle s_x s_y \rangle = \left(2v^2 + 4v^4 + (2V + 10)v^6 + \ldots\right) / \left(1 + Vv^4 + \ldots\right),$$
$$= 2v^2 + 4v^4 + 10v^6 + O(v^8). \qquad (9.15)$$

Susceptibility

Only two-point functions with $d(x, y) \leq n$ contribute to the expansion of the susceptibility in (9.10) to order n in v. Hence the susceptibility in every order n is given by a finite number of terms or diagrams. One considers all permitted diagrams for fixed lattice points x, y with $d(x, y) \leq n$ and calculates the corresponding two-point function. Adding these functions as in (9.10) leads to the high-temperature

ℓ	diagrams	number	g'_ℓ
2		2	2
4		4	4
6		3×4	
		2×2	$2V + 10$
		$2(V - 3)$	

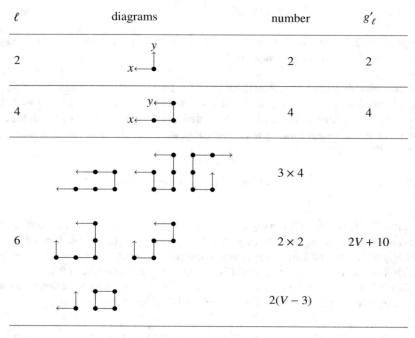

Fig. 9.5 The leading high-temperature diagrams contributing to the numerator in (9.12) for next-to-nearest neighbor points x and y

expansion of χ. To order v^{10} one finds

$$\chi(v) = \sum \chi'_\ell v^\ell = 1 + 4v + 12v^2 + 36v^3 + 100v^4 + 276v^5 + 740v^6$$
$$+ 1972v^7 + 5172v^8 + 13492v^9 + 34876v^{10} + \dots \quad (9.16)$$

For example, the term $12v^2$ is the contribution of the diagram with $g'_2 = 2$ in Fig. 9.5 plus the second diagram with $g'_2 = 1$ for which the two bonds connecting x and y point in the same direction. The sum over y in (9.10) generates a further factor 4, such that $\chi'_2 = 4 \cdot (2 + 1) = 12$. To convert the series (9.16) into a series in K,

$$\chi(K) = \sum_{\ell \geq 0} \chi_\ell K^\ell , \quad (9.17)$$

we substitute $v = \tanh K$ into (9.16) and expand the resulting expression in powers of K. This yields the following expansion in inverse powers of the temperature:

$$\chi(K) = 1 + 4K + 12K^2 + 104/3\,K^3 + 92K^4 + 3608/15\,K^5 + 3056/5\,K^6$$
$$+ 484528/315\,K^7 + 400012/105\,K^8 + 26548808/2835\,K^9$$
$$+ 107828128/4725\,K^{10} + \dots . \tag{9.18}$$

This series was calculated to 21st order in [6, 7]. Terms of even higher order are found in [8] and for generalized Ising models with higher spins in [9].[2]

Extrapolation to the Critical Point

The power series (9.18) has a finite radius of convergence $R > 0$ and thus defines an analytical function on the disk $|K| < R$. Since all coefficients χ_ℓ are positive, there must be a singularity at the point $K = R$ on the real axis. We identify this singular point with the critical value $K_c = J/T_c$. The *ratio test* yields the critical temperature

$$R = \lim_{\ell \to \infty} \frac{\chi_\ell}{\chi_{\ell+1}} = \frac{J}{T_c} . \tag{9.19}$$

More accurate values for the critical temperature and critical exponent γ are obtained by fitting the truncated power series to the expected scaling of the susceptibility near the critical point:

$$\chi(K) = \sum \chi_\ell K^\ell \propto (1 - K/K_c)^{-\gamma}$$
$$= 1 + \sum_{\ell=1}^{\infty} \frac{\gamma(\gamma+1)\cdots(\gamma+\ell-1)}{\ell!} \left(\frac{K}{K_c}\right)^\ell . \tag{9.20}$$

The ratio of coefficients defines a linear function in $1/\ell$,

$$\frac{\chi_\ell}{\chi_{\ell-1}} = \frac{1}{K_c} + \frac{\gamma-1}{K_c}\frac{1}{\ell} , \tag{9.21}$$

and the slope of this function together with its value at $1/\ell \to 0$ yields the critical exponent and the critical temperature. The ratios $\chi_\ell/\chi_{\ell-1}$ of the coefficients to order 25 and the linear fit are depicted in Fig. 9.6. The linear interpolation yields

$$T_c \approx 2.26694\,J \quad \text{and} \quad \gamma \approx \frac{1.69129}{2.26695} + 1 = 1.74606 . \tag{9.22}$$

---

[2] The coefficients of orders 20 and 21 in [7–9] do not agree with those in [6]. According to a private communication by Paolo Butera, the numbers in [6] are probably erroneous.

Fig. 9.6 The ratios $\chi_\ell/\chi_{\ell-1}$ for the high-temperature expansion (9.17) to order 25. From the slope and intersection with the vertical axis, we read off the critical exponent γ and the critical temperature T_c of the $2d$ Ising model

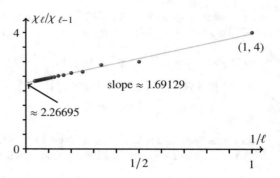

The estimates agree well with the exact values in (6.7):

$$T_c = \frac{2J}{\log(1+\sqrt{2})} \approx 2.26919\,J \quad \text{and} \quad \gamma = \frac{7}{4} = 1.75 \,. \tag{9.23}$$

The online library [9] contains further high-temperature expansions of several basic observables in Ising models with spins $1/2, 1, 3/2, 2, 5/2, 3, 7/2, 5,$ and ∞—in two dimensions on the square lattice and in three dimensions on the simple-cubic and the body-centered cubic lattices.

9.2.2   Three-Dimensional Model

High-temperature expansions for the three-dimensional Ising model to order 10 in the inverse temperature are found in the early work [10]. With refined techniques more and more higher-order terms were calculated in [11–14], and with the steady increase of computer power, this was possible for a large class of spin models and lattice field theories. The diagrammatic method has been applied to Ising-type models, and the resulting coefficients to order 25 are tabulated in [9].

Free Energy Density and Specific Heat
The finite lattice method developed in [15] avoids the tedious work of counting all the high-temperature diagrams. Nevertheless, the amount of calculation still grows exponentially with the order of the expansion. In [16] a variant of the method was applied to calculate the high-temperature expansion of the free energy density,

$$-\beta f = \log(2\cosh^3 K) + \sum_{\ell \geq 4} f'_\ell v^\ell, \quad v = \tanh K \,, \tag{9.24}$$

Table 9.2 High-temperature expansion coefficients f'_ℓ to order 46 [16]

ℓ	f'_ℓ	ℓ	f'_ℓ
2	0	26	4437596650548
4	3	28	525549581866326/7
6	22	30	6448284363491202/5
8	375/2	32	179577198475709847/8
10	1980	34	395251648062268272
12	24044	36	21093662188820520521/3
14	319170	38	126225408651399082182
16	18059031/4	40	4569217533196761997785/2
18	201010408/3	42	29159128711096862385794 0/7
20	5162283633/5	44	84107222262379235048686604/11
22	16397040750	46	14120314204713719766888210
24	266958797382		

to order 46. The expansion coefficients f'_ℓ are listed in Table 9.2. Differentiating $\beta f(K)$ twice with respect to β, one obtains the expansion of the specific heat to the same order. The corresponding expansion in K to order 12 reads

$$c = \sum \alpha_\ell K^\ell = 3K^2 + 33\,K^4 + 542\,K^6 + 123547/15\,K^8$$
$$+ 14473442/105\,K^{10} + 11336607022/4725\,K^{12} + \ldots \qquad (9.25)$$

Since this is an expansion in K^2, we fit the high-temperature series to the scaling law

$$c = \sum \alpha_{2\ell} K^{2\ell} \sim \left(1 - \frac{K^2}{K_c^2}\right)^{-\alpha}. \qquad (9.26)$$

For large ℓ the ratio of coefficients defines a linear function of $1/\ell$, similarly as for the susceptibility:

$$\frac{\alpha_{2\ell}}{\alpha_{2\ell-2}} \longrightarrow \frac{1}{K_c^2} + \frac{\alpha - 1}{K_c^2}\frac{1}{\ell}. \qquad (9.27)$$

Figure 9.7 shows the ratios for all coefficients to order 46. The three ratios involving the coefficients of lowest orders 2, 4, 6, and 8 do not fall onto a straight line and are left out in the data analysis. The linear fit included in Fig. 9.7 yields

$$T_c = 4.5102\,J \quad \text{and} \quad \alpha = 0.1226. \qquad (9.28)$$

Fig. 9.7 The ratios $\alpha_{2\ell}/\alpha_{2\ell-2}$ for the high-temperature expansion (9.25) to order 46. From the slope and intersection with the vertical axis, we read off the critical exponent α and the critical temperature T_c of the 3d Ising model

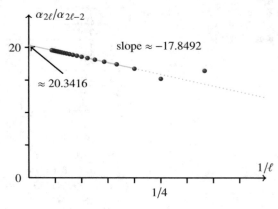

Fig. 9.8 The local estimates $T_c^{(\ell)}$ for the critical temperature given in (9.29) for the high-temperature expansion to order 46

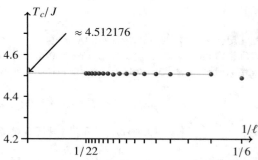

We can do even better when we estimate the critical temperature and exponent from the ratios $r_\ell = \alpha_{2\ell}/\alpha_{2\ell-2}$ of two neighboring ℓ according to

$$T_c^{(\ell)} = J\sqrt{(\ell+1)\,r_{\ell+1} - \ell r_\ell} \quad \text{and} \quad \alpha^{(\ell)} = -\frac{(\ell^2-1)r_{\ell+1} - \ell^2 r_\ell}{(\ell+1)r_{\ell+1} - \ell r_\ell} \tag{9.29}$$

and let ℓ become large. Figure 9.8 contains the estimates for T_c/J for all available orders. Leaving out the lowest five estimates which do not fall onto a straight line, the linear extrapolation leads to

$$T_c = 4.512176\,J\ . \tag{9.30}$$

Computer simulations in combination with finite size scaling analysis predict a critical temperature $T_c \approx 4.5115232\,J$ [17].

Figure 9.9 shows the local estimates $\alpha^{(\ell)}$ defined in (9.29) for ℓ up to 22. Discarding the five lowest orders in the high-temperature expansion, a linear fit yields the critical exponent

$$\alpha = 0.109385\ , \tag{9.31}$$

Fig. 9.9 The local estimates $\alpha^{(\ell)}$ for the critical exponent α according to (9.29) for the high-temperature expansion to order 46

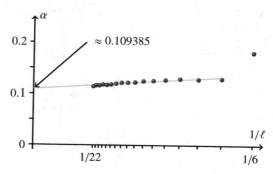

Table 9.3 Expansion coefficients of $\chi(v)$ for the Ising model on a cubic lattice (taken from [9])

ℓ	χ_ℓ'	ℓ	χ_ℓ'
1	6	14	3973784886
2	30	15	18527532310
3	150	16	86228667894
4	726	17	401225368086
5	3510	18	1864308847838
6	16710	19	8660961643254
7	79494	20	40190947325670
8	375174	21	186475398518726
9	1769686	22	864404776466406
10	8306862	23	4006394107568934
11	38975286	24	18554916271112254
12	182265822	25	85923704942057238
13	852063558		

which compares well with the precise estimate $\alpha = 0.11008(1)$ taken from [18]. Further improvement is possible if one uses the modified-ratio method and also includes corrections to scaling.

Susceptibility

The high-temperature expansion of the susceptibility for the three-dimensional Ising model on a hypercubic lattice,

$$\chi(v) = 1 + \sum_{\ell \geq 1} \chi_\ell' v^\ell , \qquad (9.32)$$

to order 20 was computed earlier on in [11] and then continued to order 25 in [9]. The known coefficients χ_ℓ' are listed in Table 9.3. The coefficient ratios of the derived series $\chi(K) = \sum \chi_\ell K^\ell$ to order 25 are depicted in Fig. 9.10. Similarly as in two dimensions, we fit this series expansion with the expected scaling law for the

Fig. 9.10 Ratios $\chi_\ell/\chi_{\ell-1}$ for the high-temperature expansion $\sum \chi_\ell K^\ell$ for the three-dimensional Ising model

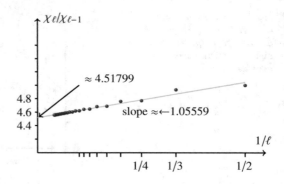

susceptibility and find the critical values

$$T_c \approx 4.51799\, J \quad \text{and} \quad \gamma \approx 1 + \frac{1.05559}{4.51799} = 1.23364 \,. \tag{9.33}$$

These values differ by less than 1 percent from the best known values listed in Table 7.3 on page 142.

9.3   Low-Temperature Expansion of Ising Models

In spin systems at zero temperature, only the configuration(s) with minimal energy is occupied. At low temperature, configurations with higher energies may occur, but they are exponentially suppressed. Thus in a low-temperature expansion, we study the deviation of the system from the state with minimal energy. The expansion corresponds to the weak-coupling expansion in field theory. Indeed, for a continuous target space, we recover the perturbation theory in quantum field theory.

The energy of the Ising model is minimal if all spins point in the direction of the magnetic field. Hence, for a positive magnetic field, the configuration

$$\omega_0 = \{s_x = 1 \mid x \in \Lambda\} \tag{9.34}$$

has the lowest energy

$$E_0 = -PJ - Vh \,. \tag{9.35}$$

The number of nearest-neighbor pairs is $P = dV$. There is one configuration with minimal energy. We reach configurations with higher energies by flipping the spins at certain lattice points. A configuration ω is uniquely characterized by the set $X(\omega) \subset \Lambda$ of lattice points with flipped spins. All spins in the complement of X are parallel to the magnetic field, and all spins in X are antiparallel. This means that the number of flipped spins is equal to the volume $|X|$ of X and the number of nearest-

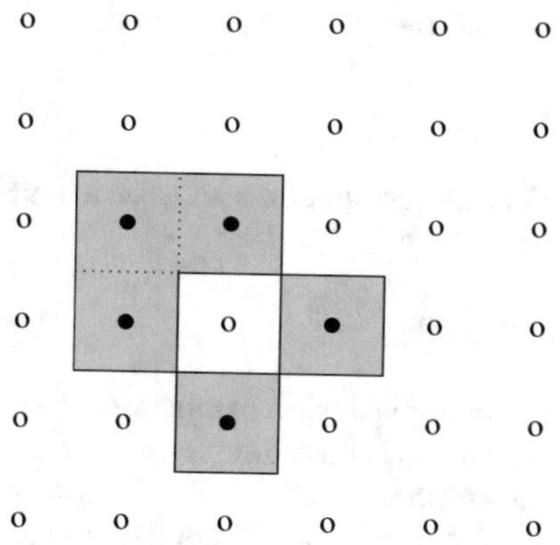

Fig. 9.11 A spin configuration is uniquely determined by the set of sites with flipped spins marked as *filled circle*

neighbor pairs with antiparallel spins is equal to the area $|\partial X|$ of its boundary. Figure 9.11 shows a configuration of the Ising model on a square lattice with $|X| = 5$ and $|\partial X| = 16$. A configuration ω is almost uniquely characterized by its polygon ∂X (in higher dimensions: polyhedron). The only ambiguity is that ω and $-\omega$ possess the same polygons—the graphical representation does not distinguish between the inner and outer part of ∂X.

9.3.1 Free Energy and Magnetization of Two-Dimensional Model

Every lattice point ● with spin antiparallel to the magnetic field contributes $2h$ to the energy. Similarly, a nearest-neighbor pair ●○ with opposite spins contributes $2J$. Hence, we have

$$E(X) = E_0 + 2h\,|X| + 2J\,|\partial X| \,, \tag{9.36}$$

and the partition function may be written as

$$Z = e^{-\beta E_0} \sum_X e^{-2\beta h\,|X| - 2\beta J\,|\partial X|} \equiv e^{-\beta E_0}\, \Xi \,, \tag{9.37}$$

where we introduced the function Ξ, which has the low-temperature expansion

$$\Xi(z, \zeta) \sum_{n,p=0}^{\infty} z^n\, u^p\, G_V(n, p) \,. \tag{9.38}$$

We assume a positive h such that the fugacity[3] satisfies

$$z = e^{-2\beta h} < 1 \, . \tag{9.39}$$

For a ferromagnetic system on a hypercubic lattice, $|\partial X|$ is an even integer, and thus we choose

$$u = e^{-4\beta J} \ll 1 \tag{9.40}$$

as small expansion parameter at low temperatures, such that the order of the series expansion is the power $p = |\partial X|/2$ of u. The combinatorial factor $G_V(n, p)$ is just the number of subsets $X \subset \Lambda$ with volume $|X| = n$ and surface area $|\partial X| = 2p$.

The sets X and X' corresponding to the configurations ω and $-\omega$ have different statistical weights, namely,

$$X : \ z^{|X|} u^{|\partial X|/2} \quad \text{and} \quad X' : \ z^{V-|X|} u^{|\partial X|/2} \, ,$$

such that in the thermodynamic limit and $h > 0$, the statistical weight of X' vanishes relative to that of X. This is why the low-temperature system shows a spontaneous magnetization for $h \neq 0$ and infinite volume. However, without magnetic field $E(X) = E(X')$ and the configurations ω and $-\omega$ have identical weights. Thus, in a finite volume, there is no magnetization for $h = 0$.

Next we must enumerate the low-temperature diagrams to obtain the combinatorial factors G_V in Eq. (9.38). Table 9.4 shows all diagrams on the square to order $p = 5$ in the expansion parameter u. Observe the close relationship with the high-temperature expansion in Table 9.1. This relation is a two-dimensional peculiarity and originates from the self-duality of the two-dimensional Ising model. From

$$\Xi = 1 + Vu^2 z + 2Vu^3 z^2 + Vu^4 \left(z^4 + 6z^3 + (V - 5)z^2/2 \right)$$
$$+ Vu^5 \left(2z^6 + 8z^5 + 18z^4 + 2(V - 8)z^3 \right) + \dots \tag{9.41}$$

follows the low-temperature expansion of the temperature-dependent contribution to the free energy density:

$$-\beta \Delta f = \frac{\log \Xi}{V} = zu^2 + 2z^2 u^3 + \left(z^4 + 6z^3 - 5z^2/2 \right) u^4$$
$$+ \left(2z^6 + 8z^5 + 18z^4 - 16z^3 \right) u^5 + \dots \tag{9.42}$$

---

[3] It is the inverse of the fugacity used in Sect. 8.7.

Table 9.4 Leading diagrams contributing to the low-temperature expansion of Ξ

p	n	diagram	number	p	n	diagram	number
2	1		V	5	3		$2V(V-8)$
3	2		$2V$	5	4		$2V$
4	2		$\frac{1}{2}V(V-5)$	5	4		$8V$
4	3		$2V$	5	4		$4V$
4	3		$4V$	5	4		$4V$
4	4		V	5	5		$8V$
				5	6		$2V$

The field dependence of the magnetization m for $h > 0$ is given by

$$m = -\frac{\partial f}{\partial h} = 1 + 2z\frac{\partial}{\partial z}(\beta \Delta f) = 1 - 2zu^2 - 8z^2 u^3 - \left(8z^4 + 36z^3 - 10z^2\right)u^4$$

$$- \left(24z^6 + 80z^5 + 144z^4 - 96z^3\right)u^5 + \dots \quad (9.43)$$

Now we switch off the h-field and remain with the spontaneous magnetization

$$m = \sum m_\ell u^\ell = 1 - 2u^2 - 8u^3 - 34u^4 - 152u^5 - \dots \quad (9.44)$$

In the zero-temperature limit, all spins are aligned in direction defined by the magnetic field before it was switched off. The systems show a spontaneous magnetization. In [19] the finite lattice method has been applied to obtain the low-temperature series for the partition function, order parameter, and susceptibility of the Ising model on the square lattice. In particular the expansion for the magnetization was calculated to order 38. Half of the known coefficients are listed in Table 9.5.

Figure 9.12 shows a plot of the spontaneous magnetization $m = \sum m_\ell u^\ell$ with the coefficients in Table 9.5. The magnetization vanishes for values u larger than the critical value of u_c.

Table 9.5 Coefficients m_ℓ and ratios of coefficients for the $2d$ Ising model [19]

ℓ	m_ℓ	$m_\ell/m_{\ell-1}$	ℓ	m_ℓ	$m_\ell/m_{\ell-1}$
0	1		10	−454378	5.16056
1	0		11	−2373048	5.22263
2	−2		12	−12515634	5.27408
3	−8	4.00000	13	−66551016	5.31743
4	−34	4.25000	14	−356345666	5.35447
5	−152	4.47059	15	−1919453984	5.38649
6	−714	4.69737	16	−10392792766	5.41445
7	−3472	4.86275	17	−56527200992	5.46093
8	−17318	4.98790	18	−308691183938	5.48046
9	−88048	5.08419	19	−1691769619240	

Fig. 9.12 The magnetization as a function of $u = e^{-4\beta J}$ in the low-temperature expansion

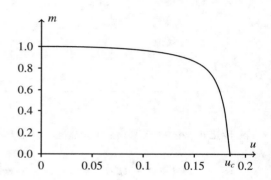

Extrapolation to the Critical Point

The low-temperature series for m has only negative coefficients m_ℓ except for m_0. Hence, if the series has a finite radius of convergence, then a singularity of $m(u)$ lies on the positive real axis at some value u_c. As earlier on we fit the low-temperature expansion to the expected scaling,

$$m = \sum m_\ell\, u^\ell \sim \left(1 - \frac{u}{u_c}\right)^\beta , \tag{9.45}$$

and analyze the series with the help of the ratio test:

$$\frac{m_\ell}{m_{\ell-1}} = \frac{1}{u_c} - \frac{1+\beta}{u_c}\frac{1}{\ell} . \tag{9.46}$$

Figure 9.13 shows the ratios of successive coefficients for the orders given in [19] and the corresponding linear fit for $\ell \geq 4$. The slope of the fitting function is approximately -6.58598, and the intersection with the ordinate is given by

Fig. 9.13 Extrapolation for the determination of the critical temperature and the critical exponent β in the low-temperature approximation

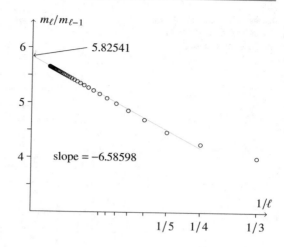

$1/u_c \approx 5.82541$. This yields the estimates

$$T_c = -\frac{4J}{\log u_c} = 2.26985\, J \quad \text{and} \quad \beta = 0.13056 \qquad (9.47)$$

for the critical temperature and critical exponent. This critical temperature is very close to the exact value in (9.23), and the exponent is near the exact value $\beta = 1/8$.

9.3.2 Three-Dimensional Model

While the high-temperature series are well-behaved, the situation at low temperatures is less satisfactory, in particular above two dimensions. With finite lattice methods, it is possible to calculate low-temperature expansions on the simple-cubic lattice in powers of the small parameter $u = e^{-4J/T}$. For example, the magnetization has been calculated to order 20 in [20, 21] and later on to order 26 in [22]. Using a modification of the shadow-lattice techniques, Vohwinkel obtained low-temperature series for the free energy, magnetization, and susceptibility [23]. The magnetization was calculated to order 32 and the corresponding coefficients m_ℓ are listed in Table 9.6. Note that in three dimensions the coefficients have alternating signs and we expect the first singularity of $m(u)$ on the negative real axis. The ratio test shows that the first singularity for the series occurs near $u \approx -0.3$. This unphysical singularity makes it difficult to apply the ratio method, and thus we use Padé approximants to analyze the low-temperature series. The $[p, q]_f$ Padé approximant of a function $f(u)$ is the ratio of a polynomial $p(u)$ of degree p and a polynomial $q(u)$ of degree q such that the series expansion of $p(u)/q(u)$ agrees with the series expansion of $f(u)$ through order $p + q$ [24]. Figure 9.14 shows the low-temperature expansion of the magnetization which becomes singular at $u \approx \pm 0.3$ together with its [5, 5] Padé approximant.

Table 9.6 Coefficients m_ℓ
and ratios of coefficients for
the $3d$ Ising model [23]

ℓ	m_ℓ	ℓ	m_ℓ
0	1	18	30371124
3	−2	19	−101585544
4	0	20	338095596
5	−12	21	−1133491188
6	14	22	3794908752
7	−90	23	−12758932158
8	192	24	42903505030
9	−792	25	−144655483440
10	2148	26	488092130664
11	−7716	27	−1650000819068
12	2326	28	5583090702798
13	−79512	29	−18918470423736
14	252054	30	64167341172984
15	−846628	31	−217893807812346
16	2753520	32	740578734923544
17	−9205800		

Fig. 9.14 Plot of the
low-temperature series of
order 32 and its [5, 5] Padé
approximant

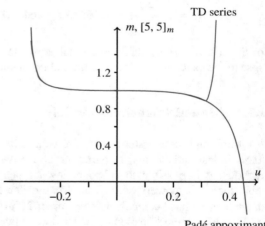

TD series

$m, [5, 5]_m$

Padé appoximant

To extract the critical point u_c and the critical exponent β, we analyze the Padé
approximants to the low-temperature series of $f(u) = m(u)/m'(u)$, which near the
critical point should behave like $(u - u_c)/\beta$, such that the critical coupling u_c is
identified with the first zero of the Padé approximant on the positive real axis, and
the inverse critical exponent is equal to the slope at the critical value u_c. The critical
values for several Padé approximants are listed in Table 9.7. The prediction for u_c
of the approximant [5, 5] comes closest to the precise number $u_c = 0.41205$ from
simulations, and for this best fit, we read off a critical exponent $\beta = 0.3063$.

Table 9.7 Critical coupling u_c and critical exponent β from various Padé approximants

$[r, s] =$	$[3, 3]$	$[4, 4]$	$[5, 5]$	$[6, 6]$	$[7, 7]$	Averages
$u_c =$	0.4177	0.4033	0.4124	0.4031	0.4098	0.4093
$\beta =$	0.3204	0.2595	0.3063	0.2586	0.2848	0.2859

Table 9.8 Critical parameters of the Ising model on a square lattice from Padé approximants

$[r, s] =$	$[2, 2]$	$[3, 3]$	$[4, 4]$	$[5, 5]$	$[6, 6]$	Exact
$u_c =$	0.166666	0.171573	0.171573	0.171573	0.171573	0.171573...
$T_c/J =$	2.232443	2.269186	2.269185	2.269185	2.269185	2.269185...
$\beta =$	0.111111	0.125000	0.125000	0.125000	0.125000	0.125

Table 9.9 Critical parameters for $3d$ Ising model from high-temperature expansions

β_c	γ	ν	η	α	From
0.22165459(10)	1.2373(2)	0.63012(16)	0.03639(15)		[29]
0.2216545(1)	1.2369(2)	0.6298(3)		0.1035(5)	[30]
0.221655(2)	1.2371(1)	0.6299(2)	0.0360(8)		[31]

9.3.3   Improved Series Studies for Ising-Type Models

Typically the Padé approximant method yields more accurate results for the critical parameters as the ratio method. For example, for the $2d$ Ising model on the square lattice, the lowest Padé approximant to $m(u)/m'(u)$, where $m(u) = \sum m_\ell u^\ell$ is the low-temperature expansion up to order 38, yields the critical parameters listed in Table 9.8. Already the fourth-order approximant agrees with the exact result in all digits given in the table. To achieve a further improvement in the precision of the estimates of the critical parameters from the analysis of extended high- or low-temperature series, one should properly allow for the expected nonanalytic correction to the leading power law behavior or thermodynamic quantities near a critical point. A singular quantity is expected to behave, in the vicinity of the critical point β_c as

$$g(\beta) \sim A_\gamma |t|^\lambda \left(1 + a_\chi |t|^{\lambda_1} + a'_\chi |t|^{2\lambda_1} + \cdots + e_\chi t + e'_\chi t^2 + \ldots \right) \qquad (9.48)$$

when $t = 1 - \beta/\beta_c \to 0$; see [25]. The critical exponent λ and the leading confluent correction exponents λ_1 are universal. The established ratio extrapolation and Padé approximant methods are generally inadequate to solve the numerical problem of extracting β_c, the critical exponent, and the leading confluent exponent. Instead one may use the differential approximant method put forward in [26]. With such improvements it is possible to find rather precise values for critical exponents for a large class of spin models in various dimensions; for a detailed discussion, see the textbook [3]. In Table 9.9 we compiled results from high-temperature expansions for the *Ising model on a simple-cubic lattice*. From scaling relations between critical exponents, one finds $\beta \approx 0.3265$. In [27] the low-temperature series for the partition function, order parameter, and susceptibility of the q-state *Potts model on the square lattice* to high order in u for $q = 2, 3, 4, \ldots, 10$ are given. The online library [9] contains high-temperature expansions of basic observables in two- and three-dimensional Ising models with spins $1/2, 1, \ldots, 5, \infty$. Besides the critical exponents, there are further universal quantities, namely, the universal amplitude ratios. Critical exponents and universal amplitude ratios for many interesting spin models are compiled in the review [28].

9.4   High-Temperature Expansions of Nonlinear O(N) Models

For $N > 1$ the high-temperature series for the nonlinear O(N) sigma model with energy function (6.18) are significantly less known as those for the Ising model. For $N = 0$ (the self-avoiding walk model), the susceptibility has been calculated to order β^{23} on the simple-cubic lattice in [32]. With the linked cluster expansion technique, the high-temperature series has been extended to order β^{23} for the nonlinear models with $N \leq 12$ in [33, 34].

Actually, the linked cluster expansion technique can be adapted to produce expansions for the general class of models with partition functions

$$Z = \int d\mu(\omega)\, e^{-\beta H + \sum h_x s_x}, \quad \text{where} \quad H = -\sum_{\langle x, y \rangle} s_x s_y, \tag{9.49}$$

contains nearest-neighbor interactions between the spins $s_x \in \mathbb{R}^N$. The $h_x \in \mathbb{R}^N$ represent an external field, and $d\mu$ is the product of O(N)-invariant single-spin measures:

$$d\mu(\omega) = \prod_x d\mu(s_x), \qquad d\mu(Rs) = d\mu(s). \tag{9.50}$$

We absorbed the nearest-neighbor coupling J in the inverse temperature β. For small β and without external field, we obtain the expansion

$$Z = 1 + \sum_{\ell=1}^{\infty} \frac{\beta^{2\ell}}{(2\ell)!} \int d\mu(\omega)\, H^{2\ell}, \tag{9.51}$$

with only even powers of H, since $d\mu$ is invariant under rotations of the individual spins. For the product measure $d\mu$, the averages on the right-hand side are expressed in terms of moments of the single-spin distribution $d\mu(s)$, and these moments are generated by $z(h) = \int d\mu(s) \exp(hs)$. For O(N)-invariant systems, the generating function depends only on the modulus of h such that for a vanishing field, the moments are totally symmetric O(N)-invariant tensors, e.g.,

$$\int d\mu(s)\, s_a s_b = \mathscr{C}_2 \delta_{ab} \equiv \mathscr{C}_2 C_{ab}$$

$$\int d\mu(s)\, s_a s_b s_c s_d = \mathscr{C}_4 \left(\delta_{ab}\delta_{cd} + \delta_{ac}\delta_{bd} + \delta_{ad}\delta_{bc}\right) \equiv \mathscr{C}_4 C_{abcd} . \qquad (9.52)$$

The totally symmetric tensor

$$C_{a_1 \dots a_{2\ell}} = \delta_{a_1 a_2}\delta_{a_3 a_4} \cdots \delta_{a_{2k-1} a_{2\ell}} + \dots \qquad (9.53)$$

contains $(2\ell - 1)!!$ terms corresponding to the possible Wick contractions. For the *Gaussian model* with normalized single-spin distribution

$$d\mu(s) = \left(\frac{\alpha}{2\pi}\right)^{N/2} e^{-\alpha s^2/2}, \quad s \in \mathbb{R}^N \qquad (9.54)$$

the coefficients of the invariant tensors are

$$\mathscr{C}_{2\ell} = \frac{1}{\alpha^\ell} . \qquad (9.55)$$

For the *nonlinear O(N) sigma model* with single spins randomly distributed on the unit sphere, the normalized single-spin measure is proportional to $\delta(s^2-1)\, d^N s$. The generating function in (7.122), normalized to $z(0) = 1$, has the Taylor expansion

$$z(h) = \sum_{\ell=0}^{\infty} \frac{\Gamma(N/2)}{\Gamma(\ell + N/2)} \frac{1}{\ell!} \left(\frac{h^2}{4}\right)^\ell , \qquad (9.56)$$

from which one extracts the coefficients

$$\mathscr{C}_{2\ell} = \frac{\Gamma(N/2)}{2^\ell \Gamma(N/2 + \ell)} = \frac{1}{N(N+2) \cdots (N + 2\ell - 2)} . \qquad (9.57)$$

9.4.1   Expansions of Partition Function and Free Energy

The term of order $\beta^{2\ell}$ in (9.51) contains sums of products of 2ℓ spins. Thus the high-temperature expansion involves products of links $s_x s_y$, and because for each site each s_x must appear an even number of times, one generates closed polygons. Since the spins also carry an internal index, we must attach an internal index to each link. A link may now be chosen several times. More precisely, a link connecting two nearest neighbors ℓ-times represents the average of $(s_x s_y)^\ell$, where x, y are the endpoints of the link. If 2ℓ links with indices $a_1, \dots, a_{2\ell}$ end at a given site (a vertex of the graph), we assign the factor $\mathscr{C}_{2\ell} C_{a_1 \dots a_{2\ell}}$ to the vertex. In addition $s_x s_y = s_x^a s_y^a$

involves a contraction over internal indices, and hence we must finally contract all internal indices occurring in the graph. For example, to $(s_x s_y)^2 (s_y s_u)^2$ we assign a graph with two lines from x to its neighbor y and two lines from y to its neighbor u. Since indices of nearest-neighbor pairs are contracted, the average yields

$$C_{ab} C_{abcd} C_{cd} = (N^2 + 2N) \mathcal{C}_2^2 \mathcal{C}_4 .$$

Below we need the following contractions of invariant tensors:

$$C_{ab} C_{ab} = N$$
$$C_{abcd} C_{abcd} = 3(N^2 + 2N)$$
$$C_{abcdef} C_{abcdef} = 15(N^3 + 6N^2 + 8N)$$
$$C_{abcd} C_{bcde} = 3(N+2) C_{ae}$$
$$C_{abcd} C_{cd} = (N+2) C_{ab} \tag{9.58}$$
$$C_{abcdef} C_{ef} = (N+4) C_{abcd}$$
$$C_{abcdef} C_{cdef} = 3(N+2)(N+4) C_{ab}$$
$$C_{ab} C_{abcdef} C_{cdef} = 3N(N+2)(N+4) .$$

Now we are ready to calculate the partition functions of O(N) models to order β^6. The emerging diagrammatic expansion is also very useful to obtain the high-temperature expansions for other objects of interest, for example, the susceptibility.

Order β^2

The second-order contribution to Z is proportional to $\langle H^2 \rangle$. Setting $\sigma_b = s_x s_y$ for a bond b with endpoints x, y, we obtain

$$\langle H^2 \rangle = \sum_b \langle \sigma_b^2 \rangle = \sum_c C_{ac} C_{ac} \mathcal{C}_2^2 = dVN \mathcal{C}_2^2 , \tag{9.59}$$

where dV comes from the sum over all links and is the number of associated graphs

$$x \bullet\!\!=\!\!=\!\!\bullet y$$

on the lattice.

Order β^4

The function H^4 is a sum of all terms $\sigma_{b_1} \sigma_{b_2} \sigma_{b_3} \sigma_{b_4}$ associated with four bonds. Only graphs with even vertices contribute and these graphs are depicted in Fig. 9.15. The number of graphs of a given type is listed below the graph. In the contribution

$$\langle H^4 \rangle = \sum \langle \sigma_b^4 \rangle + 6 \sum \langle \sigma_{b_1}^2 \, \sigma_{b_2}^2 \rangle + 24 \sum \langle \sigma_{b_1} \sigma_{b_2} \sigma_{b_3} \sigma_{b_4} \rangle , \tag{9.60}$$

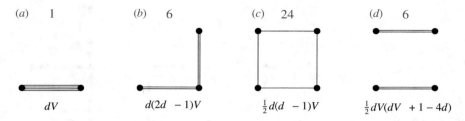

Fig. 9.15 Diagram contributing in order β^4

one only sums over distinct links. All terms in the second sum belonging to disconnected links are represented by the disconnected graph (d) Fig. 9.15. The four links in the last sum in (9.60) define the loop graph (c). Thus every term of $\langle H^4 \rangle$ is represented by a graph. The number related to a graph is the product of its multiplicity and the combinatorial factor in (9.60), and both are given in Fig. 9.15. Therefore we find

$$\langle H^4 \rangle = dV C_{abcd} C_{abcd} \, \mathscr{C}_4^2$$
$$+ 6dV(2d-1)C_{abcd}C_{ab}C_{cd} \, \mathscr{C}_4 \mathscr{C}_2^2 \qquad (9.61)$$
$$+ 12dV(d-1)C_{ab}C_{bc}C_{cd}C_{da} \, \mathscr{C}_2^4$$
$$+ 3dV(dV-4d+1)C_{ab}C_{ab}C_{cd}C_{cd} \, \mathscr{C}_2^4 \ .$$

Contracting the invariant tensors according to (9.58) leads to

$$\langle H^4 \rangle = 3dVN \left((N+2)C_4^2 + 2(2d-1)(N+2)\mathscr{C}_4\mathscr{C}_2^2 \right.$$
$$\left. +4(d-1)\mathscr{C}_2^4 + (dV-4d+1)N\mathscr{C}_2^4 \right) \ . \qquad (9.62)$$

Order β^6
The function H^6 is a sum of all terms $\sigma_{b_1}\sigma_{b_2}\sigma_{b_3}\sigma_{b_4}\sigma_{b_5}\sigma_{b_6}$ associated with six links. The following terms contribute to the partition function:

$$\langle H^6 \rangle = \sum \langle \sigma_b^6 \rangle + 15 \sum \langle \sigma_{b_1}^4 \sigma_{b_2}^2 \rangle + 90 \sum \langle \sigma_{b_1}^2 \sigma_{b_2}^2 \sigma_{b_3}^2 \rangle$$
$$+360 \sum \langle \sigma_{b_1}^2 \sigma_{b_2}\sigma_{b_3}\sigma_{b_4}\sigma_{b_5} \rangle + 120 \sum \langle \sigma_{b_1}^3 \sigma_{b_2}\sigma_{b_3}\sigma_{b_4} \rangle \qquad (9.63)$$
$$+720 \sum \langle \sigma_{b_1}\sigma_{b_2}\sigma_{b_3}\sigma_{b_4}\sigma_{b_5}\sigma_{b_6} \rangle \ .$$

The first and last sums are represented by the connected diagrams a, f and f' in Fig. 9.16. The remaining sums are represented by connected and disconnected diagrams. The disconnected ones are depicted in Fig. 9.17. The analytical expression

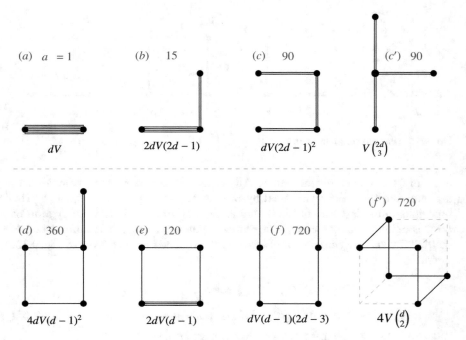

Fig. 9.16 All connected diagrams contributing to $\langle H^6 \rangle$. The numbers below the graphs count the number of diagrams. The combinatorial factors in (9.63) are also listed

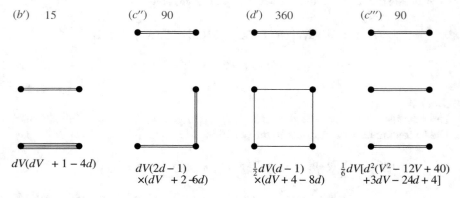

Fig. 9.17 Disconnected diagrams contributing to $\langle H^6 \rangle$. The numbers below the graphs count the number of diagrams

for $\langle H^6 \rangle$ in terms of the invariant tensors, coefficients $\mathscr{C}_\ell$, and dimension of space d is now easily found:

$$\langle H^6 \rangle = dV\, C_{abcdef} C_{abcdef}\, \mathscr{C}_6^2$$
$$+ 30dV(2d-1)C_{abcdef}C_{abcd}C_{ef}\, \mathscr{C}_6\mathscr{C}_4\mathscr{C}_2$$
$$+ 90dV(2d-1)^2 C_{ab}C_{abcd}C_{cdef}C_{ef}\, \mathscr{C}_4^2\mathscr{C}_2^2$$
$$+ 60dV(2d-1)(d-1)C_{ab}C_{abcdef}C_{cd}C_{ef}\, \mathscr{C}_6\mathscr{C}_2^3$$
$$+ 1440dV(d-1)^2 C_{ab}C_{abcd}C_{ce}C_{ef}C_{fd}\, \mathscr{C}_4\mathscr{C}_2^4$$
$$+ 240dV(d-1)C_{abcd}C_{bcde}C_{ef}C_{fa}\, \mathscr{C}_4^2\mathscr{C}_2^2$$
$$+ 720dV(d-1)(2d-3)C_{ab}C_{bc}C_{cd}C_{de}C_{ef}C_{fa}\, \mathscr{C}_2^6$$
$$+ 480dV(d-1)(d-2)C_{ab}C_{bc}C_{cd}C_{de}C_{ef}C_{fa}\, \mathscr{C}_2^6$$
$$+ 15dV(dV+1-4d)C_{abcd}C_{abcd}C_{ef}C_{ef}\, \mathscr{C}_4^2\mathscr{C}_2^2$$
$$+ 90dV(2d-1)(dV+2-6d)C_{ab}C_{abcd}C_{cd}C_{ef}C_{ef}\, \mathscr{C}_4\mathscr{C}_2^4$$
$$+ 180dV(d-1)(dV+4-8d)C_{ab}C_{bc}C_{cd}C_{da}C_{ef}C_{ef}\, \mathscr{C}_2^6$$
$$+ 15dV(d^2(V^2-12V+40)+3dV-24d+4))(C_{ab}C_{ab})^3\, \mathscr{C}_2^6 \,.$$

After contracting the invariant tensors, this becomes

$$\langle H^6 \rangle = 15dVn(N+2)(N+4)\, \mathscr{C}_6^2$$
$$+ 90dV(2d-1)N(N+2)(N+4)\, \mathscr{C}_6\mathscr{C}_4\mathscr{C}_2$$
$$+ 90dV(2d-1)^2 N(N+2)^2\, \mathscr{C}_4^2\mathscr{C}_2^2$$
$$+ 60dV(2d-1)(d-1)N(N+2)(N+4)\, \mathscr{C}_6\mathscr{C}_2^3$$
$$+ 1440dV(d-1)^2 N(N+2)\, \mathscr{C}_4\mathscr{C}_2^4$$
$$+ 720dV(d-1)N(N+2)\, \mathscr{C}_4^2\mathscr{C}_2^2$$
$$+ 720dV(d-1)(2d-3)N\, \mathscr{C}_2^6$$
$$+ 480dV(d-1)(d-2)N\, \mathscr{C}_2^6$$
$$+ 45dV(dV+1-4d)N^2(N+2)\, \mathscr{C}_4^2\mathscr{C}_2^2$$
$$+ 90dV(2d-1)(dV+2-6d)N^2(N+2)\, \mathscr{C}_4\mathscr{C}_2^4$$
$$+ 180dV(d-1)(dV+4-8d)N^2\, \mathscr{C}_2^6$$
$$+ 15dV(d^2(V^2-12V+40)+3dV-24d+4)N^3\, \mathscr{C}_2^6 \,.$$

The disconnected graphs cancel in the free energy density

$$-\beta V f_\beta = \frac{\beta^2}{2} \langle H^2 \rangle + \frac{\beta^4}{4!} \left(\langle H^4 \rangle - 3\langle H^2 \rangle^2 \right)$$

$$+ \frac{\beta^6}{6!} \left(\langle H^6 \rangle - 15\langle H^4 \rangle \langle H^2 \rangle + 30\langle H^2 \rangle^3 \right) + \cdots .$$

In particular for the nonlinear O(N) sigma model in d dimensions, we find

$$-\beta f(\beta) = \frac{d\beta^2}{2N} + \frac{d\beta^4}{4N^3} \frac{2dN + 4d - 3N - 4}{N + 2} \tag{9.64}$$

$$+ \frac{d\beta^6}{3N^5} \left(8d^2 - \frac{3d(9N^2 + 50N + 56) - 2(10N^2 + 51N + 52)}{(N + 2)(N + 4)} \right) ,$$

and for $N = 1$ we recover the leading order contributions to the Ising model free energy density in dimension d. The internal energy densities for the models in two and three dimensions have the expansions

$$u_{d=2} = \frac{2\beta}{N} + \frac{2\beta^3}{N^3} \frac{N + 4}{N + 2} - \frac{8\beta^5}{N^5} \frac{N^2 + 3N - 12}{(N + 2)(N + 4)} + \cdots$$

$$u_{d=3} = \frac{3\beta}{N} + \frac{3\beta^3}{N^3} \frac{3N + 8}{N + 2} + \frac{6\beta^5}{N^5} \frac{11N^2 + 84N + 176}{(N + 2)(N + 4)} + \cdots . \tag{9.65}$$

When one tries to calculate higher-order contributions the way we did, then the process becomes very laborious after the first few terms in the high-temperature expansion. One runs into the problem of polygon counting on the lattice. Such polygons can be generated by an n-step random walk on the lattice. The major difficulty is to ensure that no diagrams have been overlooked at each stage.

It is advantageous to consider intensive quantities, for example, the free energy density or the susceptibility, to which only connected diagrams contribute. The linked cluster expansion is a systematic method to construct all connected diagrams [33]. Each term in the expansion is represented by a graph consisting of vertices $v \in \mathcal{V}$ and lines joining them. There are internal lines $\ell \in \mathcal{L}$ and external lines (the graphs belonging to the free energy have no external lines). Lines with only one terminal point are external. We denote the number of external lines attached to a vertex v by $E(v)$. All lines ℓ have an initial point $i(\ell)$ and a final point $f(\ell)$ and the two endpoints are different. For the O(N) models, the order of all vertices, i.e., the total number of lines (internal and external) ending at v, must be even.

We associate four numbers to each graph $G \in \mathcal{G}$. These are the (topological) symmetry number $S(G)$, the lattice embedding number $I(G)$, the O(N) symmetry factor $C(G)$, and the weight $\mathring{W}(G)$. First one tabulates all graphs of a particular topological class which contribute to a given order. Figure 9.16 contains the six classes:

$$\{a\}, \quad \{b\}, \quad \{c, c'\}, \quad \{d\}, \quad \{e\} \quad \text{and} \quad \{f, f'\}.$$

In [33] the following linked cluster expansion for the coefficient in the high-T expansion of the *susceptibility* is given

$$\chi_{2k} = \sum_{G \in \mathscr{G}_{2k}} (2\kappa)^L I(G) C(G) \mathring{W}(G) \frac{E!}{S_E(G)}, \tag{9.66}$$

where the sum extends over all topologically inequivalent graphs with k internal lines. The symmetry number $S(G)$ is given by the incidence matrix associated with the graph. Although we shall not go further into algorithmic considerations, we shall use the results obtained by the linked cluster expansion in combination with efficient computer programs. We shall only consider simple hypercubic lattices.

Butera and Comi computed through order β^{21} the high-temperature expansions for the internal energy density of nonlinear O(N) models on hypercubic lattices for arbitrary N [35]. Thus they added *eight* more terms to our result in (9.65). For the ratio test, we computed the ratios of coefficients in the associated expansion (9.26) of the specific heat $c = \partial u / \partial T$. They are listed in Table 9.10. For large ℓ the ratios depend linearly on $1/\ell$ and this is clearly visible in Fig. 9.18. With the linear extrapolation (9.27), one obtains already reasonable accurate values for the critical temperatures, given in Table 9.11. For example, one could compare to the accurate value $\beta_c = 0.6862385(20)$ for the $N = 3$ model; see [36].

Unfortunately the series for the internal energy are still too short to extract accurate values for the critical exponent α. In [34] the high-temperature expansions for the susceptibility and the second correlation length

$$\mu_2(\beta) = \sum_x x^2 \langle s_0 s_x \rangle = \sum s_\ell(N) \beta^\ell \tag{9.67}$$

Table 9.10 Ratios $r_\ell = \alpha_{2\ell}/\alpha_{2\ell-2}$ for various O(N) models in three dimensions

ℓ	2	3	4	5	6
$N = 2$	2.62500	3.84921	3.45242	3.86091	4.03943
$N = 3$	1.13333	1.64021	1.43292	1.61356	1.69459
$N = 4$	0.62500	0.89583	0.76566	0.86438	0.90978
$N = 5$	0.394286	0.561031	0.470829	0.53166	0.56018
ℓ	7	8	9	10	11
$N = 2$	4.14362	4.22924	4.29748	4.35237	4.39740
$N = 3$	1.74202	1.78203	1.81484	1.84154	1.86350
$N = 4$	0.93656	0.95973	0.97927	0.99541	1.00874
$N = 5$	0.57719	0.59223	0.60525	0.61619	0.62528

Fig. 9.18 The ratios of coefficients $r_\ell = \alpha_{2\ell}/\alpha_{2\ell-2}$ in the high-temperature expansion of the specific heat for the $O(3)$ and $O(4)$ models in three dimensions

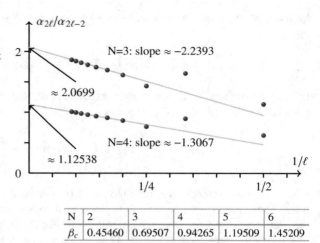

Table 9.11 Critical temperature for O(N) models in three dimensions

N	2	3	4	5	6
β_c	0.45460	0.69507	0.94265	1.19509	1.45209

have been extended to order β^{21} with the help of the (vertex renormalized) linked cluster expansion [33]. The corresponding estimates for the critical point and critical exponents γ and ν are listed in Table 9.12 . The critical exponent α_{hte} is calculated with the hyperscaling relation. Also listed in the table are the critical exponents from strong-coupling expansions and six-loop approximations, improved by the Padé-Borel resummation technique. A comparison of the numbers may indicate how uncertain they are. For small N there are precise data from lattice simulations or other methods; see the references on page 229.

9.5   Polymers and Self-Avoiding Walks

Polymers are long molecules composed of many, say n, monomers. When the interaction between the monomers is negligible, the geometric configuration is similar to a Brownian chain of n successive independent steps, made at random. This is the analogue of a random walk on a lattice. If the monomers are interacting, the problem is more difficult. If they repel, then the chain is more extended than the Brownian chain. The simplest realization for a repulsion is to imagine a random walk on a lattice in which it is forbidden to visit again a previously visited site: this is called a self-avoiding walk. The resulting model is non-Markovian since the nth step depends on the whole past of the chain. It is expected that the typical size of the chain will be larger as a function of n than that of the Brownian chain, because of this geometric repulsion.

It was shown by P. de Gennes that the problem of self-avoiding walks on the lattice may be mapped onto a singular $N = 0$ limit of the O(N) model. To perform this limit, we choose a different normalization for the spins, namely,

$$s_x s_x = N, \quad \text{for all} \quad x \in \Lambda . \tag{9.68}$$

Table 9.12 Critical temperatures, susceptibility exponent, and heat capacity exponent of $O(N)$ models on simple-cubic lattices from high-temperature expansions (hte). Also listed are the exponents obtained from strong-coupling expansions (sc), from six-loop approximations combined with the Pade-Borel resummation technique (six-loop), from renormalization group equation with full-momentum dependence of correlation function (rge), and from lattice simulations (mc)

N	1	2	3	4	6	8	10	12	From
$\beta_{hte,c}$	0.2217	0.4542	0.6930	0.9359	1.4286	1.9263	2.4267	2.929	[34]
$\beta_{mf,c}$	0.1667	0.3333	0.5000	0.6667	1.0000	1.3333	1.6667	2.000	MF
γ_{hte}	1.244	1.327	1.404	1.474	1.582	1.656	1.712	1.759	[34]
γ_{sc}	1.241	1.318	1.387	1.451	1.558	1.638	1.725	1.763	[37]
γ_{6-loop}	1.239	1.315	1.3386	1.449	1.556	1.637	1.697	1.743	[38]
ν_{hte}	0.634	0.677	0.715	0.750	0.804	0.840	0.867	0.889	[34]
ν_{sc}	0.630	0.670	0.705	0.737	0.790	0.829	0.866	0.890	[37]
ν_{6-loop}	0.631	0.670	0.706	0.738	0.790	0.830	0.859	0.881	[38]
ν_{rge}	0.632	0.674	0.715	0.754			0.889		[39]
ν_{mc}	0.630	0.672	0.711	0.749	0.818				[36, 40, 41]
α_{hte}	0.098	−0.031	−0.145	−0.250	−0.412	−0.520	−0.601	−0.667	[36, 40, 41]
α_{sc}	0.107	−0.010	−0.117	−0.213	−0.370	−0.489	−0.576	−0.643	[37]
α_{mc}	0.1101	−0.1336							[36, 40, 41]

As a consequence the coefficients in (9.57) are rescaled,

$$\mathscr{C}_{2\ell} = \frac{N^\ell}{N(N+2)\cdots(N+\ell k - 2)} , \tag{9.69}$$

and have the scaling limit

$$\lim_{N\to 0} \mathscr{C}_{2\ell} = \begin{cases} 1 \text{ for } \ell = 1 \\ 0 \text{ for } \ell > 1 . \end{cases} \tag{9.70}$$

In the previously considered expectation values $\langle H^{2\ell}\rangle$, all totally symmetric O(N)-invariant tensors are contracted. But a contraction of all indices yield at least one power of N. Actually the contraction of all indices in the terms belonging to disconnected graphs yield at least two powers of N. This means that all expectation values $\langle H^{2\ell}\rangle$ vanish and this gives the trivial result $Z = 1$. However, let us go a bit further and consider the high-temperature expansion of the spin-spin correlation function

$$G^{1,1}(x) = \langle s_0^1 s_x^1 \rangle_{N\to 0} = \lim_{N\to 0} \frac{1}{Z} \sum_{\ell=0}^{\infty} \frac{\beta^\ell}{\ell!} \int d\mu(\omega)\, s_0^1 s_x^1\, H^\ell , \tag{9.71}$$

where x is any site on the lattice. In the limit $N \to 0$, all terms containing coefficients $\mathscr{C}_4, \mathscr{C}_6, \mathscr{C}_8, \ldots$ vanish. This means that graphs with vertices that have more than two lines attached, or equivalently with several lines between nearest neighbors, do not contribute. All graphs must be connected since disconnected graphs contain at least one full contraction of indices and hence are suppressed by at least one power of N. We conclude that only nonintersecting connected graphs contribute which connect the lattice points 0 and x. Two examples are shown in Fig. 9.19. Since $\langle s^a s^b \rangle \propto \delta_{ab}$, the internal index is preserved along the line connecting 0 and x, and hence all spins on the line have the same index as the spins at the endpoints. The contribution of a graph is equal to $\ell! \mathscr{C}_2^\ell$, where 0 and x are connected by a line with length ℓ. Thus we find

$$G^{1,1}(x) = \sum_\ell c_\ell(x)\, \beta^\ell \tag{9.72}$$

where $c_\ell(x)$ is the number of self-avoiding random walks from $0 \to x$ with length ℓ. Averaging over x yields the susceptibility

$$\chi = \sum_\ell c_\ell\, \beta^\ell , \tag{9.73}$$

with coefficient c_ℓ counting the number of random walks of length ℓ beginning at the origin. The singularity of χ at the critical temperature is related to the behavior

Fig. 9.19 Only graphs
contribute in which the two
points 0 and x are connected
by a nonintersecting line on
the lattice

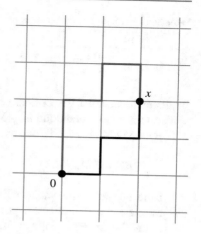

of the coefficients c_ℓ for large ℓ. With the help of Stirling's approximation for large factorials, one estimates the coefficients in the expansion (9.20) and finds

$$\chi \sim \sqrt{\frac{\gamma}{2\pi}} \left(\frac{e}{\gamma}\right)^\gamma \sum_\ell \ell^{\gamma-1} \left(\frac{\beta}{\beta_c}\right)^\ell . \tag{9.74}$$

Another quantity of interest is the mean square displacement over all ℓ-step self-avoiding walks

$$R_\ell^2 = \frac{1}{c_\ell} \sum_x x^2 c_\ell(x) . \tag{9.75}$$

Its root is the average size of a walk consisting of ℓ steps. If the *two*-point correlation function of the $N \to 0$ vector model falls off exponentially with a correlation length ξ which diverges near β_c as $(\beta_c - \beta)^{-\nu}$, this is compatible with (9.75) if, for large ℓ, $R_\ell \sim \ell^\nu$. This provides a surprising connection between the problem of self-avoiding walks and properties of N-vector models.

In two dimensions the exact critical temperature on a honeycomb lattice is known, $\beta_c^{-2} = 2 + \sqrt{2}$, and the critical exponents are [42]

$$\gamma = \frac{43}{32} \quad \text{and} \quad \nu = \frac{3}{4} . \tag{9.76}$$

The computation of c_ℓ for not too small ℓ is a formidable computational problem since, according to (9.74), the number of self-avoiding random walks grows as

$$c_\ell \sim \frac{\ell^{\gamma-1}}{\beta_c^\ell} . \tag{9.77}$$

Table 9.13 Critical temperatures for the self-avoiding random walks for $4 \leq d \leq 8$ [43]

d	4	5	6	7	8
β_c	0.147622	0.1131	0.09193	0.07750	0.06703

In three dimensions the critical exponents were calculated with a self-avoiding walk enumeration technique in [43]. From walks with up to 30 steps (e.g., $c_{30} = 270569905525454674614$), one extracts

$$\beta_c \approx 0.2134907, \quad \gamma \approx 1.1567 \quad \text{and} \quad \nu \approx 0.5875 , \tag{9.78}$$

and from self-avoiding walks with 24 or less steps in $d \geq 4$ dimensions, one can estimate the nonuniversal critical temperatures; see Table 9.13. An earlier conjecture for the exponent ν is due to the chemist Flory. The Flory exponents are

$$\nu_{\text{Flory}} = \begin{cases} 3/(2+d) & \text{for } d \leq 4 \\ 1/2 & \text{for } d > 4 . \end{cases} \tag{9.79}$$

This value is correct in $d = 2$ and $d \geq 4$ (apart from logarithmic corrections in four dimensions) and comes very close for $d = 3$, where $\nu = 0.5888$. We expect that in $d \geq 4$ dimensions the second critical exponent is $\gamma = 1$.

9.6   Problems

9.1 (High-Temperature Expansion for the $3d$ Ising Model) Examine the diagrams for the high-temperature expansion of the partition function of the three-dimensional Ising model with Hamiltonian

$$H = -J \sum_{\langle xy \rangle} s_x s_y, \quad s_x, s_y \in \{-1, 1\} .$$

The number of nearest-neighbor pairs is equal to $P = 3V$. You will find the following series expansion:

$$Z = (\cosh K)^{3V} \, 2^V \left(1 + 3Vv^4 + 22Vv^6 + \frac{1}{2} \{9V(V - 1) + 375V\} v^8 + .. \right)$$

with $v = \tanh(\beta J)$. Furthermore, determine the series expansion of $e^{-\beta f}$ (with f denoting the free energy density) to order v^8.

9.2 (Correlation Functions of O(N) Models) Compute, up to order β^4, the correlation function $\langle s(x)s(0) \rangle$ of the three-dimensional O(N) models at high temperature.

References

1. C. Domb, *Phase Transitions and Critical Phenomena*, vol. 3, ed. by C. Domb, M.S. Green (Academic Press, London, 1974)
2. G.A. Baker, *Quantitative Theory of Critical Phenomena* (Academic Press, London, 1990)
3. J. Oitmaa, C. Hamer, W. Zheng, *Series Expansion Methods for Strongly Interacting Lattice Models* (Cambridge University Press, Cambridge, 2010)
4. C. Itzykson, J.M. Drouffe, *Statistical Field Theory*, vol. 2 (Cambridge University Press, Cambridge, 1991)
5. A.J. Berlinsky, A.B. Harris, *Statistical Mechanics*, Graduate Texts in Physics (Springer, 2020)
6. M.F. Sykes, D.S. Gaunt, P.D. Roberts, J.A. Wyles, High temperature series for the susceptibility of the Ising model. I. Two dimensional lattices. J. Phys. **A5**, 624 (1972)
7. B.G. Nickel, J.J. Rehr, High-temperature series for scalar-field lattice models: generation and analysis. J. Stat. Phys. **61**, 1 (1990)
8. W.P. Orrick, B.G. Nickel, A.J. Guttmann, J.H.H. Perk, The susceptibility of the square lattice Ising model: new developments. J. Stat. Phys. **102**, 795 (2001)
9. P. Butera, M. Comi, A library of extended high-temperature expansions of basic observables for the spin S Ising models on two- and three-dimensional lattices. J. Stat. Phys. **109**, 311 (2002)
10. W.J. Camp, J.P. Van Dyke, High-temperature series for the susceptibility of the spin-s Ising model: Analysis of confluent singularities. Phys. Rev. **B11**, 2579 (1975)
11. D.S. Gaunt, M.F. Sykes, The critical exponent γ for the three-dimensional Ising model. J. Phys. **A 12**, L25 (1979)
12. G. Bhanot, M. Creutz, U. Glässner, K. Schilling, Specific heat exponent for the 3-d Ising model from a 24-th order high temperature series. Phys. Rev. **B49**, 12909 (1994)
13. A.J. Guttmann, I.G. Enting, The high-temperature specific heat exponent of the 3-dimensional Ising model. J. Phys. **A27**, 8007 (1994)
14. P. Butera, M. Comi, Extension to order β^{23} of the high-temperature expansions for the spin 1/2 Ising model on the simple-cubic and body-centered-cubic lattices. Phys. Rev. **B62**, 14837 (2000)
15. T. de Neef, I.G. Enting, Series expansions from the finite lattice method. J. Phys. **A10**, 801 (1977)
16. H. Arisue, T. Fujiwara, New algorithm of the finite lattice method for the high-temperature expansion of the Ising model in three dimensions. Phys. Rev. **E67**, 066109 (2003)
17. K. Binder, E. Luijten, Monte Carlo tests of renormalization-group predictions for critical phenomena in Ising models. Phys. Rep. **344**, 179 (2001); M. Hasenbusch, A finite size scaling study of lattice models in the 3d Ising universality class. Phys. Rev. **B82**, 174433 (2010)
18. D. Simmons-Duffin, the Lightcone Bootstrap and the spectrum of the 3d Ising CFT. JHEP **03**, 086 (2017); S. Rychkov, 3D Ising model: a view from the Conformal Bootstrap island. Comptes Rendus Physique **21**, 185 (2020)
19. I.G. Enting, A.J. Guttmann, I. Jensen, Low-temperature series expansions for the spin-1 Ising model. J. Phys. **A27**, 6987 (1994)
20. M.F. Sykes, D.S. Gaunt, J.W. Essam, C.J. Elliot, Derivation of low-temperature expansions for Ising model. VI Three-dimensional lattices-temperature grouping. J. Phys. **A6**, 1507 (1973)
21. G. Bhanot, M. Creutz, J. Lacki, Low temperature expansion for the Ising model. Phys. Rev. Lett. **69**, 1841 (1992)
22. A.J. Guttmann, I.G. Enting, Series studies of the Potts model: I. the simple cubic Ising model. J. Phys. **A26**, 807 (1993)
23. C. Vohwinkel, Yet another way to obtain low temperature expansions for discrete spin systems. Phys. Lett. **B301**, 208 (1993)
24. G.A. Baker, P. Graves-Morris, *Padé Approximants* (Cambridge University Press, Cambridge, 2010)
25. F. Wegner, Corrections to scaling laws. Phys. Rev. **B5**, 4529 (1972)

26. H.D. Hunter, G.A. Baker, Methods of series analysis. III. Integral approximant methods. Phys. Rev. **B19**, 3808 (1979)
27. P. Butera, M. Comi, Series studies of the Potts model. 2. Bulk series for the square lattice. J. Phys. **A27**, 1503 (1994)
28. A. Pelissetto, E. Vicari, Critical phenomena and renormalization-group theory. Phys. Rep. **368**, 549 (2002)
29. M. Campostrini, A. Pelissetto, P. Rossi, E. Vicari, 25th order high temperature expansion results for three-dimensional Ising like systems on the simple cubic lattice. Phys. Rev. **E65**, 066127 (2002)
30. H. Arisue, T. Fujiwara, K. Tabata, Higher orders of the high-temperature expansion for the Ising model in three dimensions. Nucl. Phys. Proc. Suppl. **129**, 774 (2004)
31. P. Butera, M. Comi, Critical universality and hyperscaling revisited for Ising models of general spin using extended high temperature series. Phys. Rev. **B65**, 144431 (2002)
32. D. MacDonald, D.L. Hunter, K. Kelly, N. Jan, Self avoiding walks in two to five dimensions: exact enumerations and series studies. J. Phys. **A25**, 1429 (1992)
33. M. Lüscher, P. Weisz, Application of the linked cluster expansion to the n-component phi^4 theory. Nucl. Phys. **B300**, 325 (1988)
34. P. Butera, M. Comi, N-vector spin models on the sc and the bcc lattices: a study of the critical behavior of the susceptibility and of the correlation length by high temperature series extended to order β^{21}. Phys. Rev. **B56**, 8212 (1997)
35. P. Butera, M. Comi, Critical specific heats of the N-vector spin models on the sc and the bc lattices. Phys. Rev. **B60**, 6749 (1999)
36. M. Campostrini, M. Hasenbusch, A. Pelissetto, P. Rossi and E. Vicari, Critical exponents and equation of state of the three dimensional Heisenberg universality class. Phys. Rev. **B65**, 144520 (2002)
37. H. Kleinert, Strong-coupling behavior of ϕ^4-theories and critical exponents. Phys. Rev. **D57**, 2264 (1998)
38. S.A. Antonenko, A.I. Sokolov, Critical exponents for $3d$ $O(n)$-symmetric models with $n > 3$. Phys. Rev. **E51**, 1894 (1995)
39. F. Benitez, J.P. Blaizot, H. Chaté, B. Delamotte, R. Méndes-Galain, N. Wschebor, Non-perturbative renormalization group preserving full-momentum dependence: implementation and quantitative evaluation. Phys. Rev. **E85**, 026707 (2012)
40. M. Hasenbusch, A finite scaling study of lattice models in the three-dimensional Ising universality class. Phys. Rev. **B82**, 174433 (2010)
41. S. Holtmann, T. Schulze, Critical behavior and scaling functions of the three-dimensional O(6) model. Phys. Rev. **E68**, 036111 (2003)
42. B. Nienhuis, Exact critical exponents of the $O(n)$ models in 2 dimensions. Phys. Rev. Lett. **49**, 1062 (1982)
43. N. Clisby, R. Liang, G. Slade, Self-avoiding walk enumeration via lattice expansion. J. Phys. **A40**, 10973 (2007)

Peierls Argument and Duality Transformations

10

In this chapter we shall present exact results which apply to many lattice models of interest. Even before the exact solution of the two-dimensional Ising model by Onsager, Peierls [1] proved the existence of two ordered phases at low temperatures. His argument can be extended to many other models with discrete target spaces. Here we present Peierls argument for the two- and three-dimensional Ising models.

We continue with the duality transformations which relate two lattice models. A duality transformation maps a system at weak coupling or low temperature into a system at strong coupling or high temperature and thus leads to new insights into the strong-coupling regime of lattice models. *Duality transformations* exist for many Abelian theories even in higher dimensions. In case the two lattice models are identical up to a rescaling of the couplings, we call the transformation *self-dual*. The two-dimensional Ising model without magnetic field is self-dual [2, 3]. For nonself-dual models, the dual theory may be considerably more complex than the original one. For example, the dual of the three-dimensional Ising model is a Z_2 gauge theory on the dual lattice [4]. In this chapter we shall study lattice systems for which the target spaces form Abelian groups. Unfortunately, it is difficult or impossible to find duality transformations for non-Abelian models. A detailed account of duality transformations in field theories and statistical mechanics is given in [5–7].

10.1 Peierls Argument

First we present the beautiful reasoning due to Peierls to prove that at sufficient low temperatures, the two-dimensional Ising model is in an ordered phase. The proof can be extended to other discrete spin models in two or more dimensions, and this will be discussed later in this section.

To begin with, we choose fixed boundary conditions and set all Ising spins at the boundary to one. The choice of nonperiodic boundary conditions will be important at a later stage. Every spin configuration ω is uniquely characterized by a set of

© The Author(s), under exclusive license to Springer Nature Switzerland AG 2021
A. Wipf, *Statistical Approach to Quantum Field Theory*, Lecture Notes
in Physics 992, https://doi.org/10.1007/978-3-030-83263-6_10

Fig. 10.1 The Peierls contours (loops) enclose regions on the lattice where the spins are -1. With the $+$ boundary conditions, the $+$ spins are outside the contours and the $-$ spins are inside

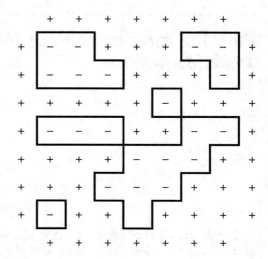

nonintersecting loops on the dual lattice,

$$\Gamma_\omega = \{\gamma_1, \gamma_2, \ldots, \gamma_n\} \, ,$$

where every loop, called Peierls contour, encloses an island with spins down; see Fig. 10.1. If $|\gamma_i|$ denotes the length of γ_i, then a configuration ω has exactly $\sum_i |\gamma_i|$ nearest-neighbor pairs with antiparallel spins. Hence its energy is given by

$$H_\Lambda(\omega) = -J \, \# \, (\text{pairs with equal spins})$$

$$+J \, \# \, (\text{pairs with unequal spins})$$

$$= -JP + 2J \sum_{\gamma_i \in \Gamma_\omega} |\gamma_i| \, ,$$

where P is the total number of nearest-neighbor pairs (cp. with (9.36) for $h = 0$). The constant contribution cancels in expectation values such that we obtain the following probability for the occurrence of a configuration ω (recall that $K = \beta J$):

$$P[\omega] = \frac{1}{Z} \exp\left(-2K \sum_{\gamma_i \in \Gamma_\omega} |\gamma_i|\right), \quad Z = \sum_\Gamma \exp\left(-2K \sum_{\gamma_i \in \Gamma} |\gamma_i|\right). \quad (10.1)$$

Lemma 10.1 (Peierls Inequality) *The probability for the occurrence of a contour γ may be bounded from above as follows:*

$$P[\gamma] \equiv P\left[\{\omega : \gamma \in \Gamma_\omega\}\right] \leq e^{-2K|\gamma|} \, . \quad (10.2)$$

Proof The left-hand side may be written as

$$\frac{1}{Z}\sum_{\omega:\gamma\in\Gamma_\omega}\exp\left(-2K\sum_{\gamma'\in\Gamma_\omega}|\gamma'|\right)=\frac{1}{Z}\,\mathrm{e}^{-2K|\gamma|}\sum_{w:\gamma\in\Gamma_\omega}\exp\left(-2K\sum_{\gamma'\in\Gamma_\omega\setminus\gamma}|\gamma'|\right)$$

$$=\frac{1}{Z}\,\mathrm{e}^{-2K|\gamma|}\sum_{\omega:\gamma\in\Gamma_{P_\gamma\omega}}\exp\left(-2K\sum_{\gamma'\in\Gamma_\omega}|\gamma'|\right),$$

where we have used the fact that if we remove γ from a contour gas Γ_ω with $\gamma\in\Gamma_\omega$, we obtain the contour gas of the configuration $P_\gamma\omega$, where $P_\gamma\omega$ originates from ω by a flip of all spins enclosed by γ. Since the last term represents a summation over a *subset* of all configurations, we have proved the inequality (10.2).

The inequality shows that the probability for the occurrence of a long contour decreases exponentially with its length, independent of the lattice size. Note that the number of horizontal and vertical edges of a contour is always even such that $|\gamma|\in\{4,6,8,\dots\}$. We now use the above inequality to estimate the probability of spin configurations with $s_x=-1$. Here it will be important to recall that we imposed fixed boundary conditions with all spins $+1$ at the boundary. Note that any x with $s_x=-1$ is enclosed by *at least* one contour.

Lemma 10.2 *The number $A(n)$ of contours of length n which enclose a certain point $x\in\Lambda$ is bounded from above according to*

$$A(n)\le\frac{n-2}{2}\,3^{n-1}\,.$$

Proof First of all, we notice that the ray $y(\lambda)=x+\lambda e_1$, $\lambda>0$, emanating from x, intersects a contour at least once. This is sketched in Fig. 10.2. Now let us consider the vertical link of a given contour γ_x enclosing x which intersects the ray at a maximal distance $\lambda_{\max}$ from x. The possible values of $\lambda_{\max}$ are $1/2,\,3/2,\dots,\,(n-3)/2$, i.e., we have $(n-2)/2$ possible values of $\lambda_{\max}$. The largest value is attained for the rectangle of height 1 and length $(n-2)/2$. If we now move along the contour, then each of the remaining $n-1$ links may "choose" between three possible directions with respect to its predecessor: left, right, or straight on. This gives a multiplicative factor of 3^{n-1} and this yields the upper bound for $A(n)$.

Theorem 10.1 *If $K>0.7$, then there exist two different Gibbs measures P_β^+, P_β^- for the Ising model on the infinite square lattice, where*

$$\langle s_x\rangle_{P_\beta^+}>0\quad\text{and}\quad\langle s_x\rangle_{P_\beta^-}<0\tag{10.3}$$

holds for all sites x.

Fig. 10.2 Estimation of the number of loops which enclose a fixed lattice point

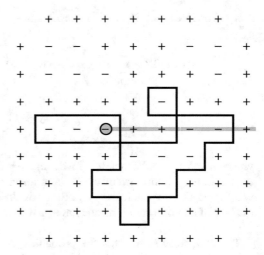

Thus we observe spontaneous magnetization at low temperatures. We know already from the exact solution or the duality arguments that spontaneous magnetization occurs for $K > K_c$ with $K_c = \frac{1}{2} \log(1 + \sqrt{2}) \approx 0.44$.

Proof Firstly, we estimate the probability of the occurrence of $s_x = -1$ when the spins on the boundary are $+1$. If $\{\gamma_x\}$ is the set of contours enclosing x, then

$$P^+[s_x = -1] \le \sum_{\gamma_x} P[\gamma] \le \sum_{n \in \mathbb{N}} A(n)\, e^{-2Kn} = \sum_{m \in \mathbb{N}} A(2m)\, e^{-4Km}$$

$$\le \sum_{m \in \mathbb{N}} (m-1) 3^{2m-1} e^{-4Km} = \frac{1}{3} \sum_{m \in \mathbb{N}} (m-1)\, e^{-\alpha m}$$

$$= \frac{1}{3} e^{-\alpha} \sum_{n \in \mathbb{N}_0} n\, e^{-n\alpha} = \frac{1}{3} e^{-\alpha} \left(-\frac{\partial}{\partial \alpha} \sum_{n \in \mathbb{N}_0} e^{-n\alpha} \right) = \frac{1}{3} \frac{y^2}{(1-y)^2} ,$$

wherein we assumed that the constant $\alpha \equiv 4K - 2 \log 3$ is positive and in the last step we defined $y = e^{-\alpha} \in (0, 1)$. Next we determine the y-values for which the probability is less than $1/2$. The upper bound $1/2$ is attained for

$$2y^2 = 3(1 - y)^2 \quad \text{or} \quad y = y_p \equiv 3 \pm \sqrt{6} .$$

We conclude that the probability for any given spin being -1 is less than $1/2$ for $y < y_p$ or equivalently for $\alpha > -\log(3 - \sqrt{6})$. With $+1$-boundary conditions, we thus observe spontaneous magnetization, if

$$\beta J > K_p, \quad K_p = \frac{1}{2} \log 3 - \frac{1}{4} \log\left(3 - \sqrt{6}\right) \approx 0.69853. \tag{10.4}$$

The estimate holds for all sites x and lattice sizes. At low temperature the system is therefore characterized by a positive magnetization $\langle s_x \rangle^+ > 0$ in the thermodynamic limit. On the other hand, setting all Ising spins at the boundary to -1, we would find a probability $P^-[s_x = 1] < 1/2$ for $K > K_p$ for all lattice sizes. This implies a negative magnetization in the thermodynamic limit. Different boundary conditions force the statistical system into different phases in the thermodynamic limit. Hence there exist at least two different equilibrium states at low temperature.

10.1.1 Extension to Higher Dimensions

To which other spin systems can one extend Peierls argument? The argument begins with a minimal energy configuration ω_0 compatible with the imposed boundary conditions. For the Ising model with $+1$-boundary conditions, this is the ordered configuration:

$$\omega_0 = \{s_x = 1 | x \in \Lambda\}. \tag{10.5}$$

Then one studies configurations with higher energies which are exponentially suppressed. Although the excited configurations carry lots of entropy—there are many of them—the energy suppression wins at sufficiently low temperatures. To generalize this type of *energy-entropy arguments*, one needs a generalization of the *Peierls contours* for other lattices and more general target spaces. Actually the Peierls argument can be extended to spin models with a discrete Abelian group G as target space. Here we focus on the Ising model in higher dimensions with target space Z_2 for which the closed contours are borderless hypersurfaces on the dual lattice. A contour thereby separates the interior from its complement, the exterior.

There exists a one-to-one relation between configurations ω with $+1$-boundary conditions and sets of nonintersecting contours. We find the following estimate for the number of contours of a given size:

Lemma 10.3 *In d dimensions the number $A(n)$ of different Peierls contours of size n fulfills the inequality*

$$\exp\left(\frac{n - 2d}{2d - 2} \log d\right) < A(n) < \frac{n - 2}{2d - 2}(6d - 9)^{n-1}. \tag{10.6}$$

Proof To prove the lower bound, we construct elongated contours of size n which enclose the site x. To that aim we consider a chain of k adjacent lattice points starting at x. Moving along the chain from one site to the next site, we jump one step toward one of the d *positive coordinate directions*. This guarantees that two chains are different if only one of their jumps is in a different direction. Clearly there exist d^{k-1} different chains of this type. We now consider the corresponding dual d-chain, the border of which is a Peierls contour which encloses x. More explicitly, the d-chain is just the union of the k elementary cubes dual to the sites along the chain. The size of such a contour is the surface area of the d-chain which is $n = (2d - 2)k + 2$, since the inner areas of the chain segments cancel. Thus the length of the chain is

$$k(n) = \frac{n - 2}{2d - 2} ,$$ (10.7)

and we obtain the lower bound

$$A(n) > d^{k(n)-1} = d^{(n-2d)/(2d-2)} = \exp\left(\frac{n - 2d}{2d - 2} \log d\right) .$$ (10.8)

To derive the upper bound, we proceed similarly as in two dimensions. Thereby we call a k-dimensional cube on the lattice k-cube. In particular two-cubes and one-cube are plaquettes and links, respectively. The corresponding objects from the dual lattice are called dual cubes, dual plaquettes and dual links. Here we are mostly dealing with dual $(d - 1)$−cubes which we call *cells*.

The ray $x + \lambda e_1$ emanating from x intersects every contour γ_x around x at least once. We focus on that cell in the contour γ_x with maximal distance $\lambda_{\max}$ from x. The possible values of $\lambda_{\max}$ are $-\frac{1}{2} + k$ with $k \in \{1, \ldots k(n)\}$ and $k(n)$ from (10.7). Thereby the largest value is realized for a column with base area of 1 and length $k(n)$ toward the e_1-direction. Beginning with the cell with maximal distance from x, we construct a Peierls contour around x by successively gluing more and more cells together. We can glue a cell to one of the $2d - 3$ free faces of an already attached cell. Actually there are three ways to glue a cell to the face of a given cell: A face is a dual $(d - 2)$-cube and defines a unique plaquette on the original lattice. The newly glued cell must be dual to one of the links forming the boundary of this plaquette. But one of the four links on the boundary of the plaquette is dual to the already attached cell and hence must be excluded. Thus we are left with three possible ways of gluing. Multiplying the combinatorial factors yields the upper bound in (10.6).

To prove spontaneous symmetry breaking at low temperatures, we assume that the constant $\alpha = 4K - 2\log(6d - 9)$ is positive. The probability for the spin at x to be -1 is bounded from above as

$$P^+[s_x = -1] \leq \sum_{m \in \mathbb{N}} A(2m) \, e^{-4Km} \leq \frac{1}{\zeta^2} \frac{y^2}{(1 - y)^2} ,$$ (10.9)

where we used the abbreviations

$$y = e^{-\alpha} \quad \text{and} \quad \zeta^2 = 3(2d - 3)(d - 1) .$$ (10.10)

This probability is less than $1/2$ for

$$y < \frac{\zeta}{\zeta + \sqrt{2}} \quad \text{or} \quad K > \frac{1}{4}\log\left(1 + \frac{\sqrt{2}}{\zeta}\right) + \frac{1}{2}\log(6d - 9) .$$ (10.11)

In two dimensions $\zeta^2 = 3$ and we recover our previous result. For $d = 3$ we have $\zeta^2 = 9/2$ and we observe two phases, if

$$K > \frac{1}{4} \log 135 \quad \text{or} \quad T < 0.1359 \, T_{\text{c,mf}} . \tag{10.12}$$

From the mean field result, we expect that K_c decreases as $1/d$. Hence the lower bounds (10.11) are by no means optimal. The inequality derived in [8] is much better in high dimensions:

$$A(n) \leq \exp\left(64n(\log d)/d\right) . \tag{10.13}$$

However, an easier way to prove the existence of an ordered low-temperature phase in higher-dimensional ferromagnetic systems makes use of the correlation inequalities in Sect. 8.6 in conjunction with the known results for the two-dimensional system.

10.2 Duality Transformation of Two-Dimensional Ising Model

In their pioneering work, Kramers and Wannier discovered a transformation which maps the 2d Ising model with couplings $(\beta, h = 0)$ into itself, but with couplings $(*\beta, h = 0)$ [2, 3]. The temperature after the transformation is a monotonously decreasing function of the temperature of the original model such that the high-temperature phase maps into the low-temperature phase and vice versa. We begin with rewriting the high-temperature series expansion of the partition function (9.6) as follows $(K = \beta J)$:

$$Z = (\cosh K)^P 2^V \sum_{G \in \mathcal{G}} v^{L(G)} = (\cosh K)^P 2^V \sum_G \prod_x v^{n_x(G)/2} , \tag{10.14}$$

with $v = \tanh(K) \ll 1$. Here $\mathcal{G}$ denotes the set of high-temperature graphs. These are diagrams with closed curves (loops) and even vertices only. $L(G)$ denotes the length of the graph G or the number of its links. Figure 10.3 shows a high-temperature graph with $L(G) = 12$. Finally, the even number $n_x(G)$ in the last representation of the partition function is equal to the number of links ending at vertex x. In two dimensions $n_x(G)$ takes the values 0, 2, or 4.

Now we shall argue that the sum (10.14) may be viewed as partition function on the *dual lattice*. The dual lattice $*\Lambda$ of a square lattice is again a square lattice with sites at the centers of the plaquettes of the original lattice. Now we assign a spin configuration $*\omega = \{s_{*x}\}$ on the dual lattice to every spin configuration $\omega = \{s_x\}$ on the lattice as follows: for two nearest neighbors $*x, *y$ on the dual lattice, we set $s_{*x} s_{*y} = -1$ if a loop of the high-temperature graph belonging to ω crosses the link between $*x$ and $*y$. Else we set $s_{*x} s_{*y} = 1$. Actually $*\omega$ and $-*\omega$ belong to the same graph. On the other hand, ω and $-\omega$ also belong to the same high-temperature

Fig. 10.3 A
high-temperature graph G of
the two-dimensional Ising
model of length $L(G) = 12$.
The dual lattice $^*\Lambda$ with its
lattice points $^*x$ is illustrated
as well

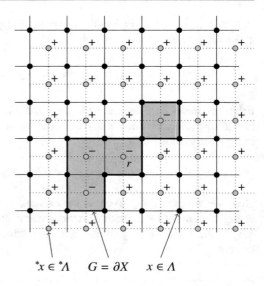

$$^*x \in {}^*\Lambda \qquad G = \partial X \qquad x \in \Lambda$$

graph, and this implies that there is a bijective mapping $\omega \rightarrow {}^*\omega$. The loops of a
graph G encircle a set X of sites on the dual lattice, and all spins s_{*x} in X have the
same sign. The spins in the complement of X have the other sign.

Let us consider the plaquette $p(x)$ on the dual lattice with vertices $^*x, {}^*y, {}^*u, {}^*v$
and center point $x \in \Lambda$. Now it is easy to see that the order of the vertex x is

$$n_x = 2 - \frac{1}{2}\left(s_{*x}s_{*y} + s_{*y}s_{*u} + s_{*u}s_{*v} + s_{*v}s_{*x}\right) \equiv 2 - \frac{1}{2}p(x) \,. \tag{10.15}$$

Inserting this result into the partition function (10.14) yields

$$Z = (\cosh K)^P 2^V \sum_{*\omega} \prod_{x} \left(v \cdot v^{-p(x)/4}\right)$$

$$= (\cosh K)^{2V}(2v)^V \sum_{*\omega} v^{-\frac{1}{2}\sum_{(*x,*y)} s_{*x}s_{*y}}$$

$$= (2\sinh K \cosh K)^V \sum_{*\omega} e^{-*\beta\,*H(s)} \,, \tag{10.16}$$

where we took into account that every dual link belongs to two plaquettes. In the
last step, we defined $v = \exp(-2^*K)$. We may rewrite this relation as follows:

$$2\sinh(2^*K) = e^{2^*K} - e^{-2^*K} = \frac{1}{v} - v = \coth K - \tanh K = \frac{2}{\sinh 2K} \,.$$

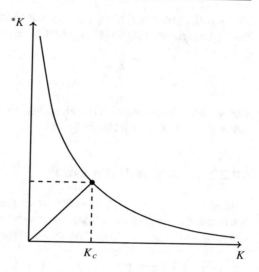

Fig. 10.4 Relation between the reduced temperature K on the lattice Λ and the reduced temperature $*K$ on the dual lattice $*\Lambda$

Hence, the duality relation takes the symmetric form

$$\sinh 2K \cdot \sinh 2{*}K = 1 .\tag{10.17}$$

This duality relation links the temperature T of the Ising model with the temperature $*T$ of the Ising model on the dual lattice. It is a *symmetric* and reciprocal relation: If K increases monotonically from 0 to ∞, then $*K$ decreases monotonically from ∞ to 0. Figure 10.4 shows the function $*K(K)$ as well as the fixed point of the map $K \to {*}K$. By using Eq. (10.17) we obtain the relation

$$(2 \sinh K \cosh K)^2 = \sinh 2K \sinh 2K \overset{(10.17)}{=} \frac{\sinh 2K}{\sinh 2{*}K} .$$

Together with (10.16) this leads to the duality relation

$$\frac{Z(K)}{(\sinh 2K)^{V/2}} = \frac{Z(*K)}{(\sinh 2{*}K)^{V/2}} .\tag{10.18}$$

Let us now assume that there is a critical coupling K_c where the free energy density (in the thermodynamic limit) is singular. Then the duality relation (10.18) implies that there must exist another singularity at $*K_c$. Hence, in the case of there being a unique critical point, it would be located at $K_c = {*}K_c$. Then K_c is a solution of

$$\sinh 2K_c = \pm 1 \implies K_c = \pm\frac{1}{2}\log\left(1 + \sqrt{2}\right) \approx \pm 0.4407 .$$

The negative solution thereby corresponds to the antiferromagnetic case $J < 0$. For the ferromagnetic Ising model, we obtain the critical temperature

$$T_c = \frac{2J}{\log\left(1 + \sqrt{2}\right)} \approx 2.2692J \ . \tag{10.19}$$

For systems with several critical points, the duality relation leads to relations between pairs of critical couplings only.

10.2.1 An Algebraic Derivation

Now we present a second, more algebraic, derivation of the duality relation which generalizes more easily to higher dimensions and to other spin systems. We rewrite the Ising model partition function in d dimensions as follows:

$$Z = \sum_\omega \prod_{\langle x,y\rangle} \left(\cosh K + \sinh K \, s_x s_y\right) = \sum_\omega \prod_{\langle x,y\rangle} \sum_{k=0,1} c_k(K) \left(s_x s_y\right)^k \ , \tag{10.20}$$

where we defined $c_0(K) := \cosh K$ and $c_1(K) := \sinh K$. At this point we are lead to introduce a field which assigns to every link either 0 or 1:

$$k : \langle x, y\rangle \to k_{xy} \in \{0, 1\} \ .$$

For a fixed configuration of link variables $\{k_{xy}\}$, we obtain the following contribution to the partition function:

$$\sum_\omega \prod_{\langle x,y\rangle} c_{k_{xy}}(K) \left(s_x s_y\right)^{k_{xy}} = \prod_{\langle x,y\rangle} c_{k_{xy}}(K) \sum_\omega \prod_{\langle x,y\rangle} (s_x s_y)^{k_{xy}}$$

$$= \prod_{\langle x,y\rangle} c_{k_{xy}}(K) \sum_\omega \prod_x s_x^{\partial k(x)} \ , \tag{10.21}$$

where $\partial k(x)$ is the sum of all variables defined on links ending at x:

$$\partial k(x) = \sum_{y:\langle y,x\rangle} k_{xy} = \sum_{\ell:x\in\partial\ell} k_\ell \ .$$

The operator ∂ represents the discrete version of the divergence and in d dimensions takes the values $\{0, 1, \ldots, 2d\}$. For any integer n, we have

$$\sum_{s=-1,1} s^n = 2\delta_2(n), \qquad \delta_2(n) = \begin{cases} 1 \ n \text{ even} \\ 0 \ n \text{ odd} \, , \end{cases} \tag{10.22}$$

such that the sum over all spin configurations in (10.21) is performed easily and leads to the following representation of the partition function in arbitrary dimensions:

$$Z = 2^V \sum_{\{k\}} \prod_\ell c_{k_\ell}(K) \prod_x \delta_2(\partial k(x)) .$$ (10.23)

In two dimensions we assign a spin configuration $^*\omega$ on the dual lattice to each configuration $k = \{k_\ell\}$ on the links as follows: If the link $\langle {}^*x, {}^*y \rangle$ between two nearest neighbors $^*x$ and $^*y$ on the dual lattice crosses the link ℓ, we set

$$k_\ell = \frac{1}{2}(1 - s_{*x}s_{*y}) .$$ (10.24)

We recover the relation (10.15) wherein n_x is equal to the divergence of k at x:

$$\partial k(x) = 2 - \frac{1}{2}p(x) .$$

Since $p(x) \in \{-4, 0, 4\}$, the right-hand side is always even and all δ_2-constraints in (10.23) are fulfilled. This means that the transformation (10.24) yields all link configurations $\{k\}$ that satisfy the δ_2-constraints. The sum over link configurations turns into the sum over spin configurations on the dual lattice. Since every link corresponds to exactly one link on the dual lattice, the product over all ℓ becomes the product over all nearest-neighbor pairs $\langle {}^*x, {}^*y \rangle$. Thus (10.23) can be written as

$$Z = 2^V \sum_{*\omega} \prod_{\langle {}^*x, {}^*y \rangle} c_{(1-s_{*x}s_{*y})/2}(K) .$$ (10.25)

Rewriting $c_k(K)$ according to

$$c_{k_\ell}(K) \overset{(10.24)}{=} (\cosh K \sinh K)^{1/2} \exp\left(-\tfrac{1}{2}s_{*x}s_{*y} \log \tanh K\right)$$

and inserting this into (10.25) gives

$$Z = (2\cosh K \sinh K)^V \sum_{*\omega} \exp\left(-\frac{1}{2}\log\tanh K \sum_{\langle {}^*x, {}^*y \rangle} s_{*x}s_{*y}\right)$$

$$= (\sinh 2{}^*K)^{-V} \sum_{*\omega} \exp\left({}^*K \sum_{\langle {}^*x, {}^*y \rangle} s_{*x}s_{*y}\right) ,$$ (10.26)

where $^*K$ is related to K as in (10.17). Thus we recovered our previous result.

10.2.2 Two-Point Function

In order to interpret the dual spins s_{*x}, we slightly modify the previous derivation to determine the *two-point function* of the dual model:

$$\langle s_{*x} s_{*y} \rangle = \frac{1}{*Z} \sum_{*\omega} s_{*x} s_{*y} \exp\left({}^*K \sum_{\langle *u, *v \rangle} s_{*u} s_{*v} \right) .$$

The partition function $^*Z$ in the denominator has already been dualized, and we focus on the numerator,

$$Z_{*x*y} = \sum_{*\omega} s_{*x} s_{*y} \exp\left({}^*K \sum_{\langle *u, *v \rangle} s_{*u} s_{*v} \right)$$

$$= \sum_{*\omega} s_{*x} s_{*y} \prod_{\langle *u, *v \rangle} \sum_{k=0,1} c_k({}^*K)\, (s_{*u} s_{*v})^k \ ,$$

where the product extends over all nearest-neighbor pairs on $^*\Lambda$. Thus we have

$$Z_{*x*y} = 2^V \sum_{\{k\}} \prod_{*\ell} c_{k*\ell}({}^*K) \prod_{*u} \delta_2 \left(\delta_{*u*x} + \delta_{*u*y} + \partial k({}^*u) \right) . \tag{10.27}$$

We now wish to find a representation for the configurations $\{k\}$ on the dual links such that all δ_2-constraints in (10.27) are fulfilled. We proceed as follows: We connect the points $^*x$ and $^*y$ by an arbitrary path $\mathscr{C}_{*x*y}$ on the dual lattice as illustrated in Fig. 10.5. We then choose the following representation for the dual-link variables:

$$k_{*\ell} = \begin{cases} \frac{1}{2}(1 - s_x s_y) & {}^*\ell \notin \mathscr{C}_{*x*y} \\ \frac{1}{2}(1 + s_x s_y) & {}^*\ell \in \mathscr{C}_{*x*y} . \end{cases}$$

In other words if the link $\langle x, y \rangle$ on the lattice intersects the path $^*\mathscr{C}_{*x*y}$ on the dual lattice, then we modify the transformation rule. According to this representation, ∂k is an even number for all lattice points $^*\Lambda$, except for $^*x$, $^*y$, where it is odd. Hence, we may cast the representation (10.27) in the form

$$Z_{*x*y} = 2^V \sum_{\omega} \prod_{\langle x, y \rangle \in \mathscr{S}_{*x*y}} c_{(1+s_x s_y)/2}({}^*K) \prod_{\langle x, y \rangle \notin \mathscr{S}_{*x*y}} c_{(1-s_x s_y)/2}({}^*K) \ ,$$

where $\mathscr{S}_{*x*y}$ denotes the set of links on the lattice Λ which intersect the path $\mathscr{C}_{*x*y}$. With a calculation similar to the one above (10.26), we arrive at

$$Z_{*x*y} = (\sinh 2K)^{-V} \sum_{\omega} \exp\left(\sum_{\langle x, y \rangle} K_{xy} s_x s_y \right) . \tag{10.28}$$

Fig. 10.5 A path on the dual lattice connecting the arguments $^*x$ and $^*y$ of the two-point correlation function

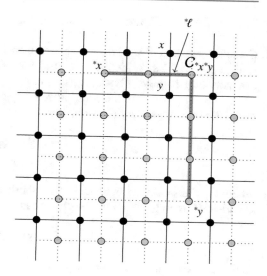

Note that the couplings in the exponent depend on the nearest-neighbor pair: if the link between a pair does not intersect the path $\mathscr{C}_{^*x^*y}$, then the corresponding coupling is given by βJ. If the link between the pair does intersect the path $\mathscr{C}_{^*x^*y}$, then the corresponding coupling $-\beta J$ is negative and antiferromagnetic. In summary, the numerator $Z_{^*x^*y}$ itself is a partition function with ferromagnetic as well as antiferromagnetic couplings between nearest neighbors. Thus the correlation function $\langle s_{^*x} s_{^*y} \rangle$ of the dual Ising model is the ratio of two partition functions on Λ. One contains ferromagnetic and antiferromagnetic couplings, whereas the other one includes only ferromagnetic couplings.

10.2.3 Potts Models

The two-dimensional standard q-state Potts model with energy function (6.10) is self-dual, and the duality relation between the couplings reads

$$e^{-^*K_p} = \frac{1 - e^{-K_p}}{1 + (q-1)e^{-K_p}}, \quad K_p = \beta J_p , \tag{10.29}$$

see the problem on p. 257. A particular simple proof based on the random-bond model was given in [9]. On a finite lattice, the duality transformation can be extended to Potts models subject to cyclic boundary conditions,

$$s_{x+Ne_i} = s_x + c_i, \quad c_i \bmod q , \tag{10.30}$$

and this generalization may be utilized to study universal aspects of phase transitions in three-dimensional gauge theories [10].

10.2.4 Curl and Divergence on a Lattice

Consider a hypercubic lattice Λ in d dimensions. Suppose that there are some statistical variables k_ℓ defined on the oriented links of Λ. If we change the orientation of ℓ, then k_ℓ changes its sign, $k_{\langle x,y\rangle} = -k_{\langle y,x\rangle}$. The k_ℓ should belong to an Abelian group with the addition as group operation. The circulation along the perimeter of a elementary plaquette p is

$$\mathrm{d}k(p) = \sum_{\ell\in\partial p} k_\ell \,, \tag{10.31}$$

where the links ℓ on the boundary of p inherit the orientation of the plaquette. If k is curl-free, $\mathrm{d}k = 0$, then it is (locally) the gradient of a function φ:

$$k_{\langle y,x\rangle} = \varphi(y) - \varphi(x) \,. \tag{10.32}$$

In this case the "integral" along any contactable loop $\mathscr{C}$ on the lattice vanishes:

$$\oint_\mathscr{C} k = \sum_{\ell\in\mathscr{C}} k_\ell = 0 \,. \tag{10.33}$$

The links must have the same orientation as the loop. Besides the curl we can define a discrete version of the divergence on the lattice. The divergence of k is

$$(\partial x)(x) = \sum_{\ell:x\in\partial\ell} k_\ell \,, \tag{10.34}$$

where all links are emanating from x. For a divergence-free field the flux through elementary cubes of the dual lattice vanish.

We do not want to go any further at this point. However, if one goes beyond hypercubic lattices in higher dimensions, it is useful to know some basic facts about the difference calculus on lattices. A comprehensive and exhaustive representation based on simplices, chains, border and co-border operators, and Stokes theorem is contained in [11].

10.3   Duality Transformation of Three-Dimensional Ising Model

The three-dimensional Ising model is not self-dual—the application of the duality transformation yields the Z_2 lattice gauge theory. Thus we cannot predict its critical point form duality alone. But a duality relating two different models can be useful for studying the excitations in the high- and low-temperature regimes since it is still true that the transformation relates the high-temperature phase of one model to the low-temperature phase of the other model and vice versa.

Our point of departure is the representation (10.23) of the partition function,

$$Z = 2^V \sum_{\{k\}} \prod_\ell c_{k_\ell}(K) \prod_x \delta_2 \left(\partial k(x)\right) , \qquad (10.35)$$

which holds in all dimensions. In three dimensions the divergence at a given site

$$(\partial k)(x) = \sum_{\ell : x \in \partial \ell} k_\ell \qquad (10.36)$$

is the sum of six terms—one term for every link ending at x—and assumes the values $\{0, 1, \ldots, 6\}$. Again we can fulfill the δ_2-constraints on $\partial k(x)$ in (10.35) by introducing suitable variables on the dual lattice $*\Lambda$. The dual of a hypercubic lattice is again a hypercubic lattice with sites in the centers of the elementary cells of the original lattice. Nearest neighbors of $*\Lambda$ sit in adjacent cells of Λ. Every link ℓ crosses exactly one plaquette $*p_\ell$ of the *dual lattice* as depicted in Fig. 10.6. We now assign to every link $*\ell$ of the dual lattice a variable $U_{*\ell}$ with values in the Abelian group $Z_2 = \{-1, 1\}$. Next we map a link configuration $\{U_{*\ell}\}$ on the dual lattice to a link configuration $\{k_\ell\}$ on the original lattice as follows: If $*p_\ell$ is the plaquette dual to ℓ, then we set

$$k_\ell = \frac{1}{2}\left(1 - U_{*p_\ell}\right), \quad \text{where} \quad U_{*p} = \prod_{*\ell \in \partial *p} U_{*\ell} \qquad (10.37)$$

is the product of the link variables of the four edges bounding the dual plaquette. Now we shall prove that the divergence of k only takes the values 0, 2, 4, and 6 and thus satisfies all δ_2-constraints in (10.35).

Fig. 10.6 Every link ℓ of the lattice crosses exactly one plaquette $*p_\ell$ of the dual lattice, and a dual plaquette has four dual links as edges

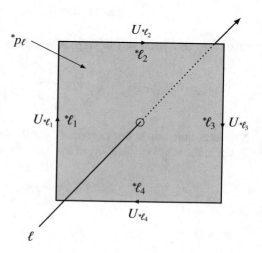

To calculate the divergence of k at x, we consider the elementary cube c_x of the dual lattice with center x. The six links ending at x intersect the six plaquettes on the boundary of c_x, such that the divergence in (10.36) takes the form

$$\partial k(x) = 3 - \frac{1}{2} \sum_{*p \in \partial c_x} U_{*p} \,. \tag{10.38}$$

The plaquette variables U_{*p} have the values 1 or -1, and if all dual-link variables $U_{*\ell}$ on the edges of the cube c_x are 1, then the divergence at x is zero. In contrast, if we change the sign of one link variable, then exactly two plaquette variables change sign such that the sum changes by a multiple of 4 or the divergence by a multiple of 2. This then proves that ∂k takes the values in $\{0, 2, 4, 6\}$ and thus fulfills the δ_2-constraints.

We now rewrite the product over all links in (10.35) as a product over all dual plaquettes. Since k in (10.37) satisfies all δ_2-constraints in (10.35), we find

$$Z = 2^V \sum_{\{k(U)\}} \prod_{\{*p\}} c_{\frac{1}{2}(1 - U_{*p})}(K) \,. \tag{10.39}$$

Below we shall see that there are many different configurations $\{U_{*\ell}\}$ belonging to the same configuration $\{k_\ell\}$. Configurations which are mapped into the same k are called equivalent. By construction equivalent configurations have the same weight c in (10.39), and the sum over $\{k\}$ in the partition function (10.35) becomes a sum over equivalence classes. This is indicated by the sum over $\{k(U)\}$ in (10.39). Using the relation

$$c_{k_\ell} = \cosh K \, e^{k_\ell \log \tanh K}$$

$$\overset{(10.37)}{=} (\cosh K \sinh K)^{1/2} \exp\left(-\tfrac{1}{2} \log \tan K U_{*p_\ell}\right)$$

we end up with the following representation of the partition function:

$$Z = 2^V (\cosh K \sinh K)^{3V/2} \sum_{\{k(U)\}} \exp\left(*K \sum_{\{*p\}} \prod_{*\ell \in \partial *p} U_{*\ell}\right) \,. \tag{10.40}$$

Note that the sum in the exponent extends over all plaquettes of the dual lattice. Besides, the relation $*K(K)$ is of the same form as in two dimensions:

$$*K = -\frac{1}{2} \log \tanh K \,. \tag{10.41}$$

10.3.1 Local Gauge Transformations

Let us now determine the set of configurations $\{U_{*\ell}\}$ which are mapped into the same configuration $\{k_\ell\}$ by the mapping (10.37). Multiplying the group elements $U_{*\ell}$ on all links $*\ell$ ending at a fixed $*x$ with -1 does not change $\{k_\ell\}$ and $\{U_{*p_\ell}\}$ since every dual plaquette contains either two or none of these links. This operation may be performed at each site of the dual lattice independently such that there are 2^{V^*} equivalent configurations in each class. Thus we conclude that the variables $\{k_\ell\}$ and $\{U_{*p}\}$ do not change under so-called gauge transformations of the dual-link variables:

$$U_{\langle *x, *y\rangle} \longrightarrow U'_{\langle *x, *y\rangle} = g_{*x} U_{\langle *x, *y\rangle} g_{*y}^{-1}, \qquad g : {}^*\Lambda \longrightarrow Z_2 = \{-1, 1\}. \qquad (10.42)$$

They are examples of gauge-invariant variables. Other gauge invariant objects are the so-called Wilson loop variables: for any closed path (loop) $*\mathscr{C}$ on the dual lattice, the gauge-invariant loop variable is

$$W({}^*\mathscr{C}) = \prod_{\ell \in {}^*\mathscr{C}} U_{*\ell} \in Z_2. \qquad (10.43)$$

Now the question arises how to perform the sum in (10.40) over gauge-nonequivalent configurations $\{U_{*\ell}\}$, i.e., configurations with different $k(U)$. There are two possible approaches, e.g., one *fixes the gauge* and picks from each gauge class

$$\{U_{\langle *x, *y\rangle}\} \sim \{g_{*x} U_{\langle *x, *y\rangle} g_{*y}^{-1}\} \equiv {}^g U_{\langle *x, *y\rangle} \qquad (10.44)$$

one representative and sums over these representatives. Since equivalent configurations give the same contribution, it does not matter which representatives are picked. Alternatively, one simply sums over *all* configurations $\{U_{*\ell}\}$ in (10.40). Of course, we overcount, but since every class contains the same number of configurations, this overcounting results in the same factor 2^{V^*}, independently of k. In doing so, we find for cubic lattices with $V = V^*$ the result

$$Z = (\cosh K \sinh K)^{3V/2} \cdot \sum_{\{U\}} \exp\left({}^*K \sum_{\{*p\}} \prod_{*\ell \in \partial *p} U_{*\ell}\right). \qquad (10.45)$$

Since $\{g_{*x}\}$ entering a gauge transformation ${}^g U$ is a site-dependent lattice field, we call the gauge transformation *local*, and theories which admit a local (space-dependent) symmetry are called gauge theories. What we have proven then is that the three-dimensional Ising model is dual to a Z_2 *gauge theory*. This duality holds in both directions and the transformation is idempotent [6].

Finally, we may cast the gauge transformation (10.42) into a form that emphasizes the connection with electrodynamics in three Euclidean spacetime dimensions. We therefore write

$$U_{\langle *x, *y\rangle} = \exp\left(i\pi A_{\langle *x, *y\rangle}\right), \quad g_{*x} = \exp\left(i\pi \lambda_{*x}\right),$$

where the variables $A_{\langle *x, *y\rangle}$ and λ_{*x} belong to the additive group $\mathbb{Z}_2 = \{0, 1\}$. The gauge transformation (10.42) for the gauge potential A assumes the well-known form

$$A_{\langle *x, *y\rangle} \longrightarrow A'_{\langle *x, *y\rangle} = A_{\langle *x, *y\rangle} + (\lambda_{*x} - \lambda_{*y}) . \tag{10.46}$$

Since the last term between brackets represents the discretized gradient of the lattice field λ, this formula is just the lattice version of the well-known gauge transformation $A'_\mu = A_\mu + \partial_\mu \lambda$ in electrodynamics.

10.4 Duality Transformation of Three-Dimensional Z_n Gauge Model

Here we extend the results of the previous section to Z_n models. This time the point of departure is the Z_n gauge theories which are mapped into Z_n spin models by the duality transformation. We shall use a slightly different method which emphasizes the close relationship between finite Fourier transformations and duality transformations [12]. The link variables of the Z_n gauge theory are elements of the multiplicative cyclic group, the elements of which can be written as

$$U_\ell = e^{2\pi i \theta_\ell / n}, \quad \theta_\ell \in \{0, 1, \ldots, n-1\} . \tag{10.47}$$

The θ_ℓ are in the additive group of integers with addition performed modulo n. As in the previous section, we introduce the plaquette variables,

$$U_p = \prod_{\ell \in \partial p} U_\ell = e^{2\pi i \theta_p / n}, \quad \theta_p = \sum_{\ell \in \partial p} \theta_\ell , \tag{10.48}$$

where the orientation of a link on the boundary of a plaquette is inherited from the orientation of the plaquette. The real and gauge invariant energy function contains the sum over all plaquettes:

$$S = \sum_p \left(1 - \Re U_p\right) = \sum_p \left(1 - \cos \frac{2\pi \theta_p}{n}\right) . \tag{10.49}$$

Here we interpret the energy function as Euclidean action of a lattice gauge theory with coupling constant g related to β according to $\beta = 1/g^2$ (in this section β is not

$1/kT$). In the partition function, one sums over the $3V$ link variables:

$$Z(\beta) = \sum_{\{\theta_\ell\}} e^{-\beta S} = e^{-3\beta V} \sum_{\{\theta_\ell\}} \prod_p \exp\left(\beta \cos \frac{2\pi\theta_p}{n}\right) . \tag{10.50}$$

To extend the duality transformation from the Ising model to Z_n models, we perform for each plaquette a finite Fourier transform on $\mathbb{Z}_n$:

$$\exp\left(\beta \cos \frac{2\pi\theta}{n}\right) = \sum_{k=0}^{n-1} c_k(\beta) \cos\left(\frac{2\pi\theta k}{n}\right) . \tag{10.51}$$

At a fixed plaquette the Fourier coefficients are given by the inverse transformation

$$c_k(\beta) = \frac{1}{n} \sum_{\theta=0}^{n-1} \exp\left(\beta \cos \frac{2\pi\theta}{n}\right) \cos\left(\frac{2\pi\theta k}{n}\right) . \tag{10.52}$$

Inserting the Fourier representation into (10.50) yields

$$Z(\beta) = e^{-3\beta V} \sum_{\{\theta_\ell\}} \prod_p \sum_{k_p=0}^{n-1} c_{k_p}(\beta) \exp\left(\frac{2\pi i \theta_p k_p}{n}\right)$$

$$= e^{-3\beta V} \sum_{\{\theta_\ell\}} \sum_{\{k\}} \left(\prod_p c_{k_p}(\beta)\right) \exp\left(\sum_p \frac{2\pi i \theta_p k_p}{n}\right) . \tag{10.53}$$

Writing every plaquette variable θ_p as a sum of link variables as in (10.48), the exponent takes the form

$$\frac{2\pi i}{n} \sum_\ell \theta_\ell \sum_{p:\ell\in\partial p} k_p , \tag{10.54}$$

and the sum over θ_ℓ leads to the constraint

$$\sum_{p:\ell\in\partial p} k_p = 0 \bmod n , \tag{10.55}$$

where the summation symbol is defined to include the sign corresponding to the relative orientation of p and ℓ. The sum extends over the plaquettes of the staple belonging to ℓ, i.e. the plaquettes with a boundary containing the link ℓ. The staple of ℓ in Fig. 10.7 consists of the four plaquettes $p, \ldots, p''''$. To summarize, the sum

Fig. 10.7 The link ℓ is at the boundary of the four plaquettes p, p', p'', and p''' defining the staple belonging to the link. The links $*\ell$, $*\ell'$, $*\ell''$, and $*\ell'''$ dual to these plaquettes form a loop which encircles ℓ and form the boundary of the plaquette $*p$ dual to ℓ

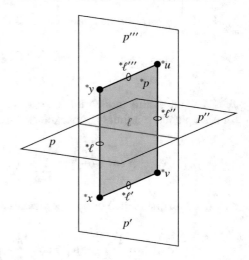

over the configurations $\{\theta_\ell\}$ can be calculated and yields

$$Z(\beta) = \mathrm{e}^{-3\beta V} \sum_{\text{constrained } \{k\}} \prod_p c_{k_p} . \tag{10.56}$$

To solve for the constraints, it is convenient to switch to the dual lattice. With each link ℓ and plaquette p of the original lattice, we associate the dual plaquette $*p$ and dual link $*\ell$ of the dual lattice. One may write $k_p = k_{*\ell}$ such that the constraint (10.55) can be written as

$$\sum_{*\ell \in \partial *p} k_{*\ell} = 0 \bmod n . \tag{10.57}$$

We conclude that the configurations $\{k_{*\ell}\}$ on the dual lattice have no circulation and thus can be written as gradients of Z_n-spin configurations on the dual lattice,

$$k_{*\ell} = s_{*x} - s_{*y} , \tag{10.58}$$

where $*x$ and $*y$ are the two endpoints of $*\ell$ and the difference is modulo n. In terms of the unconstrained spin variables $\{s_{*x} | *x \in *\Lambda\}$, the partition function takes the form

$$Z(\beta) = \mathrm{const} \times \sum_{\{s\}} \mathrm{e}^{-*\beta *S[s]} . \tag{10.59}$$

Table 10.1 The Boltzmann weights $\exp[-{}^*\beta\,{}^*s(k)]$ for the Z_2, Z_3, and Z_4 spin models dual to gauge theories in three dimensions

n	$k = 0$	$k = 1$	$k = 2$	$k = 3$
2	$e^\beta + e^{-\beta}$	$e^\beta - e^{-\beta}$		
3	$e^\beta + 2e^{-\beta/2}$	$e^\beta - e^{-\beta/2}$	$e^\beta - e^{-\beta/2}$	
4	$e^\beta + 2 + e^{-\beta}$	$e^\beta - e^{-\beta}$	$e^\beta - 2 + e^{-\beta}$	$e^\beta - e^{-\beta}$

The action (energy) of the dual spin model is the sum of nearest-neighbor terms,

$$
{}^*S[s] = \sum_{{}^*\ell} {}^*s\,(k_{{}^*\ell})\,, \qquad k_{{}^*\ell} = s_{{}^*x} - s_{{}^*y}, \quad {}^*\ell = \langle {}^*x, {}^*y \rangle\,, \tag{10.60}
$$

and the nearest-neighbor interaction is given by the Boltzmann weights

$$
e^{-{}^*\beta\,{}^*s(k)} = c_k(\beta) = \frac{1}{n} \sum_{\ell=0}^{n-1} e^{\beta \cos(2\pi\ell/n)} \cos\left(2\pi\ell\,\frac{k}{n}\right), \quad k \in \mathbb{Z}_n\,. \tag{10.61}
$$

Note that the contributions with ℓ and $n - \ell$ are identical. The Boltzmann weights of the Z_2, Z_3, and Z_4 spin models are listed in Table 10.1.

For $n \leq 4$ we find the energy functions of the clock (planar Potts) models; see Sect. 6.2.2. In particular, for $n = 2$ we recover the nearest-neighbor interaction of the two-state clock model,

$$
{}^*\beta\,{}^*s(k) = \tilde{c}_0(\beta) - {}^*\beta \cos\left(\frac{2\pi k}{2}\right), \qquad {}^*\beta = \frac{1}{2} \log \coth \beta\,, \tag{10.62}
$$

which is equivalent to the ubiquitous Ising model, together with the already known dual coupling. Recall that k is the difference of neighboring spin variables. Similarly, the Z_3 gauge theory is dual to the three-state clock model with nearest-neighbor interaction

$$
{}^*\beta\,{}^*s(k) = \tilde{c}_0(\beta) - {}^*\beta \cos\left(\frac{2\pi k}{3}\right), \qquad \beta^* = \frac{2}{3} \log\left(\frac{e^\beta + 2e^{-\beta/2}}{e^\beta - e^{-\beta/2}}\right), \tag{10.63}
$$

where $k \in \{0, 1, 2\}$. Finally, the Z_4 gauge theory is dual to the four-state clock model with nearest-neighbor interaction given by

$$
{}^*\beta\,{}^*s(k) = c_0(\beta) - {}^*\beta \cos\left(\frac{2\pi k}{4}\right), \qquad {}^*\beta = \log \coth \frac{\beta}{2}\,, \tag{10.64}
$$

which is equivalent to two copies of the Ising model. For $n > 4$ the dual models are not clock models but belong to the class of generalized Z_n spin models with energy functions (6.15).

The three-dimensional Ising model and more generally the n-state Potts model show a phase transition from an ordered low-temperature phase into a disordered

Table 10.2 The critical couplings $^*\beta_c$ of the planar Potts models [13] and the corresponding critical couplings β_c of the dual Z_n spin systems in three dimensions

	$n = 2$	$n = 3$	$n = 4$
$^*\beta_c$(Potts models)	0.2217	0.367	0.4434
β_c(Z_n gauge models)	0.7613	1.084	1.5226

high-temperature phase at a critical point β_c. From the duality map, we conclude that the dual gauge theory also shows a transition at $^*\beta_c = {}^*\beta(\beta_c)$. For small n the critical temperatures of the clock models are known [13], and the resulting critical couplings of the dual gauge theories are listed in Table 10.2.

10.4.1 Wilson Loops

Let us discuss the gauge-invariant Wilson loops as correlation function. Recall that the Wilson loop $W_\mathscr{C}$ is the product of the link variables U_ℓ around a closed curve $\mathscr{C}$, where the orientation of ℓ is inherited from the orientation of the loop $\mathscr{C}$. Thus

$$W_\mathscr{C} = \prod_{\ell \in \mathscr{C}} U_\ell = \exp\left(\frac{2\pi \mathrm{i}}{n} \sum_{\ell \in \mathscr{C}} \theta_\ell \right) . \tag{10.65}$$

The expectation value of Wilson loops

$$\langle W_\mathscr{C} \rangle = \frac{1}{Z(\beta)} \sum_{\theta_\ell} W_\mathscr{C}[\theta]\, e^{-\beta S[\theta]} \tag{10.66}$$

can be dualized similarly as the partition function. The constraint (10.55) is thereby changed to

$$\sum_{p:\ell \in \partial p} k_p + \delta_{\ell,\mathscr{C}} = 0 \bmod n , \tag{10.67}$$

where the additional term is 1 if $\ell \in \mathscr{C}$ and else is 0. On the dual lattice, this additional term leads to the modified constraint

$$\sum_{^*\ell \in \partial^* p} k_{^*\ell} + \mathrm{linking}(\partial^* p, \mathscr{C}) = 0 \bmod n , \tag{10.68}$$

where linking$(\partial^* p, \mathscr{C}) = 1$ if the boundary of the plaquette $^*p$ and the loop $\mathscr{C}$ are linked and otherwise is 0. Consider now an arbitrary surface $\mathscr{S}$ bounded by $\mathscr{C}$ and made up of plaquettes $p_1, \ldots, p_r$. Let $^*\ell_1, \ldots, {}^*\ell_r$ be the variables on links dual to these plaquettes. They intersect the surface $\mathscr{S}$ such that the constraints are solved by

$$k_{*\ell} = s_{*x} - s_{*x} + \text{intersect}(*\ell, \mathscr{S}) \, . \tag{10.69}$$

This then leads to the following result

$$\langle W_{\mathscr{C}} \rangle = \left\langle \prod_{*\ell \cap \mathscr{S} \neq 0} \frac{c_{*\ell+1}}{c_{*\ell}} \right\rangle \, , \tag{10.70}$$

where the expectation value is evaluated in the dual spin model with Boltzmann weights (10.61). In particular, when $\mathscr{C}$ is the boundary of a single plaquette with dual link $*\ell$, we obtain

$$\langle W_p \rangle = \left\langle \frac{c_{*\ell+1}}{c_{*\ell}} \right\rangle \, . \tag{10.71}$$

Since the gauge action (energy) is proportional to the sum of the W_p we interpret $\langle W_p \rangle$ as average action density given by the derivative of the free energy density with respect to the coupling β. This means that an n'th-order phase transition is a discontinuity in the $(n-1)$'th derivative of $\langle W_p \rangle(\beta)$.

10.4.2 Duality Transformation of U(1) Gauge Model

Numerical and theoretical arguments indicate the absence of phase transitions in the three-dimensional $U(1)$ lattice gauge theory [12, 14, 15]. The model shows confinement for all values of the coupling β, analogous to the expected behavior of non-Abelian gauge theory in four dimensions. The theory is defined as the $n \to \infty$ limit of the Z_n gauge theories. In this limit the discrete link variables

$$\frac{2\pi\theta_\ell}{n} \quad \text{with} \quad \theta_\ell \in \mathbb{Z}_n \tag{10.72}$$

turn into real variables θ_ℓ with values in $[0, 2\pi)$ and the link variables $\exp(i\theta_\ell)$ into elements of the Abelian group $U(1)$. The lattice action becomes

$$S = \sum_p \left(1 - \cos\theta_p\right), \quad \theta_p = \sum_{\ell \in \partial p} \theta_\ell \, , \tag{10.73}$$

and it is invariant under local gauge transformations

$$\theta_{\langle x,y\rangle} \longrightarrow \theta_{\langle x,y\rangle} + \lambda_x - \lambda_y, \quad \lambda_{x,y} \in [0, 2\pi) \, . \tag{10.74}$$

The sum over θ_ℓ in the Z_n-partition function (10.50) becomes an integral such that the partition function of the $U(1)$ model takes the form

$$Z(\beta) = \int \prod_\ell d\theta_\ell \, e^{-\beta S} \, . \tag{10.75}$$

with the action (10.73). The dual of the $U(1)$ gauge theory is the $n \to \infty$ limit of the spin systems dual to the Z_n gauge models. In particular the sum in (10.52) turns into an integral which can be expressed in terms of modified Bessel functions:

$$c_k(\beta) = \frac{1}{2\pi} \int_0^{2\pi} d\theta \, e^{\beta \cos\theta} \cos(k\theta) = I_k(\beta), \quad k \in \mathbb{Z} \, . \tag{10.76}$$

Now the sum defining the constraint (10.55) must vanish in $\mathbb{Z}$ and not only in $\mathbb{Z}_n$ such that the partition function has the dual representation

$$Z(\beta) = \text{const} \times \sum_{\{s\}} \prod_{\langle *x, *y\rangle} I_{s*_x - s*_y}(\beta) \, , \tag{10.77}$$

where one averages over all spin configuration $\{s*_x \in \mathbb{Z}\}$ defined on the sites of the dual lattice. For weak couplings one can use the approximation

$$I_k(\beta) \longrightarrow \frac{e^\beta}{\sqrt{2\pi\beta}} \, e^{-k^2/4\beta} \quad \text{for} \quad \beta \to \infty \tag{10.78}$$

to further simplify the partition function. Various expressions for the partition function in the weak-coupling regime and physical pictures of the phases have been obtained in [16].

10.5   Duality Transformation of Four-Dimensional Z_n Gauge Model

The dual of an Abelian lattice gauge theory in four dimensions is again an Abelian lattice gauge theory [4]. In particular, the Z_2, Z_3, and Z_4 systems are self-dual and show one phase transition at their self-dual points. Z_n gauge theories based on the Wilson action (10.49) are no longer self-dual for $n \geq 5$ and show two phase transitions with a massless phase appearing between the strong- and weak-coupling phases [17–19]. However, systems with gauge-invariant Villain action [20] instead of the Wilson action are self-dual for all n.

To find the dual of the Z_n gauge theory with Wilson action, we proceed as in Sect. 10.4 up to the result (10.56). To solve the constraint (10.55), we again switch to the dual lattice. In four dimensions the dual of a plaquette p is a plaquette $*p$ on the dual lattice, and the dual of a link ℓ is a cube $*c$ on the dual lattice. We set

$k_p = \theta_{*p}$. The constraint (10.55) on the original lattice translates into

$$\sum_{*p \in \partial^*c} \theta_{*p} = 0 \bmod n \tag{10.79}$$

on the dual lattice. Since the configuration $\{\theta_{*p}\}$ has no circulation, it can be written as (generalized) gradient of a configuration defined on the dual links,

$$\theta_{*p} = \sum_{*\ell \in \partial^*p} \theta_{*\ell} , \tag{10.80}$$

where the summation symbol is defined to include the sign corresponding to the relative orientation of $*p$ and $*\ell \in \partial^*p$. In terms of the unconstrained link variables $\{\theta_{*\ell}\}$, the partition function takes the form

$$Z(\beta) = \text{const} \times \sum_{\{\theta_{*\ell}\}} e^{-*\beta \, {}^*S} , \tag{10.81}$$

where the action contains a sum over the plaquettes of the dual lattice:

$$^*S = \sum_{*p} {}^*s \left(\theta_{*p} \right) . \tag{10.82}$$

The contribution of a single plaquette to the Boltzmann weight is

$$e^{-*\beta \, {}^*s(\theta)} = \sum_{\ell=0}^{n-1} e^{\beta \cos(2\pi \ell/n)} \cos\left(2\pi \ell \frac{\theta}{n} \right) . \tag{10.83}$$

From Table 10.1 we can read off the single plaquette action of the dual Z_2, Z_3, and Z_4 gauge models. They define the Wilson actions

$$^*S(\theta_p) = \text{const}(\beta) - \sum_{*p} \cos\left(\frac{2\pi \theta_{*p}}{n} \right), \quad n = 2, 3, 4 , \tag{10.84}$$

on the dual lattice with dual couplings $^*\beta = {}^*\beta(\beta)$ given Table 10.1. We conclude that the Z_2, Z_3, and Z_4 gauge models with Wilson action are self-dual in four dimensions. If any of these models possess one critical coupling, then it must be the self-dual coupling defined by $^*\beta = \beta$. The self-dual couplings for the Z_2, Z_3, and Z_4 gauge theories are listed in Table 10.3.

Probably the most attractive feature of duality transformations is the relation between strong-coupling and weak-coupling regimes or between high-temperature and low-temperature phases of dual theories. For small couplings we can trust weak-coupling perturbation theory, and with duality we may use the perturbative

Table 10.3 Self-dual
couplings $\beta(^*\beta = \beta)$ of the
Z_n gauge theories in four
dimensions

	$n = 2$	$n = 3$	$n = 4$
$\beta_{\text{self–dual}}$	$\frac{1}{2}\log(1 + \sqrt{2})$	$\frac{2}{3}\log(1 + \sqrt{3})$	$\log(1 + \sqrt{2})$
	≈ 0.44069	≈ 0.67004	≈ 0.88137

results to study the non-perturbative sector of the dual theory. Further results on dualities, in particular on the representation of various correlation functions in the dual model, can be found in the reviews cited at the end of this chapter.

10.6   Problems

10.1 (Self-duality of Potts Chain in Magnetic Field) We wish to show that the q-state Potts chain with partition function (8.34) admits a duality transformation. To this aim we consider the auxiliary function

$$\tilde{Z}_\Lambda(\zeta, z) = ((\zeta - 1)(z - 1))^{-N/2} Z_\Lambda(\zeta, z)$$

$$= q' \left(\frac{\zeta - 1}{z - 1} \right)^{N/2} + \left(\frac{\lambda_+^2}{(\zeta - 1)(z - 1)} \right)^{N/2} + \left(\frac{\lambda_-^2}{(\zeta - 1)(z - 1)} \right)^{N/2}$$

depending on the couplings $z = e^{2\beta h}$ and $\zeta = e^{\beta J}$.

1. Show that $\tilde{Z}$ is invariant under duality transformation

$$(^*\zeta - 1)(z - 1) = q, \quad (^*z - 1)(\zeta - 1) = q$$

according to

$$\tilde{Z}_\Lambda (\zeta, z) = \tilde{Z}_\Lambda (^*\zeta, ^*z) .$$

Actually all three terms between brackets in the right-hand side of the formula for $\tilde{Z}_\Lambda$ are invariant.

2. Show that the duality relations relating the old to the new couplings define two broken linear Möbius transformations

$$^*\zeta = \frac{z + q - 1}{z - 1} \quad \text{and} \quad ^*z = \frac{\zeta + q - 1}{\zeta - 1} ,$$

and these transformations map circles to circles.

3. The partition function of the *Ising chain* is given by

$$Z_\Lambda(\zeta, z) = \left(\frac{(\zeta - 1)(z - 1)}{(^*\zeta - 1)(^*z - 1)} \right)^{N/2} Z_\Lambda(^*\zeta, ^*z),$$

and in Sect. 8.7.1 we determined the Lee-Yang zeros of the partition function in the complex fugacity plane. Use the duality transformation to show that the zeros of the partition function in the complex ζ-plane (for fixed $\beta h \in \mathbb{R}$) are on the imaginary axis. Compare your results with [21].

10.2 (Self-duality of Two-Dimensional Potts Models) Prove that the standard q-state Potts model without external field is self-dual in two dimensions. Its energy function is given in (6.10) wherein $h = 0$. Show that the relation between the couplings of the Potts model and dual Potts model is given by (10.29). Use this result to find the critical temperature of the two-dimensional Potts models.

10.3 (Dual of Z_6 Gauge Theory in Three Dimensions) Calculate the dual model of the Z_6 gauge theory with Wilson action in three dimensions. Show that the dual model belongs to the class of *generalized* Potts models.

References

1. R. Peierls, Statistical theory of adsorption with interaction between the adsorbed atoms. Proc. Camb. Phil. Soc. **32**, 471 (1936)
2. H.A. Kramers, G.H. Wannier, Statistics of the two-dimensional ferromagnet. Part I. Phys. Rev. **60**, 252 (1941)
3. H.A. Kramers, G.H. Wannier, Statistics of the two-dimensional ferromagnet. Part II. Phys. Rev. **60** (1941) 263
4. F. Wegner, Duality in generalized Ising models and phase transitions without local order parameter. J. Math. Phys. **12**, 2259 (1971)
5. L. Kadanoff, Lattice Coulomb representations of two-dimensional problems. J. Phys. **A11**, 1399 (1978)
6. R. Savit, Duality in field theory and statistical systems. Rev. Mod. Phys. **52**, 453 (1980)
7. C. Gruber, A. Hintermann, D. Merlini, *Group Analysis of Classical Lattice Systems*. Lecture Notes in Physics, vol. 60 (Springer, Berlin, 1977)
8. J.L. Lebowitz, A.E. Mazel, Improved Peierls argument for high-dimensions Ising models. J. Stat. Phys. **90**, 1051 (1998)
9. F.Y. Wu, The Potts model. Rev. Mod. Phys. **54**, 235 (1982) [Erratum-ibid **55**, 315 (1983)]
10. L. von Smekal, Universal aspects of QCD-like theories. Nucl. Phys. Proc. Suppl. **228**, 179 (2012)
11. W. Schwalm, B. Moritz, M. Giona, M. Schwalm, Vector difference calculus for physical lattice models. Phys. Rev. **E59**, 1217 (1999)
12. G. Bhanot, M. Creutz, Phase diagram of Z(N) and U(1) gauge theories in three dimensions. Phys. Rev. **D21**, 2892 (1980)
13. W.J. Blöte, R.H. Swendsen, First order phase transitions and the three state Potts model. J. of Appl. Phys. **50**, 7382 (1979)
14. A.M. Polyakov, Compact gauge fields and the infrared catastrophe. Phys. Lett. **B59**, 82 (1975)
15. A.M. Polyakov, Quark confinement and topology of gauge groups. Nucl. Phys. **B120**, 429 (1977)
16. T. Banks, R. Myerson, J. Kogut, Phase transitions in Abelian lattice gauge theories. Nucl. Phys. **B129**, 493 (1977)
17. J.M. Drouffe, Series analysis in four-dimensional Z_n lattice gauge systems. Phys. Rev. **D18**, 1174 (1978)

18. M. Creutz, L. Jacobs, C. Rebbi, Experiments with a gauge invariant Ising system. Phys. Rev. Lett. **42**, 1390 (1979)
19. A. Ukawa, P. Windey, A.H. Guth, Dual variables for lattice gauge theories and the phase structure of Z(N) systems. Phys. Rev. **D21**, 1023 (1980)
20. J. Villain, Theory of one- and two-dimensional magnets with an easy magnetization plane. II. The planar, classical, two-dimensional magnet. J. de Phys. **36**, 581 (1975)
21. F.Y. Wu, Self-dual property of the Potts model in one dimension. cond-mat/9805301

Renormalization Group on the Lattice

11

Previously we considered a variety of equilibrium systems which undergo second-order phase transitions. In this chapter we will show how the idea of scaling leads to a universal theory of critical phenomena, and we will derive some exact results for order-disorder transitions.

Simulations of systems with second-order transition characterized by an order parameter show typical configurations in the high- and low-temperature phases and in the vicinity of a critical point. Characteristic configurations of the two-dimensional Ising model near the critical temperature are depicted in Fig. 11.1. At high temperature the spins are randomly oriented, and there is only a short-range correlation between the fluctuating spins. With increasing temperature the system becomes more and more disordered, the already small regions of aligned spins become even smaller, and the correlation length tends to zero. At low temperature, on the other hand, we typically see aligned spins within macroscopic *domains* and some few, finite regions with an opposite alignment. With decreasing temperature the fluctuations over the mean field become more and more decorrelated, and as a result the correlation length decreases when we cool the system. When we approach the critical point by heating the system in the low-temperature phase or by cooling it in the high-temperature phase, then long-wavelength excitations are most easily excited and dominate the properties in the critical region. At the critical point, there are domains of arbitrary size, and we may not be able to distinguish between images taken at different length scales. The observed scale invariance is rather unusual since specific physical systems usually have specific scales. Therefore, in the vicinity of a critical point, these specific scales must in some sense become irrelevant.

What happens to typical configurations in the high- and low-temperature phases when we gradually change the scale (and not the temperature) by some coarse-graining procedure, for example, by decreasing the resolution of our microscope? In the disordered high-temperature phase, the small domains become even smaller, and this has the same effect as an increase in temperature. In contrast, in the ordered low-temperature phase, the typical size of the small domains with nonaligned spins

A. Wipf, *Statistical Approach to Quantum Field Theory*, Lecture Notes in Physics 992, https://doi.org/10.1007/978-3-030-83263-6_11

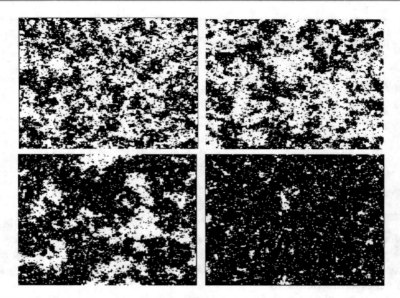

Fig. 11.1 Typical configurations from Monte Carlo simulations above and below the critical temperature $T_c = 2.2692$. The configurations belong to ensembles with $T = 2.5$ (upper left), $T = 2.4$ (upper right), $T = 2.3$ (lower left), and $T = 2.2$ (lower right)

becomes even smaller, and this effect is similar to the one obtained by lowering the temperature of the system. In either case a change in scale with decimation factor $b > 1$ decreases the correlation length and drives the system away from its critical point. Only at criticality, where the correlation length diverges, does a typical configuration not change under coarse graining. These considerations lead to the natural question whether a scale transformation is indeed equivalent to a change in temperature and of further coupling constants. Thereby the word equivalent means that the partition function and correlation functions do not alter.

K. Wilson was awarded the Nobel Prize for his major contributions to the physics of scale transformations or more generally of renormalization group (RG) transformations. This non-perturbative approach to the description of critical phenomena has become a very powerful tool in statistical physics as well as in quantum field theory. Early contributions to quantum field theory and particle physics have been obtained by Stueckelberg, Peterman, Gell-Man, Low, and Brezin. Some ten years later, Kadanov [1], Fisher [2, 3], and Wilson [4–6] developed the renormalization group method in statistical systems. The textbooks [7–11] present the renormalization group approach within quantum field theory and statistical physics with applications to critical phenomena. The following sections are devoted to the following problem: What general properties of the system may be extracted from RG transformations?

There are many ways to construct scale transformations or more general RG transformations. Examples of real space RG transformations are the cumulant method, finite-cluster method, Migdal-Kadanov transformation, and Monte Carlo renormalization. In particular, the latter method yields precise values for the critical exponents and will be discussed. Alternatively one may set up the RG transformation in momentum space with continuous momenta in the Brillouin zone. The momentum-space RG transformation may be performed by integrating out the high momentum modes. Since high momenta correspond to *short* length scales, this amounts to a coarse graining in real space. In momentum space the rescaling factor may be continuous and arbitrarily close to 1. Hence we may consider infinitesimal transformations. Examples of momentum-space RG transformations include the ε expansion, Callan-Symanzik equation, and functional renormalization group equations. The latter can be formulated in the continuum and will be discussed in Chap. 12.

11.1 Decimation of Spins

The decimation of spins can be done exactly for the Ising chain with external field. The "thinned out" system is identical to the original system at different temperature and with different external field. However, for spin systems in two and more dimensions, each decimation of spins generates new terms in the energy function, and the iterated decimation cannot be performed analytically as for the Ising chain.

11.1.1 Ising Chain

We consider the partition function of the periodic Ising chain with N spins and energy proportional to

$$- \beta H = K \sum_{x=1}^{N} s_x s_{x+1} + h \sum_{x=1}^{N} s_x \quad \text{with} \quad K = \beta J \ . \tag{11.1}$$

We assume N to be an even number. Summing only over every second spin ($b = 2$), i.e. over spins at even lattice points, the partition function takes the form

$$Z(N, K, h) = \sum_{s_1, s_2, \ldots} e^{K s_1 s_2 + \frac{1}{2} h(s_1 + s_2)} e^{K s_2 s_3 + \frac{1}{2} h(s_2 + s_3)} \times \ldots$$

$$= \sum_{s_1, s_2, \ldots} e^{K(s_1 s_2 + s_2 s_3) + \frac{1}{2} h(s_1 + 2 s_2 + s_3)} \times \ldots$$

$$= \sum_{s_1, s_3, \ldots} \left(e^{(K + \frac{1}{2} h)(s_1 + s_3) + h} + e^{-(K - \frac{1}{2} h)(s_1 + s_3) - h} \right) \cdots \ . \tag{11.2}$$

Thus, after decimation we obtain an Ising-type system with spins located at the odd lattice points. We now may introduce new coupling constants K', h' as well as a function $g(K, h)$ such that

$$e^{(K+\frac{1}{2}h)(s_1+s_3)+h} + e^{-(K-\frac{1}{2}h)(s_1+s_3)-h} = e^{2g(K,h)}e^{K's_1s_3+\frac{1}{2}h'(s_1+s_3)} . \tag{11.3}$$

We shall calculate the new couplings and g at a later stage. Inserting the expression (11.3) for every factor in (11.2) yields the partition function on the thinned out lattice with couplings K', h':

$$Z(N, K, h) = e^{Ng} \sum_{s_1, s_3, \dots} e^{K's_1s_3+\frac{1}{2}h'(s_1+s_3)} e^{K's_3s_5+\frac{1}{2}h'(s_3+s_5)} \times \dots$$

$$= e^{Ng} Z\left(\frac{N}{2}, K', h'\right) . \tag{11.4}$$

We summarize this remarkable result: the decimation of spins reproduces the Ising model on the thinned out chain with twice the lattice spacing. The energy on the coarser lattice has the same functional form as on the microscopic lattice:

$$\beta H \longrightarrow \beta'H' - g(K, h)N, \quad -\beta'H' = K' \sum_{x' \text{ odd}} s_{x'}s_{x'+2} + h' \sum_{x' \text{ odd}} s_{x'} . \tag{11.5}$$

We have used a very particular *decimation procedure* to get to this transformation of the energy function. Other decimation procedures, where the set of degrees of freedom after decimation is not necessarily a subset of the original degrees of freedom, shall be discussed below.

In order to extract the new coupling constants, we evaluate Eq. (11.3) for three different configurations of the two spins (s_1, s_3). We have

$$(s_1, s_3) = (1, 1) : \qquad 2e^h \cosh(2K + h) = e^{2g}e^{K'+h'}$$

$$(s_1, s_3) = (-1, -1) : \qquad 2e^{-h} \cosh(2K - h) = e^{2g}e^{K'-h'}$$

$$(s_1, s_3) = (1, -1) : \qquad 2\cosh(h) = e^{2g}e^{-K'} .$$

Solving these equations for K', h' and g, we obtain the map

$$K \xrightarrow{R_2} K' = \frac{1}{4}\log\frac{\cosh(2K + h)\cosh(2K - h)}{\cosh^2 h}$$

$$h \xrightarrow{R_2} h' = h + \frac{1}{2}\log\frac{\cosh(2K + h)}{\cosh(2K - h)} \tag{11.6}$$

$$g(K, h) = \frac{1}{8}\log\left(16\cosh(2K + h)\cosh(2K - h)\cosh^2 h\right) .$$

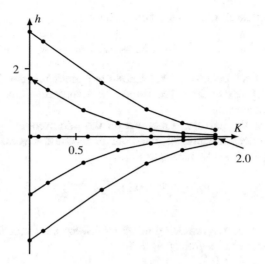

Fig. 11.2 The flow of the couplings (K, h) under repeated application of the decimation transformation R_2 for the Ising chain. For the chain the exact transformation is known and reproduces the energy function with nearest-neighbor coupling. The same flow expressed in terms of the variables e^{-2h} and e^{-2K} has been obtained in [12]

$$
\begin{array}{c}
\bullet \; K \; \bullet \; K \; \bullet \; K \; \bullet \; K \; \bullet \; K \; \bullet \; K \; \bullet \; K \; \bullet \; K \; \bullet \; K \; \bullet \; K \; \bullet \; K \; \bullet \\
R_2 \downarrow \\
\bullet \quad K' \quad \bullet \quad K' \quad \bullet \quad K' \quad \bullet \quad K' \quad \bullet \quad K' \quad \bullet \quad K' \quad \bullet \\
R_2 \downarrow \\
\bullet \qquad K'' \qquad \bullet \qquad K'' \qquad \bullet \qquad K'' \qquad \bullet
\end{array}
$$

Fig. 11.3 Decimation of degrees of freedom by successive blocking transformations. Each decimation R_2 reproduces the energy function with different couplings and doubles the distance between nearest neighbors

Figure 11.2 shows the trajectories of the couplings in the (K, h)-plane after the repeated application of the decimation transformation R_2. The decimation doubles the distance between neighboring lattice sites as sketched in Fig. 11.3, and this scale factor appears as index of R. The initial couplings in the plot are $K = 2$ and $h \in \{\pm 0.2, \pm 0.05, 0\}$. The series of points

$$
(K, h) \xrightarrow{R_2} (K', h') \xrightarrow{R_2} (K'', h'') \xrightarrow{R_2} (K''', h''') \xrightarrow{R_2} \dots
$$

in the plane of couplings is *attracted* to the axis $K = 0$. This means that the nearest-neighbor coupling K decreases for every decimation of spins. If the magnetic field of the microscopic system is zero, then it stays zero since the Z_2 symmetry without field is preserved by the decimation map. Thus the line $h = 0$ is a trajectory of the renormalization group (RG). More generally, if R_b is a renormalization group transformation with scaling factor b, then we have

$$R_b \circ R_b = R_{b^2} .$$

(11.7)

Note that there is no inverse of R_b, since we cannot reverse the procedure of integrating out degrees of freedom. Hence the transformations R_b form only a semigroup and not a group.

Let us see how the partition functions on the microscopic lattice and on the coarse-grained lattices are related when R_2 is repeated several times. For two decimations the relation (11.4) leads to

$$Z(N, K, h) = e^{Ng(K,h)} e^{\frac{1}{2}Ng(K',h')} Z\left(\frac{N}{4}, K'', h''\right).$$

(11.8)

Further iterations yield the following relation between the free energy densities of the microscopic and coarse-grained systems:

$$f(K, h) = -\frac{1}{\beta}\left(g(K, h) + \frac{1}{2}g(K', h') + \frac{1}{2^2}g(K'', h'') + \frac{1}{2^3}g(K''', h''') + \ldots\right).$$

(11.9)

Note that the function g has the same form for every iteration step. For simplicity we consider the decimation without magnetic field:

$$K' = R_2(K) = \frac{1}{2}\log\cosh(2K), \quad g = \frac{1}{4}\log\left(4\cosh(2K)\right).$$

(11.10)

Only the couplings $K = 0$ and $K = \infty$ are inert under decimation—they are *fixed points* of the RG transformation R_2. These two couplings represent the high-temperature and low-temperature fixed points of the Ising chain.

The previous derivation shows that the correlation between two spins defined on both the microscopic and diluted lattice is the same before and after decimation:

$$\frac{1}{Z(N, K)} \sum_{\Omega} s_{x'} s_{y'} \exp\left(K \sum_{\langle u,v \rangle} s_u s_v\right)$$

$$= \frac{1}{Z(\frac{1}{2}N, K')} \sum_{\Omega'} s_{x'} s_{y'} \exp\left(K' \sum_{\langle u',v' \rangle} s_{u'} s_{v'}\right).$$

In the decimation (11.4), the sites x' and y' must be odd lattice points. Now we rescale the coarse-grained lattice such that the distance between neighboring sites shrinks from 2 to 1. Thus, if two points are separated by a distance $2n$ on the microscopic lattice, then they are separated by a distance n on the rescaled coarse-grained lattice. It follows that the two-point correlation

$$\langle s_x s_y \rangle \sim e^{-|x-y|/\xi}, \quad |x-y| \gg \xi \tag{11.11}$$

falls off faster after decimation and rescaling. We conclude that every transformation R_2 halves the correlation length ξ, i.e.,

$$\xi' = \frac{\xi}{2} . \tag{11.12}$$

According to (8.18) the correlation length diverges at the low-temperature fixed point and vanishes at the high-temperature fixed point. The low-temperature fixed point is a critical point, whereas at the high-temperature fixed point, the interaction vanishes. The RG trajectories flow into the trivial fixed point with $\xi = 0$ since the coupling constant K and correlation length ξ both decrease with every decimation of spins.

11.1.2 The Two-Dimensional Ising Model

Let us consider the two-dimensional Ising model without external field, given by

$$\beta H = -K \sum_{\langle x, y \rangle} s_x s_y . \tag{11.13}$$

We shall see that the decimation generates, besides nearest-neighbor interactions, interactions between next-nearest neighbors. Iterating the decimation generates interactions between widely separated spins.

Again we consider the sum over all configuration in the partition function. In the decimation we sum over the spins marked by open circles in Fig. 11.4 and thus construct an effective spin model with spins located at the filled circles. These sites define the coarse-grained lattice. For example, the contribution of the spin located

Fig. 11.4 In one decimation step, one sums over the spins on the lattice sites marked with open circles

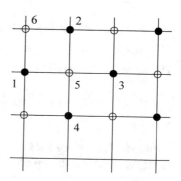

at lattice site denoted by 5 is given by

$$\sum_{s_5=\pm1} e^{K s_5(s_1+s_2+s_3+s_4)} = e^{K(s_1+s_2+s_3+s_4)} + e^{-K(s_1+s_2+s_3+s_4)} . \tag{11.14}$$

The most general Boltzmann weight involving s_1, s_2, s_3, and s_4 compatible with the symmetries of the system has the form

$$e^{2g} \exp \left(\frac{1}{2} K_1' \underbrace{(s_1s_2 + s_2s_3 + s_3s_4 + s_4s_1)}_{\text{NN}} + K_2' \underbrace{(s_1s_3 + s_2s_4)}_{\text{nNN}} + K_3' \underbrace{(s_1s_2s_3s_4)}_{\text{Q}} \right) ,$$

where (NN) and (nNN) denote nearest neighbors and next-nearest neighbors and (Q) represents squares. The last two expressions are equal for all configurations of spins s_1, s_2, s_3, and s_3 if the following independent equations are satisfied:

$$\begin{aligned}
(s_1, s_2, s_3, s_4) &= (1, 1, 1, 1): & 2\cosh(4K) &= e^{2g}e^{2K_1'+2K_2'+K_3'} \\
(s_1, s_2, s_3, s_4) &= (1, -1, -1, -1): & 2\cosh(2K) &= e^{2g}e^{-K_3'} \\
(s_1, s_2, s_3, s_4) &= (1, 1, -1, -1): & 2 &= e^{2g}e^{-2K_2'+K_3'} \\
(s_1, s_2, s_3, s_4) &= (1, -1, 1, -1): & 2 &= e^{2g}e^{-2K_1'+2K_2'+K_3'} .
\end{aligned}$$

Solving these equations for the new couplings, we obtain the relations

$$\begin{aligned}
K_1' &= 2K_2' = \tfrac{1}{4} \log\cosh(4K) \\
K_3' &= \tfrac{1}{8} \log\cosh(4K) - \tfrac{1}{2} \log\cosh(2K) \\
g &= \tfrac{1}{16} (\log\cosh(4K) + \log\cosh(2K) + 8\log 2) .
\end{aligned} \tag{11.15}$$

We get such contributions from all spins on the open circles. Thereby a term like $\exp(K_1's_1s_2/2)$ occurs in the sum over s_6 as well. Denoting the spin configuration on the coarse-grained lattice by ω', we can write

$$Z(V, K) = Z'\left(\frac{V}{2}, K'\right) = \sum_{\omega'} e^{-(\beta H)'(\omega')} . \tag{11.16}$$

The so-called *Landau-Ginzburg-Wilson* energy function

$$-(\beta H)' = Vg + K_1' \sum_{\text{NN}} s_{x'}s_{y'} + K_2' \sum_{\text{nNN}} s_{x'}s_{y'} + K_3' \sum_{\text{Q}} s_{x'}s_{y'}s_{u'}s_{v'} , \tag{11.17}$$

where x', y', u', v' are sites on the dilute lattice, is not of the same form as the energy function of the microscopic system.

If we iterated the decimation without approximation, we would generate more and more terms involving interactions between widely separated spins. To proceed analytically one needs some sort of truncation. In a first attempt, we could set the coupling constants $K'_2 = K'_3$ equal to zero, but this approximation would be quite insufficient. This becomes apparent by inspecting the fixed points which are simply $K_1 = 0$ and $K_1 = \infty$ as in the one-dimensional model. Thus we would *not* observe any phase transition. A better approximation is to set $K'_3 = 0$ and to count next-nearest neighbors as nearest neighbors. With this truncation the partition function becomes

$$Z(V, K) = e^{Vg} \sum_{\omega'} \exp\left(K' \sum_{\langle x', y'\rangle} s_{x'} s_{y'} \right), \quad K' = K'_1 + K'_2 . \tag{11.18}$$

Inserting the results in (11.15), we find the renormalization group map

$$K \longrightarrow K'(K) = \frac{3}{8} \log \cosh 4K , \tag{11.19}$$

which is depicted in Fig. 11.5. The map has fixed points at $K = 0$, $K = \infty$ and at

$$K^* = 0.50698 . \tag{11.20}$$

This is reasonably close to the known critical point $K_c = 0.4407$. Actually the fixed point is unstable: starting at $K \neq K^*$, the flow drives the system either to the high-temperature fixed point at $K = 0$ or the low-temperature fixed point at $K = \infty$.

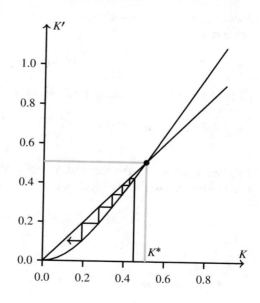

Fig. 11.5 The fixed point of the truncated block-spin transformation (11.19). The nontrivial fixed point at $K^* = 0.50698$ is unstable

11.2   Fixed Points

We now turn toward a more general discussion of the RG method. For this purpose we consider a d-dimensional lattice model, characterized by its coupling constants

$$K = \{K_A | A \subset \Lambda\} = (K_1, K_2, \dots) \,, \tag{11.21}$$

where we enumerated the subsets of the lattice.[1] The set of couplings should be complete in the following sense: A RG transformation that substitutes b^d microscopic degrees of freedom on the lattice Λ by one degree of freedom on the coarser lattice Λ' leads to an *energy function* with the same kind of interactions as the energy function of the original system. Thus, starting with the energy function of the form

$$H(\omega) = -\sum_{A \subset \Lambda} K_A s_A, \quad s_A = \prod_{x \in A} s_x \,, \tag{11.22}$$

the renormalized energy function on Λ' has the same form, up to an additive extensive constant,

$$H(\omega) \longrightarrow H'(\omega') - Vg(K), \quad H'(\omega') = -\sum_{A \subset \Lambda'} K'_A S_A, \tag{11.23}$$

where A denotes the same set as in (11.21). We thereby assume a set $\{A\}$ to exist on the original lattice as well as on the diluted lattice. Furthermore we assume the reduced degrees of freedom $S_{x'}$ to exhibit the same algebraic properties as the s_x.

Note that the constant term $Vg(K)$ in (11.23) occurs in all RG transformations. A dilution of the system changes the couplings according to the *renormalization group map*

$$K'_i = R_i(K_1, K_2, \dots) \,, \tag{11.24}$$

where the partition function remains unchanged, i.e,

$$e^{-F(V,K)} = \sum_{\omega \in \Omega} e^{-H(\omega)} = e^{Vg(K)} \sum_{\omega' \in \Omega'} e^{-H'(\omega')} = e^{Vg(K)-F(V',K')} \,. \tag{11.25}$$

Now we assume that the thermodynamic limit $V \to \infty$ exists. Hence, the free energy densities of the two systems are related by the recursion relation

$$f(K) = b^{-d} f(K') - g(K), \quad V = b^d V' \tag{11.26}$$

---

[1] $x \in A$ may appear several times, similarly as in Sect. 8.6.

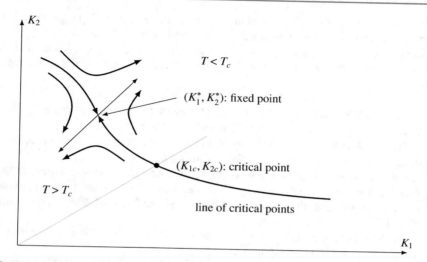

Fig. 11.6 The fixed point K^* has one relevant and one irrelevant direction

in the thermodynamic limit. This relation is quite familiar since it occurred in (11.5) in connection with the Ising chain. We have already argued that a *fixed point* K^* of the RG transformation either defines a critical point of the system characterized by a divergent correlation length $\xi_c = \infty$ or belongs to a noninteracting system with vanishing correlation length $\xi = 0$. The converse needs not be true, i.e., there may exist critical points which are *no* fixed points. Unfortunately, only simple models as the Ising chain may be described by a *finite* number of couplings. However, it is plausible to assume that the couplings K_A corresponding to long-range interactions are suppressed and thus may be neglected. Hence, in practice, one works with a finite number of coupling constants $\{K_1, \ldots, K_n\}$.

Now let us consider a two-dimensional space of couplings (K_1, K_2), where we denote the critical couplings by $K_c = (K_{1c}, K_{2c})$. The generic case as illustrated in Fig. 11.6 shows a *line of critical points*, where K^* lies on this curve. This may be explained as follows: since we absorbed the temperature in the couplings, a change of temperature in

$$(K_{1c}, K_{2c}) = (\beta J_1, \beta J_2)$$

is equivalent to a dilation in the space of coupling $\{K_i\}$. Thus, when we lower the temperature, then we move away from $K = 0$ on a ray with constant ratio K_2/K_1 and expect to find a critical point at some critical temperature T_c. This temperature will depend on this ratio K_2/K_1, and when the ratio changes, then the point

$$(K_{1c}, K_{2c}) = \left(\frac{J_1}{T_c}, \frac{J_2}{T_c}\right)$$

describes a curve in the (K_1, K_2)-plane. If we admit more couplings, we expect to find critical hypersurfaces on which $\xi = \infty$ instead of critical lines.

Let us now see how the properties of a statistical system are related to its critical points, critical surfaces, or fixed points. Some simple properties of the RG flow are:

- Near the critical surface, a RG trajectory moves away from the surface, since $\xi = \infty$ on the surface and every RG transformation reduces ξ.
- A RG transformation cannot change the phase of the system, since a dilution of the system cannot generate order from disorder and vice versa.
- If we start above the critical temperature, then the system should evolve to the free fixed point, characterized by $T = \infty$. In contrast, if we start below the critical temperature, then we should end up at the "ground state" fixed point $T = 0$.
- If the flow begins exactly on the critical surface, then it stays on this surface, since $\xi = \infty$ implies $\xi' = \xi/b = \infty$.
- It is the exception and not the rule that a critical point is a fixed point. In general we expect a finite set of isolated fixed points. Only in lower-dimensional systems there may be an infinite set of fixed points (cp. Sect. 12.5.1).

11.2.1 The Vicinity of a Fixed Point

Let $K^* = (K_1^*, K_2^*, \dots)$ be a fixed point of the RG transformation:

$$K^* = R(K^*) . \tag{11.27}$$

To examine the RG flow in the vicinity of K^*, we write $K = K^* + \delta K$ and linearize the flow around the fixed point:

$$K_i' = K_i^* + \delta K_i' = R_i(K_j^* + \delta K_j) = K_i^* + \frac{\partial R_i}{\delta K_j}\Big|_{K^*} \delta K_j + O(\delta K^2) .$$

We read off the linearized RG transformation:

$$\delta K_i' = \sum_j M_i{}^j \delta K_j, \qquad M_i{}^j = \frac{\partial R_i}{\partial K_j}\Big|_{K^*} . \tag{11.28}$$

This linear map determines the flow in the vicinity of the critical point. It is characterized by the eigenvalues and left eigenvectors of the associated matrix M:

$$\sum_j \Phi_\alpha^j M_j^i = \lambda_\alpha \Phi_\alpha^i = b^{y_\alpha} \Phi_\alpha^i . \tag{11.29}$$

If K^* lies on a critical surface, then a subset of eigenvectors span the space tangential to the critical surface at K^*. Note that the eigenvalue λ_α has been substituted by b^{y_α}. This is justified, since we have

$$\lambda_\alpha(b)\lambda_\alpha(b) = \lambda_\alpha(b^2)$$

by virtue of the semigroup property of the RG transformation. Thus every eigenvalue λ_α defines a critical exponent y_α which will enter our discussion of scaling relations below. We now consider the new variables

$$g_\alpha = \sum_i \Phi_\alpha^i \delta K_i \ . \tag{11.30}$$

These are projections of δK onto the eigenvectors Φ_α. We have

$$g'_\alpha = \sum_i \Phi_\alpha^i \delta K'_i = \sum_{ij} \Phi_\alpha^i M_i{}^j \delta K_j = \sum_j b^{y_\alpha} \Phi_\alpha^j \delta K_j = b^{y_\alpha} g_\alpha \ . \tag{11.31}$$

We now return to the recursion relation (11.26) for the free energy density. The contribution $g(K)$ thereby originates from integrating out the short-range fluctuations and thus represents a *smooth function*. Thus the singular part of the free energy density fulfills the homogeneous relation

$$f_s(K) = b^{-d} f_s(K') \ . \tag{11.32}$$

We linearize the flow around the fixed point and obtain the following scaling behavior for the singular part of the free energy density:

$$f_s(K^* + \delta K) = b^{-d} f_s(K^* + \delta K') \ . \tag{11.33}$$

We now omit the argument K^* and write

$$f_s(K^* + \delta K) \equiv f_s(g_1, g_2, \ldots), \qquad \delta K \stackrel{(11.30)}{=} \delta K(g) \ .$$

Performing ℓ iteration steps, we find

$$f_s(g_1, g_2, \ldots) = b^{-d\ell} f_s\left(b^{\ell y_1} g_1, b^{\ell y_2} g_2, \ldots\right) \ . \tag{11.34}$$

Depending on the sign of the exponent y_α, we observe a qualitatively different scaling behavior:

- For $y_\alpha > 0$ the deviation g_α increases continuously, and the flow moves the point $K^* + g_\alpha$ away from the fixed point K^*. This is called a *relevant perturbation*.

- For $y_\alpha < 0$ the deviation g_α decreases, and the flow carries the point $K^* + g_\alpha$ toward the fixed point K^*. This corresponds to an *irrelevant perturbation*.
- Deviations with $y_\alpha = 0$ are called *marginal*.

Physical quantities corresponding to relevant perturbations are, e.g., the temperature or the (dimensionless) external field:

$$t = \frac{T - T_c}{T_c} \equiv g_1 \quad \text{and} \quad h = g_2 . \tag{11.35}$$

Let us slightly reinterpret these findings: the RG transformation acts on the space of coupling constants or equivalently on the space of interactions which enter the energy function. In general, this is an ∞-dimensional space. Let us now consider again the general class of energy functions (11.22), i.e.,

$$H = -\sum_{A \subset \Lambda} K_A s_A \equiv -\sum K_i O_i . \tag{11.36}$$

An expansion of the Hamiltonian around the fixed point, $H = H^* + \delta H$, yields

$$H^* = -\sum_i K_i^* O_i, \quad \delta H = -\sum \delta K_i O_i = -\sum_\alpha g_\alpha Q_\alpha . \tag{11.37}$$

Applying the linearized RG transformation ℓ times changes H according to

$$H^* + \delta H \longrightarrow H^* - \sum g_\alpha' Q_\alpha \longrightarrow H^* - \sum g_\alpha'' Q_\alpha \longrightarrow \cdots$$
$$\longrightarrow H^* - \sum_\alpha b^{\ell y_\alpha} g_\alpha Q_\alpha ,$$

where the Q_α are called *scaling operators* and the g_α *scaling fields*. Operators with positive y_α are relevant, operators with $y_\alpha < 0$ are irrelevant, and operators with $y_\alpha = 0$ are marginal. The relevant operators of the Ising model are its energy H and its average field $\sum s_x$.

11.2.2 Derivation of Scaling Laws

Let us assume that $g_1 = t$ and $g_2 = h$ in (11.35) are relevant couplings and $g_3, g_4, \ldots$ are irrelevant ones. Moreover, we choose ℓ such that

$$b^{y_1 \ell} = \frac{1}{t} \quad \text{or} \quad b^\ell = t^{-1/y_1} . \tag{11.38}$$

Then the scaling relation (11.34) yields

$$f_s(K^* + \delta K) \equiv f_s(t, h, g_3, \dots) = t^{d/y_1} f_s\left(1, \frac{h}{t^{y_2/y_1}}, \frac{g_3}{t^{y_3/y_1}} \dots\right). \qquad (11.39)$$

Similarly, setting $b^\ell = h^{-1/y_2}$, we arrive at

$$f_s(t, h, g_3, \dots) = h^{d/y_2} f_s\left(\frac{t}{h^{y_1/y_2}}, 1, \frac{g_3}{h^{y_3/y_2}}, \dots\right). \qquad (11.40)$$

Note that the arguments

$$\frac{g_i}{t^{y_i/y_1}} \xrightarrow{t \to 0} 0 \quad \text{and} \quad \frac{g_i}{h^{y_i/y_2}} \xrightarrow{h \to 0} 0, \qquad i = 3, 4, \dots \qquad (11.41)$$

of the singular part f_s of the free energy density vanish at the fixed point, since the exponents of t and h are negative.

Now we can relate the thermodynamic critical exponents to the exponents of the linearized RG transformation by differentiating the singular part of the free energy density with respect to the relevant couplings t and h. At this point it is useful to collect the relevant thermodynamic quantities as introduced in Sect. 7.3:

$$\text{Magnetization:} \qquad m(t, h) = \langle s_x \rangle = -\frac{\partial f}{\partial h} \qquad (11.42)$$

$$\text{Susceptibility:} \qquad \chi(t, h) = \beta \sum_x \langle s_0 s_x \rangle_c = -\frac{\partial^2 f}{\partial h^2} \qquad (11.43)$$

$$\text{Internal energy density:} \quad u(t, h) = \lim_{\Lambda \to \mathbb{Z}^d} \frac{1}{V} \langle H \rangle = \frac{\partial(\beta f)}{\partial \beta} \qquad (11.44)$$

$$\text{Specific heat:} \qquad c(t, h) = \frac{\partial u}{\partial T} = -\beta^2 \frac{\partial u}{\partial \beta} = -T \frac{\partial^2 f}{\partial T^2}. \qquad (11.45)$$

Their singular behavior is characterized by the critical exponents α, β, γ, and δ:

$$c(t, 0) \sim E_\pm |t|^{-\alpha} \quad , \quad m(t, 0) \sim B t^\beta \qquad (11.46)$$

$$\chi(t, 0) \sim A_\pm |t|^{-\gamma} \quad , \quad m(0, h) \sim |h|^{1/\delta} \text{sign}(h). \qquad (11.47)$$

The correlation length and two-point function define the critical exponents η and ν:

$$\text{Correlation length:} \qquad \xi^{-1} = -\lim_{|x| \to \infty} \frac{1}{|x|} \log\langle s_0 s_x \rangle_c \sim |t|^\nu \qquad (11.48)$$

$$\text{Green's function:} \qquad \langle s_0 s_x \rangle \sim \frac{1}{|x|^{d-2+\eta}}. \qquad (11.49)$$

Since the specific heat is proportional to the second t-derivative of f, we conclude

$$f_s \sim |t|^{2-\alpha} . \tag{11.50}$$

A comparison with (11.39) immediately results in $2 - \alpha = d/y_1$. A similar reasoning for the other derivatives of f yields further relations between the critical exponents β, γ, and δ and the relevant exponents y_1 and y_2. To relate ν and η to the relevant exponents, we must allow for an inhomogeneous external field $h(x)$, similarly as in the Ornstein-Zernike extension of Landau's theory. Finally one ends up with the following important relations between $\alpha, \beta, \gamma, \delta, \nu, \eta$ and y_1, y_2:

$$2 - \alpha = \frac{d}{y_1} \quad , \quad \beta = \frac{d - y_2}{y_1}$$

$$\gamma = \frac{2y_2 - d}{y_1} \quad , \quad \frac{1}{\delta} = \frac{d - y_2}{y_2} \tag{11.51}$$

$$\nu = \frac{1}{y_1} \quad , \quad d - 2 + \eta = 2(d - y_2) .$$

Since the six critical exponents only depend on the two relevant exponents y_1 and y_2 and the dimension of the system, we find the following scaling relations between the critical exponents:

$$\gamma = \nu(2 - \eta) \quad \text{(Fisher)}$$

$$\alpha + 2\beta + \gamma = 2 \quad \text{(Rushbrooke)}$$

$$\gamma = \beta(\delta - 1) \quad \text{(Widom)} \tag{11.52}$$

$$\nu d = 2 - \alpha \quad \text{(Josephson, "hyperscaling relation")} .$$

The critical exponents of the two- and three-dimensional Ising models and the three-dimensional Heisenberg model are found in Table 11.1 As expected the scaling relations are fulfilled. In $d \neq 4$ they are not fulfilled for the mean field exponents in the last row of Table 11.1.

The critical exponents do not depend on the microscopic details of the interactions, since the correlation length diverges at the critical point. This *universality* of the critical behavior arises from the singular part of the free energy which is

Table 11.1 Critical exponents for the Ising model and the Heisenberg model, cp. Table 7.2

	α	β	γ	δ	η	ν
$d = 2$ Ising	0	1/8	7/4	15	1/4	1
$d = 3$ Ising	0.11	0.32	1.24	4.8	0.05	0.63
Class, Heisenberg $d = 3$	−0.12	0.36	1.37	4.6	0.04	0.7
MFA, arbitrary d	0	1/2	1	3	0	1/2

independent of the (infinitely many) *irrelevant* couplings $g_i, i \geq 3$. The critical exponents of a system are fixed by the space dimension, the number of components of the order parameter, and the symmetry of the interaction. Two models which share these properties belong to the same universality class.

11.3 Block-Spin Transformation

The Monte Carlo renormalization group (MCRG) method has been developed by Ma, Swendsen and others [13–15]. This powerful method is based on the block-spin transformation which maps a microscopic system to a coarse-grained system. It is a generalization of the previously considered decimation procedure. We present the transformation for two-dimensional Ising-type models on a square lattice with general energy function

$$\beta H = - \sum_{A \subset \Lambda} K_A s_A, \qquad s_A = \prod_{x \in A} s_x, \qquad s_x \in \{-1.1\}, \qquad (11.53)$$

and subject to periodic boundary conditions. We absorb the inverse temperature β in the couplings K_A and we write H instead of βH.

We partition the lattice into blocks of size b^2 and assign a *block spin* to the spins in every block. Figure 11.7 shows a possible blocking with rescaling factor of $b = 2$. Let us denote by $x = (x_1, x_2)$ with $x_i \in \{1, \dots, N\}$ the sites on a square lattice Λ. Then a block consisting of b^2 lattice points is mapped onto one lattice point x' of the coarse-grained lattice,

$$x'(x) = \big(x_1'(x_1), x_2'(x_2)\big) = \Big(\mathrm{ceil}\,\frac{x_1}{b}, \mathrm{ceil}\,\frac{x_2}{b}\Big), \qquad (11.54)$$

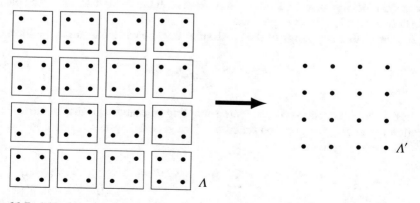

Fig. 11.7 A blocking transformation which combines four microscopic spins to one block spin

where ceil(x) is the smallest integer not less than x. If $b = 2$, then all sites $(1, 1)$, $(1, 2)$, $(2, 1)$, and $(2, 2)$ are mapped onto the site $(1, 1)$ on the diluted lattice Λ'. Thereby the lattice shrinks by the factor $1/b$.

A *block-spin transformation* maps a microscopic configuration $\omega = \{s_x\}$ with a certain probability onto a configuration $\omega' = \{S_{x'}\}$ on the coarser lattice and is characterized by its *kernel* $T(\omega', \omega)$ according to

$$\omega' = \sum_{\omega \in \Omega} T(\omega', \omega) \,, \tag{11.55}$$

where $T(\omega', \omega)$ is the probability that ω is mapped to ω'. This blocking kernel should satisfy the conditions

$$0 \leq T(\omega', \omega) \leq 1 \quad \text{and} \quad \sum_{\omega'} T(\omega', \omega) = 1 \,. \tag{11.56}$$

The Boltzmann factors of the microscopic configurations determine the Boltzmann factors of the block-spin configurations via

$$e^{-H'(\omega')} = \sum_{\omega \in \Omega} T(\omega', \omega)\, e^{-H(\omega)} \,. \tag{11.57}$$

The previously considered decimation of the Ising chain is a particular blocking transformation with kernel

$$T(\omega', \omega) = \prod_{x' \in \Lambda'} \delta\left(s_{2x'}, S_{x'}\right) \,,$$

i.e., two neighboring spins are mapped to one block spin. Thus, decimation is a particular blocking, where all but one spin from each block are discarded. The block spin is then equal to the one spin left.

Assuming that the energy of the block spins may be written again in the form

$$H'(\omega') = -\sum_{A' \subset \Lambda'} K_{A'} S_{A'} \tag{11.58}$$

we can derive a recursion relation for the coupling constants. Because of the second relation in (11.56), the partition function of the blocked system is equal to that of the microscopic system:

$$Z'_{H'} = \sum_{\omega' \in \Omega'} e^{-H'(\omega')} = \sum_{\omega' \in \Omega'} \sum_{\omega \in \Omega} T(\omega', \omega) e^{-H(\omega)} = \sum_{\omega \in \Omega} e^{-H(\omega)} = Z_H \,. \tag{11.59}$$

The connected correlation functions of the s_A are obtained by differentiating $\log Z$ with respect to the corresponding couplings, for example,

$$\langle s_A \rangle = \frac{\partial \log Z}{\partial K_A}$$

$$\langle s_A; s_B \rangle \equiv \langle s_A s_B \rangle - \langle s_A \rangle \langle s_B \rangle = \frac{\partial \langle s_A \rangle}{\partial K_B} . \tag{11.60}$$

The correlation functions of the blocked system are given by

$$\langle S_{A'} \rangle' \equiv \frac{\sum_{\omega'} S_{A'} e^{-H'(\omega')}}{\sum_{\omega'} e^{-H'(\omega')}} = \frac{\sum_{\omega'} S_{A'} \sum_{\omega} T(\omega', \omega) e^{-H(\omega)}}{\sum_{\omega} e^{-H(\omega)}} . \tag{11.61}$$

The last relation shows how one can calculate correlations of the block spins from the microscopic ensemble. To proceed we introduce the derivative of $\langle S_{A'} \rangle'$ in the blocked system with respect to the coupling constant K_B of the microscopic system:

$$\begin{aligned}
T_{A'B} &= \frac{\partial \langle S_{A'} \rangle'}{\partial K_B} = \frac{\partial}{\partial K_B} \frac{\sum_{\omega'} S_{A'} \sum_{\omega} T(\omega', \omega) e^{-H(\omega)}}{\sum_{\omega} e^{-H(\omega)}} \\
&= \frac{\sum_{\omega} \left\{ \sum_{\omega'} S_{A'} T(\omega', \omega) \right\} s_B e^{-H(\omega)}}{\sum_{\omega} e^{-H(\omega)}} - \langle S_{A'} \rangle' \langle s_B \rangle .
\end{aligned}$$

This yields

$$T_{A'B} = \left\langle s_B \sum_{\omega'} S_{A'} T(\omega', \omega) \right\rangle - \langle S_{A'} \rangle' \langle s_B \rangle , \tag{11.62}$$

and with the help of (11.61), all expectation values on the right-hand side can be calculated from the microscopic ensemble and the blocking kernel. We use the chain rule to find the alternative expression

$$T_{A'B} = \frac{\partial \langle S_{A'} \rangle'}{\partial K_B} = \sum_{C'} \frac{\partial \langle S_{A'} \rangle'}{\partial K_{C'}} \frac{\partial K_{C'}}{\partial K_B} = \sum_{C'} \langle S_{A'}; S_{C'} \rangle' \frac{\partial K_{C'}}{\partial K_B} . \tag{11.63}$$

Again the connected correlation functions on the right-hand side can be extracted from the microscopic ensemble. Thus, from the two expressions (11.62) and (11.63) for $T_{A'B}$, we can read off how the couplings of the blocked system vary when we vary the microscopic couplings.

In order to illustrate the MCRG transformation we return to the two-dimensional Ising model and combine four spins to one block spin as indicated in Fig. 11.7. For the blocking kernel, we choose the *majority rule*: four spins $\{s_x\}$ are mapped onto one block spin $S_{x'}$ according to

$$T(\omega', \omega) = \prod_{x' \in \Lambda'} t\left(S_{x'}, \sum_{x \in x'} s_x\right) \tag{11.64}$$

with

$$\sum s_x > 0 \Longrightarrow S_{x'} = 1 \qquad \text{with probability } 1$$

$$\sum s_x < 0 \Longrightarrow S_{x'} = -1 \qquad \text{with probability } 1 \tag{11.65}$$

$$\sum s_x = 0 \Longrightarrow \begin{cases} S_{x'} = +1 & \text{with probability } 1/2 \\ S_{x'} = -1 & \text{with probability } 1/2 \, . \end{cases}$$

Let us determine the RG transformation for the Ising model without external field on a small 4×4 lattice. The $2^{16} = 65536$ microscopic configurations are easily generated such that (11.57) allows for a direct calculation of the energy of the coarse-grained system. In two dimensions H' has no the same functional form as H, but on the blocked 2×2 lattice, there are only three different types of interactions: between nearest neighbors, next-nearest neighbors, and all four spins (for simplicity we drop the prime at the couplings):

$$\begin{aligned} H' = - &K_1 \sum_{x'} S_{x'} \left(S_{x'+(1,0)} + S_{x'+(0,1)}\right) \\ &- K_2 \sum_{x'} S_{x'} \left(S_{x'+(1,1)} + S_{x'+(1,-1)}\right) \\ &- K_3 \sum_{x'} S_{x'} S_{x'+(1,0)} S_{x'+(0,1)} S_{x'+(1,1)} \\ &- K_4 \sum_{x'} 1 \, . \end{aligned} \tag{11.66}$$

Note that we deliberately overcounted, since, for instance, the two terms in the second row, which are identical on a 2×2 lattice, are different on larger lattices. The $2^4 = 16$ configurations on the coarse-grained lattice fall into one of the four classes listed in Table 11.2, and members of the same class share the same Boltzmann factor. Hence we end up with four equations for the couplings K_1, K_2, K_3, K_4 in Eq. (11.66). In each class we pick the first representative and obtain the equations

$$e^{-H'(\omega_i')} = \sum_{\omega} T(\omega_i', \omega) \, e^{-H(\omega)} \equiv e^{c_i}, \qquad i = 1, 2, 3, 4 \, . \tag{11.67}$$

Table 11.2 Grouping of the blocked configurations on the 2×2 lattice into classes. Configurations in the same class have equal weight

Class	Configurations	$-H'$
$\mathscr{C}_1$	++ −− ++ −−	$8K_1 + 8K_2 + 4K_3 + 4K_4$
$\mathscr{C}_2$	−+ +− ++ ++ +− −+ −− −− ++ ++ −+ +− −− −− +− −+	$-4K_3 + 4K_4$
$\mathscr{C}_3$	++ +− −− −+ −− +− ++ −+	$-8K_2 + 4K_3 + 4K_4$
$\mathscr{C}_4$	+− −+ −+ +−	$-8K_1 + 8K_2 + 4K_3 + 4K_4$

From Table 11.2 we can find the couplings of block spins in terms of these constants as follows:

$$K_1 = \frac{1}{16}(c_1 - c_4)$$

$$K_2 = \frac{1}{32}(c_1 - 2c_3 + c_4)$$

$$K_3 = \frac{1}{32}(c_1 - 4c_2 + 2c_3 + c_4)$$

$$K_4 = \frac{1}{32}(c_1 + 4c_2 + 2c_3 + c_4) .$$

The iteration of the RG transformation is then performed by using the couplings K_i again on a 4×4 lattice. Figure 11.8 shows the projection of some renormalization group trajectories onto the K_1, K_2-plane. These trajectories are obtained with the program rengroupis2d.c on p. 286 and are based on the discussion found at http://www-zeuthen.desy.de/~hasenbus/lecture.html. Points in the space of couplings that flow into the fixed point define the *critical surface*, and the intersection of this surface with the line $(K_1, 0, 0)$ yields the critical coupling of the Ising model with nearest-neighbor interaction. Besides the trivial high-temperature fixed point at $K^* = 0$, we find a nontrivial fixed point at

$$K^* = (K_1^*, K_2^*, K_3^*) = (0.302796, 0.104246, 0.023298) . \tag{11.68}$$

The numerical value of the critical coupling of the $2d$ Ising model with nearest-neighbor interactions lies at

$$K_{1,c}^{(1)} \approx 0.458961 ,$$

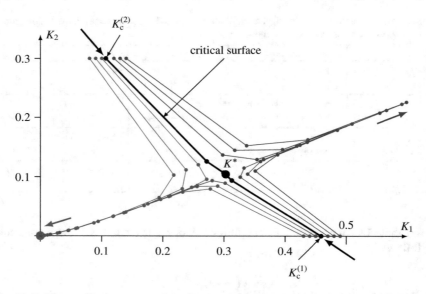

Fig. 11.8 Renormalization group flow in the (K_1, K_2)−plane. Besides the trivial high-temperature fixed point at $K = 0$, we find a nontrivial fixed point. The intersection of the critical surface with the K_1−axis yields the critical temperature of the Ising model with nearest-neighbor interaction

and this value can be compared with the exact result $\frac{1}{2} \log(1 + \sqrt{2}) \approx 0.4407$. In order to obtain the critical exponents, we examine the linearized RG transformation close to the fixed point K^*, i.e., the matrix

$$T_{ab} = \frac{\partial K'_a}{\partial K_b}\bigg|_{K^*} . \tag{11.69}$$

The corresponding difference quotients as calculated with `rengroupis2d.c` read

$$T = \begin{pmatrix} 1.37342\ 0.50762\ 0.06027 \\ 1.66800\ 0.79017\ 0.00564 \\ 0.55608\ 0.22621\ 0.20882 \end{pmatrix} . \tag{11.70}$$

Next we have to diagonalize this nonsymmetric matrix in order to determine the critical exponents:

$$\sum_a \Phi^i_a T_{ab} = \lambda^i \Phi^i_b .$$

The eigenvalues and left eigenvectors are

$$\lambda_1 = 2.0658 , \quad \Phi_1 = (0.9268, 0.3743, 0.0312)$$
$$\lambda_2 = 0.1890 , \quad \Phi_2 = (0.3027, 0.1010, -0.9477) \tag{11.71}$$
$$\lambda_3 = 0.1176 , \quad \Phi_3 = (-0.7695, 0.4185, 0.4825) ,$$

and yield the following exponents y_i in $\lambda_i = b^{y_i}$:

$$\frac{1}{\nu} = y_1 \approx 1.0467, \quad y_2 \approx -2.4037, \quad y_3 \approx -3.0880 . \tag{11.72}$$

Without external field the Ising model has only one relevant coupling t and infinitely many irrelevant couplings with negative exponents. The truncated model has just two irrelevant couplings. With $h = 0$ we may confirm only the relation $\nu = 1/y_1$ in (11.51). We see that the result (11.72) for ν approximates the exact value $\nu = 1$ rather well in contrast to the mean field approximation, which predicts $\nu_{\mathrm{mf}} = 0.5$.

11.4 Continuum Limit of Noninteracting Scalar Fields

We begin our discussion of the continuum limit by reconsidering the simple free scalar field theory. For that purpose we introduce explicitly a *lattice spacing a* (earlier denoted by ε) and study the *continuum limit* $a \to 0$ of the two-point function. Thereby we should distinguish between sites, parameters, and fields of the system on the unit lattice and on the lattice with lattice spacing a. We use the following conventions: n, k, and m_L refer to the dimensionless sites, momenta, and mass on the unit lattice and x, p, and m to the corresponding dimensionful quantities on the lattice with spacing a. We begin with the Fourier representation of the two-point function on the unit lattice:

$$G_L(n) = \frac{1}{(2\pi)^d} \int_{-\pi}^{\pi} d^d k \, \frac{e^{ikn}}{m_L^2 + \hat{k}^2}, \quad \hat{k}_\mu = 2 \sin \frac{k_\mu}{2} . \tag{11.73}$$

Setting $n = x/a$, $k = pa$, and $m_L = am$, we obtain

$$G_L(x) = \left(\frac{a}{2\pi}\right)^d \frac{1}{a^2} \int_B d^d p \, \frac{e^{ipx}}{m^2 + \hat{p}^2}, \quad \hat{p}_\mu = \frac{2}{a} \sin \frac{ap_\mu}{2} , \tag{11.74}$$

where the momentum is inside the Brillouin zone $B = [-\pi/a, \pi/a]^d$ which becomes $\mathbb{R}^d$ in the continuum limit. The Green's function satisfies the difference equation

$$-\sum_\mu \left(G_L(x + ae_\mu) - 2G_L(x) + G_L(x - ae_\mu) \right) + (am)^2 G_L(x) = \delta_{x,0} .$$

The rescaled dimensionful two-point function

$$G_a(x) = \frac{1}{a^{d-2}} G_L(x) = \frac{1}{(2\pi)^d} \int_B d^d p \, \frac{e^{ipx}}{m^2 + \hat{p}^2} \tag{11.75}$$

satisfies the rescaled difference equation

$$-\sum_\mu \frac{1}{a^2} \left(G_a(x + ae_\mu) - 2G_a(x) + G_a(x - ae_\mu) \right) + m^2 G_a(x) = \frac{1}{a^d} \delta_{x,0} \ .$$

In the continuum limit, the right-hand side approaches the Dirac delta function,

$$\frac{1}{a^d} \delta_{x,0} \ \overset{a \to 0}{\longrightarrow} \ \delta(x) \ , \tag{11.76}$$

such that the difference equation turns into the linear *differential* equation

$$\left(-\Delta + m^2 \right) G(x) = \delta^d(x), \quad \text{where} \quad \lim_{a \to 0} G_a = G \ . \tag{11.77}$$

It is just the defining equation for the Green function of the operator $-\Delta + m^2$ on the Euclidean space $\mathbb{R}^d$. In the same limit, the Fourier representation (11.75) becomes

$$\frac{1}{(2\pi)^d} \int_{\mathbb{R}^d} d^d p \, \frac{e^{ipx}}{m^2 + p^2} \ . \tag{11.78}$$

Several remarks are in order at this point:

- The rescaling of the two-point function in (11.75) corresponds to a *field renormalization*. The mass parameter m_L and the lattice field ϕ_L in the lattice action

$$S = \frac{1}{2} \sum_{\langle n,m \rangle} (\phi_L(m) - \phi_L(n))^2 + \frac{m_L^2}{2} \sum_n \phi_L^2(n)$$

$$= \frac{1}{2} \sum_{x \in (a\mathbb{Z})^d} \frac{a^d}{a^{d-2}} \left(\sum_{\mu=1}^d \frac{(\phi_L(x + ae_\mu) - \phi_L(x))^2}{a^2} + \frac{m_L^2}{a^2} \phi_L^2(x) \right) \ . \tag{11.79}$$

  are *dimensionless*. In the continuum limit, the difference quotients become partial derivatives, and, if we rescale the field according to

$$\phi_L(n) \longrightarrow \phi(x) = \frac{1}{a^{(d-2)/2}} \phi_L(an) \ . \tag{11.80}$$

then the Riemann sum turns into an integral over $\mathbb{R}^d$, such that the lattice action approaches the continuum action of the Klein-Gordon field

$$S = \frac{1}{2} \int d^d x \left((\nabla\phi(x), \nabla\phi(x)) + m^2\phi^2(x) \right), \quad \text{where} \quad m = \frac{m_L}{a}.$$

In d spacetime dimensions, a scalar field has mass-dimension $[\text{mass}]^{(d-2)/2}$ such that the naive rescaling (11.80) just restores the correct mass-dimension of ϕ. The rescaling implies a rescaling of the k-point function

$$G_L(n_1, \ldots, n_k) \longrightarrow G_L(x_1, \ldots, x_k) = a^{-k(d-2)/2} G(an_1, \ldots, an_k).$$
$$(11.81)$$

- For large arguments the two-point functions on the lattice and in the continuum fall off exponentially:

$$G_L(n) \xrightarrow{m_L|n| \gg 1} e^{-m_L|n|} \quad , \quad G(x) \xrightarrow{m|x| \gg 1} e^{-m|x|}. \qquad (11.82)$$

- The lattice Green function becomes approximately rotationally invariant for large arguments $|n| \gg 1/m_L$, and this is needed to obtain an SO(d)-invariant correlation function in the continuum limit. Hence, although the lattice regularization breaks the rotation invariance, the rotational symmetry is restored in the continuum limit.

11.4.1 Correlation Length for Interacting Systems

For interacting theories the correlation length on a lattice or in the continuum is intimately related to a mass parameter, similarly as for a noninteracting theory. To explain this connection, we recall the definition of a mass and a correlation length:

1. The dimensionless *bare mass* m_L directly appears in the lattice action.
2. The dimensionless *correlation length* ξ_L is defined via the two-point function

$$\frac{1}{\xi_L} = -\lim_{|n| \to \infty} \frac{\log G_L(n)}{|n|}. \qquad (11.83)$$

For the free field, we simply have $\xi_L = 1/m_L$, but in general ξ_L depends on all bare couplings of the lattice theory.
3. A particle described by the field ϕ is characterized by a dimensional *physical mass* m as measured in experiments.

4. The choice of bare parameters determines the correlation length ξ_L in lattice units from which we extract the physical mass according to

$$m = \frac{1}{\xi} = \frac{1}{a\,\xi_L(m_L, \dots)} . \tag{11.84}$$

Thus, the *lattice spacing* a depends on the fixed physical mass m, the dimensionless mass parameter m_L, and possible further bare couplings.

We may interpret the connection between the bare and the physical mass as follows: Firstly, we may set the lattice spacing a equal to some arbitrarily chosen physical distance. If ϕ describes a particle of mass m, then the product $\xi_L = 1/am$ is identified with its Compton wavelength in units of a. When we specify a and m, then we fix the bare parameter $m_L(\xi_L)$ provided the remaining bare parameters are specified as well. A change of the (unobservable) lattice spacing a is compensated by a change of the (unobservable) bare parameter m_L whereby physical quantities remain unchanged. Thus the trajectory $m_L(a)$ represents a *curve of constant physics*.

To reach the continuum limit $a \to 0$, we should adjust the free lattice parameters such that the correlation length ξ_L is very large compared to the lattice spacing or equivalently that $\xi \gg a$. This means that we perform the limits $\xi_L \to \infty$ or $m_L \to 0$ so that the dimensional correlation functions have a well-defined continuum limit characterized by a correlation length $\xi > 0$. More generally, for each free lattice bare parameter, we identify a physical quantity (mass, vacuum expectation value, decay constant, etc.) which should be reached in the continuum limit. The a-dependence of the bare parameters can be fixed by the requirement that the physical quantities do not change. This step also relates the lattice spacing to a physical scale.

In *MC simulations* one typically chooses a set of bare parameters and extracts the corresponding lattice spacing for a prescribed physical mass m according to (11.84). Thereby the correlation length should be much larger than the lattice spacing in order to avoid lattice effects and much smaller than the volume in order to avoid finite volume effects. This means that the inequality

$$1 \ll \xi_L \ll N \tag{11.85}$$

should always be satisfied. Currently, high-performance computer clusters admit simulations of scalar field theories on lattices with up to 128^4 sites.

11.5 Continuum Limit of Spin Models

In the vicinity of a critical point (curve, surface) of a general spin model, the correlation length ξ_L becomes large. Thus, near criticality the correlations extend over very many lattice spacings such that the discreteness of the lattice becomes irrelevant. This is the reason why certain lattice models at criticality can be interpreted as Euclidean quantum field theories or quantum field theories at finite temperature. At finite temperature we perform the limit $a \to 0$ such that

$$T = \frac{1}{a N_d} \tag{11.86}$$

is fixed. Away from criticality the lattice model is viewed as lattice regularized theory. Performing a continuum limit means then that we approach a critical point along a particular curve in the space of couplings. Close to a critical point, the correlation length in units of a diverges as

$$\frac{\xi}{a} = \xi_L = \kappa (\beta_c - \beta)^{-\nu} \tag{11.87}$$

with critical exponent ν. For a *fixed correlation length*, this equation determines the lattice spacing $a = a(\beta)$, and the continuum limit is reached for $\beta \to \beta_c$ with ξ being fixed. For a given lattice spacing, the dimensionless parameter β can be expressed in terms of the dimensional length ξ. The physical mechanism that transforms a dimensionless parameter into a dimensional parameter is called *dimensional transmutation*. It is observed in many relevant field theories with dimensionless couplings.

We consider a correlation function with long-range behavior

$$\langle O(n)O(m) \rangle \sim e^{-m_{LO}|n-m|} , \tag{11.88}$$

wherein m_{LO} acts as screening parameter. Since the distance $|n - m|$ is measured in units of the lattice spacing, the correlation function yields the screening mass in units of $a(\beta)$. Universality implies that near a critical point

$$m_O \, a(\beta) = m_{LO} = \kappa_O \, (\beta_c - \beta)^{\nu} , \qquad \beta \uparrow \beta_c , \tag{11.89}$$

such that the product $m_O \xi$ attains a constant value in the vicinity of the critical point:

$$m_O \xi = \kappa_m \kappa . \tag{11.90}$$

The dimensionless numbers κ and κ_m are accessible to numerical simulations. Thus we can "measure" the screening mass m_O in units of the free parameter $1/\xi$.

11.6 Programs for Chap. 11

The program `rengroupis2d.c` computes the trajectories of the block-spin transformation of the $2d$ Ising model on a 4×4 lattice. The blocking kernel is based on the *majority rule* with block-spin Hamiltonian (11.66). The trajectories in Fig. 11.8 were computed with this program. When one runs the program, it first wants to know an initial K_1-value and then computes ig points on the renormalization group trajectory beginning at $(K_1, 0, 0, 0)$. One also may change the initial couplings in line 10 of the code. With a fine-tuning, we localized the nontrivial fixed point K^* listed in lines 11 and 12. The two critical points in Fig. 11.8 are localized at

$$K_c^{(1)} = (0.4589605276967553, 0, 0, 0)$$

$$K_c^{(2)} = (0.106770494, 0.3, 0, 0) .$$

To compute the critical exponents, one needs to set ig=1 in line 21 and uncomment the lines beginning with X.

```
1    /* Programm rengroupis2d.c */
2    /* computes trajectories of MC_RG transformation*/
3    /* numerically. Blocking according to majority rule.*/
4    #include <stdio.h>
5    #include <stdlib.h>
6    #include <math.h>
7    #include <string.h>
8    #include "constrenising2.h"
9    #include "stdrenising.h"
10   double k1,k2=0.3,k3=0,k4=0;
11   /*X double k1=0.30279597088,k2=0.10424577852;*/
12   /*X double k3=0.02329831163,k4=2.66479130426;*/
13   int main(void){
14     /*X k1o=k1;k2o=k2;k3o=k3;*/
15     /*X puts("delta = ");scanf("%lf",&delta);*/
16     /*X k1=k1+delta;*/
17     conf=1<<V; /* number of configurations */
18     neighbors();
19     puts("K1 = ");scanf("%lf",&k1);
20     printf("(%1.4f, %1.4f)",k1,k2);
21     for (ig=0;ig<20;ig++){
22       c1=0;c2=0;c3=0;c4=0;
23       for (i=0;i<conf;i++){
24         /* binary code of i = configurations */
25         for (p=0;p<V;p++){
26           s[p]=(i>>p)%2;s[p]=2*s[p]-1;
27         };
28         h1=0;h2=0;h3=0;
29         for (p=0;p<V;p++){
30           h1=h1+s[p]*(s[nr[p]]+s[no[p]]);
31           h2=h2+s[p]*(s[nro[p]]+s[nru[p]]);
32           h3=h3+s[p]*s[nr[p]]*s[no[p]]*s[nro[p]];
33         };
34         boltz=exp(k1*h1+k2*h2+k3*h3);
35         blockspin(s);
36         for (p=0;p<VB;p++){
37           kc1[p]=bs[p]*k11[p];
```

```
38              kc2[p]=bs[p]*kl2[p];
39              kc3[p]=bs[p]*kl3[p];
40              kc4[p]=bs[p]*kl4[p];
41          };
42          if ((kc1[0]>=0)&&(kc1[1]>=0)&&(kc1[2]>=0)&&(kc1[3]>=0)){
43              if (kc1[0]*kc1[1]*kc1[2]*kc1[3]==0) c1=c1+0.5*boltz;
44              else c1=c1+boltz;}
45          if ((kc2[0]>=0)&&(kc2[1]>=0)&&(kc2[2]>=0)&&(kc2[3]>=0)){
46              if (kc2[0]*kc2[1]*kc2[2]*kc2[3]==0) c2=c2+0.5*boltz;
47              else c2=c2+boltz;}
48          if ((kc3[0]>=0)&&(kc3[1]>=0)&&(kc3[2]>=0)&&(kc3[3]>=0)){
49              if (kc3[0]*kc3[1]*kc3[2]*kc3[3]==0) c3=c3+0.5*boltz;
50              else c3=c3+boltz;}
51          if ((kc4[0]>=0)&&(kc4[1]>=0)&&(kc4[2]>=0)&&(kc4[3]>=0)){
52              if (kc4[0]*kc4[1]*kc4[2]*kc4[3]==0) c4=c4+0.5*boltz;
53              else c4=c4+boltz;}
54      };
55      l1=log(c1);l2=log(c2);l3=log(c3);l4=log(c4);
56      k1=(l1-l4)/16;
57      k2=(l1-2*l3+l4)/32;
58      k3=(l1-4*l2+2*l3+l4)/32;
59      k4=(l1+4*l2+2*l3+l4)/32;
60      printf("%1.4f,%1.4f)",k1,k2);
61  };
62  /*X printf("[%1.8f,%1.8f,%1.8f]\n",*/
63  /*X (k1-k1o)/delta,(k2-k2o)/delta,(k3-k3o)/delta);*/
64  printf("\n");
65  return 0;
66 }
```

The following header file defines the constants and variables. The arrays kl1[VB], ... represent the four distinct classes of configurations on the blocked lattice.

```
1  /* header file constrenising2.h*/
2  #define N 4 /* lattice length*/
3  #define V (N*N) /* number of lattice points*/
4  #define VB (V/4) /* volume of blocked lattice*/
5  short x,y,xm,xp,ym,yp;
6  short s[V],nr[V],no[V],nro[V],nru[V];
7  short bs[VB],kc1[VB],kc2[VB],kc3[VB],kc4[VB];
8  short kl1[VB]={1,1,1,1},kl2[VB]={1,1,-1,1};
9  short kl3[VB]={-1,-1,1,1},kl4[VB]={-1,1,1,-1};
10 unsigned int ig,i,il,j,jl,conf;
11 unsigned short p,q;
12 double k1o,k2o,k3o,delta,c1,c2,c3,c4,l1,l2,l3,l4,boltz;
13 int h1,h2,h3;
```

The header file `stdrenising.h` determines the nearest neighbors and next-nearest neighbors of a given lattice point and provides the block-spin configuration $bs[VB]$ of a given spin configuration $s[V]$.

```
1   /* header file stdrenising.h */
2   /* provides (next-)nearest neighbours and block spins */
3   void neighbors(void){
4     for (il=0;il<V;il++){
5       y=il/N;x=il-y*N;
6       xp=x+1,yp=y+1,ym=y-1;
7       nr[il]=y*N+xp%N;
8       no[il]=(yp%N)*N+x;
9       nro[il]=(yp%N)*N+xp%N;
10      nru[il]=((ym+N)%N)*N+xp%N;
11    };
12  }
13  void blockspin(short *s)
14  {
15    for (il=0;il<VB;il++){
16      p=(2*il)/N;jl=p*N+2*il;
17      bs[il]=s[jl]+s[jl+1]+s[jl+N]+s[jl+N+1];
18    };
19  }
```

11.7 Problems

11.1 (Scaling Relations) Starting from the scaling relation for the singular part of the free energy,

$$f_s(t, h) \sim b^{-d} f_s \left(b^{y_1} t, b^{y_2} h\right) ,$$

derive the four relations in (11.51) which contain α, β, γ, and δ.

11.2 (Decimation for spin−1 Ising chain) Generalize the decimation procedure to the spin−1 Ising chain with energy function

$$H = -J \sum_x s_x s_{x+s} - \sum_x C, \quad s_x = \pm 1, 0 .$$

New interactions are generated by the decimation. Show that if all possible even interactions between nearest neighbors are included in the energy function,

$$H = -J \sum_x s_x s_{x+1} - K \sum_x s_x^2 s_{x+1}^2 - D \sum_x s_x^2 - \sum_x D$$

a set of consistent recursion equations for the couplings is obtained. The recursion equations take a simple form in terms of the parameters

$$x = e^{-\beta(J+K+D/2)}, \quad y = e^{-2\beta J} \quad \text{and} \quad z = e^{-\beta(J+K+D)} .$$

Study the trajectories of the renormalization group transformation.

11.3 (Decimation of $2d$ **Ising Model)** Rewrite the program `rengroupis2d.c` on p. 286 by using the decimation procedure in place of the majority rule to define the blocking kernel. Calculate the couplings K_1^*, K_2^*, K_3^*, K_4^* at the nontrivial fixed point and the corresponding critical exponents. Compare with the flow obtained with the majority rule and depicted in Fig. 11.8.

References

1. L.P. Kadanoff, Scaling laws for Ising models near T_c. Physica **2**, 263 (1966)
2. M.E. Fisher, The renormalization group in the theory of critical behavior. Rev. Mod. Phys. **46**, 597 (1974)
3. M.E. Fisher, Renormalization group theory: its basis and formulation in statistical physics. Rev. Mod. Phys. **70**, 653 (1998)
4. K.G. Wilson, Renormalization group and critical phenomena. I. Renormalization group and the Kadanoff scaling picture. Phys. Rev. **B4**, 3174 (1971)
5. K.G. Wilson, Renormalization group and critical phenomena. II. Phase-space cell analysis of critical behavior. Phys. Rev. **B4**, 3184 (1971)
6. K.G. Wilson, The renormalization group and critical phenomena. Rev. Mod. Phys. **55**, 583 (1983)
7. N.N. Bogoliubov, D.V. Shirkov, *Introduction to the Theory of Quantized Fields* (Interscience, New York, 1959)
8. E. Brezin, J.C Le Guillou, J. Zinn-Justin, Field theoretical approach to critical phenomena, in *Phase Transitions and Critical Phenomena*, vol. 6, ed. by C. Domb, M.S. Green (Academic, London, 1976), p. 125
9. P. Pfeuty, G. Toulouse, *Introduction to the Renormalization Group and to Critical Phenomena* (Wiley, New York, 1977)
10. D.J. Amit, *Field Theory, the Renormalization Group and Critical Phenomena* (World Scientific, Singapore, 1993)
11. J.I. Binney, N.J. Dowrick, A.J. Fisher, M.E.J. Newmann, *The Theory of Critical Phenomena. An Introduction to the Renormalization Group* (Clarendon Press, Oxford, 1992)
12. D.R. Nelson, M.E. Fisher, Soluble renormalization groups and scaling fields for low-dimensional Ising systems. Ann. Phys **91**, 226 (1975)
13. R.H. Swendsen, Monte Carlo renormalization group. Phys. Rev. Lett. **42**, 859 (1979)
14. R.H. Swendsen, Monte Carlo calculation of renormalized coupling parameters. Phys. Rev. Lett. **52**, 1165 (1984)
15. S.K. Ma, Renormalization group by Monte Carlo methods. Phys. Rev. Lett. **37**, 461 (1976)

Functional Renormalization Group

12

The functional renormalization group (FRG) is a particular implementation of the renormalization group concept which combines the functional methods of quantum field theory with the renormalization group idea of Kenneth Wilson. It interpolates smoothly between the known microscopic laws and the complex macroscopic phenomena in physical systems. It is a momentum-space implementation of the renormalization group idea and can be formulated directly for a continuum field theory—no lattice regularization is required. In most approaches one uses a scale-dependent Schwinger functional or scale-dependent effective action. The scale parameter acts similarly as an adjustable screw of a microscope. For large values of a momentum scale k or equivalently for a high resolution of the microscope, one starts with the known microscopic laws. With decreasing scale k or equivalently with decreasing resolution of the microscope, one moves to a coarse-grained picture adequate for macroscopic phenomena. The flow from microscopic to macroscopic scales is given by a conceptionally simple but technically demanding flow equation for scale-dependent functionals. A priori the method is non-perturbative and does not rely on an expansion in a small coupling constant.

The flow of the scale-dependent Schwinger functional $W_k[j]$ is determined by the Wilson-Polchinski functional renormalization group equation. Actually the flow equations used by K. Wilson [1] with a specific cutoff procedure (the incomplete integration) is equivalent to the equation of J. Polchinski [2] containing an arbitrary cutoff function, as was observed in [3].

In this chapter we will use the flow equation for the scale-dependent effective action $\Gamma_k[\varphi]$. Apart from a cutoff term, it is just the Legendre transform of the scale-dependent Schwinger functional. The flow of Γ_k from microscopic to macroscopic scales is determined by the flow equation [4] due to C. Wetterich. The flow interpolating between the classical action $S[\varphi]$ and the full effective action $\Gamma[\varphi]$ is depicted in Fig. 12.1. To actually calculate Γ_k, one incorporates quantum fluctuations between a momentum scale k and a large cutoff scale Λ. For large k near the cutoff, one should recover the classical action $S[\varphi]$. With decreasing scale

© The Author(s), under exclusive license to Springer Nature Switzerland AG 2021
A. Wipf, *Statistical Approach to Quantum Field Theory*, Lecture Notes
in Physics 992, https://doi.org/10.1007/978-3-030-83263-6_12

Fig. 12.1 Sketch of the renormalization group flow in theory space. Each axis labels a different operator which may enter the effective action, e.g., φ^2, φ^4, $(\partial\varphi)^2$, etc. For a given initial condition at the cutoff scale Λ, the functional renormalization group equation determines the flow of Γ_k. Different regulator functions R_k lead to different trajectories in theory space, but in principle all trajectories end at the full quantum effective action Γ

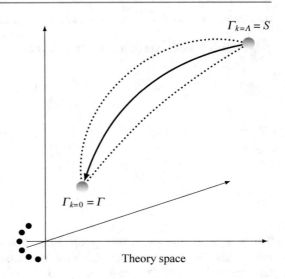

long-range effects are included, and for $k \to 0$ one recovers the complete effective action containing quantum fluctuations on all momentum scales. In recent years the functional renormalization group has been applied to many physical systems of interest: in particle physics to the theory of strong interaction, the electroweak phase transition, and the gauge hierarchy problem; in gravity to the asymptotic safety scenario; and in condensed matter theory to a unified description of classical bosons, the Hubbard model, disordered systems, liquid He^4, frustrated magnets, nucleation processes, superconductivity, and nonequilibrium systems. In nuclear physics it has been applied to effective models and the equation of state of nuclear matter and, finally, in atomic physics to investigate ultracold atoms. Useful reviews on various aspects and applications of the functional renormalization group method are contained in [5–11].

12.1 Scale-Dependent Functionals

To find an effective average action, we begin with the generating functional of the Euclidean n-point correlation functions

$$Z[j] = \int \mathscr{D}\phi\, e^{-S[\phi]+(j,\phi)}, \quad (j,\phi) = \int d^d x\, j(x)\phi(x)\,. \tag{12.1}$$

Its logarithm defines the Schwinger functional $W[j] = \log Z[j]$ which generates all connected correlation functions. The Legendre transform of $W[j]$ is the effective action (cp. Sect. 5.2)

$$\Gamma[\varphi] = (j,\varphi) - W[j] \quad \text{with} \quad \varphi(x) = \frac{\delta W[j]}{\delta j(x)} \tag{12.2}$$

which generates the one-particle irreducible correlation functions. The last equation in (12.2) determines $j[\varphi]$ which must be inserted into the right-hand side of the first equation. The functional Γ encodes all properties of the underlying quantum field theory in a most economic way.

In order to introduce scale-dependent functionals, we add a scale-dependent *IR-cutoff* term ΔS_k to the classical action in the functional integral (12.1) and obtain the scale-dependent generating functional

$$Z_k[j] = \int \mathscr{D}\phi \, e^{-S[\phi]+(j,\phi)-\Delta S_k[\phi]} \, . \tag{12.3}$$

The corresponding scale-dependent Schwinger functional $W_k[j]$ is given by

$$W_k[j] = \log Z_k[j] \, . \tag{12.4}$$

As regulator we choose a quadratic functional with a momentum-dependent mass,

$$\Delta S_k[\phi] = \frac{1}{2} \int \frac{\mathrm{d}^d p}{(2\pi)^d} \, \phi^*(p) R_k(p) \phi(p) \equiv \frac{1}{2} \int_p \phi^*(p) R_k(p) \phi(p) \, , \tag{12.5}$$

such that the flow equation will have a one-loop structure. We impose the following natural conditions on the cutoff function $R_k(p)$:

- For $k \to 0$ and fixed momentum p, the function should vanish such that we recover the conventional effective action for $k \to 0$. Hence we demand

$$R_k(p) \xrightarrow{k \to 0} 0 \quad \text{for fixed} \quad p \, . \tag{12.6}$$

- When k approaches the large cutoff scale Λ, then we should recover the classical theory. With the help of the saddle point approximation, one shows that the scale-dependent effective action $\Gamma_{k \to \Lambda}$ defined below tends to the classical action if

$$R_k \xrightarrow{k \to \Lambda} \infty \, . \tag{12.7}$$

- The cutoff function must regularize the theory in the IR, and this is the case if

$$R_k(p) > 0 \quad \text{for} \quad p \to 0 \, . \tag{12.8}$$

In many cases one demands $R_k(p) \longrightarrow k^2$ for small momenta p and in addition sends the cutoff to infinity. Possible cutoffs are

The exponential regulator: $\qquad R_k(p) = \dfrac{p^2}{e^{p^2/k^2} - 1} \, , \tag{12.9}$

Fig. 12.2 Plots of the exponential cutoff function $R_k(p^2)$ and its derivative $k\partial_k R_k(p^2)$. The cutoff function vanishes for $k \to 0$ and fixed p^2, it becomes large for $k \to \infty$, and it is positive for small arguments p^2

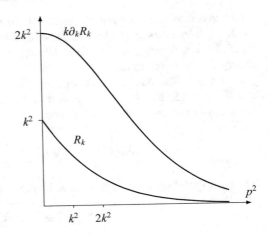

The optimized regulator:

$$R_k(p) = (k^2 - p^2)\,\theta\left(k^2 - p^2\right) , \qquad (12.10)$$

The quartic regulator:

$$R_k(p) = k^4/p^2 , \qquad (12.11)$$

The sharp regulator:

$$R_k(p) = \frac{p^2}{\theta\left(p^2 - k^2\right)} - p^2 , \qquad (12.12)$$

The Callan–Symanzik regulator:

$$R_k(p) = k^2 . \qquad (12.13)$$

The exponential cutoff function and its derivative are plotted in Fig. 12.2.

Similarly as for the conventional effective action, one introduces the average field of the cutoff theory with external source:

$$\varphi(x) = \frac{\delta W_k[j]}{\delta j(x)} . \qquad (12.14)$$

Note that for a fixed source the average field depends on the cutoff function and vice versa for a fixed average field the source depends on the cutoff function.

Next we perform a modified Legendre transformation and define the *scale-dependent effective action*

$$\Gamma_k[\varphi] = (j, \varphi) - W_k[j] - \Delta S_k[\varphi] . \qquad (12.15)$$

Here it is understood that (12.14) has been solved for $j = j[\varphi]$ and the solution is used on the right-hand side. Note that (12.15) is not the Legendre transform of $W_k[j]$ because of the additional term ΔS_k and hence does not need to be convex. But for $k \to 0$ the additional term vanishes and $\Gamma_{k\to 0}$ must become convex. Now we vary the effective average action with respect to φ to derive its equation of motion for the average field:

$$\frac{\delta \Gamma_k}{\delta \varphi(x)} = \int \frac{\delta j(y)}{\delta \varphi(x)} \varphi(y) + j(x) - \int \frac{\delta W_k[j]}{\delta j(y)} \frac{\delta j(y)}{\delta \varphi(x)} - \frac{\delta \Delta S_k[\varphi]}{\delta \varphi(x)} \,.$$

With (12.14) the first term on the right-hand side cancels against the third term and

$$\frac{\delta \Gamma_k}{\delta \varphi(x)} = j(x) - \frac{\delta}{\delta \varphi(x)} \Delta S_k[\varphi] = j(x) - (R_k \varphi)(x) \,. \tag{12.16}$$

12.2 Derivation of the Flow Equation

Now we derive the flow equation for the scale-dependent effective action. Its argument φ is kept fixed such that its conjugated current defined in (12.16) depends on the scale. Differentiating Γ_k in (12.15) results in

$$\partial_k \Gamma_k = \int d^d x \, \partial_k j(x) \varphi(x) - \partial_k W_k[j] - \int \frac{\partial W_k[j]}{\partial j(x)} \partial_k j(x) - \partial_k \Delta S_k[\varphi] \,,$$

where the variation of W_k yields two contributions: the second term on the right-hand side comes from the scale dependence of the parameters in W_k and the third term from the scale dependence of its argument j. Hence, in the partial derivative $\partial_k W_k[j]$, one varies the parameters and not the argument. With (12.14) the first and third terms cancel and

$$\partial_k \Gamma_k = -\partial_k W_k[j] - \partial_k \Delta S_k[\varphi]$$

$$= -\partial_k W_k[j] - \frac{1}{2} \int d^d x d^d y \, \varphi(x) \partial_k R_k(x, y) \varphi(y) \,. \tag{12.17}$$

Clearly, the partial derivative of W_k in (12.4) is given by

$$\partial_k W_k[j] = -\frac{1}{2} \int d^d x d^d y \, \langle \phi(x) \partial_k R_k(x, y) \phi(y) \rangle_k \,,$$

and it relates to the connected two-point function

$$G_k^{(2)}(x, y) \equiv \frac{\delta^2 W_k[j]}{\delta j(x) \delta j(y)} = \langle \phi(x) \phi(y) \rangle_k - \varphi(x) \varphi(y) \tag{12.18}$$

as follows:

$$\partial_k W_k[j] = -\frac{1}{2} \int d^d x d^d y \, \partial_k R_k(x, y) G_k^{(2)}(y, x) - \partial_k \Delta S_k[\varphi]$$

$$= -\frac{1}{2} \text{tr} \left(\partial_k R_k \, G_k^{(2)} \right) - \partial_k \Delta S_k[\varphi] \,. \tag{12.19}$$

Now we insert this simple result into the flow equation (12.17) and find

$$\partial_k \Gamma_k = \frac{1}{2} \int d^d x d^d y \, \partial_k R_k(x, y) \, G_k^{(2)}(y, x) \,. \tag{12.20}$$

It is known that the second functional derivatives of a convex functional and its Legendre transform define two operators which are inverses of each other; see Eq. (5.57) on p. 100. But since Γ_k is only the modified Legendre transform of W_k, we must first calculate the corrections to the quoted result. With

$$\varphi(x) = \frac{\delta W_k[j]}{\delta j(x)} \quad \text{and} \quad j(x) = \frac{\delta \Gamma_k}{\delta \varphi(x)} + \int d^d y \, R_k(x, y)\varphi(y)$$

we find the following relation between the second derivatives:

$$\delta(x - y) = \int d^d z \, \frac{\delta \varphi(x)}{\delta j(z)} \frac{\delta j(z)}{\delta \varphi(y)} = \int d^d z \, G_k^{(2)}(x, z) \left\{ \Gamma_k^{(2)} + R_k \right\} (z, y) \,. \tag{12.21}$$

Between the curly brackets, there appears the second functional derivative of Γ_k

$$\Gamma_k^{(2)}(x, y) = \frac{\delta^2 \Gamma_k}{\delta \varphi(x) \delta \varphi(y)} \,. \tag{12.22}$$

We conclude that the expression between the curly brackets in (12.21) is the inverse of the connected two-point function G_k. In operator notation this identity reads

$$G_k^{(2)} = \frac{1}{\Gamma_k^{(2)} + R_k} \,. \tag{12.23}$$

Inserting this result into (12.20) provides us with the flow equation for the scale-dependent effective action [4]

$$\partial_k \Gamma_k[\varphi] = \frac{1}{2} \text{tr} \left(\frac{\partial_k R_k}{\Gamma_k^{(2)}[\varphi] + R_k} \right) \,. \tag{12.24}$$

The closed *Wetterich equation* (12.24) is a nonlinear functional integro-differential equation which contains the full propagator. It is the starting point for many applications in various branches of theoretical physics.

In passing we note that the intermediate result (12.19) is just the *Polchinski equation* for the scale-dependent Schwinger functional:

$$\partial_k W_k = -\frac{1}{2}\mathrm{tr}\left(\partial_k R_k G_k^{(2)}\right) - \frac{1}{2}\left(G_k^{(1)}, \partial_k R_k G_k^{(1)}\right) . \tag{12.25}$$

On the right-hand side, the scale-dependent connected one- and two-point functions appear. These are just the first and second functional derivatives of $W_k[j]$ with respect to its argument.

Both the Polchinski equation (12.25) and Wetterich equation (12.24) are exact functional renormalization group equations. They are related by a (modified) Legendre transformation and as such they are equivalent. The Polchinski equation has a more simple structure since the Schwinger functional and its derivatives appear at most quadratically in this equation. This is not the case for the Wetterich equation in which the second functional derivative of the effective action occurs in the denominator. But it is exactly this property which stabilizes the flow when one tries to actually solve the flow equation in some approximation. This explains why Polchinski's equation is favored in structural investigations, whereas Wetterich's equation is mainly used in explicit calculations.

In applications the flow equation must be truncated which means that it is projected onto some finite-dimensional subspace. Unfortunately it is a highly nontrivial task to find some controlled error estimate for the flow. Typically one improves the truncation in successive steps by including more and more running couplings to see how quickly the flow stabilizes. This gives a first impression about the stability and quality of the flow. In addition one may compare the flows for different regulator functions in a given truncation scheme. For a "good truncation," the resulting couplings in the infrared should vary little with the regulator function. The most difficult part in any truncation is to include all relevant degrees of freedom in the infrared. If the effective action at the cutoff is quadratic,

$$\Gamma_\Lambda[\varphi] = \frac{1}{2}\int d^dx\, \varphi(-\Delta + m_\Lambda^2)\varphi ,$$

then the unique solution of the FRG equation reads

$$\Gamma_k[\varphi] = \Gamma_\Lambda[\varphi] + \frac{1}{2}\log\det\left(\frac{-\Delta + m_\Lambda^2 + R_k}{-\Delta + m_\Lambda^2 + R_\Lambda}\right) = \Gamma_\Lambda[\varphi] + \int d^dx\, u_{k,\Lambda} .$$

$$\tag{12.26}$$

For the optimized cutoff in (12.10), the integrand $u_{k,\Lambda}$ entering the last term is

$$\text{Two dimensions:} \quad u_{k,\Lambda} = \frac{1}{8\pi}\left(m_\Lambda^2 \log\frac{m_\Lambda^2 + \Lambda^2}{m_\Lambda^2 + k^2} + k^2 - \Lambda^2\right),$$

$$\text{Three dimensions:} \quad u_{k,\Lambda} = \frac{1}{6\pi^2}\left(m_\Lambda^3 \arctan\frac{m_\Lambda(k-\Lambda)}{m_\Lambda^2 + k\Lambda}\right.$$
$$\left. + m_\Lambda^2(\Lambda - k) + \frac{k^3 - \Lambda^3}{3}\right),$$

$$\text{Four dimensions:} \quad u_{k,\Lambda} = \frac{1}{64\pi^2}\left(m_\Lambda^4 \log\frac{m_\Lambda^2 + k^2}{m_\Lambda^2 + \Lambda^2}\right.$$
$$\left. + m_\Lambda^2\left(\Lambda^2 - k^2\right) + \frac{k^4 - \Lambda^4}{2}\right).$$

12.3 Functional Renormalization Applied to Quantum Mechanics

Before applying the flow equation to quantum field theory, we study approximate solutions for the ubiquitous anharmonic oscillator with classical Euclidean action

$$S[\omega] = \int d\tau \left(\frac{1}{2}\dot{q}^2 + V(q)\right), \tag{12.27}$$

depending on a classical potential V. Many techniques and approximations used in field theories can nicely be illustrated and checked against semi-analytic results for this simple quantum-mechanical model. Here we are mainly interested in the *effective potential* and consider the following low energy approximation to the effective average action

$$\Gamma_k[\omega] = \int d\tau \left(\frac{1}{2}\dot{q}^2 + u_k(q)\right) \tag{12.28}$$

with scale-dependent effective potential u_k. We have neglected higher derivative terms or mixed terms of the form $q^n\dot{q}^m$. The truncation (12.28) is the local potential approximation (LPA). It is the leading order in a systematic gradient expansion of the effective action.

On the right-hand side of the flow equation (12.24), the second functional derivative of Γ_k enters, which in the LPA has the simple form $\Gamma_k^{(2)} = -\partial_\tau^2 + u_k''(q)$. In order to find the flow projected onto the effective potential, it suffices to consider a constant q, for which

$$\int d\tau \, \partial_k u_k(q) = \frac{1}{2} \int d\tau d\tau' \partial_k R_k(\tau - \tau') \left(-\partial_\tau^2 + u_k''(q) + R_k \right)^{-1} (\tau' - \tau)$$

$$= \frac{1}{2} \int d\tau \int_{-\infty}^{\infty} \frac{dp}{2\pi} \frac{\partial_k R_k(p)}{p^2 + u_k''(q) + R_k(p)} . \tag{12.29}$$

To proceed we must choose a regulator function which conforms to the above-mentioned conditions. It has been argued in [12, 13] that the regulator function in (12.10) is optimal since it improves the stability properties of the flow equation. Its derivative is $\partial_k R_k(p) = 2k\theta(k^2 - p^2)$ such that for a constant q, we obtain the following nonlinear partial differential equation for the effective potential:

$$\partial_k u_k(q) = \frac{1}{\pi} \frac{k^2}{k^2 + u_k''(q)} . \tag{12.30}$$

Note that the minimum of $u_k(q)$ cannot be the ground state energy but differs by a q-independent contribution. This already becomes clear by studying the free particle case for which $u_\Lambda = 0$ and (12.30) yields $u_k = (k - \Lambda)/\pi$. In order to extract the true ground state energy from u_k, we perform a subtraction to avoid the buildup of unphysical zero-point energy contributions [10]. The free particle limit fixes this subtraction in the flow equation and we end up with

$$\partial_k u_k = \frac{1}{\pi} \left(\frac{k^2}{k^2 + u_k''(q)} - 1 \right) = -\frac{1}{\pi} \frac{u_k''(q)}{k^2 + u_k''(q)} . \tag{12.31}$$

Let us consider an even potential at the cutoff scale. Then the right-hand side of (12.31) is an even function at the cutoff, and the solution of the flow equation will be even at all scales. To proceed we make a polynomial ansatz for the potential,

$$u_k(q) = \sum_{n=0,1,2\ldots} \frac{1}{(2n)!} a_{2n}(k) \, q^{2n} , \tag{12.32}$$

where the effects of the short-wavelength fluctuations are encoded in scale-dependent couplings a_{2n}. Inserting the second derivative of u_k with respect to q into (12.31) and comparing coefficients in a series expansion in powers of q^2 yields an infinite set of coupled ordinary differential equations,

$$\frac{da_0}{dk} = -\frac{1}{\pi} a_2 \Delta_0, \qquad \Delta_0 = \frac{1}{k^2 + a_2} ,$$

$$\frac{da_2}{dk} = -\frac{k^2}{\pi} a_4 \Delta_0^2 ,$$

$$\frac{da_4}{dk} = -\frac{k^2 \Delta_0^2}{\pi} \left(a_6 - 6a_4^2 \Delta_0 \right) , \tag{12.33}$$

$$\frac{da_6}{dk} = -\frac{k^2 \Delta_0^2}{\pi} \left(a_8 - 30a_4a_6\Delta_0 + 90a_4^3\Delta_0^2 \right) ,$$

$$\vdots \quad \vdots$$

where the dots indicate the equations for the higher coefficients. As initial conditions we use the parameters a_{2n} at the cutoff—the parameters in the classical potential.

We first project the flow onto the space of fourth-order polynomials and hence impose $a_6 = 0$ in Eq. (12.33). Using the standard notation

$$a_0 = E, \quad a_2 = \omega^2 \quad \text{and} \quad a_4 = \lambda , \tag{12.34}$$

we find the following truncated system of flow equations:

$$\frac{dE_k}{dk} = -\frac{\omega_k^2}{\pi}\Delta_0, \qquad \frac{d\omega_k^2}{dk} = -\frac{k^2\lambda_k}{\pi}\Delta_0^2, \qquad \frac{d\lambda_k}{dk} = \frac{6k^2\lambda_k^2}{\pi}\Delta_0^3 . \tag{12.35}$$

With the `octave` program on p. 325, we solved these differential equations subject to the initial conditions $E_\Lambda = 0$, $\omega_\Lambda = 1$ and a varying quartic coupling λ at the cutoff scale. The scale-dependent couplings E_k and ω_k^2 are depicted in Fig. 12.3. They hardly change for $k \gg \omega$, and only when the scale parameter is comparable

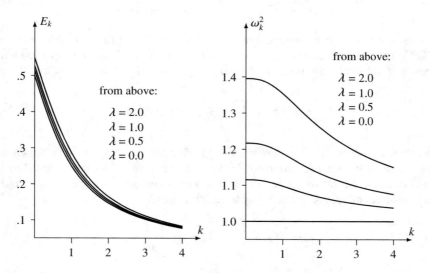

Fig. 12.3 The flow of the couplings E_k and ω_k^2 of the anharmonic oscillator for the cutoff values $E_\Lambda = 0$ and $\omega_\Lambda^2 = 1$. The values at $k = 0$ contain quantum corrections from all length scales

Table 12.1 Energies of the ground state and first excited state of the anharmonic oscillator with varying λ extracted from the flow in the LPA projected onto polynomials of order 4 and order 12. We used the optimized regulator and the Callan–Symanzik regulator $R_k(p) = k^2$. The fifth and last columns contain the "exact values." The energies and couplings are given in units of $\hbar\omega$

	Ground state energy				Energy of first excited state			
Cutoff	Optimal order 4	Optimal order 12	Callan order 4	Exact result	Optimal order 4	Optimal order 12	Callan order 4	Exact result
$\lambda = 0$	0.5000	0.5000	0.5000	0.5000	1.5000	1.5000	1.5000	1.5000
$\lambda = 1$	0.5277	0.5277	0.5276	0.5277	1.6311	1.6315	1.6307	1.6313
$\lambda = 2$	0.5506	0.5507	0.5504	0.5508	1.7324	1.7341	1.7314	1.7335
$\lambda = 3$	0.5706	0.5708	0.5703	0.5710	1.8177	1.8207	1.8159	1.8197
$\lambda = 4$	0.5885	0.5889	0.5882	0.5891	1.8923	1.8968	1.8898	1.8955
$\lambda = 5$	0.6049	0.6054	0.6045	0.6056	1.9593	1.9652	1.9562	1.9637
$\lambda = 6$	0.6201	0.6207	0.6196	0.6209	2.0205	2.0278	2.0168	2.0260
$\lambda = 7$	0.6343	0.6350	0.6336	0.6352	2.0771	2.0857	2.0728	2.0836
$\lambda = 8$	0.6476	0.6484	0.6469	0.6487	2.1299	2.1397	2.1250	2.1374
$\lambda = 9$	0.6602	0.6611	0.6594	0.6614	2.1794	2.1905	2.1741	2.1879
$\lambda = 10$	0.6721	0.6732	0.6713	0.6735	2.2263	2.2385	2.2205	2.2357
$\lambda = 20$	0.7694	0.7714	0.7679	0.7719	2.5994	2.6209	2.5898	2.6166

to the characteristic scale ω of the oscillator do the couplings begin to flow. For a positive $\omega = \omega_{k=0}$, the effective potential is minimal at the origin such that the *ground state* energy is just $E_0 = \min(u_{k=0})$. The energy of the *first excited state* is given by the curvature of the effective potential at its minimum:

$$E_1 = E_0 + \sqrt{u''_{k=0}(0)} = E_0 + \omega_{k=0} . \tag{12.36}$$

The second and sixth columns in Table 12.1 contain the energies of the ground state and first excited state for varying values of the quartic coupling λ of the anharmonic oscillator. The fifth and last columns contain the "exact values" for E_0 and E_1 obtained by a numerical diagonalization of the matrix-Schrödinger operator of the anharmonic oscillator with SLAC lattice derivative [14, 15]. A comparison reveals that the simple projection of the LPA flow onto polynomials of order 4 already leads to rather accurate values for the energies.

12.3.1   Projection onto Polynomials of Order 12

To judge the quality of the polynomial truncation, we also calculated the flow projected onto even polynomials of order 12 with the `octave` program on p. 325. The corresponding energies are listed in the columns 3 and 7 in Table 12.1, and they are almost identical to the values obtained for the projection onto fourth-order polynomials. For positive ω^2 we do not gain much by including higher-order terms in the polynomial truncation.

12.3.2 Changing the Regulator Function

In order to study the dependence of the flow on the regulator function, we now use the momentum-independent Callan–Symanzik regulator $R_k(p) = k^2$. After subtracting the unphysical zero-point energy contributions, the flow equation for the effective potential (12.29) takes the form

$$\partial_k u_k = \frac{1}{2} \left(\frac{k^2}{k^2 + u_k''(q)} \right)^{1/2} - \frac{1}{2} . \tag{12.37}$$

As before we choose an even potential at the cutoff such that u_k stays even at all scales. Expanding u_k in even powers of q and comparing coefficients, we find the following equations for the flow projected onto fourth-order polynomials:

$$\frac{da_0}{dk} = \frac{1}{2} \left(k \Delta_0^{1/2} - 1 \right) ,$$

$$\frac{da_2}{dk} = -\frac{k}{4} a_4 \Delta_0^{3/2} , \tag{12.38}$$

$$\frac{da_4}{dk} = \frac{9k}{8} a_4^2 \Delta_0^{5/2} .$$

Similarly as for the optimized regulator, we calculated the energies of the ground state and first excited state with the `octave` program on p. 325, and the results are collected in the fourth and second to last columns in Table 12.1. The values are almost as accurate as those obtained with the optimized cutoff function.

12.3.3 Solving the Flow Equation for Non-convex Potentials

The classical potential with negative ω^2 shows a local maximum at the origin and two minima at $\pm v$, where $v^2 = -6\omega^2/\lambda$. The flow equation (12.30) reveals that u_k changes most rapidly near positions where its curvature is minimal which happens at maxima of u_k. The denominator $k^2 + u_k''$ in the equation is positive for large scales, and the structure of the flow equation ensures that it remains positive during the flow. It follows from the flow equation (12.31) that $u_k(q)$ increases with decreasing k if $u_k''(q)$ is positive and it decreases if $u_k''(q)$ is negative. This implies that a double-well potential flattens when it flows to the infrared and finally becomes convex for $k \to 0$. This is expected on general grounds and has been discussed in the context of the Wetterich equation in [16]. Figure 12.4 shows the solution of the partial differential equation (12.31) for a classical double-well potential with couplings $\omega^2 = -1$ and

Fig. 12.4 Flow of the scale-dependent effective potential in the local potential approximation. At the cutoff the flow begins with a non-convex double-well potential and for small scales ends up with a convex potential $u_{k=0}$. Depicted is the solution of (12.31) with initial couplings $\omega^2 = -1$ and $\lambda = 1$

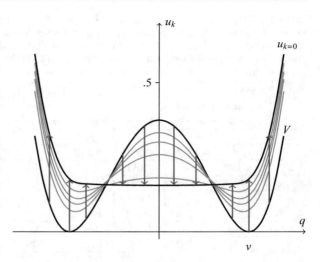

$\lambda = 1$. Here we see explicitly how the potential becomes convex at small scales. The parabolic differential equation (12.31) has been solved numerically with the octave function flowpde on p. 327. For the numerical treatment, we discretized *coordinate space* and replaced the continuous position variable q by equidistant points $q_1, \ldots, q_N$ in a suitable chosen interval $[-L, L]$. Then the function $u_k(q)$ of the two variables k and q becomes a vector-valued function $u_k(q_i)$ of the continuous flow parameter k. At the same time, the partial differential equation (12.31) turns into a system of N ordinary differential equations

$$\partial_k u_k(q_i) = -\frac{1}{\pi} \frac{(\Delta u_k)(q_i)}{k^2 + (\Delta u_k)(q_i)} , \tag{12.39}$$

where Δ is a discretized second derivative. In the function flowpde, we use

$$(\Delta u_k)(q_i) = \frac{u_k(q_{i+1}) + u_k(q_{i-1}) - 2u_k(q_i)}{(\Delta q)^2}$$

for which an *even* classical potential flows into an even effective potential.

The energies of the ground state and first excited state of the anharmonic oscillators with double-well potential are given in Table 12.2. We see that the projection of the flow onto fourth-order polynomials yields inaccurate results for weak couplings. The reason is that for small λ/ω^3 the potential near the minimum of the classical potential changes rapidly and this happens away from the origin where the polynomial approximation applies. The numerical solution of the flow equation (12.31) does better. But for decreasing λ or equivalently increasing barrier of the classical potential, it becomes more and more difficult to extract reliable results from the LPA. For large barriers the low-lying energies come in almost degenerate

Table 12.2 Energies of the ground state and first excited state for the anharmonic oscillator with $\omega^2 = -1$ and varying λ. All energies are calculated with the optimized regulator. For small λ the polynomial approximation of order 4 fails. Also shown are the energies calculated from the numerical solution of the partial differential equation (12.31). In the fifth and last column, we listed the "exact values." Energies and couplings are given in units of $\hbar\omega$

	Ground state energy				Energy of first excited state			
	Optimal order 4	Optimal order 12	pde	Exact	Optimal order 4	Optimal order 12	pde	Exact
$\lambda = 1$			−0.8732	−0.8556			−0.7887	−0.8299
$\lambda = 2$		−0.2474	−0.2479	−0.2422		0.0049	0.0063	−0.0216
$\lambda = 3$	0.2473	−0.0681	−0.0679	−0.0652	−0.2241	0.3514	0.3500	0.3307
$\lambda = 4$	−0.0186	0.0286	0.0290	0.0308	0.3511	0.5753	0.5755	0.5598
$\lambda = 5$	0.0654	0.0949	0.0953	0.0967	0.5835	0.7455	0.7462	0.7324
$\lambda = 6$	0.1234	0.1457	0.1461	0.1472	0.7509	0.8842	0.8851	0.8723
$\lambda = 7$	0.1688	0.1871	0.1876	0.1885	0.8851	1.0021	1.0030	0.9909
$\lambda = 8$	0.2063	0.2223	0.2228	0.2236	0.9987	1.1052	1.1061	1.0944
$\lambda = 9$	0.2671	0.2530	0.2535	0.2543	1.1863	1.1972	1.1981	1.1866
$\lambda = 10$	0.2386	0.2803	0.2808	0.2816	1.0978	1.2805	1.2814	1.2701
$\lambda = 20$	0.4536	0.4632	0.4639	0.4643	1.7866	1.8638	1.8648	1.8538

pairs, and the splitting of the doublet is induced by instanton effects. To detect these exponentially suppressed non-perturbative effects, one needs to go beyond the leading order LPA [17].

Comparison with Weak-Coupling Perturbation Expansion

From ordinary weak-coupling stationary perturbation theory for the anharmonic oscillator, one obtains a series expansion for the ground state energy in powers of λ/ω^3. The leading terms are [18]

$$E_0 = \frac{\omega}{2} + \frac{3\omega}{4}\varepsilon - \frac{21\omega}{8}\varepsilon^2 + \dots, \quad \varepsilon = \frac{\lambda}{24\omega^3}. \tag{12.40}$$

We wish to extract a similar weak-coupling expansion from the truncated flow equations (12.35). For the harmonic oscillator with a vanishing λ, the coupling ω_k is scale independent and $\lambda_k = 0$ for all k. This points to the following series expansion for the (dimensionless) scale-dependent couplings in powers of the dimensionless small parameter $\varepsilon \ll 1$:

$$E_k/\omega = \alpha_0 + \alpha_1\varepsilon + \alpha_2\varepsilon^2 + \dots$$
$$\omega_k^2/\omega^2 = 1 + \beta_1\varepsilon + \beta_2\varepsilon^2 + \dots \tag{12.41}$$
$$\lambda_k/24\omega^3 = \varepsilon + \gamma_2\varepsilon^2 + \dots .$$

Inserting these expansions into the truncated flow equations (12.35) and comparing coefficients of $\varepsilon^0, \varepsilon^1$, and ε^2 leads to simple flow equations for the coefficient functions $\alpha_0, \alpha_1, \alpha_2, \beta_1, \beta_2$, and γ_2. For large scales the coefficients must vanish and this initial condition determines the solution. Details of the calculation are found in problem 12.1.

It is easier to find an explicit solution of these equations if we first neglect the running of the anharmonic coupling and set $\lambda_k = \lambda$ or equivalently $\gamma_2 = 0$. For this crude truncation, the coefficient functions α_0, α_1 are given in (12.114) and α_2 in (12.115). The ground state energy is obtained if we set $x = k/\omega = 0$ in these results, such that

$$E_0 = \frac{\omega}{2} + \frac{3\omega}{4}\varepsilon - \frac{21\omega}{8}\kappa\varepsilon^2 + \dots, \quad \kappa = \frac{8\pi^2 + 29}{14\pi^2} = 0.7813 . \tag{12.42}$$

Already from our crude approximation, we have obtained the correct coefficient $3/4$ for the first-order contribution. The coefficient of the second-order contribution is off by the factor $\kappa \approx 0.78$. Even for a running λ_k, we can integrate the differential equations for the coefficients analytically. The small-x expansions of the coefficients are given in (12.114) and lead to

$$E_0 = \frac{\omega}{2} + \frac{3\omega}{4}\varepsilon - \frac{21\omega}{8}\kappa'\varepsilon^2 + \dots, \quad \kappa' = \frac{10\pi^2 - 29}{7\pi^2} = 1.0088 . \tag{12.43}$$

We see that the simple local potential approximation projected onto polynomials of degree 4 reproduces the second-order coefficient in the weak-coupling expansion very well. It exceeds the correct value by less than 1%.

12.4 Scalar Field Theory

A quantum-mechanical system can be interpreted as one-dimensional field theory, and this explains why the previous methods and results are easily extended to field theories—at least to field theories without local symmetries. In this section we study the flow equation for a scalar field theory in d dimensions with Euclidean action

$$\mathscr{L} = \frac{1}{2}(\partial_\mu \phi)^2 + V(\phi) . \tag{12.44}$$

First we consider the local potential approximation of the effective average action

$$\Gamma_k[\varphi] = \int d^d x \left(\frac{1}{2}(\partial_\mu \varphi)^2 + u_k(\varphi) \right) . \tag{12.45}$$

Its second functional derivative $\Gamma_k^{(2)} = -\Delta + u_k''(\varphi)$ enters the flow equation. Projecting the flow onto constant average fields, we arrive at the following flow equation for the scale-dependent effective potential:

$$\partial_k u_k(q) = \frac{1}{2} \int \frac{d^d p}{(2\pi)^d} \frac{\partial_k R_k(p)}{p^2 + u_k''(q) + R_k(p)} . \tag{12.46}$$

For the optimized regulator, the integral can be calculated in closed form, and the result contains the volume of the d-dimensional ball divided by $(2\pi)^d$:

$$\mu_d = \frac{\pi^{d/2}}{(2\pi)^d \Gamma(d/2+1)} = \frac{1}{(4\pi)^{d/2}\Gamma(d/2+1)} . \tag{12.47}$$

The flow equation has the simple form

$$\partial_k u_k(\varphi) = \mu_d \frac{k^{d+1}}{k^2 + u_k''(\varphi)} , \tag{12.48}$$

where a prime denotes the derivative with respect to the field. For $d = 1$ we recover the flow equation for the anharmonic oscillator (12.30). Inserting the polynomial ansatz (12.32) for an even potential into the flow equation and comparing coefficients, we end up with similar flow equations as in quantum mechanics,

$$\frac{da_0}{dk} = +\mu_d k^{d+1} \Delta_0, \qquad \Delta_0 = \frac{1}{k^2 + a_2} ,$$

$$\frac{da_2}{dk} = -\mu_d k^{d+1} \Delta_0^2 a_4 ,$$

$$\frac{da_4}{dk} = -\mu_d k^{d+1} \Delta_0^2 \left(a_6 - 6a_4^2 \Delta_0 \right) , \tag{12.49}$$

$$\frac{da_6}{dk} = -\mu_d k^{d+1} \Delta_0^2 \left(a_8 - 30a_4 a_6 \Delta_0 + 90a_4^3 \Delta_0^2 \right) ,$$

$$\vdots \quad \vdots$$

where the geometric factor μ_d was introduced in (12.47).

12.4.1 Fixed Points

To localize the fixed points of RG flow in the local potential approximation, we introduce the dimensionless field and potential,

$$\varphi = k^{(d-2)/2} \sqrt{\mu_d} \, \chi \quad \text{and} \quad u_k(\varphi) = k^d \mu_d v_k(\chi) \tag{12.50}$$

and rewrite the flow equation for $u_k(\varphi)$ in terms of these rescaled quantities,

$$k\partial_k v_k + dv_k - \frac{d-2}{2}\chi v'_k = \frac{1}{1 + v''_k} , \qquad (12.51)$$

where a prime is now the derivative with respect to the dimensionless field χ. At a fixed point of the flow, the first term on the left-hand side vanishes, such that a fixed point potential v_* satisfies the following second-order differential equation:

$$dv_* - \frac{d-2}{2}\chi v'_* = \frac{1}{1 + v''_*} . \qquad (12.52)$$

In any dimension this equation has the constant solution $dv_* = 1$ corresponding to a trivial Gaussian fixed point. Does it also possess other regular solutions corresponding to non-Gaussian fixed points? The answer to this question depends on the dimension d of spacetime.

For an even classical potential, v_k is even as well and we can set

$$v_k(\chi) = w_k(\varrho), \quad \text{with} \quad \varrho = \frac{\chi^2}{2} . \qquad (12.53)$$

The flow equation for $w_k(\varrho)$ takes the form

$$k\partial_k w_k(\varrho) + dw_k(\varrho) - (d-2)\varrho w'_k(\varrho) = \frac{1}{1 + w'_k(\varrho) + 2\varrho w''_k(\varrho)} , \qquad (12.54)$$

where the prime denotes the derivative with respect to ϱ. Note that in two dimensions a classical scalar field is dimensionless such that the last term on the left-hand side of the associated fixed point equation

$$dw_*(\varrho) - (d-2)\varrho w'_*(\varrho) = \frac{1}{1 + w'_*(\varrho) + 2\varrho w''_*(\varrho)} \qquad (12.55)$$

is absent. This property is the main reason why two-dimensional scalar field theories admit infinitely many fixed point solutions [19]. Actually the same happens for two-dimensional Yukawa theories with scalars and fermions in interaction [20].

In a polynomial truncation to order m, we expand the dimensionless potential in powers of ϱ:

$$w^{(m)} = \sum_{n=0}^{m} c_n \varrho^n . \qquad (12.56)$$

The corresponding flow equations for the couplings c_n read

$$k\partial_k c_0 = -dc_0 + \Delta_0, \qquad \Delta_0 = (1 + c_1)^{-1},$$

$$k\partial_k c_1 = -2c_1 - 6c_2\Delta_0^2,$$

$$k\partial_k c_2 = (d - 4)c_2 - 15c_3\Delta_0^2 + 36c_2^2\Delta_0^3,$$

$$k\partial_k c_3 = (2d - 6)c_3 - 28c_4\Delta_0^2 + 180c_2c_3\Delta_0^3 - 216c_2^3\Delta_0^4,$$

$$k\partial_k c_4 = (3d - 8)c_4 - 45c_5\Delta_0^2 + (336c_2c_4 + 225c_3^2)\Delta_0^3$$
$$- 1620c_2^2c_3\Delta_0^4 + 1296c_2^4\Delta_0^5.$$

$$\vdots \qquad \vdots$$

Scalar Fields in Three Dimensions

Many three-dimensional field theories admit nontrivial fixed points and the scalar theory is no exception. As we cannot solve the fixed point equation (12.55) exactly, we have to find ways to get at least approximate solutions. We use the polynomial truncation which leads to equations (12.57) with vanishing left-hand sides. Thus we find m algebraic equations for the $m + 1$ fixed point couplings:

$$0 = f_0(c_0^*, c_1^*) = f_1(c_1^*, c_2^*) = \cdots = f_{m-1}(c_1^*, \ldots, c_m^*).$$

The functions are just the right-hand sides of the system of equations (12.57) for $d = 3$. They are polynomials in $c_0^*, c_2^*, \ldots, c_m^*$ and $\Delta_0 = 1/(1 + c_1^*)$. Since the slope at the origin c_1^* is the only non-polynomial variable, we solve the system for $c_0^*, c_2^*, c_3^*, \ldots, c_m^*$ in terms of c_1^*. Algebraic computer programs[1] find the solution for polynomials of order 42. In the intermediate manipulation, it is useful to introduce $c_1^* - 1$ as new variable. The explicit expressions for the lowest couplings read

$$c_0^* = \frac{1}{3}\frac{1}{1 + c_1^*}$$

$$c_2^* = -\frac{c_1^*(1 + c_1^*)^2}{3}$$

$$c_3^* = \frac{c_1^*(1 + c_1^*)^3(1 + 13c_1^*)}{45}$$

$$c_4^* = -\frac{c_1^{*2}(1 + c_1^*)^4(1 + 7c_1^*)}{21}, \tag{12.58}$$

---

[1] The calculation in this section was performed with REDUCE 3.8.

Table 12.3 With $n!$ multiplied coefficients c_n^* of two polynomial approximations to the fixed point solution. The lowest coefficients do not change much when one increases the polynomial degree from $m = 20$ to $m = 42$

	c_0^*	c_1^*	c_2^*	c_3^*	c_4^*	c_5^*	c_6^*
$m = 20$	0.409534	−0.186066	0.082178	0.018981	0.005253	0.001104	−0.000255
$m = 42$	0.409533	−0.186064	0.082177	0.018980	0.005252	0.001104	−0.000256
	c_7^*	c_8^*	c_9^*	c_{10}^*	c_{11}^*	c_{12}^*	c_{13}^*
$m = 20$	−0.000526	−0.000263	0.000237	0.000632	0.000438	−0.000779	−0.002583
$m = 42$	−0.000526	−0.000263	0.000236	0.000629	0.000431	−0.000799	−0.002643
	c_{14}^*	c_{15}^*	c_{16}^*	c_{17}^*	c_{18}^*	c_{19}^*	c_{20}^*
$m = 20$	−0.002029	0.007305	0.028778	0.034696	−0.077525	−0.381385	0.000000
$m = 42$	−0.002216	0.006677	0.026544	0.026320	−0.110498	−0.517445	−0.587152

$$\vdots \quad \vdots$$

$$c_m^* = c_1^{*2}(1 + c_1^*)^m P_{m-3}(c_1^*) \, ,$$

where P_k is a polynomial of order k. We recover the trivial solution with $c_1^* = 0$ and

$$c_0^* = \frac{1}{3}, \quad 0 = c_2^* = c_3^* = c_4^* = \ldots \qquad (12.59)$$

corresponding to the Gaussian fixed point $w_*' = 0$. To find an approximate nontrivial fixed point solution, we truncate the tower of algebraic equations and set $c_m^* = 0$ which leads to $P_{m-3}(c_1^*) = 0$. Generically, the polynomial P_k has several real roots and we must pick a particular one. We choose roots c_1^* such that for large m they converge to a fixed value. For this choice of slopes at the origin, the approximating polynomials converge to a power series with maximal radius of convergence. For example, for polynomials of orders 20 and 42, we find the two roots $c_1^* = -0.186066$ and $c_1^* = -0.186041$ in agreement with [21]. Inserting the solution for $m = 20$ and for $m = 42$ into (12.58) yields the coefficients of the polynomials given in Table 12.3.

In Fig. 12.5 we plotted the polynomials of orders 10, 20, 30, and 40 which approximate the fixed point solution in the local potential approximation. The potentials are compared with the numerical solution of the fixed point equation.

Numerical Solution

When we try to solve the fixed point equation by numerical means, we generically run into a singularity at a finite value of ϱ. At the singular point, the potential and its derivative are both finite but its higher derivatives are divergent. A solution

Fig. 12.5 Fixed point solution w_* of a scalar field theory with even potentials in three dimensions. Depicted are the numerical solution to the differential equation (12.60) and results of polynomial approximations with polynomials of degrees 10, 20, 30, and 40

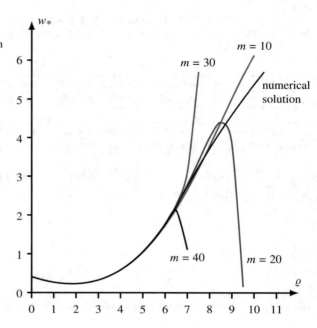

of the fixed point equation

$$w''_* = \frac{1}{2\varrho} \frac{1}{d w_* - (d-2) \varrho w'_*} - \frac{1+w'_*}{2\varrho}, \quad (d=3), \tag{12.60}$$

may become singular when the denominator $3w_*(\varrho) - \varrho w'_*(\varrho)$ vanishes. Regularity at the origin demands that the "initial values" $w_*(0)$, $w'_*(0)$ are related as

$$w'_*(0) = c^* \quad \text{and} \quad w_*(0) = \frac{1}{3}\frac{1}{1+c^*}. \tag{12.61}$$

This is just the first relation in (12.58). From the polynomial truncation, we already know that the slope at the origin is $w'_*(0) \approx -0.18606$. We applied the shooting method with a seventh-order Runge-Kutta integrator and determined the slope for which the numerical solution is regular for $\varrho \leq 120$. The resulting slope $w'_*(0) \approx -0.186064249376$ is almost identical with the slope -0.186064279993 of the approximating polynomial of degree 42.

12.4.2 Critical Exponents

Let us study the solution to the flow equation in the vicinity of a prescribed fixed point solution w_*. Thus we write $w_k = w_* + \delta_k$ and linearize the flow equation

(12.54) in the small perturbation δ_k. Using the fixed point equation of w_*, we end up with the following linear differential equation for the small fluctuation δ_k:

$$k\partial_k \delta_k = -d\delta_k + (d-2)\varrho\delta_k' - \left(dw_* - (d-2)\varrho w_*'\right)^2 \left(\delta_k' + 2\varrho\delta_k''\right) . \quad (12.62)$$

To make progress we insert the polynomial approximation for the fixed point solution and make a polynomial ansatz for the perturbation:

$$\delta_k(\varrho) = \sum_{n=0}^{m-1} d_n \varrho^n \quad \varrho = \frac{\chi^2}{2} . \quad (12.63)$$

The resulting linear system for the coefficients d_m has the form

$$k\partial_k \begin{pmatrix} d_0 \\ d_1 \\ \vdots \\ d_{m-1} \end{pmatrix} = M\left(c_0^*\right) \begin{pmatrix} d_0 \\ d_1 \\ \vdots \\ d_{m-1} \end{pmatrix} \quad (12.64)$$

and the critical exponents are identified with the eigenvalues of the m-dimensional matrix M. With an algebraic computer program, we calculated the eigenvalues for polynomial truncations up to order 46. The results are listed in Table 12.4. The Z_2-symmetric model has two negative exponents $\omega_0 = -3$ and $\omega_1 = -1/\nu$. The former corresponds to the trivial scaling of the ground state energy and is unrelated to the critical behavior. The remaining exponents $\omega_2, \omega_3, \ldots$ are all positive. We see that the smallest exponents extracted from the polynomial truncations converge with increasing m and can be extrapolated to $m = \infty$. The LPA prediction obtained in this way $\nu = 0.649562$ should be compared with the accurate prediction $\nu = 0.630$

Table 12.4 The five smallest nontrivial eigenvalues of the stability matrix corresponding to the Wilson–Fisher fixed point. The exponents are calculated for different polynomial truncations up to degree $m = 46$ of the LPA. The eigenvalue $\omega_0 = -3$ of the scaling operator 1 is not listed

m	$\nu = -1/\omega_1$	ω_2	ω_3	ω_4	ω_5
10	0.648617	0.658053	2.985880	7.502130	17.913494
14	0.649655	0.652391	3.232549	5.733445	9.324858
18	0.649572	0.656475	3.186784	5.853987	9.141093
22	0.649554	0.655804	3.170538	5.977066	8.522811
26	0.649564	0.655629	3.182910	5.897290	8.844632
30	0.649562	0.655791	3.180847	5.903039	8.907607
34	0.649561	0.655749	3.178636	5.922910	8.702583
38	0.649562	0.655731	3.180577	5.908885	8.814225
42	0.649562	0.655755	3.180216	5.909910	8.847386
46	0.649562	0.655746	3.179541	5.915754	8.738608

of the high-temperature expansion; see Table 9.9. To obtain more accurate results, one needs to go beyond the local potential approximation.

12.5   Linear O(N) Models

The previous results are readily extended to the linear O(N) models in d dimensions. So let us assume that the scalar field ϕ has N real components and that the interaction term in the Lagrangian density

$$\mathscr{L} = \frac{1}{2} \left(\partial_\mu \phi \right)^2 + V(\phi), \tag{12.65}$$

is O(N) invariant. This means that the potential depends only on the modulus of $\phi \in \mathbb{R}^N$. In the fixed point analysis, we use the dimensionless field χ and the dimensionless potential v_k in (12.50). For the O(N) models, it is convenient to introduce the invariant composite field

$$\varrho = \frac{1}{2} \sum_{i=1}^{N} \chi_i^2 \, , \tag{12.66}$$

since the invariant scaling potential is a function of ϱ only, $v_k(\chi) = w_k(\varrho)$. For several components the flow equation (12.51) generalizes to

$$k\partial_k v_k(\chi) + d v_k - \frac{d-2}{2} \chi v_k' = \text{tr} \left(\frac{1}{1 + \partial_i \partial_j v_k(\chi)} \right) . \tag{12.67}$$

For $v_k = w_k(\varrho)$ the matrix in the denominator becomes $\partial_i \partial_j v_k = \delta_{ij} w_k' + \chi_i \chi_j w_k''$ and has just two eigenvalues: the single eigenvalue $w_k' + 2\varrho w_k''$ and the highly degenerate eigenvalue w_k'. Hence for multicomponent fields with O(N) invariant interaction, the flow equation (12.54) translates into

$$k\partial_k w_k + d w_k - (d-2) \varrho w_k' = \frac{N-1}{1 + w_k'} + \frac{1}{1 + w_k' + 2\varrho w_k''} \, , \tag{12.68}$$

where a prime denotes the derivative with respect to the invariant field ϱ. The first term on the right-hand side is easily recognized as the contribution of the $N-1$ Goldstone modes. The last term is related to the massive radial mode. For large N the Goldstone modes give the main contribution to the flow equation.

To study the critical behavior, we linearize the flow equation about a fixed point solution w_* and hence set $w_k = w_* + \delta_k$. The fluctuation δ_k obeys the linear differential equation

$$k\partial_k \delta_k = -d\delta_k + (d-2) \varrho \delta_k' - \frac{(N-1)\,\delta_k'}{(1 + w_*')^2} - \frac{\delta_k' + 2\varrho \delta_k''}{(1 + w_*' + 2\varrho w_*'')^2} \, . \tag{12.69}$$

Table 12.5 The slope of the fixed point solution at the origin and the three smallest nontrivial eigenvalues of the stability matrix corresponding to the Wilson–Fisher fixed point of the O(N) models. The eigenvalue $\omega_0 = -3$ of the scaling operator 1 is not listed. The numbers are calculated with a polynomial truncation of degree 40

N	1	2	3	100	1000
$-w'_*(0)$	0.186064	0.230186	0.263517	0.384172	0.387935
$\nu = -1/\omega_1$	0.64956	0.70821	0.76113	0.99187	0.99923
ω_2	0.6556	0.6713	0.6990	0.97218	0.99844
ω_3	3.1798	3.0710	3.0039	2.98292	2.99554

Now we proceed exactly in the same way as for the one-component model without Goldstone modes considered in the previous section. The fixed point solution and the critical exponents in the polynomial truncation are computed with an algebraic computer program. We used polynomials of degree 40 to determine the eigenvalues of the stability matrix listed in Table 12.5. For a given N, we must find polynomial approximations which for small values of ϱ converge to a fixed point solution. For this purpose one picks that root of the polynomial $P_{m-3}(c_1^*)$ for which the roots converge with increasing degree of the polynomials. This amounts to a fine-tuning of the slope at the origin. The correct slopes are calculated beforehand and are used in the algebraic computer program.

Actually we can do better and extend the polynomial truncation to much higher order. For example, for the Z_2-model we obtain the value $\nu = 0.649\,561\,776$ from a truncation to polynomials of degree $m = 60$. With the help of a conformal mapping, one can further extend the polynomial truncation to order 75 which yields the more accurate value[2] $\nu = 0.649\,561\,773\,880....$ Plots of the scaling potentials w_* of various O(N) models and more critical exponents are found in [21].

Looking at the exponents in the table, one may conjecture that for large N the critical exponents converge to $\omega_n^\infty = 2n - 3$. Since it was shown in [22] that the LPA is exact in the large N limit for the effective potential, we expect that the ω_n^∞ are the exact critical exponents in this limit. In the following section, we shall see that this is indeed the case. Actually the exponents ω_n converge to their limiting values ω_n^∞ as

$$\omega_n = \omega_n^\infty + \frac{\delta_n}{N} + \dots . \tag{12.70}$$

Linear fits in the small parameter $1/N$ to the slope as well as the critical exponent ν for $N = 100,\ 500,$ and 1000, as depicted in Fig. 12.6, yield the asymptotic formulas

$$w'_*(0) \approx -0.3881 + 0.4096/N, \quad \nu \approx 0.9998 - 0.9616/N . \tag{12.71}$$

---

[2] Private communication by Daniel Litim.

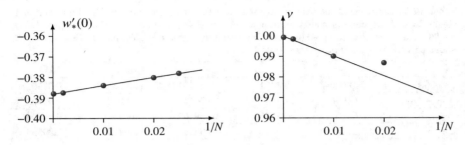

Fig. 12.6 The slope $w'_*(0)$ and the critical exponent ν for large values of N. Fitting the data points for $N \geq 100$ with a linear function in $1/N$ yields (12.71). The interpolating linear functions are plotted in the figure

Table 12.6 Critical exponents ν and $\omega \equiv \omega_2$ from derivative expansion (de) [26], BMW approximation [27], and lattice simulation combined with the high-T expansion [28–30]; see Table 9.12

N	1	2	3	4	10	100
ν_{de}	0.631	0.666	0.704	0.739	0.881	0.990
ν_{bmw}	0.632	0.674	0.715	0.754	0.889	0.990
$\nu_{mc,hte}$	0.630	0.672	0.711	0.749	0.867	
ω_{bmw}	0.78	0.75	0.73	0.72	0.80	1.00
ω_{mc}	0.832	0.785	0.773	0.765		

The number -0.3881 is close to that of the large N expansion, which is $-0.3883\ldots$ Our prediction for ν is also close to the exact result $\nu \approx 1 - 1.081/N$ derived in [23]. Notice that the subleading terms are very small for the N used in the fit. To make further progress, one must go beyond the LPA. In a next step in the derivative expansion, one would include a momentum- and field-dependent wave function renormalization and terms containing derivatives of the fields. Also, to determine critical exponents like the anomalous dimension, one should allow for a momentum dependence of the correlation functions. An approximation which keeps the momentum dependence in the two-point function has been proposed in [24]. This approximation was used to obtain better values for critical exponents of the O(N) models. Table 12.6 contains the results obtained with such approximations. They are compared with the exponents extracted from simulations and high-temperature expansions. Including the momentum dependence has become more important over the last years. This leads to a higher numerical effort, and with accurate pseudo-spectral methods, one can find global solutions of a broad class of partial differential equations that are encountered in next-to-leading-order approximations [25].

12.5.1 Large N Limit

For a large number of field components N, we only keep the terms of order N on the right-hand side of the flow equation (12.68). In the resulting differential equation

$$k\partial_k w_k = (d-2)\,\varrho w_k' - dw_k + \frac{N}{1+w_k'} \tag{12.72}$$

the contribution of the radial mode is missing. Thus in the large N limit, only the fluctuating Goldstone modes are responsible for a nontrivial flow to the infrared. To obtain a differential equation which is linear in the first derivatives and hence can be solved with the methods of characteristics, we differentiate (12.72) with respect to the field ϱ. Introducing the dimensionless scale parameter $t = \log(k/\Lambda)$, we obtain the flow equation for $w'(t,\varrho) = w_k'(\varrho)$:

$$\partial_t w' = (d-2)\,\varrho w'' - 2w' - \frac{N}{(1+w')^2} w'' \,. \tag{12.73}$$

Before we present and discuss the analytical solutions of this differential equation, we study the fixed point solutions and calculate the critical exponents.

Fixed Point Analysis

It is not difficult to solve the ordinary differential equation for a fixed point solution $w_*'(\varrho)$. The general solution depends on a free parameter c and solves

$$\frac{1}{\sqrt{\pm w'}}\frac{\varrho}{N} = H_\pm(w_*') + c, \qquad \pm w_*' \geq 0 \,, \tag{12.74}$$

where we introduced the functions

$$H_+(w') = \frac{1}{\sqrt{w'}}\frac{3w'+2}{2w'+2} + \frac{3}{2}\arctan\left(\sqrt{w'}\right) \,, \tag{12.75}$$

$$H_-(w') = \frac{1}{\sqrt{-w'}}\frac{3w'+2}{2w'+2} - \frac{3}{2}\operatorname{arctanh}\left(\sqrt{-w'}\right) \,. \tag{12.76}$$

Setting $\varrho = 0$ in (12.74) yields the slope at the origin, $w_*'(0) = -0.3883467189\ldots$, and this result can be compared with the values in Table 12.5. All solutions vanish at $\varrho_0 = N$. An expansion around this node reveals that only the solution with $c = 0$ is analytic. Solutions with positive c are globally defined and solutions with negative c are multivalued. We plotted solutions for different values of c as functions of ϱ/N in Fig. 12.7. To find all critical exponents, it is more convenient to use the fixed point

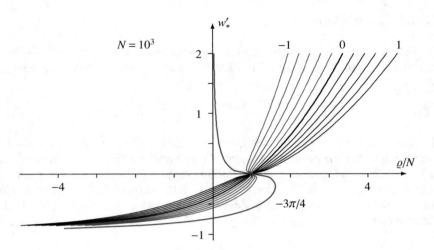

Fig. 12.7 One parametric family of derived fixed point potentials in the large N limit. Depicted are solutions with c-values between -1 and 1 and the particular solution with $c = -3\pi/4$ which touches the w'_* axis. All solutions have a node at $\varrho = N$. The solutions with $c \neq 0$ are nonanalytic and have zero slope w''_* at the node

equation for w_* in place of w'_*:

$$(d - 2)\varrho w'_* - dw_* + \frac{N}{(1 + w'_*)} = 0 . \tag{12.77}$$

To linearize the flow equation (12.72) near a fixed point, we set $w \approx w_*(\varrho) + e^{\omega t}\delta(\varrho)$. The small fluctuations obey the eigenvalue problem

$$(\omega + d)\delta = (d - 2)\varrho\delta' - \frac{N}{(1 + w'_*)^2}\delta' . \tag{12.78}$$

Following [31] we now express the fluctuation in terms of the fixed point solution. To that aim we use the ϱ-derivative of the fixed point equation,

$$(d - 2)\varrho w''_* - 2w'_* - \frac{N}{(1 + w'_*)^2}w''_* \tag{12.79}$$

to simplify the right-hand side of (12.78) as follows:

$$(\omega + d)\,\delta = \frac{2\omega'_*}{\omega''_*}\,\delta' . \tag{12.80}$$

This equation is easily solved and leads to

$$w(t, \varrho) \approx w_*(\varrho) + \text{const} \times e^{\omega t} \, w'_*(\varrho)^{(\omega+d)/2} \ . \tag{12.81}$$

If the solutions $w(t, \varrho)$ and w_* have Taylor expansions in powers of ϱ with a finite radius of convergence, the result (12.81) implies that the eigenvalues of the stability matrix must be quantized:

$$\omega \in \{2n - d \mid n = 0, 1, 2, \ldots\} \ . \tag{12.82}$$

In three dimensions there exist two relevant perturbations. The eigenvalue -3 belongs to the scaling operator 1 and the eigenvalue -1 yields the critical exponent $\nu = 1$. Note that already in the polynomial truncation, we have anticipated this simple result for the exponents in the large N limit.

12.5.2 Exact Solution of the Flow Equation

The most general solution of the first-order evolution equation (12.73) was derived in [32, 33] with the method of characteristics. It is the solution (12.74) with the constant c replaced by an arbitrary function of $F(e^{2t} w')$ and thus has the form

$$H_\pm(w') - \frac{1}{\sqrt{\pm w'}} \frac{\varrho}{N} = F(e^t w'), \quad \pm w' > 0 \ , \tag{12.83}$$

where the functions $H_\pm$ have been defined in (12.75) and (12.76). To prove that this is indeed a solution of the flow equation (12.73), one differentiates with respect to ϱ and t, solves the resulting equations for $\partial_t w'$ and w'', and inserts the solutions back into the flow equation. The free function F is fixed by initial conditions. Here it is the requirement that at the cutoff scale Λ, the effective potential u_k is equal to the classical potential $V(\phi)$. Recalling the relation between the dimensionful and dimensionless fields and potentials, this means

$$w'_\Lambda(\varrho) = \frac{1}{\Lambda^2} V'(\tilde{\varrho}), \qquad \tilde{\varrho} = \frac{1}{2}\phi^2 = \mu_d \Lambda \varrho \ , \tag{12.84}$$

Now we evaluate equations (12.83) at the cutoff scale where ϱ is a known function of w'_Λ, determined by solving

$$w'_\Lambda(\varrho) = \frac{1}{\Lambda^2} V'(\mu_3 \Lambda \varrho), \qquad \mu_3 = \frac{1}{6\pi^2} \ , \tag{12.85}$$

for ϱ in terms of w'_Λ. Collecting our results we end up with the following rather simple solution of the flow equation:

$$H_\pm(w') - \frac{1}{\sqrt{\pm w'}} \frac{\varrho}{N} = H_\pm \left(e^{2t} w'\right) - \frac{e^{-t}}{\sqrt{\pm w'}} \frac{\varrho \left(e^{2t} w'\right)}{N}, \quad \pm w' > 0 . \quad (12.86)$$

Note that ϱ on the left-hand side is an independent parameter, whereas the function $\varrho(e^{2t} w')$ on the right-hand side is obtained by inverting (12.85). Also note that the dependence on the number of fields only enters via ϱ/N. Thus we may set $N = 1$ if we use ϱ/N in place of ϱ as field variable.

Let us consider a theory with a quartic classical potential:

$$V(\tilde{\varrho}) = \frac{\tilde{\lambda}_\Lambda}{2} (\tilde{\varrho} - \tilde{\kappa}_\Lambda)^2 . \quad (12.87)$$

At the cutoff scale, it is defined in the regime with spontaneous symmetry breaking, with the minimum of the potential at $\tilde{\kappa}_\Lambda$. The corresponding w'_Λ in (12.84) reads

$$w'_\Lambda(\varrho) = \lambda_\Lambda (\varrho - \kappa_\Lambda) \quad \text{with} \quad \lambda_\Lambda = \mu_3 \frac{\tilde{\lambda}_\Lambda}{\Lambda}, \quad \kappa_\Lambda = \frac{1}{\mu_3} \frac{\tilde{\kappa}_\Lambda}{\Lambda} . \quad (12.88)$$

This means that for a quartic potential,

$$\varrho \left(e^{2t} w'\right) = \frac{e^{2t} w'}{\lambda_\Lambda} + \kappa_\Lambda , \quad (12.89)$$

and this result is inserted into the right-hand side of (12.86). To actually calculate the scale-dependent potential, we prescribe at every scale the values of w' and determine the corresponding values of the composite field ϱ from the mappings

$$\frac{\varrho}{N} = \frac{e^{-t}}{N} \left(\frac{e^{2t} w'}{\lambda_\Lambda} + \kappa_\Lambda\right) + \sqrt{\pm w'} \left(H_\pm(w') - H_\pm \left(e^{2t} w'\right)\right), \quad \pm w' > 0 . \quad (12.90)$$

For small w' the right-hand side converges to $(1 - e^{-t})$ such that a node ϱ_0 of w' flows according to

$$\varrho_0(t) = N + e^{-t} (\kappa_\Lambda - \kappa_{\text{crit}}), \quad \kappa_{\text{crit}} = N . \quad (12.91)$$

If we tune the cutoff parameter κ_Λ to κ_{crit}, then the node of the dimensionless w' becomes scale invariant. Any other choice of this parameter leads to an instability

Fig. 12.8 Flow of the *dimensionless* effective potential with cutoff parameters $\lambda_\Lambda = 1$ and $\kappa_\Lambda = 1$. For this fine-tuned value of κ_Λ, the potential flows into the regular fixed point solution with $c = 0$. Shown are potentials with scale parameter $k/\Lambda = 1$, 0.95, 0.8, and 0.6. All dimensionless potentials have their minimum at $\varrho/N = 1$

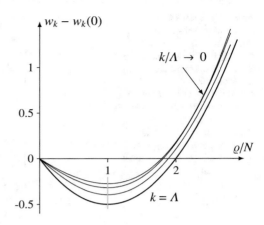

since the $\varrho_0(t) - \varrho_0(0)$ grows exponentially fast. For small scales equation (12.91) reads

$$\frac{\varrho}{N} = e^{-t}\left(\frac{\kappa_\Lambda}{N} - 1\right) + \sqrt{\pm w'}\, H_\pm(w'), \quad t << -1 . \tag{12.92}$$

As expected, this is only a solution of the fixed point equation if $\kappa_\Lambda = \kappa_{\text{crit}}$. For this fine-tuned value, the potential flows into the fixed point solution with $c = 0$. Figure 12.8 shows the flow of the dimensionless potential with critical coupling κ_{crit}.

Symmetry Breaking

To study the phases of the O(N) models for large N, we return to the dimensionful effective potential

$$\frac{u_k(\tilde{\varrho})}{\Lambda^3} = \mu_3 e^{3t} w_k\left(\frac{e^{-t}\tilde{\varrho}}{\mu_3 \Lambda}\right) . \tag{12.93}$$

In the far infrared u_k is minimal at

$$\frac{\tilde{\varrho}_0(t)}{\mu_3 \Lambda} = e^t \varrho_0(t) \rightarrow \kappa_\Lambda - \kappa_{\text{crit}} \equiv \delta\kappa_\Lambda \quad \text{for} \quad t \rightarrow -\infty . \tag{12.94}$$

Thus for any *positive* control parameter $\delta\kappa_\Lambda$, the system flows into the ordered phase with broken O(N) symmetry. When we lower the scale, the minimum moves inward and the normalized field settles at $\delta\kappa_\Lambda$. The flow of the potential with cutoff parameter $\kappa_\Lambda/\kappa_{\text{crit}} = 1.5$ is depicted in Fig. 12.9. For any *negative* control parameter $\delta\kappa_\Lambda$, the minimum of the full effective potential is always at the origin. Even if the classical potential has a minimum at $\kappa_\Lambda < \kappa_{\text{crit}}$, the system flows into the disordered O(N) symmetric phase as is shown in Fig. 12.10. In other words if the symmetry breaking of the classical theory is not strong enough, the quantum fluctuations drive the system into the symmetric phase. No secondary minimum develops, and hence

Fig. 12.9 Flow of the *dimensionful* effective potential for $\kappa_\Lambda/\kappa_{\mathrm{crit}} = 1.5$ and $\lambda_\Lambda = 1$. With decreasing k the minimum moves inward until it settles as $\delta\kappa_\Lambda$. For $k = 0$ the potential is flat between 0 and $\delta\kappa_\Lambda$. Shown are potentials with k/Λ values 1, 0.95, 0.9, 0.8, 0.6, and 0. Since $\kappa_\Lambda > \kappa_{\mathrm{crit}}$, the system ends up in the phase with broken O(N)

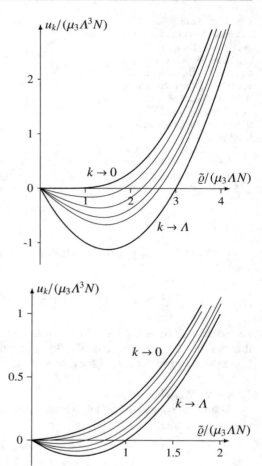

Fig. 12.10 Flow of the *dimensionful* effective potential for $\kappa_\Lambda/\kappa_{\mathrm{crit}} = 0.5$ and $\lambda_\Lambda = 1$. Due to quantum fluctuations, the minimum of the classical potential moves to the origin. From below to above, the scale parameter k/Λ has the values 1, 0.95, 0.9, 0.8, 0.6, and 0. Since $\kappa_\Lambda < \kappa_{\mathrm{crit}}$, the system ends up in the symmetric phase

there is a second-order phase transition at $\kappa_\Lambda = \kappa_{\mathrm{crit}}$. Clearly there is just one relevant coupling, namely, κ_Λ. Only the control parameter $\delta\kappa_\Lambda$ must be tuned to fix the qualitative behavior of the quantum system in the infrared. Since in the large N limit the LPA for the effective potential becomes exact, our conclusions will not be modified if one considers other truncations.

12.6 Wave Function Renormalization

In a full next-to-leading-order approximation in the derivative expansion, one would include a wave function renormalization $Z_k(p, \varphi)$ depending on the scale, fields, and momenta. Typically this leads to nonlinear parabolic partial differential RG equations which are not always easy to solve. Here we neglect the field and momentum dependence and consider a scale-dependent wave function renormalization Z_k in the effective action

$$\Gamma_k[\varphi] = \int \mathrm{d}^d x \left(\frac{1}{2} Z_k (\partial_\mu \varphi)^2 + u_k(\varphi) \right) . \tag{12.95}$$

Thereby the scale-dependent constant Z_k is the field-dependent wave function renormalization at zero field, $Z_k = Z_k(\phi = 0)$. The variation of Γ_k contains the derivatives of the couplings in the effective potential as well as the derivative of the wave function renormalization,

$$\partial_k \Gamma_k = \int \mathrm{d}^d x \left(\frac{1}{2} (\partial_k Z_k) (\partial_\mu \varphi)^2 + \partial_k u_k(\varphi) \right) , \tag{12.96}$$

and its second functional derivative with respect to the field is

$$\Gamma_k^{(2)} = -Z_k \Delta + u_k''(\varphi) . \tag{12.97}$$

The flow equation simplifies if we use $Z_k R_k$ in place of R_k in the regulator term ΔS_k. This is also suggested by preserving the invariance under field rescalings. With this parametrization the flow equation in the truncation (12.95) reads

$$\partial_k \Gamma_k = \frac{1}{2} \mathrm{tr} \left(\frac{\partial_k (Z_k R_k)}{Z_k (p^2 + R_k) + u_k''} \right) . \tag{12.98}$$

As earlier on we employ the optimized regulator in (12.10) for which the flow equation for the effective potential takes the form

$$\partial_k u_k = \frac{\mathscr{Z}_k}{Z_k k^2 + u_k''}, \quad \text{where} \quad \mathscr{Z}_k = \frac{\mu_d}{d+2} \partial_k \left(k^{d+2} Z_k \right) . \tag{12.99}$$

The geometric constant μ_d was introduced earlier in (12.47). Without wave function renormalization, the differential equation simplifies to the flow equation (12.48). Inserting the series expansion

$$u_k(\varphi) = a_0 + \sum_{n=1}^{\infty} \frac{a_n(k)}{n!} \varphi^n \tag{12.100}$$

and comparing coefficients, we obtain the flow equations for the couplings a_n. The lowest couplings obey

$$\frac{\mathrm{d}a_0}{\mathrm{d}k} = \mathscr{L}_k \Delta_0, \qquad \Delta_0 = \frac{1}{Z_k k^2 + a_2}$$

$$\frac{\mathrm{d}a_1}{\mathrm{d}k} = -\mathscr{L}_k \Delta_0^2 a_3$$

$$\frac{\mathrm{d}a_2}{\mathrm{d}k} = -\mathscr{L}_k \Delta_0^2 (a_4 - 2\Delta_0 a_3^2)$$

$$\frac{\mathrm{d}a_3}{\mathrm{d}k} = -\mathscr{L}_k \Delta_0^2 \left(a_5 - 6\Delta_0 a_3 a_4 + 6\Delta_0^2 a_3^3 \right) \tag{12.101}$$

$$\frac{\mathrm{d}a_4}{\mathrm{d}k} = -\mathscr{L}_k \Delta_0^2 \left(a_6 - 8\Delta_0 a_3 a_5 - 6\Delta_0 a_4^2 + 36\Delta_0^2 a_3^2 a_4 - 24\Delta_0^3 a_3^4 \right)$$

$$\frac{\mathrm{d}a_5}{\mathrm{d}k} = -\mathscr{L}_k \Delta_0^2 \left(a_7 - 10\Delta_0 a_3 a_6 - 20\Delta_0 a_4 a_5 + 60\Delta_0^2 a_3^2 a_5 \right.$$
$$\left. + 90\Delta_0^2 a_3 a_4^2 - 240\Delta_0^3 a_3^3 a_4 + 120\Delta_0^4 a_3^5 \right)$$

$$\frac{\mathrm{d}a_6}{\mathrm{d}k} = -\mathscr{L}_k \Delta_0^2 \left(a_8 - 12\Delta_0 a_3 a_7 - 30\Delta_0 a_4 a_6 - 20\Delta_0 a_5^2 + 90\Delta_0^2 a_3^2 a_6 \right.$$
$$+ 90\Delta_0^2 a_4^3 + 360\Delta_0^2 a_3 a_4 a_5 - 480\Delta_0^3 a_3^3 a_5 - 1080\Delta_0^3 a_3^2 a_4^2$$
$$\left. + 1800\Delta_0^4 a_3^4 a_4 - 720\Delta_0^5 a_3^6 \right) .$$

In leading order $Z_k = 1$ and we recover the flow equations without wave function renormalization in (12.49).

12.6.1 RG Equation for Wave Function Renormalization

To extract the scale dependence of Z_k in the flow equation

$$\int \mathrm{d}^d x \left(\frac{1}{2} (\partial_k Z_k) \, (\partial_\mu \varphi)^2 + \partial_k u_k(\varphi) \right) = \frac{1}{2} \mathrm{tr} \left(\frac{\partial_k (Z_k R_k)}{Z_k (p^2 + R_k) + u_k''(\varphi)} \right) \tag{12.102}$$

we project the right-hand side onto the operator $(\partial \varphi)^2$. Clearly, this is only possible if we admit an inhomogeneous average field for which p^2 and $u_k''(\varphi)$ do not commute. But it is sufficient to expand the full propagator on the right-hand side in powers of the field up to second order. For this expansion we set $u_k''(\varphi) =$

$a_2 + \Delta u_k'' + \ldots$ and expand in powers of the field-dependent $\Delta u_k''$:

$$\frac{1}{Z_k(p^2 + R_k) + u_k''(\varphi)} = \sum_{n=0}^{\infty}(-)^n G_0 \left(\Delta u_k''(\varphi)G_0\right)^n . \quad (12.103)$$

For the optimized regulator, the free propagator

$$G_0 = \frac{1}{Z_k(p^2 + R_k) + a_2} . \quad (12.104)$$

becomes p-independent below the scale k:

$$G_0 \left(|p| < k\right) = \Delta_0 = \frac{1}{Z_k k^2 + a_2} . \quad (12.105)$$

In order to project the right-hand side of (12.102) onto the operator $(\partial\varphi)^2$, we only need to consider terms quadratic in φ. With $\Delta u_k'' = a_3\varphi + a_4\varphi^2/2 + \ldots$, we obtain

$$\int d^d x\, \partial_k Z_k\, (\partial_\mu\varphi)^2 = \text{tr} \left\{\partial_k(Z_k R_k) \left(a_3^2\, G_0\varphi G_0\varphi G_0 - \frac{a_4}{2}G_0\varphi^2 G_0\right)\right\}_{(\partial\varphi)^2} . \quad (12.106)$$

In momentum space the left-hand side takes the form

$$\partial_k Z_k \int_p \varphi(-p)p^2\varphi(p), \quad \text{where} \quad \int_p \equiv \int \frac{d^d p}{(2\pi)^d} . \quad (12.107)$$

We distinguish between a function in x space and its Fourier transform in p space via their arguments. Inserting the Fourier transform of the free propagator

$$G_0(x) = \int_p e^{ipx} G_0(p) \quad (12.108)$$

and that of the average field into the last term in (12.106) yields

$$-\frac{a_4}{2}\varphi^2(p=0) \int_p \partial_k \left(Z_k R_k(p)\right) G_0^2(p) .$$

Only the field at zero momentum enters here, and hence no term like the one in (12.107) is generated. We conclude that for systems with even potentials for a series expansion at the origin (12.100), the wave function renormalization Z_k does not flow in the truncation considered here. For more general potentials or expansions of even potentials around a nonvanishing vacuum expectation value, $a_3 \neq 0$, and the second to last term in (12.106) reads in momentum space

$$a_3^2 \int_p \int_q \partial_k \left(Z_k R_k(q)\right) G_0^2(q)\varphi(-p)G_0(p+q)\varphi(p) \,. \tag{12.109}$$

Its projection onto $(\partial\varphi)^2$ is

$$\frac{1}{2}a_3^2 \int_q \partial_k \left(Z_k R_k(q)\right) G_0^2(q)\Delta_q G_0(q) \int_p \varphi(-p)p^2\varphi(p) \,. \tag{12.110}$$

Using these results in the projected flow equation (12.106) finally yields

$$\partial_k Z_k = \frac{a_3^2}{2} \int_q \partial_k \left(Z_k R_k(q)\right) G_0^2(q)\Delta_q G_0(q) \,. \tag{12.111}$$

For the optimized regulator, the integrand contains products of distributions, and for this reason we expand the q-integral in (12.109) directly in powers of p. The detailed calculation can be found in the Appendix to this section. The final answer is

$$k\partial_k Z_k = -\mu_d \, k^{d+2} \left(Z_k a_3 \Delta_0^2\right)^2 \,, \tag{12.112}$$

with μ_d from (12.47). The differential equations (12.99) and (12.112) yield the flow of the effective potential and scale-dependent wave function renormalization in the next-to-leading-order approximation. Actually the anomalous dimension

$$\eta = -k\partial_k \log Z_t \tag{12.113}$$

of the O(N) model vanishes in the large N limit. But there exist perturbatively non-renormalizable but asymptotically safe field theories with a large anomalous dimension at the non-Gaussian fixed point. For example, the Gross-Neveu model in three dimensions is asymptotically safe and has a large anomalous dimension [34]. Gross-Neveu models for different numbers of fermion fields are introduced and discussed in Chap. 17.

12.7 Outlook

It has already been stressed that functional renormalization group equations have been applied to a variety of quantum and statistical systems. In this chapter we could only give an introduction into this powerful method with applications to simple quantum-mechanical systems and scalar field theories. For a further reading, I refer to the textbooks and reviews [5–11]. Here it suffices to mention some interesting recent developments. Of course the flow equations can be formulated for and applied to fermionic systems as well. Thereby one can dynamically bosonize the emerging four-fermion operators with the Hubbard-Stratonovich trick [34–36]. For gauge theories with local gauge symmetries, the background field method has been adjusted to calculate the flow for the coupled systems of gauge, matter, and ghost fields [9, 37–40]. A similar technique has been used to study gravity theories with local diffeomorphism invariance and to argue that gravity very probably is an asymptotically safe theory with a nontrivial UV fixed point [41–43]. Also for many supersymmetric quantum field theories, a manifest supersymmetric flow can be constructed in superspace [20, 31, 44, 45]. Supersymmetry relates the regulator functions of the bosons and fermions and thus gives a new perspective on regulator functions in theories with interacting bosons and fermions. One can investigate for which parameter region the quantum systems are supersymmetric and for which region supersymmetry is broken. Implementing spacetime symmetries is not so much an issue for functional renormalization group flow equations as it is for a lattice regularization.

The flexible functional method offers great potential for theoretical advances in both hot and dense QCD, gravity, supersymmetry, as well as many-body physics. The method is somehow complementary to the ab initio lattice approach. In cases where a lattice regularization based on a positive Boltzmann factor fails, for example, for gauge theories at finite density, the functional method may work. Thus it is probably a good strategy to consider both methods when it comes to properties of strongly coupled systems under extreme conditions.

12.8 Programs for Chap. 12

The octave program in Listing 12.1 calculates the flow of the couplings E_k, ω_k^2, and λ_k in the LPA approximation and the truncation with $a_6 = 0$. The flow begins at $k = 10^5$, but only the values of the couplings for $k = 2$ down to $k = 0$ are plotted.

Listing 12.1 Flow of couplings in polynomial LPA

```
 1  function x=truncflowanho_lambda
 2  # calculates the flow for even potentials projected onto
 3  # quartic polynoms. At the cutoff E=0. The program asks
 4  # for lambda at the cutoff in units of omega. The running
 5  # of the effective couplings  in the infrared is plotted.
 6  # couplLambda=[0;-1;0];        # for 4th order polynomial
 7  couplLambda=[0;-1;0;0;0;0;0];  # for 12th order polynomial
 8  Nk=100000;disp=20;
 9  couplLambda(3)=lambda;
10  k=linspace(10000,0,Nk);
11  #[coupl]=lsode("flowOpt4",couplLambda,k);
12  #[coupl]=lsode("flowCallan4",couplLambda,k);
13  [coupl]=lsode("flowOpt12",couplLambda,k);
14  xh=coupl(Nk-disp:Nk,:,:);
15  kh=k(Nk-disp:Nk);
```

```
16  plot(kh,xh(:,1),kh,xh(:,2),kh,xh(:,3));
17  legend('E','omega**2','lambda');
18  printf("E0 = \t %4.4f\n",coupl(Nk,1));
19  printf("E1 = \t %4.4f\n",coupl(Nk,1)+sqrt(coupl(Nk,2)));
20  endfunction
```

Listing 12.2 Called by program 12.1: optimal regulator, fourth-order polynomial

```
1  function coupldot=flowOpt4(coupl,k)
2  ksquare=k*k;
3  P=1/(ksquare+coupl(2));xh=ksquare*P/pi;
4  coupldot(1)=xh-1/pi;
5  coupldot(2)=-xh*P*coupl(3);
6  coupldot(3)=-6*coupldot(2)*P*coupl(3);
7  endfunction
```

Listing 12.3 Called by program 12.1: Callan-Symanzik regulator and fourth-order polynomial

```
1  function coupldot=flowCallan4(coupl,k)
2  P=1/(k*k+coupl(2));
3  rootP=k*sqrt(P);
4  coupldot(1)=0.5*(rootP-1);
5  coupldot(2)=-0.25*rootP*P*coupl(3);
6  coupldot(3)=-4.5*coupldot(2)*P*coupl(3);
7  endfunction
```

The functions defined in Listings 12.2, 12.3, and 12.4 are called by program 12.1 to calculate the flow for different regulators and truncation orders.

Listing 12.4 Called by program 12.1: optimal regulator and 12th-order polynomial

```
1   function coupldot=flowOpt12(coupl,k)
2   a0=coupl(1);a2=coupl(2);a4=coupl(3);a6=coupl(4);
3   a8=coupl(5);a10=coupl(6);a12=coupl(7);
4   ks=k*k;numflow1
5   P=1/(ks+a2);P2=P*P;P2pi=ks*P2/pi;P3=P*P2;P4=P*P3;
6   a4s=a4*a4;a4q=a4s*a4s;a6s=a6*a6;
7   coupldot(1)=-a2*P/pi;
8   coupldot(2)=-a4;
9   coupldot(3)=-a6+6*a4s*P;
10  coupldot(4)=-a8+30*a4*a6*P-90*a4s*a4*P2;
11  coupldot(5)=-a10+(56*a4*a8+70*a6s)*P\
12  -1260*a6*a4s*P2+2520*a4q*P3;
13  coupldot(6)=-a12+(90*a4*a10+420*a6*a8)*P\
14   -(3780*a8*a4s+9450*a6s*a4)*P2\
15  +75600*a4s*a4*a6*P3-113400*a4q*a4*P4;
16  coupldot(7)=(132*a4*a12+924*a8*a8+990*a10*a6)*P\
17   -(8910*a10*a4s+83160*a4*a6*a8+34650*a6s*a6)*P2\
18  +(332640*a8*a4s*a4+1247400*a4s*a6s)*P3\
19  -6237000*a4q*a6*P4+7484400*a4q*a4s*(P4*P);
20  coupldot(2:7)=P2pi*coupldot(2:7);
21  endfunction
```

The `octave` function 12.5 solves the partial differential equation for the scale-dependent effective potential in (12.31). It calls the function defined in Listing 12.6.

Listing 12.5 Flow of effective potential in LPA

```
1  function flowpde(a4)
2  # Solves the partial differential equation for the
3  # flow of the effective potential by rewriting it
4  # as a system of coupled ode's. Calls fa.m
5  global Nq;
6  global dqsquareinv;
7  global W;
8  Nq=151;L=5;Nk=800;
9  q=linspace(-L,L,Nq);
10 a2=-1;
11 qsquare=q.*q;
12 dq=2*L/(Nq-1);
13 dqsquareinv=1/(dq*dq);
14 k=linspace(800,0,Nk);
15 V=0.5*a2*qsquare+a4*qsquare.*qsquare/24;
16 Veff=lsode("fa",V,k);
17 u=Veff(Nk,:);
18 plot(q,V,'r',q,u,'b')
19 legend('Vclass','Veff');
20 [umin,index]=min(u);
21 upp=(u(index+1)+u(index-1)-2*u(index))*dqsquareinv;
22 printf("\nE0 = %4.4f\nE1 = %4.4f \n",umin,umin+sqrt(upp));
23 endfunction
```

Listing 12.6 Called by program 12.5: right-hand side of ode

```
1  function xdot=fa(V,x)
2  global dqsquareinv;
3  global Nq;
4  global W;
5  xs=x*x;
6  W=dqsquareinv*(shift(V,1)+shift(V,-1)-2*V);
7  xdot=(xs./(xs+W)-1)/pi;
8  xdot(1)=0;
9  xdot(Nq)=0;
10 endfunction
```

12.9 Problems

12.1 (Weak-Coupling Expansion for Anharmonic Oscillator) In this exercise we solve the truncated flow equations (12.35) for weak couplings $\lambda/\omega^3 \ll 1$. First we introduce dimensionless variables:

$$x = \frac{k}{\omega}, \quad e_k = \frac{E_k}{\omega}, \quad o_k = \frac{\omega_k}{\omega} \quad \text{and} \quad \ell_k = \frac{\lambda_k}{24\omega^3} .$$

(a) Show that the truncated flow equations for the dimensionless variables read

$$\frac{de_k}{dx} = -\frac{o_k^2}{\pi} \frac{1}{x^2 + o_k^2}, \quad \frac{do_k^2}{dx} = -\frac{\ell_k}{\pi} \frac{24x^2}{(x^2 + o_k^2)^2}, \quad \frac{d\ell_k}{dx} = \frac{144\ell_k^2}{\pi} \frac{x^2}{(x^2 + o_k^2)^3}.$$

(b) For the harmonic oscillator with vanishing λ, we have $e_k = 1/2$, $o_k = 1$, and $\ell_k = 0$. The initial conditions are $e_{x \to \infty} = 0$, $o_{x \to \infty} = 1$, and $\ell_{x \to \infty} = \lambda/24\omega^3$. Thus for weak coupling the expansion has the form

$$e_k = \alpha_0 + \alpha_1 \varepsilon + \alpha_2 \varepsilon^2 + \dots$$
$$o_k^2 = 1 + \beta_1 \varepsilon + \beta_2 \varepsilon^2 + \dots$$
$$\ell_k = \varepsilon + \gamma_2 \varepsilon^2 + \dots$$

with $\varepsilon = \lambda/24\omega^3 \ll 1$ and scale-dependent coefficient functions. Show that these functions satisfy the differential equations

$$\dot{\alpha}_0 = -\frac{P}{\pi}, \quad \dot{\alpha}_1 = -\frac{x^2 P^2}{\pi} \beta_1, \quad \dot{\alpha}_2 = \frac{x^2 P^2}{\pi} \left(\beta_1^2 P - \beta_2 \right)$$

$$\dot{\beta}_1 = -\frac{24x^2 P^2}{\pi}, \quad \dot{\beta}_2 = \frac{24x^2 P^2}{\pi} (2\beta_1 P - \gamma_2), \quad \dot{\gamma}_2 = 144\frac{x^2 P^3}{\pi},$$

where we defined $P \equiv 1/(1+x^2)$. The coefficient functions vanish for $x \to \infty$, and the first differential equation has the solution $\alpha(x) = 1/2 - \arctan(x)/\pi$.

(c) Calculate β_1, γ_2, and β_2 with an algebraic computer program. Insert the solutions to find the functions α_2 and α_3.

(d) Prove that the coefficient functions have the following small-x expansions:

$$\alpha_0(x) \sim \frac{1}{2} - \frac{1}{\pi}x + \frac{1}{3\pi}x^3 + \dots$$

$$\alpha_1(x) \sim \frac{3}{4} - 2\pi x^3 + \dots$$

$$\alpha_2(x) \sim -\frac{3}{8}\left(10 - \frac{29}{\pi^2}\right) + 8\left(3 - \frac{4}{\pi^2}\right)x^3 + \dots \qquad (12.114)$$

$$\beta_1(x) \sim 6 - \frac{8}{\pi}x^3 + \dots$$

$$\beta_2(x) \sim -12\left(3 - \frac{8}{\pi^2}\right) + \frac{168}{\pi}x^3 + \dots$$

$$\gamma_2(x) \sim -9 + \frac{48}{\pi}x^3 + \dots$$

(e) If we make a cruder truncation with $\lambda_k = \lambda$ of equivalently with $\gamma_2 = 0$, how do the results change? Prove that in this case

$$\alpha_2(x) \sim -\frac{3}{16}\left(8 + \frac{29}{\pi^2}\right) + \frac{1}{\pi^3}(15\pi^2 + 16)x^3 + \dots \quad (12.115)$$

$$\beta_2(x) \sim -2\left(3 + \frac{16}{\pi^2}\right) + \frac{96}{\pi}x^3 + \dots$$

12.2 (Fixed Point Solution and Critical Exponents) Write a program with your favorite algebraic computer system to find the fixed point solution and the critical exponents of the Z_2 scalar field theory. You should be able to reproduce the plots in Fig. 12.5 and the critical exponents in Table 12.4.

Appendix: A Momentum Integral

In this appendix we calculate the $O(p^2)$ contribution to the integral

$$F(p) = \int d^d q \, \partial_k \{Z_k R_k(q)\} \, G_0^2(q) \, G_0(p + q) \quad (12.116)$$

for the optimized regulator function (12.10). The integrand is only nonzero for $q^2 \leq k^2$ and in this region

$$\partial_k \{Z_k R_k(q)\} \, G_0^2(q) = \left(\left(k^2 - q^2\right) \partial_k Z_k + 2k Z_k\right) \Delta_0^2 . \quad (12.117)$$

To proceed we need to consider two cases: $|p + q| \leq k$ and $|p + q| > k$ separately.

The Case $|p + q| < k$

This is the region located inside of both spheres in Fig. 12.11 where the Green function $G_0(q) = G_0(p + q) = \Delta_0$ is independent of the integration variable q. Let us decompose this variable as $q = q_\parallel + q_\perp$, where $q_\parallel$ is parallel and $q_\perp$ perpendicular to the fixed momentum p. Then the integral has the form

$$I_1 = \text{Vol}\,(S_{d-2}) \, \Delta_0^3 \int dq_\parallel \int d|q_\perp| \, |q_\perp|^{d-2} \left(\left(k^2 - q^2\right) \partial_k Z_k + 2k Z_k\right) , \quad (12.118)$$

Fig. 12.11 Sketch of the
integration regions in
momentum space. Only
momenta inside of the ball
centered at the origin
contribute to the integral. The
Green functions are constant
in the gray region inside of
both spheres. Inside the
sickle-shaped light-gray
region, $G_0(p, q)$ is
momentum dependent. The
two spheres intersect at the
lower-dimensional sphere
defined by $\{q_\parallel^*, q_\perp^*\} =$
$\{-p/2, k^2 - p^2/4\}$

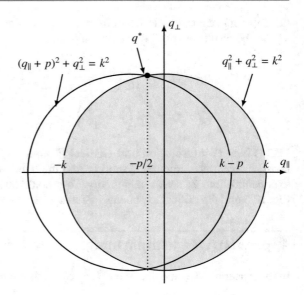

where $q^2 = q_\parallel^2 + q_\perp^2$. The volume of the unit sphere $S_{d-2} \subset \mathbb{R}^{d-1}$ originates from
the integration over the directions of $q_\perp$. Now we split the integration domain inside
of both spheres, the region marked gray in Fig. 12.11, into two spherical caps:

$$-k \le q_\parallel < -\frac{p}{2} \quad \text{and} \quad 0 \le q_\perp^2 \le k^2 - q_\parallel^2$$

$$-\frac{p}{2} \le q_\parallel < k - p \quad \text{and} \quad 0 \le q_\perp^2 \le k^2 - (p + q_\parallel)^2 .$$

The domain of integration in (12.118) is just the union of these two caps. After a
shift $q_\parallel \to q_\parallel - p$ in the integral over the second cap, we are left with the one-
dimensional integral

$$\frac{\text{Vol}\,(S_{d-2})\,\Delta_0^3}{d-1} \int_{p/2}^{k} dq_\parallel \left\{ \left(\frac{4k^2 - 4q_\parallel^2}{d+1} + 2q_\parallel p - p^2 \right) \partial_k Z_k + 4k Z_k \right\}$$

$$\times \left(k^2 - q_\parallel^2 \right)^{(d-1)/2} .$$

Its second derivative with respect to p at $p = 0$ is

$$\frac{d^2 I_1}{dp^2} \bigg|_{p=0} = -k^d V(B_d)\,\Delta_0^3\, \partial_k Z_k . \tag{12.119}$$

The Case $|p + q| < k$

This is the sickle-shaped region inside of the sphere centered at the origin but outside the displaced sphere, marked light-gray in Fig. 12.11. The integral over this region can be written as a difference of two integrals as follows:

$$
I_2 = \Delta_0^2 \left(\int_{-p/2}^{k} dq_{\parallel} \int_0^{\sqrt{k^2 - q_{\parallel}^2}} dq_{\perp} \, h\left(q_{\parallel}, q_{\perp}\right) \right.
$$

$$
\left. - \int_{-p/2}^{k-p} dq_{\parallel} \int_0^{\sqrt{k^2 - (q_{\parallel} + p)^2}} dq_{\perp} \, h\left(q_{\parallel}, q_{\perp}\right) \right) .
$$

The integrand of both integrals is

$$
h(q_{\parallel}, q_{\perp}) = \frac{\left(k^2 - q_{\parallel}^2 - q_{\perp}^2\right) \partial_k Z_k + 2k Z_k}{Z_k(q_{\parallel} + p)^2 + Z_k q_{\perp}^2 + a_2} .
$$

It is convenient to shift the variable $q_{\parallel}$ in the second integral by $-p$ such that

$$
I_2 = \Delta_0^2 \left(\int_{-p/2}^{k} dq_{\parallel} \int_0^{\sqrt{k^2 - q_{\parallel}^2}} dq_{\perp} \, h\left(q_{\parallel}, q_{\perp}\right) \right.
$$

$$
\left. - \int_{p/2}^{k} dq_{\parallel} \int_0^{\sqrt{k^2 - q_{\parallel}^2}} dq_{\perp} \, h\left(q_{\parallel} - p, q_{\perp}\right) \right) .
$$

The second derivative of this integral with respect to p at $p = 0$ is given by

$$
\left. \frac{d^2 I_2}{dp^2} \right|_{p=0} = k^d V(B_d) \, \Delta_0^4 \left(a \partial_k Z_k + k^2 Z_k \partial_k Z_k - 2k Z_k^2 \right) . \tag{12.120}
$$

Adding the two results (12.119) and (12.120) leads to the simple expression,

$$
\left. \frac{d^2 F}{dp^2} \right|_{p=0} = -2k^{d+1} V(B_d) \, \Delta_0^4 \, Z_k^2 \tag{12.121}
$$

for the curvature of $F(p)$ in (12.116) at the origin in momentum space.

References

1. K.G. Wilson, The renormalization group: Critical phenomena and the Kondo problem. Rev. Mod. Phys. **47**, 773 (1975)
2. J. Polchinski, Renormalization and effective lagrangians. Nucl. Phys. **B231**, 269 (1984)
3. T.R. Morris, On truncations of the exact renormalization group. Phys. Lett. **B334**, 355 (1994)
4. C. Wetterich, Exact evolution equation for the effective potential. Phys. Lett. **B301**, 90 (1993)
5. K. Aoki, Introduction to the nonperturbative renormalization group and its recent applications. Int. J. Mod. Phys. **B14**, 1249 (2000)
6. C. Bagnus, C. Bervillier, Exact renormalization group equations: an introductory review. Phys. Rep. **348**, 91 (2001)
7. J. Berges, N. Tetradis, C. Wetterich, Nonperturbative renormalization flow in quantum field theory and statistical physics. Phys. Rep. **363**, 223 (2002)
8. J. Polonyi, Lectures on the functional renormalization group methods. Central Eur. J. Phys. **1**, 1 (2003)
9. J. Pawlowski, Aspects of the functional renormalisation group. Ann. Phys. **322**, 2831 (2007)
10. H. Gies, Introduction to the functional RG and applications to gauge theories, in *Renormalization Group and Effective Field Theory Approaches to Many-Body Systems*, ed. by A. Schwenk, J. Polonyi. Lecture Notes of Physics, vol. 852 (Springer, Berlin, 2012)
11. P. Kopietz, L. Bartosch, F. Schütz, *Introduction to the Functional Renormalization Group*. Lecture Notes of Physics, vol. 798 (Springer, Berlin, 2010)
12. D. Litim, Optimization of the exact renormalization group. Phys. Lett. **B486**, 92 (2000)
13. D. Litim, Optimized renormalization group flows. Phys. Rev. **D64**, 105007 (2001)
14. S.D. Drell, M. Weinstein, S. Yankielowicz, Variational approach to strong coupling theory. 1. ϕ^4 theory. Phys. Rev. **D14**, 487 (1976)
15. D. Lange, A. Kirchberg, A. Wipf, From the dirac operator to Wess-Zumino models on spatial lattices. Ann. Phys. **316**, 357 (2005)
16. D.F. Litim, J.P. Pawlowski, L. Vergara, Convexity of the effective action from functional flows. ArXiv:hep-th/0602140
17. D. Zappala, Improving the renormalization group approach to the quantum-mechanical double well potential. Phys. Lett. **A290**, 35 (2001)
18. C.M. Bender, T.T. Wu, Anharmonic oscillator. Phys. Rev. **D184**, 1231 (1969)
19. T. Morris, The renormalization group and two-dimensional multicritical effective scalar field theory. Phys. Lett. **B345**, 139 (1995)
20. F. Synatschke, H. Gies, A. Wipf, Phase diagram and fixed point structure of two dimensional N=1 Wess-Zumino model. Phys. Rev. **D80**, 085007 (2009)
21. D.F. Litim, Critical exponents from optimized renormalization group flows. Nucl. Phys. **B631**, 128 (2002)
22. M. D'Attanasio, T.R. Morris, Large N and the renormalization group. Phys. Lett. **B409**, 363 (1997)
23. M. Moshe, J. Zinn-Justin, Quantum field theory in the large N limit: a review. Phys. Rep. **385**, 69 (2003)
24. J.P. Blaizot, R. Mendéz-Galain, N. Wschebor, A new method to solve the non perturbative renormalization group equations. Phys. Lett. **B632**, 571 (2006)
25. J. Borchardt, B. Knorr, Solving functional flow equations with pseudo-spectral methods. Phys. Rev. **D94**, 025027 (2016)
26. V. Von Gersdorff, C. Wetterich, Nonperturbative renormalization flow and essential scaling for the Kosterlitz-Thouless transition. Phys. Rev. **B64**, 054513 (2001)
27. F. Benitez, J.P. Blaizot, H. Chaté, B. Delamotte, R. Méndez-Galain, N. Wschebor, Nonperturbative renormalization group preserving full-momentum dependence: implementation and quantitive evaluation. Phys. Rev. **E85**, 026707 (2012)
28. M. Hasenbusch, Finite scaling study of lattice models in the three-dimensional Ising universality class. Phys. Rev. **B82**, 174433 (2010)

29. M. Campostrini, M. Hasenbusch, A. Pelissetto, P. Rossi, E. Vicari, Critical exponents and equation of state of the three dimensional Heisenberg universality class. Phys. Rev. **B65**, 144520 (2002)
30. S. Holtmann, T. Schulze, Critical behavior and scaling functions of the three-dimensional O(6) model. Phys. Rev. **E68**, 036111 (2003)
31. D. Litim, M. Mastaler, F. Synatschke-Czerwonka, A. Wipf, Critical behavior of supersymmetric O(N) models in the large-N limit. Phys. Rev. **D84**, 125009 (2011)
32. N. Tetradis, C. Wetterich, Critical exponents from effective average action. Nucl. Phys. **B422**, 541 (1994)
33. N. Tetradis, D. Litim, Analytical solutions of exact renormalization group equations. Nucl. Phys. **B464**, 492 (1996)
34. J. Braun, H. Gies, D. Scherer, Asymptotic safety: a simple example. Phys. Rev. **D83**, 085012 (2011)
35. H. Gies, C. Wetterich, Renormalization flow of bound states. Phys. Rev. **D65**, 065001 (2002)
36. J. Braun, Fermion interaction and universal behavior in strongly interacting theories. J. Phys. **G39**, 033001 (2012)
37. M. Reuter, C. Wetterich, Effective average action for gauge theories and exact evolution equations. Nucl. Phys. **B417**, 181 (1994)
38. U. Ellwanger, Flow equations and BRS invariance for Yang-Mills theories. Phys. Lett. **B335**, 364 (1994)
39. H. Gies, Running coupling in Yang-Mills theory: a flow equation study. Phys. Rev. **D66**, 025006 (2002)
40. D. Litim, J. Pawlowski, Wilsonian flows and background fields. Phys. Lett. **B546**, 279 (2002)
41. M. Reuter, Nonperturbative evolution equation for quantum gravity. Phys. Rev. **57**, 971 (1998)
42. M. Reuter, M. Niedermeier, The asymptotic safety scenario in quantum gravity. Living Rev. Relat. **9**, 5 (2006)
43. M. Reuter, F. Saueressing, Functional renormalization group equations, asymptotic safety and quantum Einstein gravity. arXiv:0708.1317
44. M. Bonini, F. Vian, Wilson renormalization group for supersymmetric gauge theories and gauge anomalies. Nucl. Phys. **B532**, 473 (1998)
45. F Synatschke, J. Braun, A. Wipf, N=1 Wess Zumino model in d=3 at zero and finite temperature. Phys. Rev. **D81**, 125001 (2010)

Lattice Gauge Theories

13

According to present-day knowledge, *all fundamental interactions* in nature are described by *gauge theories*. The best known example is electrodynamics with Abelian symmetry group U(1). In contrast, the electroweak and the strong interactions are modeled by gauge theories with the non-Abelian symmetry groups SU(2)×U(1) and SU(3), respectively. In a sense, general relativity also represents a non-Abelian gauge theory, albeit with a non-compact symmetry group [1, 2]. Continuum gauge theories are dealt with in many excellent textbooks [3–6], and it is useful, but not a necessity, that the reader has some basic knowledge of these theories. For those who are not acquainted with continuum gauge theories, we summarized those properties and concepts which are needed in the remaining chapters of this book.

The first systematic investigation of a lattice gauge theory goes back to F. Wegner [7]. He examined Ising-like systems with *local* Z_2-invariance. In Chap. 10 we already encountered the three-dimensional Z_2 gauge theory as dual of the Ising model and introduced the object that is known today in a more general setting as *Wilson loop*. Three years after Wegner's contribution, K. Wilson formulated non-Abelian gauge theories on a spacetime lattice as possible discretization and regularization of continuum gauge theories [8]. Thereby he replaced the Lie-algebra valued continuum gauge field by link variables with values in a compact Lie group in a way that the discretized theory possesses a gauge symmetry for any size of the lattice spacing. In a naive continuum limit where the lattice spacing tends to zero, the lattice action turns into the Yang-Mills action for the continuum field. He formulated a useful and often used criterion for confinement based on the Wilson loop: if the logarithm of the Wilson loop shows an *area-law* behavior, then charged particles are confined.

By working on a discrete spacetime, the path integral becomes finite-dimensional and can be evaluated by stochastic simulation techniques such as the Monte Carlo method. Shortly after the seminal work of Wilson, the first numerical simulations of pure lattice gauge theories in three and four dimensions were performed. At

© The Author(s), under exclusive license to Springer Nature Switzerland AG 2021
A. Wipf, *Statistical Approach to Quantum Field Theory*, Lecture Notes
in Physics 992, https://doi.org/10.1007/978-3-030-83263-6_13

first, this was done for the finite gauge group Z_2 in [9] and later on for non-Abelian lattice gauge theories with gauge groups SU(2) as well as SU(3) in [10, 11]. Observable quantities such as particle masses and decay widths are calculated with the Monte Carlo method. Thereby gauge field configurations are generated with probabilities proportional to e^{-S}, where S is the lattice action. When one includes fermions, then the calculations are often expensive and can require the use of the largest computers available. The simulations typically utilize algorithms based upon molecular dynamics. For more introductory material on lattice gauge theories as presented in this and the following chapters, I recommend the textbooks [12–15] and the classic review papers [16–18].

13.1   Continuum Gauge Theories

In this section we summarize the relevant properties of gauge theories in the continuum. Thereby the emphasis is put on the underlying structures and principles. We begin with the extremely successful *gauge principle*. For that purpose we consider a scalar field ϕ with values in a vector space V with scalar product denoted by $(., .)$. The scalar product is left invariant when its arguments are transformed with an element Ω of some symmetry group G. The most important example is $V = \mathbb{C}^n$ with Hermitian scalar product. Choosing an orthonormal basis ϕ is identified with its components

$$\phi = \begin{pmatrix} \phi_1 \\ \phi_2 \\ \vdots \\ \phi_n \end{pmatrix}, \qquad \phi^\dagger = \left(\phi_1^*, \phi_2^*, \ldots, \phi_n^*\right) \tag{13.1}$$

and the scalar product of two fields is

$$(\phi, \chi) = \sum_{a=1}^{n} \phi_a^* \chi_a . \tag{13.2}$$

It is invariant under a simultaneous $U(n)$ transformation of its arguments:

$$(\phi, \chi) = (\Omega\phi, \Omega\chi), \quad \Omega \in U(n) . \tag{13.3}$$

Let us now assume that the components of ϕ are free fields with equal masses, such that they all obey the Klein-Gordon equation

$$\left(\Box + \frac{m^2 c^2}{\hbar^2}\right) \phi_a = 0 . \tag{13.4}$$

We follow the habit in high energy physics and use natural units $\hbar = c = 1$. The field equations (13.4) are the Euler-Lagrange equations of the Lorentz-invariant action

$$S = \int d^d x \, \mathscr{L}(\phi, \partial_\mu \phi) \quad \text{with} \quad \mathscr{L} = \left(\partial_\mu \phi, \partial^\mu \phi\right) - m^2 \left(\phi, \phi\right) \tag{13.5}$$

Clearly, the Lagrangian density $\mathscr{L}$ is invariant under *global* $U(n)$ transformations

$$\phi(x) \longrightarrow \phi'(x) = \Omega \phi(x), \quad \Omega \in U(n), \tag{13.6}$$

since the scalar product is invariant. The transformation $\phi \to \Omega \phi$ is called *global gauge transformation* since it does not depend on the spacetime point. Invariant Lagrangians can be constructed for symmetry groups G which leave a scalar product invariant.

However, the Lagrangian density is not invariant under *local* gauge transformations, given by

$$\phi(x) \longrightarrow \phi'(x) = \Omega(x)\phi(x), \quad \Omega(x) \in G, \tag{13.7}$$

since the derivatives in (13.5) act on a spacetime-dependent $\Omega(x)$. But it is possible to extend a global symmetry to a local symmetry by coupling the charged scalar field ϕ to a *gauge potential* A_μ. In the minimal coupling, one replaces the partial derivative ∂_μ by the *covariant derivative*

$$D_\mu(A) = \partial_\mu - ig A_\mu . \tag{13.8}$$

Here g is the constant which couples the gauge field and the matter field. The two objects on the right-hand side should transform identically under Lorentz transformation. This means that the A_μ are components of a vector field A. Now we impose the important condition that $D_\mu(A)\phi$ transforms exactly like the field ϕ under gauge transformation:

$$D_\mu(A')\phi'(x) = \Omega(x) \, D_\mu(A)\phi(x) . \tag{13.9}$$

Inserting the transformation of ϕ, this condition is equivalent to

$$D_\mu(A') = \Omega \, D_\mu(A)\Omega^{-1} . \tag{13.10}$$

and this transformation rule for the covariant derivative is fulfilled if

$$A_\mu \to A'_\mu = \Omega A_\mu \Omega^{-1} - \frac{i}{g}\partial_\mu \Omega \, \Omega^{-1} . \tag{13.11}$$

For notational simplicity we did not write the spacetime dependence of the fields and gauge transformation. If $\Omega(x)$ takes its values from a Lie group G, then $\partial_\mu \Omega \Omega^{-1}$ is in the Lie algebra $\mathbf{g}$ of the group. The transformation formula suggests that $\mathrm{i} A_\mu$ should be in the Lie algebra as well such that all three terms in (13.11) are (up to a factor i) Lie-algebra valued. For the gauge group $U(n)$, the Lie algebra contains all anti-Hermitian matrices.

The antisymmetric *field strength tensor* is defined as the commutator of two covariant derivatives:

$$F_{\mu\nu}(A) = \frac{\mathrm{i}}{g}[D_\mu(A), D_\nu(A)] = \partial_\mu A_\nu - \partial_\nu A_\mu - ig[A_\mu, A_\nu] . \tag{13.12}$$

The transformation rule (13.10) for the covariant derivatives implies that it transforms according to the adjoint representation of the gauge group:

$$F_{\mu\nu}(x) \longrightarrow \Omega(x) F_{\mu\nu}(x) \Omega^{-1}(x) . \tag{13.13}$$

Similarly as in electrodynamics, we square the field strength tensor to obtain a Lorentz-invariant contribution to the Lagrangian density. Since $F^{\mu\nu} F_{\mu\nu}$ is not gauge invariant—it transforms according to the adjoint representation—we take its trace and end up with the following Lorentz and gauge-invariant Lagrangian density:

$$\mathscr{L}(\phi, A_\mu) = -\frac{1}{4}\mathrm{tr}\left(F_{\mu\nu} F^{\mu\nu}\right) + \left(D_\mu\phi, D^\mu\phi\right) - m^2(\phi, \phi) = \mathscr{L}(\phi', A'_\mu) \tag{13.14}$$

for the coupled system of scalar fields and gauge fields. The first Yang-Mills term generalizes the Maxwell term in electrodynamics. The remaining terms are obtained from (13.5) by the substitution $\partial_\mu \to D_\mu$.

Let us specialize to the Abelian gauge group U(1) for which the components A_μ of the vector potential are real fields. For an Abelian gauge theory, the field strength is gauge invariant, since Ω in (13.13) commutes with $F_{\mu\nu}$, and no trace is needed in (13.14). Thus we obtain the Lagrangian density of the vector potential in electrodynamics coupled to a charged scalar field. Upon quantization the theory based on (13.14) describes photons and charged scalar particles in interaction. In Chap. 15 we shall see how one couples fermions to the gauge field. If ϕ is replaced by the electron field, then the resulting quantized gauge theory is *quantum electrodynamics*, one of the most successful theories in physics.

The electroweak interaction, described by the *Salam-Weinberg model*, is based on the non-Abelian gauge group SU(2)×U(1). In this theory the scalar field ϕ is a complex doublet and transforms under both factors of the gauge group. The field ϕ is very important since after condensation it is responsible for the masses of elementary particles. Finally, the theory of strong interaction, quantum chromodynamics, is a gauge theory with gauge group SU(3). The matter sector of this theory contains quarks—fermions which are confined within baryons and mesons and which cannot be liberated. Many quantitative results about this strongly interacting theory stem from numerical simulations of the corresponding lattice gauge theory.

Let us discuss in more detail the minimal coupling of matter fields to the gauge field. In general the vector space V carries a representation R of the gauge group, in which case the matter field transforms according to this representation, $\phi' = R(\Omega)\phi$. For example, for the gauge group SU(2), a real field ϕ with three components could transform according to the three-dimensional triplet representation. The important covariance condition (13.6) is automatically fulfilled, if we choose

$$D_\mu \phi = \partial_\mu \phi - ig R_*(A_\mu)\phi \tag{13.15}$$

as covariant derivative. Here R_* is the representation of the Lie algebra, induced by the group representation R. Thus the commutator of covariant derivatives in the representation R is given by

$$i[D_\mu, D_\nu] = g R_*(F_{\mu\nu}) . \tag{13.16}$$

Certain properties of quantized gauge theories may depend on the representation of the gauge group. For example, locally the gauge groups SU(2) and SO(3) are identical, but this does not imply that the corresponding quantized gauge theories show the same phases and phase structure [19].

Finally let us summarize the main ingredients of a gauge theory. A gauge theory is determined by:

- The gauge group G,
- The matter fields including the representations under which they transform,
- The *universal* coupling constant g.

Unfortunately that is not all. We also must specify the masses and self-couplings of the matter fields. The Salam-Weinberg theory contains additional parameters, for example, the elements of the KMS matrix.

13.1.1 Parallel Transport

A lattice gauge theory has parallel transporters between neighboring sites as fundamental fields. These transporters are already defined in the continuum theory, and their definition is carried over to the corresponding lattice theory. Here we consider a scalar field in the defining representation, i.e., $\phi \to \Omega\phi$. The field in the continuum theory is called *covariantly constant* if

$$D_\mu \phi = 0, \quad i.e., \quad \partial_\mu \phi = ig A_\mu \phi . \tag{13.17}$$

With (13.12) this implies the integrability condition

$$0 = \left[D_\mu, D_\nu\right] \phi = -\mathrm{i}g F_{\mu\nu} \phi \; .$$

In a non-Abelian gauge theory, $F_{\mu\nu}(x)$ is Lie-algebra valued and $\phi(x)$ is a vector with n components. For a generic field strength, these n algebraic equations have no nontrivial solution.

Let us instead study the equation of covariant constancy *along a path* $\mathscr{C}_{yx}$ from x to y. The parametrized path $x(s)$ with $s \in [0, 1]$ fulfills

$$x(0) = x \quad \text{and} \quad x(1) = y \; .$$

The field ϕ is covariantly constant along the path if

$$0 = \dot{x}^\mu D_\mu \phi = \frac{\mathrm{d}\phi(s)}{\mathrm{d}s} - \mathrm{i}g A_\mu\left(x(s)\right) \dot{x}^\mu(s)\,\phi(s) \; , \tag{13.18}$$

where we used the shorthand notation $\phi(s)$ for $\phi(x(s))$. If we interpret the curve parameter s as time, then the equation above is a time-dependent Schrödinger equation with time-dependent "Hamiltonian" $\sim A_\mu\left(x(s)\right) \dot{x}^\mu(s)$. Hence, the solution of the ordinary differential equation may be written as

$$\phi(s) = \mathscr{P} \exp\left(\mathrm{i}g \int_0^s \mathrm{d}u\, A_\mu\left(x(u)\right) \dot{x}^\mu(u)\right) \phi(x) \; , \tag{13.19}$$

where $\mathscr{P}$ indicates the ordering with respect to the curve parameter s. For non-Abelian fields this *path ordering* is necessary since the Lie-algebra valued Hamiltonians at different points on the curve do not commute. Evidently, path ordering is obsolete for Abelian gauge groups.

Setting $s = 1$ we obtain the *parallel transporter* along the curve $\mathscr{C}_{yx}$:

$$\phi(y) = U\left(\mathscr{C}_{yx}; A\right) \phi(x), \quad U\left(\mathscr{C}_{yx}; A\right) = \mathscr{P} \exp\left(\mathrm{i}g \int_0^1 \mathrm{d}s\, A_\mu \dot{x}^\mu\right) \; . \tag{13.20}$$

The components of the gauge potential define a one-form, $A = A_\mu dx^\mu$, and the line integral in the exponent is just the integral of this one-form along the path $\mathscr{C}_{yx}$. Thus for an arbitrary path $\mathscr{C}$, we may write

$$U(\mathscr{C}; A) = \mathscr{P} \exp\left(\mathrm{i}g \int_{\mathscr{C}} A\right) \; . \tag{13.21}$$

For a Lie-algebra valued potential, this parallel transporter along $\mathscr{C}$ is an element of the gauge group.

Composition of Paths
If the path $\mathscr{C}$ connects x and y and $\mathscr{C}'$ connects y and z, then the composite path $\mathscr{C}' \circ \mathscr{C}$ (first $\mathscr{C}$ and afterward $\mathscr{C}'$) connects x and z. The parallel transporter along the composite path is just the product of the individual transporters:

$$U(\mathscr{C}' \circ \mathscr{C}; A) = U(\mathscr{C}'; A)U(\mathscr{C}; A) . \tag{13.22}$$

This property follows from the uniqueness of the solution of the system of ordinary differential equation (13.18) for fixed endpoints.

Stokes' Theorem
The parallel transport between two points depends on the path. If $\mathscr{C}'$ and $\mathscr{C}''$ are two different paths from x to y, then $\mathscr{C} = \mathscr{C}'^{-1} \circ \mathscr{C}''$ is a loop beginning and ending at x. For an Abelian theory, Stokes' theorem implies

$$U(\mathscr{C}; A) = \exp\left(\mathrm{i}g \oint_{\mathscr{C}=\partial\mathscr{S}} A \right) = \exp\left(\mathrm{i}g \int_{\mathscr{S}} F \right) , \tag{13.23}$$

where $\mathscr{S}$ is a surface enclosed by $\mathscr{C}$ and F denotes the field strength two-form

$$F = \frac{1}{2} F_{\mu\nu}\, dx^{\mu} \wedge dx^{\nu} = dA . \tag{13.24}$$

Unfortunately, no comparable simple generalization for non-Abelian gauge theories is known, although a non-Abelian Stokes' theorem exists in the mathematical literature [20, 21]. It seems there are not many interesting applications of this theorem in physics.

Gauge Transformation
Parallel transporters are not gauge invariant; they are only gauge covariant. We shall prove that under a gauge transformation, a transporter along $\mathscr{C}_{yx}$ transforms as

$$U(\mathscr{C}_{yx}; A') = \Omega(y)\, U(\mathscr{C}_{yx}; A)\, \Omega^{-1}(x) . \tag{13.25}$$

It immediately follows that for an arbitrary loop, the trace of the associated transporter

$$\mathrm{tr}\, U(\mathscr{C}, A') = \mathrm{tr}\, U(\mathscr{C}, A) \tag{13.26}$$

represents a gauge-invariant quantity—the *Wilson loop*. The set of all Wilson loops form an overcomplete system of gauge-invariant functions on configuration space [22].

Let us now prove the important result (13.25). The parallel transport belonging to the right-hand side of (13.25) is given by

$$\phi'(x(s)) = \Omega(x(s)) \, \mathscr{P} \exp\left(\mathrm{i}g \int_0^s A_\mu \dot{x}^\mu \mathrm{d}s\right) \Omega^{-1}(x)\, \phi(x) \,. \qquad (13.27)$$

Differentiating with respect to the parameter s, we obtain the Schrödinger equation

$$\frac{\mathrm{d}\phi'(s)}{\mathrm{d}s} = \left(\partial_\mu \Omega \Omega^{-1} + \mathrm{i}g\Omega A_\mu \Omega^{-1}\right)\big|_{x(s)} \dot{x}^\mu \, \phi'(s) = \mathrm{i}g A'_\mu(s)\dot{x}^\mu(s)\, \phi'(s) \,,$$

which has the unique solution

$$\phi'(x(s)) = \mathscr{P} \exp\left(\mathrm{i}g \int_0^s \mathrm{d}s\, A'_\mu \dot{x}^\mu\right) \phi(x) \,. \qquad (13.28)$$

A comparison of (13.27) with (13.28) for $s = 1$ yields the transformation rule (13.25).

Matter Fields
The matter field $\phi(x)$ transforms under a gauge transformation into $\Omega(x)\phi(x)$. It follows immediately that the bilinear expression

$$\big(\phi(y), U(\mathscr{C}_{yx}; A)\phi(x)\big) \qquad (13.29)$$

is gauge invariant. However, for a nonvanishing $F_{\mu\nu}$, this expression depends on the path from x to y. A complete list of gauge-invariant composite fields in a theory with scalar and gauge fields is presented in [23].

From this point on, we switch to the *Euclidean formulation* of gauge theories. After a Wick rotation from the Lorentzian to the Euclidean signature, the Lagrangian density of the Higgs model (13.14) transforms into

$$\mathscr{L}_E = \frac{1}{4}\mathrm{tr}\, F^{\mu\nu} F_{\mu\nu} + \big(D_\mu\phi, D^\mu\phi\big) + m^2\big(\phi, \phi\big) \,. \qquad (13.30)$$

The connection between gauge potential and field strength as well as the definitions of the covariant derivatives remains unchanged. In Euclidean theory the metric is $\delta_{\mu\nu}$ and this implies, for example, $F_{\mu\nu} = F^{\mu\nu}$ or $F^{\mu\nu} F_{\mu\nu} = \sum F_{\mu\nu}^2$.

13.2 Gauge-Invariant Formulation of Lattice Higgs Models

In this section we will formulate a theory with *local* gauge invariance on a lattice Λ with lattice sites $\{x\}$ and lattice spacing a. First we discretize a scalar field theory such that the resulting theory on the lattice remains invariant under *global* gauge transformations $\phi_x \to \Omega\phi_x$. The kinetic part of the lattice Lagrangian will contain nearest-neighbor terms

$$(\phi_x, \phi_y) \ . \tag{13.31}$$

These terms are invariant under global, but not under *local*, transformations,

$$\phi_x \longrightarrow \phi'_x = \Omega_x\,\phi_x, \qquad x \in \Lambda, \quad \Omega_x \in G \ , \tag{13.32}$$

since the field at different sites enters the scalar product (13.31). Before comparing the field at different points, we should parallel transport it from one point to the other. This suggests that the correct nearest-neighbor coupling reads

$$(\phi_x, U_{\langle x,y\rangle}\phi_y) \ , \tag{13.33}$$

where $U_{\langle x,y\rangle}$ is the parallel transporter from y to its neighbor x. This expression is invariant under local gauge transformations if the transporter transforms as expected:

$$U_{\langle x,y\rangle} \longrightarrow U'_{\langle x,y\rangle} = \Omega_x U_{\langle x,y\rangle}\Omega_y^{-1} \ . \tag{13.34}$$

In general the parallel transport depends on the chosen path. On the lattice we may think of $U_{\langle x,y\rangle}$ as the parallel transporter along the link,

$$U_{\langle x,y\rangle} = P\exp\left(ig\int_y^x A_\mu(z)\mathrm{d}z^\mu\right) = \mathbb{1} + ig(x-y)^\mu A_\mu(x) + O(a^2) \ , \tag{13.35}$$

although this interpretation is not of much relevance on the lattice. Inserting this expansion into the difference between the field at a given lattice site and on a neighboring site, parallel transported to the given site, we obtain the expansion

$$\phi_x - U_{\langle x,y\rangle}\phi_y = (x-y)^\mu D_\mu\phi(x) + O(a^2) \ . \tag{13.36}$$

Taking the square of this relation and bearing in mind that nearest neighbors are separated by the lattice spacing a yields (no sum over μ)

$$a^2\left(D_\mu\phi(x), D_\mu\phi(x)\right) = (\phi_x, \phi_x) + (\phi_y, \phi_y)$$
$$-2\Re\left(\phi_x, U_{\langle x,y\rangle}\phi_y\right) + O(a^3) \ . \tag{13.37}$$

These considerations suggest that we choose the parallel transporters $U_{\langle x,y\rangle}$ between neighboring sites rather than the vector potential $A_\mu(x)$ as the basic object for constructing an invariant action and path integral. Since all relevant symmetry

Fig. 13.1 The elementary variables are the parallel transporters $U_{\langle x,y\rangle}$ between nearest neighbors. The parallel transporter along a path $\mathscr{C}_{yx}$ from x to y is the product of elementary variables on the oriented links along the path. A plaquette is characterized by four sites x, y, u, and v

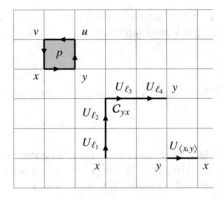

groups are compact,[1] this means that the dynamical variables on the lattice $\{U_{x,\mu}\}$ take their values in a *compact* group. This will be relevant for the functional integrals of lattice gauge theories.

13.2.1 Wilson Action of Pure Gauge Theories

Our first task is to construct an action that describes the dynamics of the gauge variables $U_{\langle x,y\rangle} \in G$ defined on the oriented links of the lattice. A gauge field configuration on the lattice is a map from the set of directed links E into G:

$$U : E = \{\ell\} \longrightarrow G. \tag{13.38}$$

The parallel transporter $U(\mathscr{C})$ along an arbitrary path $\mathscr{C}$ on the lattice is the product of the elementary transporters assigned to the directed links along the path:

$$U(\mathscr{C}_{yx}) = U_{\ell_n} \cdots U_{\ell_3} U_{\ell_2} U_{\ell_1} ,$$

cf. Fig. 13.1. Similarly as in the continuum, we have

$$U^{-1}(\mathscr{C}) = U(\mathscr{C}^{-1}) \quad \text{e.g.} \quad U(\ell^{-1}) = U(\ell) . \tag{13.39}$$

The Lagrangian density of a continuum Yang-Mills theory contains the square of the field strength. The field strength is interpreted as curvature and determines how much the matter field changes (per area) when it is parallel transported along an infinitesimally small loop. Naturally, there is no infinitesimal transport on a lattice. The most elementary transport along a closed path results from the product of the transporters along the edges of an elementary plaquette, which is defined by its

---

[1] Gravity is an exception.

corners x, y, u, and v:

$$U_p = U_{\langle x,v \rangle} U_{\langle v,u \rangle} U_{\langle u,y \rangle} U_{\langle y,x \rangle} \,, \tag{13.40}$$

see Fig. 13.1. Similarly as in Sect. 10.4, we use the notation

$$U_p = \prod_{\ell \in \partial p} U_\ell \,. \tag{13.41}$$

This notation does not reveal the dependence of U_p on x. Sometimes it is convenient to use an alternative notation: if x and $x + e_\mu$ are neighboring sites, then we denote the parallel transporter from x to $x + e_\mu$ by $U_{x,\mu}$. Similarly, a plaquette is characterized by a point and two (positive) directions, $p = (x, \mu, \nu)$, and we can write $U_p = U_{x,\mu\nu}$. The plaquette variable $U_{x,\mu\nu}$ transforms under gauge transformations according to

$$U_{x,\mu\nu} \longrightarrow \Omega_x U_{x,\mu\nu} \Omega_x^{-1} \,. \tag{13.42}$$

To motivate Wilson's choice of a gauge-invariant and real lattice action, we introduce a small lattice spacing a and write

$$U_{x,\mu} = e^{igaA_\mu(x)} \,, \tag{13.43}$$

where the interpolating field $A_\mu(x) = A_\mu^a(x) T_a$ takes its values in the Lie algebra. With the help of the Baker-Hausdorff formula [24]

$$e^A e^B = e^{A+B+\frac{1}{2}[A,B]+\dots} \tag{13.44}$$

we rewrite the product of parallel transporters around a plaquette p as follows:

$$\begin{aligned} U_{x,\mu\nu} &= e^{-igaA_\nu(x)} e^{-igaA_\mu(x+ae_\nu)} e^{igaA_\nu(x+ae_\mu)} e^{igaA_\mu(x)} \\ &= e^{-igaA_\nu(x)} e^{-igaA_\mu(x)-iga^2\partial_\nu A_\mu(x)+O(a^3)} \\ &\qquad e^{igaA_\nu(x)+iga^2\partial_\mu A_\nu(x)+O(a^3)} e^{igaA_\mu(x)} \\ &= e^{iga^2\left(\partial_\mu A_\nu(x)-\partial_\nu A_\mu(x)-ig[A_\mu(x),A_\nu(x)]\right)+O(a^3)} = e^{iga^2 F_{\mu\nu}(x)+O(a^3)} \,. \end{aligned} \tag{13.45}$$

Only terms up to order a^2 are considered in the exponent. This yields

$$U_{x,\mu\nu} + U_{x,\mu\nu}^{-1} \approx 2 \cdot \mathbb{1} - ga^4 F_{\mu\nu}^2(x) + O(a^6) \,. \tag{13.46}$$

Since the relevant term is $O(a^4)$, you would think that all terms up to this order in the exponent will contribute. However, for $U = \exp(iT)$, we obtain

$$U + U^{-1} = 2 - T^2 + O(T^4) \,.$$

It follows that only the *leading* term of order a^2 in T contributes to the fourth-order term in $U + U^{-1}$. This naive continuum limit motivates the following choice for the action of a Euclidean gauge theory with unitary gauge group:

$$S_{\text{gauge}} = \beta \sum_p \text{tr} \left\{ \mathbb{1} - \frac{1}{2} \left(U_p + U_p^\dagger \right) \right\}, \quad \beta = \frac{2}{ng^2}. \quad (13.47)$$

The sum extends over all $Vd(d-1)/2$ elementary plaquettes of the lattice, and the factor $1/n$ in the gauge coupling β takes into account that the link variables are n-dimensional matrices. The nonnegative action is minimal for configurations with $U_p = \mathbb{1}$ for all p. These "vacuum configurations" have zero action. For real or pseudo-real groups like SU(2), the trace of U is real, and the plaquette action simplifies to $\beta \, \text{tr} \, (\mathbb{1} - U_p)$. For the finite gauge group Z_n, we recover the action (10.49).

The *Wilson action* (13.47) approximates the continuum Yang-Mills action up to an error of order $O(a^2)$, similarly to the trapezoidal rule which approximates an integral up to order $O(a^2)$. A natural question is whether there is something like Simpson's rule with an error of order $O(a^4)$ in lattice gauge theory. Such improvements of the Wilson action have been suggested. One improvement based on the renormalization group method was given by K. Wilson himself [25]. Another improvement program based on perturbation theory was initiated by K. Symanzik [26] and further developed in [27]. A lattice action which gives two-loop scaling was constructed in [28].

Now we can write down a gauge-invariant action for the Higgs model on a lattice. For the gauge fields, we choose the Wilson action (13.47). For the scalar field, we take the right-hand side of (13.37), summed over all lattice sites, since this sum approximates the kinetic term in (13.30). Thereby the terms (ϕ_x, ϕ_x) and (ϕ_y, ϕ_y) can be absorbed in the potential term. Hence as gauge-invariant interaction between the gauge field and scalar field, we take

$$S_{\text{m}} = -\kappa \sum_{\langle x, y \rangle} \Re \left(\phi_x, U_{\langle x, y \rangle} \phi_y \right) \quad (13.48)$$

with hopping parameter κ. Thus we end up with the Yang-Mills-Higgs action

$$S_{\text{YMH}}(U, \phi) = S_{\text{gauge}}(U) + S_{\text{m}}(U, \phi) + \sum_x V(\phi_x) \quad (13.49)$$

with a G-invariant potential for the scalar field, $V(\Omega \phi) = V(\phi)$.

We have argued that on a lattice the $\{U_\ell\}$ should be taken as dynamical variables. Thus, the formal integration over all gauge potentials in the continuum theory turns into an integration over the group valued fields on the links, and we end up with the following functional representation of the partition function:

$$Z(\beta, \kappa) \quad \int \prod_x \mathrm{d}\phi_x \prod_\ell \mathrm{d}U_\ell \, \mathrm{e}^{-S_{\mathrm{YMH}}(U,\phi)} . \tag{13.50}$$

Since the resulting lattice theory should be gauge invariant, we better choose an invariant integration:

$$\mathrm{d}\left(\Omega\phi\right) = \mathrm{d}\phi, \quad \mathrm{d}\left(\Omega'U\Omega^{-1}\right) = \mathrm{d}U . \tag{13.51}$$

Invariant measures on compact groups exist and are studied in Chap. 14.

13.2.2 Strong- and Weak-Coupling Limits of Higgs Models

Here we will consider the four simplifying limits of the Higgs model in which $\beta \to 0$ or ∞ and $\kappa \to 0$ or ∞. Along the way we will learn about selecting gauges.

Vanishing β and Unitary Gauge
In the strong-coupling limit, the Wilson term is absent and it is a trivial limit of the theory. This is most easily seen in the unitary gauge which can be achieved for all models for which the scalar field can be gauge transformed into a fixed direction defined by a constant unit vector ϕ_0:

$$\phi_x = \varrho_x \Omega_x \phi_0 \quad \text{with} \quad \Omega_x \in G . \tag{13.52}$$

This condition fixes the gauge transformations $\{\Omega_x\}$ completely in case the little group of ϕ_0

$$H = \{\Omega \in G \,|\, \Omega\phi_0 = \phi_0\} \tag{13.53}$$

is a trivial subgroup $H = \{\mathbb{1}\}$ of the gauge group. Now we use the gauge invariance of the action and measure to rewrite the partition function as

$$Z = \int \prod_x \mathrm{d}\phi_x \prod_\ell \mathrm{d}U_\ell \, \mathrm{e}^{-S_{\mathrm{gauge}}(U')-S_{\mathrm{m}}(U', \varrho\phi_0)-V(\varrho)} \tag{13.54}$$

with gauge transformed link variables $U'_{\langle x,y \rangle} = \Omega_x^{-1}U_{\langle x,y \rangle}\Omega_y$. But since $\mathrm{d}U_\ell$ is gauge invariant, we may skip the prime at U' in the integrand. Since the integrand only depends on ϱ and U, we may change variables and find

$$Z = \left(\mathrm{Vol}_{G/H}\right)^V \int \prod_x \mathrm{d}\varrho_x \, \varrho_x^{n-1} \prod_\ell \mathrm{d}U_\ell \, \mathrm{e}^{-S_{\mathrm{gauge}}(U)-S_{\mathrm{m}}(U, \rho\phi_0)-V(\varrho)} , \tag{13.55}$$

where the first factor contains the volume of the coset space G/H and n denotes the number of ϕ-components. If we freeze the length ϱ of the scalar field by setting $\exp(-V(\varrho)) = \delta(\varrho - 1)$, we are dealing with gauged nonlinear sigma models. For these models the integration over the radial modes $\{\varrho_x\}$ yields 1, and for $\beta = 0$ the partition function Z factors into a product of independent terms and thus is trivial,

$$
Z \xrightarrow{\beta \to 0} \left(\text{Vol}_{G/H}\right)^V \left(\int dU \, e^{\kappa(\phi_0, U\phi_0)}\right)^{|E|} , \tag{13.56}
$$

where $|E|$ is the number of links on the lattice. The free energy density is an analytic function of κ given by a simple group integral.

Infinite β and Axial Gauge on Periodic Lattices
Again we impose periodic boundary conditions in all directions and thus may view the hypercubic lattice as torus. In the limit $\beta = \infty$, only vacuum configurations with $U_p = \mathbb{1}$ for all plaquettes contribute to the path integral. Then the system is most comprehensible in an axial-like gauge where one uses the gauge freedom to choose as many $U's$ as possible to be the unit matrix. In a first step of the gauge fixing, all time-like link variables $U_{x,0}$ for $t > 1$ are gauged to the unit element. These are the thick vertical links in Fig. 13.2. The remaining time-like link variables at $t = 1$ cannot be gauged to the unit element, and in the axial gauge, they are equal to the *Polyakov loops*

$$
\mathscr{P}_x \equiv \prod_{t=1}^{N} U_{(t,x),0} \stackrel{\text{gf}}{=} U_{(1,x),0} . \tag{13.57}
$$

Fig. 13.2 In the axial-like gauge, the variables on the thick marked links are fixed to the unit element

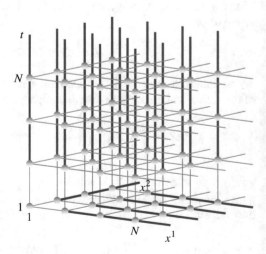

These particular Wilson loops wind around the periodic time direction and will be relevant at finite temperatures. The gauge transformations Ω_x in the time slice $t = 1$ are not used for this first gauge fixing step. In a second step, we use these gauge transformations to fix all variables on links in the x_1-direction and with $x_1 > 1$ in the time slice $t = 1$ to the unit element. These are the thick links at the bottom of the cube in Fig. 13.2 pointing toward the x_1-direction. The gauge transformations Ω_x in the slice $t = x_1 = 1$ are not used for this second gauge fixing step but in the third step where one fixes the variables in this slice with $x_2 > 1$ to the unit element. After these three steps, all variables on the thick links in Fig. 13.2 are gauge fixed to the unit element. This is almost a complete gauge fixing—only the gauge transformation at the point $(1, 1, 1)$ remains unfixed and acts as a residual *global* gauge symmetry. After gauge fixing the number of links with $U_{x,\mu} = \mathbb{1}$ is

$$N^2(N - 1) + N(N - 1) + (N - 1) = N^3 - 1 \tag{13.58}$$

in three dimensions and more generally $V - 1$ in arbitrary dimensions. This is exactly the number of gauge transformations $\{\Omega_x\}$ minus the residual gauge transformation.

In the weak-coupling limit $\beta \to \infty$, only vacuum configurations contribute to the path integral. Let us see how gauge-fixed vacuum configuration looks like in three dimensions.

Considering the time-space plaquettes of a gauge-fixed vacuum configuration, we see that *all* link variables in x_1- and x_2-directions are t-independent. Considering the space plaquettes in the slice $t = 1$, we conclude that in this slice the link variable in x_2-direction is also x_1-independent. But since the spatial link variables are time-independent, we see that *all* link variables in x_2-direction are t- and x_1-independent. Thus we are left with the nontrivial variables $U_{x,0}$ in the slice $t = 1$, the variables $U_{x,1}$ in the slice $x_1 = 1$, and the variables $U_{x,2}$ in the slice $x_2 = 1$. Recalling that all other variables are the identity, one concludes that the variables in each slice are constant. The same reasoning applies in d dimensions and one is left with just d constant group elements. In four dimensions,

$$
\begin{aligned}
U_{x,0} &\equiv \mathscr{P} & \text{if}\quad x &= (1, x_1, x_2, x_3) \\
U_{x,1} &\equiv U_1 & \text{if}\quad x &= (t, 1, x_2, x_3) \\
U_{x,2} &\equiv U_2 & \text{if}\quad x &= (t, x_1, 1, x_3) \\
U_{x,3} &\equiv U_3 & \text{if}\quad x &= (t, x_1, x_2, 1) \, .
\end{aligned}
\tag{13.59}
$$

Actually for a gauge theory with open boundary conditions, we can gauge fix *all* link variables of a vacuum configuration to the unit element such that the Higgs model reduces to a spin model for the scalar field:

$$Z \xrightarrow{\beta \to \infty} C \int \prod_x d\phi_x \exp\left(-\kappa \sum_{\langle x,y \rangle} (\phi_x, \phi_y)\right) . \tag{13.60}$$

Vanishing κ

In this limit the matter field decouples and we are left with a pure gauge theory. This is a nontrivial limit and much of the rest of the present chapter concerns this limit.

The Limit $\kappa \to \infty$

This limit is most easily studied in the unitary gauge. In this gauge we require that $(\phi_0, U\phi_0)$ takes its maximum value on each link which implies $U\phi_0 = \phi_0$. If the little group H of ϕ_0 is trivial, then all link variables must be the identity. If we freeze the length of the scalar field, then no degrees of freedom are left. If the little group is nontrivial, then the limiting theory is a pure gauge theory with gauge group H.

13.3 Mean Field Approximation

First we discuss the mean field approximation for pure gauge theories. Our starting point is the variational characterization of the free energy

$$F = \inf_P \left(\int dP(U)\, S_{\text{gauge}}(U) - S_B(P) \right) . \tag{13.61}$$

In the mean field approximation, as described in Chap. 7, $dP(U)$ is assumed to be a product measure with associated Boltzmann entropy S_B. We use the Wilson action S_{gauge} in the variational principle. In spite of a priori difficulties, such as Elitzur's theorem discussed in Sect. 13.5, this simple approximation allows for a first exploration of the phase diagram. We choose the non-invariant product measure

$$dP(U) = \prod_\ell dv_\ell(U_\ell), \quad dv(U) = p(U)dU , \tag{13.62}$$

where dU denotes the invariant Haar measure on the gauge group (as discussed in Chap. 14) and p is the probability density

$$p(U) = \frac{1}{z(h)} \exp\left(\frac{h}{n} \Re\text{tr}\, U \right) = p\left(\Omega U \Omega^{-1} \right) , \tag{13.63}$$

normalized with the single-link partition function

$$z(h) = e^{w(h)} = \int dU \exp\left(\frac{h}{n} \Re\text{tr}\, U \right) = 1 + O(h^2) . \tag{13.64}$$

It follows from (13.62) that the mean link variable is proportional to the unit matrix [29]

$$\langle U \rangle_h = \int dv(U)\, U = w'(h)\mathbb{1} . \tag{13.65}$$

The Boltzmann entropy is the number of links dV times the single-link entropy

$$s_{\mathrm{B}} = -\int d\nu \log p = \log z - \frac{h}{n}\int dU p(U)\,\mathrm{tr}\,(U) = w(h) - hw'(h)\ . \quad (13.66)$$

For the product measure, the expectation value of S_{gauge} is proportional to the number of plaquettes $Vd(d-1)/2$ times the average single plaquette action:

$$\langle U_p\rangle_h = \int d\nu_1(U_1)\cdots d\nu_4(U_4)\,\Re\mathrm{tr}\,(U_p) = nw'^4(h)\ . \quad (13.67)$$

Inserting the average action and entropy into (13.61), one obtains the mean field free energy per link,

$$\frac{f_{\mathrm{mf}}}{d} = \inf_h\left(hw'(h) - w(h) - \frac{1}{2}\beta n(d-1)\,w'^4(h)\right)\ , \quad (13.68)$$

where the function $w(h)$ is the single-link free energy defined in (13.64) and hence a convex function with a minimum at the origin where it behaves as h^2. Since the term $w'^4(h)$ behaves as h^4, the minimization always leads to the solution $h=0$, and this minimum is unique for small β. As β increases, this minimum does not remain the lowest one.

13.3.1 Z_2 Gauge Model

For the gauge group Z_2, the single-site free energy is $w(h) = \log\cosh(h)$ and

$$\frac{f_{\mathrm{mf}}}{d} = \inf_h\left(h\tanh(h) - \log\cosh(h) - \frac{1}{2}\beta(d-1)\tanh^4(h)\right)\ . \quad (13.69)$$

The behavior of the free energy density for different values of β is displayed in Fig. 13.3. At the critical value $(d-1)\,\beta_* \approx 1.3776$, the absolute minimum suddenly changes its location as depicted in Fig. 13.3, and this jump signals a first-order transition at β_*. In four dimensions we find $\beta_* = 0.4592$ to be compared with the exact value 0.44069 known from self-duality; see Table 10.3. The transition is known to be of first order as predicted by the mean field approximation. In three dimensions the mean field yields $\beta_* = 0.6888$ to be compared with the value 0.7613, known from duality; see Table 10.2. But since the three-dimensional Z_2 gauge theory is dual to the Ising model, the transition is in fact a second-order transition.

Fig. 13.3 The free energy per link f_{mf}/d of the Z_2 gauge theory in the mean field approximation. The minimum jumps at $(d-1)\beta_* \approx 1.3776$

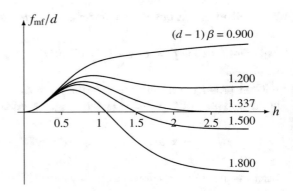

13.3.2 U(1) Gauge Theory

For this gauge theory, $z(h) = I_0(h)$, and the corresponding mean field free energy density looks similar as in Fig. 13.3 and signals a first-order transition at $(d-1)\beta_* = 3.6460$. This yields the mean field critical gauge coupling $\beta_* = 1.215$ in four dimensions. High-precision MC simulations, supplemented by a finite size analysis, indeed spotted a weakly first-order transition at $\beta_* = 1.011133$ [30, 31].

13.3.3 SU(n) Gauge Theories

The upper critical dimension for non-Abelian gauge theories should be four where the theory becomes asymptotically free. Above the critical dimension, we can trust the mean field approximation. To determine f_{mf} of SU(n) gauge theories, we need the single-link partition function, which is a sum of products of modified Bessel functions [32, 33]:

$$z_{\mathrm{SU(n)}}(h) = \sum_{m \in \mathbb{Z}} \det \begin{pmatrix} I_m & I_{m+1} & \cdots & I_{m+n-1} \\ I_{m-1} & I_m & \cdots & I_{m+n-2} \\ \vdots & \vdots & & \vdots \\ I_{m-n+1} & I_{m-n+2} & \cdots & I_m \end{pmatrix} \left(\frac{h}{n} \right) . \qquad (13.70)$$

The infinite sums entering $w = \log z$ and its derivative (here one uses Bessel function identities) can be evaluated with octave. The resulting $f_{\mathrm{mf}}(h)$ of the gauge group SU(2) in the vicinity of the critical gauge coupling, $(d-1)\beta_* = 4.2394$, is depicted in Fig. 13.4. As expected, the mean field approximation predicts a first-order transition at some critical gauge coupling β_*. In five dimensions $\beta_* \approx 1.0598$, which agrees well with numerical simulations which find a first-order transition at $\beta_* = 0.82$. In four dimensions the mean field approximation fails. There is no evidence for any transition in numerical simulations.

Fig. 13.4 The free energy per link f_{mf}/d of the SU(2) gauge theory in the mean field approximation. The minimum jumps at $(d-1)\beta_* \approx 4.2394$

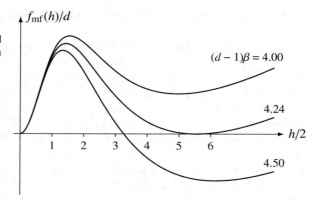

Fig. 13.5 The free energy f_{mf}/d per link of the gauged Ising model in the mean field approximation. Shown are the energies for three pairs of critical parameters (κ_*, β_*) for which the two minima are degenerate. The first-order transition becomes weaker and finally becomes a second-order transition when κ_* increases

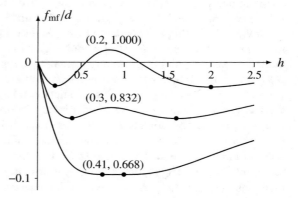

13.3.4 Higgs Model

For a lattice Higgs model with normalized scalar field, the mean field free energy per link is

$$\frac{f_{mf}}{d} = \inf_h \left((h - \kappa)w'(h) - w(h) - \frac{1}{2}\beta n(d-1)\, w'^4(h) \right). \tag{13.71}$$

For small h it behaves as $-2c\kappa h + ch^2 + O(h^4)$ with some positive constant c such that the origin ceases to be a minimum for a nonvanishing hopping parameter κ. With increasing κ the two minima of the free energy approach each other, and the first-order transition becomes weaker as depicted in Fig. 13.5. At the critical point $(\kappa_*, \beta_*) \approx (0.41, 0.668)$, the two minima merge to one minimum.

13.4   Expected Phase Diagrams at Zero Temperature

Consider a lattice Higgs model with scalar field in a fundamental representation of the gauge group. For $\kappa = 0$ the model reduces to a pure gauge theory, and for a discrete gauge group, we expect a phase transition at some critical hopping parameter. In three dimensions it is a second-order transition and in four dimensions a first-order transition. For a non-Abelian gauge theory, there is no such transition in either three or four dimensions. For $\beta = \infty$ the Higgs model collapses to a spin model with global symmetry. Generically there will be a second-order phase transition from a symmetric phase at large κ to a spontaneously broken phase at small κ. For $\beta = 0$ the theory is analytic in κ and there is no phase transition.

For nonzero but small β, Osterwalder and Seiler proved that the expansion in powers of β exists with a finite radius of convergence and that in this strong-coupling phase, the theory is confining [34]. Their proof is combinatoric in nature. It is reassuring though not very surprising that this is the case since most theories are analytic in the high-temperature phase. Later on we will show that the Wilson loop obeys an area law in the strong-coupling region. Finally, for $\kappa = \infty$ the Higgs model has no dynamics if the unitary gauge is a complete gauge fixing. If it is not a complete gauge fixing, then one is left with a pure gauge theory with gauge group $H = \{\Omega \in G \mid \Omega \phi_0 = \phi_0\}$. Hence we do not expect a phase transition at $\kappa = \infty$ for a trivial or non-Abelian H. On the other hand, if the little group H is discrete, then there will be a second-order transition in three dimensions and a first-order transition in four dimensions. There exists a generalization of the Osterwalder-Seiler theorem due to Fradkin and Shenker, which states that for a scalar field in a fundamental representation, the Higgs model is analytic for sufficiently large κ [35]. The two theorems together imply that in the coupling constant plane, there is always a path from a point $\beta, \kappa \ll 1$ deep in the confinement region, where the Wilson loop obeys an area law, to a point $\beta, \kappa \gg 1$ in the Higgs phase where the Wilson loop obeys a perimeter law, such that the expectation values of all *local*, gauge-invariant observables vary analytically along the path. This means that the Higgs phase and the confining phase are smoothly connected—similarly as the gaseous and liquid phases of water are smoothly connected over the "top" of the critical point. When the Higgs fields transform according to a non-fundamental representation, then a phase boundary may exist. This is the case for SU(n) with all Higgs fields in the adjoint representation and for U(1) with all Higgs fields in the charge-n ($n > 1$) representation.

The *SU(2) Higgs model* with fundamental scalars has been studied by Monte Carlo techniques, and the situation is depicted schematically in Fig. 13.6. One finds that for intermediate values of β, certain observables vary rapidly with the coupling, but still analytically in the thermodynamic limit. This is interpreted as a crossover [36]. For larger values of β, a line of probably first-order transitions is seen. At $\kappa = \infty$, marked by a dot in the figure, the transition must be second order. For a further discussion and an interpretation of these results, you may consult the textbook [37].

Fig. 13.6 Sketch of phase diagram for SU(2) gauge plus fundamental scalars. The Osterwalder-Seiler-Fradkin-Shenker theorem says that local observables are analytic in the gray region

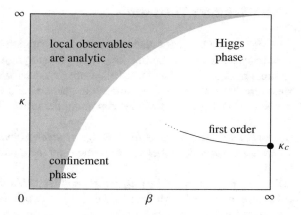

Fig. 13.7 Schematic plot of the phase diagram of the Z_2 gauged Ising model. The *dashed lines* represent a first-order phase transition, and the *continuous* line represents a second-order transition. The critical point is marked with C and the triple point by T

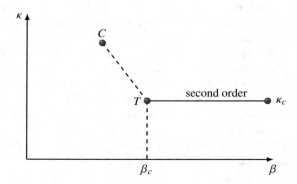

The Z_2 theory has a similar phase diagram. A crucial difference is that at $\kappa = 0$ the resulting pure gauge model shows a first-order transition at some critical β. The phase diagram is shown schematically in Fig. 13.7. The three transition lines meet at the triple point and split the phase space into three regions. Again the phase diagram includes familiar limits. In the limit $\beta \to \infty$, the gauge degrees of freedom disappear, and the system reduces to that of a four-dimensional Ising model with a second-order transition which is seen in the expectation value $\langle s_x U_{\langle x,y \rangle} s_y \rangle$. This second-order line is of Ising type all the way to the triple point T. On the $\kappa = 0$ axis, the system reduces to a pure gauge theory, and the expectation value of the plaquette variable $\langle U_p \rangle$ shows a first-order transition. This behavior continues all the way up to the triple point. A first-order line seen in both expectation values is then formed and ends with the critical point C [38].

13.5   Elitzur's Theorem

The phenomenon of spontaneously broken symmetries happens in large macro-scopic systems where the breaking of a global symmetry involves a macroscopic number of degrees of freedom. This is not the case for a local gauge symmetry. The quantum fluctuations tend to smear the ground state wave function of the system homogeneously over the whole orbit under the group. This results in

Theorem 13.1 (Elitzur) *A local gauge symmetry cannot break spontaneously. The expectation value of any gauge-non-invariant local observable must vanish.*

Elitzur's original proof in [39] which applies to Abelian gauge theories but was later extended to non-Abelian models [40]. The proofs of Elitzur's theorem are all based on the fact that inequalities which hold for any field configuration continue to hold after integrating with respect to a positive measure. In fact, positivity of the measure and gauge invariance are sufficient to prove the theorem. The theorem means that there is no analog of a magnetization: expectation values of a spin or link variables are zero, even if we introduce an external field (which explicitly breaks gauge invariance) and then carefully take first the infinite volume limit and then the $h \to 0$ limit. We must look, instead, to gauge-invariant observables which are unaffected by gauge transformations. These can be constructed by taking parallel transporters around closed loops, known as Wilson loops.

Elitzur's theorem raises the question of whether the Higgs mechanism, which gives masses to the fermions and gauge bosons of the standard model, may perhaps not work. As demonstrated in [23], such fears are ungrounded, since the physical phenomena which are associated with the Higgs mechanism can be recovered in an approach that uses gauge-invariant fields only. The masses are extracted from expectation values of gauge-invariant combinations of the Higgs and gauge fields, without any need of introducing a nonzero expectation value of the Higgs field. In particular the electroweak phase transition can be described in purely gauge-invariant terms [41]. For example, the expectation value of (ϕ, ϕ) exhibits a "jump" along the phase transition line in parameter space where the electroweak phase transition occurs.

13.5.1 Proof for Pure Z_2 Gauge Theory

First we consider the simplest possible gauge theory, the pure lattice gauge theory with gauge group Z_2 and link variables $U_\ell \in \{-1, 1\}$ on a finite lattice Λ. We couple the system to an external field h and hence add a source term to the Wilson action:

$$S_{\text{gauge},\Lambda}(U) = -\beta \sum_p U_p - h \sum_\ell U_\ell . \qquad (13.72)$$

In expectation values one sums over all elements of the configuration space $\Omega = \{U_\ell | \ell \in E\}$. Now we prove

Theorem 13.2 *The magnetization* $\langle U_\ell \rangle$ *converges to zero,*

$$\lim_{h \downarrow 0} \langle U_\ell \rangle_\Lambda (h) = 0 \,, \tag{13.73}$$

uniformly in Λ *and* β. *This also holds in the thermodynamic limit.*

Proof We bound the numerator and denominator in the expectation value

$$\langle U_{\ell_0} \rangle_\Lambda (h) = \frac{\sum_\Omega U_{\ell_0} \exp\left(-S_{\text{gauge},\Lambda}\right)}{\sum_\Omega \exp\left(-S_{\text{gauge},\Lambda}\right)} \,. \tag{13.74}$$

Let x denote an endpoint of the link ℓ_0 under consideration, and perform a local gauge transformation at x with $\Omega_x = \pm 1$. Only variables defined on the $2d$ links ending at this point are affected, $U_\ell \to U'_\ell = \Omega_x U_\ell$ if $x \in \partial \ell$, and this is depicted in Fig. 13.8. Both the original configuration $\{U_\ell\}$ and the transformed configuration $\{U'_\ell\}$ are already contained in Ω such that the partition function in the denominator can be written as

$$Z = \frac{1}{2} \sum_{\Omega_x = \pm 1} \sum_\Omega \exp\left(\beta U_p + h \sum_{\ell : x \notin \partial \ell} U_\ell\right) \exp\left(h \sum_{\ell : x \in \partial \ell} \Omega_x U_\ell\right) \,,$$

where we took into account that the plaquette variables are gauge invariant. Thus we obtain

$$Z = \sum_\Omega \exp\left(\beta U_p + h \sum_{\ell : x \notin \partial \ell} U_\ell\right) \cosh\left(h \sum_{\ell : x \in \partial \ell} U_\ell\right) .. \tag{13.75}$$

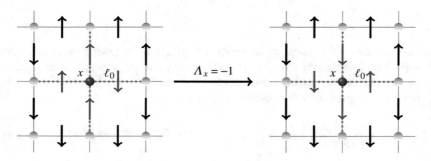

Fig. 13.8 Gauge transformation on the central site. All spins U_ℓ on the links touching the central site (dotted lines in the figure) are flipped

Repeating the argument for the numerator N in (13.74) yields

$$N = \sum_{\Omega} \exp\left(\beta U_p + h \sum_{\ell:x\notin\partial\ell} U_\ell\right) U_{\ell_0} \sinh\left(h \sum_{\ell:x\in\partial\ell} U_\ell\right) . \tag{13.76}$$

The argument of the hyperbolic functions varies between $-2dh$ and $2dh$ such that for all configurations

$$\cosh(\ldots) \geq e^{-2dh} \quad \left|U_{\ell_0}\sinh(\ldots)\right| \leq \sinh(2dh) . \tag{13.77}$$

This yields the bound

$$\left|\langle U_{\ell_0}\rangle_\Lambda(h)\right| \leq e^{2dh}\sinh(2dh) \xrightarrow{h\downarrow 0} 0 , \tag{13.78}$$

uniformly in Λ and β. If one couples the Z_2 gauge field minimally to a scalar field and adds a symmetry breaking term $h\sum_x \phi_x$, then one can prove

$$\langle\phi_x\rangle(h) \xrightarrow{h\downarrow 0} 0 , \tag{13.79}$$

uniformly in Λ and the couplings β and κ; see Problem 13.1.

13.5.2 General Argument

Let $\Phi = \{U_\ell, \phi_x, \ldots\}$ denote the collection of fields of a gauge theory and $\int \mathscr{D}\phi \ldots$ the invariant integration. We study the expectation value of a local and non-invariant function $O(\Phi)$ of the fields. Local means that it only depends on a finite number of variables $\{\Phi'\}$ defined on the sites and links in some finite area A. Non-invariant means that $\mathscr{O}$ has no invariant component, i.e.,

$$\int \prod_{x\in\mathscr{R}} d\Omega_x\, \mathscr{O}\left(^\Omega\Phi\right) = 0 , \tag{13.80}$$

where $^\Omega\Phi$ denotes the gauge transformed configuration. Following [40] we decompose the set of fields $\{\Phi\}$ into $\{\Phi'\}\cup\{\Phi''\}$. Now we average over the subgroup of gauge transformations which leave the $\{\Phi''\}$ invariant. The invariance of the action without source term and of the integration implies

$$\langle\mathscr{O}(\Phi)\rangle_{\lambda,J} = \frac{1}{Z_{\Lambda,J}} \int \mathscr{D}\Phi \prod_{x\in\mathscr{R}} d\Omega_x\, e^{-S[\Phi]+(J,\,^\Omega\Phi')+(J,\Phi'')}\, \mathscr{O}\left(^\Omega\Phi\right) . \tag{13.81}$$

Since $\{\Phi'\}$ contains a finite and Λ-independent number of degrees of freedom

$$\left| e^{(J,\Phi')} - 1 \right| \leq \varepsilon(J) \xrightarrow{J \downarrow 0} 0 \,,$$

and this bound is uniform in Λ and in Φ, at least if the latter is a compact variable. Inserting

$$e^{(J,\,^{\Omega}\Phi')} = 1 + \left(e^{(J,\,^{\Omega}\Phi')} - 1 \right)$$

into (13.81), the contribution of the first term vanishes because of (13.80). The contribution of the second term vanishes in the limit $\lim_{J\downarrow 0} \lim_{\Lambda \to \mathbb{Z}^d}$. This proves (together with a similar bound on the partition function) that

$$\langle \mathcal{O}(\Phi) \rangle = \lim_{J\downarrow 0} \lim_{\Lambda \to \mathbb{Z}^d} \langle \mathcal{O}(\Phi) \rangle_{\Lambda, J} = 0 \,. \tag{13.82}$$

The reasoning in the proofs hinges very much on the *local* gauge invariance and on the positivity of the measure.

13.6 Observables in Pure Gauge Theories

For a gauge-invariant lattice action, the normalized probability measure

$$d\mu[U] = \frac{1}{Z} \, e^{-S_{\mathrm{gauge}}(U)} \prod_{\ell \in E} dU_\ell, \quad Z = \int e^{-S_{\mathrm{gauge}}(U)} \prod_{\ell \in E} dU_\ell \tag{13.83}$$

which defines the functional integral, is gauge invariant. The dimensionality of the integral on a finite d-dimensional lattice is $dV \dim(G)$. For example, for the pure SU(2) gauge theory on a hypercubic 16^4-lattice, one is confronted with a 786,432-dimensional integral. According to Elitzur's theorem, it is only reasonable to consider expectation values of *gauge-invariant* quantities. For a pure gauge theory, the gauge-invariant functions of the link variables are given by *traces* of parallel transporters along closed paths (loops). Thus, we define

$$W[\mathscr{C}] = \mathrm{tr} \left(U_{\ell_n} \cdots U_{\ell_1} \right), \quad \mathscr{C} = \ell_n \circ \cdots \circ \ell_1 \,. \tag{13.84}$$

In most works $W[\mathscr{C}]$ is called Wilson loop. Sometimes the product of the U's along the loop, i.e., the argument of the trace in (13.84), is called a Wilson loop. It also occurs that a Wilson loop is defined as the expectation value of $W[\mathscr{C}]$. With a *Wilson loop*, we will associate the gauge-invariant quantity W defined in (13.84).

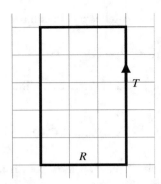

Fig. 13.9 Wilson loop belonging to a rectangular loop with edge lengths R and T. From its expectation value, one may extract the string tension between two static charges

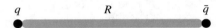

Fig. 13.10 Between a static quark-antiquark pair, a flux tube develops and leads to a linearly rising potential

13.6.1 String Tension

Let $W[R, T]$ denote the Wilson loop associated with a plane rectangular loop of edge lengths R and T, respectively, as illustrated in Fig. 13.9. The function

$$V_{q\bar{q}}(R) = -\lim_{T \to \infty} \frac{1}{T} \log \langle W[R, T] \rangle \qquad (13.85)$$

is interpreted as potential energy of a static $q\bar{q}$-pair separated by a distance R. From the static potential, we can extract the *string tension*

$$\sigma = \lim_{R \to \infty} \frac{V_{q\bar{q}}(R)}{R} . \qquad (13.86)$$

The justification of this interpretation will be given at a later stage.

A positive string tension means that the potential energy of the static charges increases linearly with increasing distance R. The potential becomes infinitely large for asymptotically separated charges, and hence this state does not exist in a dynamical theory. The formation of a flux tube of constant energy density between the charges, as depicted in Fig. 13.10, could explain the linear rising potential: the energy of the tube is proportional to its length R and would give rise to a constant force between the charges. On the other hand, a vanishing string tension would imply that the force between the charges decreases with increasing separation. If it decreases sufficiently fast, then a finite amount of energy suffices to separate the two charges. Color-electric flux tubes have indeed been observed in lattice simulations [42].

We assumed that the test particles are infinitely heavy charged objects without their own dynamics. In the real world with dynamical matter, the energy between the departing charges increases only as long as the potential energy is smaller than the energy necessary for pair production from vacuum. In the string picture, the pair production gives rise to *string breaking*. If pair production sets in, then the created pairs screen the test charges, and finally we observe two neutral states departing from each other. Even without dynamical matter, string breaking happens if the test particles are in particular representations of the gauge group or if the gauge group has a trivial center [43].

K. Wilson proposed the *area* or *perimeter law* for the expectation value of the Wilson loop as a criterion for confinement in lattice gauge theories:

$$\langle W[R, T]\rangle \sim e^{-T V_{q\bar{q}}(R)} \sim e^{-\sigma \times \text{area}} \qquad \text{confinement}$$

$$\langle W[R, T]\rangle \sim e^{-T V_{q\bar{q}}(R)} \sim e^{-\gamma \times \text{perimeter}} \qquad \text{deconfinement} . \quad (13.87)$$

For a plane rectangular loop, the area and perimeter are TR and $2(T + R)$, respectively. E. Seiler and C. Borgs proved that the static potential increases monotonously, i.e., $V'_{q\bar{q}} \geq 0$. They also showed that this increase is at most linear, $V''_{q\bar{q}} \leq 0$ [44, 45]. Hence, for large separation of the charges, the static potential should have the form

$$V_{q\bar{q}}(R) \sim \sigma R + \text{const.} - \frac{c}{R} + o\left(R^{-1}\right) , \qquad (13.88)$$

where c is a universal and positive constant. The term $-c/R$ is the so-called Lüscher term and originates from the quantum fluctuations of the flux tube connecting the two static charges [46].

Let us finally *motivate* why $V_{q\bar{q}}$ represents a static $q\bar{q}$-potential. In quantum electrodynamics the phase factor entering the functional integral in the presence of an external four-current density is modified to

$$\exp(\mathrm{i}S) \longrightarrow \exp\left(\mathrm{i}S + \mathrm{i}\int \mathrm{d}^4x \, j^\mu A_\mu\right) . \qquad (13.89)$$

We parametrize the world line $\mathscr{C}$ of an electrically charged point particle by $z^\mu(\tau)$ with time-like four-velocity $\dot{z}^\mu$. The four-current density is

$$j^\mu(x) = g \int_{\mathscr{C}} \mathrm{d}\tau \, \dot{z}^\mu(\tau)\delta^4\big(x - z(\tau)\big)$$

and gives rise to the additional phase factor

$$\exp\left(i \int d^4x j^\mu A_\mu\right) = \exp\left(ig \int_{\mathscr{C}} dz^\mu A_\mu(z)\right) = \exp\left(ig \int_{\mathscr{C}} A\right), \quad (13.90)$$

where the integral is taken along the particle path $\mathscr{C}$. We obtain the Euclidean version by the substitutions $dz^0 \to -idz^0$ and $A_0 \to iA_0$. The Wick rotation transforms the phase (13.90) again into a phase. If we choose for $\mathscr{C}$ a loop representing the world lines of a heavy test particle and its antiparticle put into the system at a given time and removed at a later time, then the partition function in presence of the particles reads

$$\frac{1}{Z} \int \mathscr{D}A_\mu \exp\left(-S_E[A] + ig \oint_{\mathscr{C}} A\right) = \left\langle \exp\left(ig \oint_{\mathscr{C}} A\right)\right\rangle = \langle W[\mathscr{C}]\rangle .$$
$$(13.91)$$

This is just the continuum result for the expectation value of the Wilson loop.

13.6.2 Strong-Coupling Expansion for Pure Gauge Theories

The Wilson action of a pure lattice gauge theory

$$S_{\text{gauge}} = \beta \sum_p \text{tr}\left(\mathbb{1} - \Re U_p\right), \quad \beta = \frac{2}{ng^2}, \quad (13.92)$$

contains only the bare coupling constant g as free parameter. The theory may be considered as classical spin system with inverse temperature $1/kT \propto 1/g^2$. Thus the perturbation theory in g^2 corresponds to the low-temperature expansion in the spin model, and the strong-coupling limit $g \gg 1$ corresponds to the high-temperature limit of the spin model. We now argue that expectation values of Wilson loops,

$$\langle W[\mathscr{C}]\rangle = \frac{\int \prod_\ell dU_\ell\, W[\mathscr{C}]\, e^{-S_{\text{gauge}}(U)}}{\int \prod_\ell dU_\ell\, e^{-S_{\text{gauge}}(U)}} \quad (13.93)$$

obey an area law in the strong-coupling regime. To perform the strong-coupling expansion, we expand the exponential of the lattice action in powers of β. Then the problem is to calculate integrals over the group,

$$I^{a_1\cdots a_p,c_1\cdots c_q}_{b_1\cdots b_p,d_1\cdots d_q} = \int_G dU\, U^{a_1}_{b_1} \cdots U^{a_p}_{b_p} U^{\dagger c_1}_{d_1} \cdots U^{\dagger c_q}_{d_q}, \quad (13.94)$$

where the Haar measure is normalized to 1, $\int dU = 1$. For the cases $(m, n) = (1, 0)$ and $(1, 1)$, the answers derive from the Peter-Weyl theorem (cf. Sect. 14.3.1):

$$\int_G dU\, U^a_b = 0 \quad , \quad \int_G dU\, U^a_b U^{\dagger c}_d = \frac{1}{\dim(U)}\, \delta^a_d \delta^c_b \,. \tag{13.95}$$

The integral

$$\int_G dU\, U^a_b U^c_d \tag{13.96}$$

is zero if the tensor product of the defining representation with itself does not contain the trivial singlet representation. It vanishes for the groups SU(n) with $n \geq 3$. More generally, the center of SU(n) is $\mathbb{Z}_n$, and as a result the integral (13.94) is nonvanishing only if $p = q + kn$ with integer k. Note that the integral (13.96) does not vanish for SU(2) and G_2.

In the expectation value (13.93), the constant contribution $\beta \sum_p \mathrm{tr}\,\mathbb{1}$ to the Wilson action cancels and hence will be dropped such that

$$e^{-S_{\text{gauge}}(U)} = \prod_p \left(1 + \beta\, \Re U_p + O(\beta^2)\right) \,.$$

Because of (13.95) the terms of order β do not contribute to the partition function:

$$Z = 1 + O(\beta^2) = 1 + O(1/g^4) \,. \tag{13.97}$$

The numerator in (13.93) has the strong-coupling expansion

$$\int \prod_\ell dU_\ell\, W[\mathscr{C}] \prod_p \left(1 + \beta\, \Re U_p + O(\beta^2)\right) \,. \tag{13.98}$$

Consider a loop $\mathscr{C} = \ell_n \circ \cdots \circ \ell_2 \circ \ell_1$ as depicted in Fig. 13.11. Because of (13.95) every link variable must appear at least twice in the integrand. Hence only those products of plaquettes contribute that define a surface with boundary $\partial\mathscr{C}$. Besides a connected surface with boundary $\partial\mathscr{C}$, the plaquettes may define further surfaces without boundaries. If the integral (13.96) vanishes, then the plaquettes of the surface with boundary $\partial\mathscr{C}$ must have opposite orientation to $\partial\mathscr{C}$. Since every plaquette contributes a factor β, the surface with *minimal* area A gives the leading order contribution. Hence, in leading order the expectation value (13.93) is given by

$$\langle W[\mathscr{C}]\rangle \sim \left(c^2\beta\right)^A = \exp\left(-A \log \frac{ng^2}{c^2}\right) \,. \tag{13.99}$$

For a rectangular Wilson loop, the minimal area is $A = RT$. The constant c originates from the group integrations and depends on the gauge group.

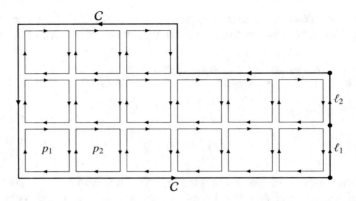

Fig. 13.11 Filling of a loop $\mathscr{C}$ with plaquettes $p_1, p_2, \ldots$

Let us briefly comment on the subleading terms in the strong-coupling expansion. Only products of plaquettes which define a surface with boundary $\mathscr{C}$ contribute to the numerator in (13.93). The contributions of the additional closed surfaces that are not connected to the surface enclosed by $\mathscr{C}$ cancel with the corresponding contributions in the denominator. An area A may self-intersect or may be tangent to a closed surface. In any case, we expect a series expansion of the form

$$\langle W[\mathscr{C}]\rangle = \sum_{A:\,\partial A=\mathscr{C}} \mathrm{e}^{-\sigma A} \tag{13.100}$$

with string tension σ. Actually this area law should hold for all couplings and in particular for weak couplings corresponding to small lattice sizes. Unfortunately, up to now there exists no proof of this expectation.

13.6.3 Glueballs

Gluons, the "photons" of quantum chromodynamics, carry color charge and hence interact strongly with each other. Thus we may expect bound states of several gluons even in the absence of matter. These states, called *glueballs*, are characterized by three quantum numbers J^{PC}: the total angular momentum J and the discrete quantum numbers P and C. The latter characterize the behavior under parity transformation and charge conjugation. Gluons have spin 1 such that glueballs carry an integer spin. We neglect the quark exchange between gluons and consider pure gauge theories.

The energy of excited states is extracted from the exponential tails of suitable two-point functions. Let $G(\tau)$ be the Euclidean two-point function of a gauge-invariant operator $\hat{\mathcal{O}}$ (an observable) given by the vacuum expectation value

$$G_{\mathrm{E}}(\tau) = \langle 0|T\,\hat{\mathcal{O}}(\tau)\hat{\mathcal{O}}(0)|0\rangle = \langle 0|\hat{\mathcal{O}}\mathrm{e}^{-\tau\hat{H}}\,\hat{\mathcal{O}}|0\rangle = \sum_n \left|\langle n|\hat{\mathcal{O}}|0\rangle\right|^2 \mathrm{e}^{-\tau E_n} ,$$

(13.101)

where the $|n\rangle$ form a complete set of eigenstates of the Hamiltonian $\hat{H}$ with energies E_n. For large Euclidean times, the two-point function behaves as

$$G_{\mathrm{E}}(\tau) \longrightarrow \left|\langle 0|\hat{\mathcal{O}}|0\rangle\right|^2 + \left|\langle 1|\hat{\mathcal{O}}|0\rangle\right|^2 \mathrm{e}^{-E_1\tau}\left(1 + \mathcal{O}\left(\mathrm{e}^{-\tau(E_2-E_1)}\right)\right) .$$

(13.102)

We see that the excited state with lowest energy and nonvanishing matrix element $\langle 1|\hat{\mathcal{O}}|0\rangle$ determines the large time behavior. This matrix elements can only be nonzero if the states $\hat{\mathcal{O}}|0\rangle$ and $|1\rangle$ have identical quantum numbers with respect to all conserved charges. Thus, to calculate the mass of a glueball with specified quantum numbers, we must pick a suitable $\hat{\mathcal{O}}$ which projects onto a subsector specified by the quantum numbers J^{PC}.

In a pure gauge theory, every gauge-invariant operator is given in terms of Wilson loops in the framework of the functional integral quantization. Hence we consider a sum of Wilson loops

$$\mathcal{O} = \sum \alpha_i W[\mathscr{C}_i] \quad \text{with} \quad \langle \mathcal{O}(\tau)\mathcal{O}(0)\rangle = \sum_{i,j} \bar{\alpha}_i\alpha_j \left\langle W[\mathscr{C}_i^\tau]W[\mathscr{C}_j]\right\rangle .$$

(13.103)

Thereby $\mathscr{C}^\tau$ denotes the loop $\mathscr{C}$ shifted by τ lattice points in the Euclidean time direction. To project onto states with zero momentum, one often averages over the spatial directions of the lattice. The eigenvalues of parity P and charge conjugation C are ± 1, and it is not difficult to implement the corresponding projections. It is not so easy to project onto a subspace with fixed angular momentum since a lattice theory is invariant only under a finite subgroup of the rotation group.

Cubic Group

The symmetry transformations of a lattice with fixed point form a finite subgroup of the rotation group. In case of a hypercubic lattice, it is the cubic group, i.e., one of the platonic groups. There are three types of symmetry axis: the axes going through the centers of opposite faces of the cube, the axes going through the centers of opposite edges, and the body diagonals as depicted in Fig. 13.12. Hence the order of the group is $1 + F/2 \times 3 + E/2 \times 1 + V/2 \times 2 = 24$, where F, E, and V denote the numbers of faces, edges, and vertices of the cube. The group is isomorphic to

Fig. 13.12 The symmetry operations of a hypercubic lattice have three types of symmetry axis: the three axes going through the centers of opposite faces, the six axes going through the centers of opposite edges, and the four body diagonals

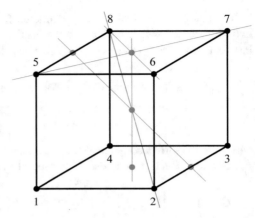

Table 13.1 Character table of the five irreducible representations of the cubic group

Class	# elements	A_1	A_2	E	T_1	T_2
e	1	1	1	2	3	3
$\mathscr{C}_2$	3	1	1	2	-1	-1
$\mathscr{C}_3$	8	1	1	-1	0	0
$\mathscr{C}_2'$	6	1	-1	0	1	-1
$\mathscr{C}_4$	6	1	-1	0	-1	1

the octahedral group $\mathscr{O}_{24}$ or the permutation group $\mathscr{S}_4$ and contains five conjugacy classes:

1. The trivial class with neutral element e,
2. The class $\mathscr{C}_2$ with the π-rotations about the axes connecting opposite faces,
3. The class $\mathscr{C}_3$ with the $2\pi/3$-rotations about the body diagonals,
4. The class $\mathscr{C}_2'$ with the π-rotations about the axes connecting opposite edges,
5. The class $\mathscr{C}_4$ of the $\pm\pi/2$-rotations about the axes connecting opposite faces.

There are as many irreducible representations as there are conjugacy classes, and according to *Burnside's theorem*, the sum of squares of the dimensions of these representations is equal to the order of the group. Hence, the cubic group has five irreducible representations and the sum of the squares of their dimensions is 24. There are the ubiquitous one-dimensional trivial representation A_1, a second one-dimensional representation A_2, a two-dimensional representation E, and two three-dimensional representations T_1 and T_2. The latter consists of the symmetry transformations of the cube. Table 13.1 contains the characters of the five irreducible representations on the conjugacy classes e, $\mathscr{C}_2$, $\mathscr{C}_3$, $\mathscr{C}_2'$, and $\mathscr{C}_4$.

Projecting on Fixed Quantum Numbers
Now we are ready to construct an irreducible representation of parity, charge conjugation, and the cubic group from a given Wilson loop $W[\mathscr{C}]$. Let $\mathscr{C}_{a,P,g}$

denote the loop resulting from $\mathscr{C}$ by a translation on the lattice with $a \in \mathbb{Z}_3$, a parity transformation $P \in \{\mathbb{1}, -\mathbb{1}\}$, and a group transformation $g \in \mathcal{O}_{24}$. Then we introduce

$$\mathscr{W}_{\theta PC}[\mathscr{C}] = \sum_a \sum_{P,g} (-1)^P \chi_\theta(g) \left(W[\mathscr{C}_{a,P,g}] + (-)^C W^*[\mathscr{C}_{a,P,g}] \right) , \qquad (13.104)$$

where θ denotes one of the five irreducible representations of the cubic group. This combination of Wilson loops has quantum numbers θ^{PC} and a vanishing spatial momentum. Simple loops may lead to a vanishing $\mathscr{W}_{\theta PC}[\mathscr{C}]$ for some representations.

The infinitely many irreducible representations θ_ℓ of the rotation group are labeled by the angular momentum $\ell = 0, 1, 2, \ldots$ and have dimension $2\ell + 1$. In general, the irreducible representation θ_ℓ branches into several irreducible representations of the subgroup $\mathcal{O}_{24}$. The branching rules are obtained by comparing the characters of the rotation group with those of $\mathcal{O}_{24}$. In the rotation group, the character $\chi_\ell(\phi)$ of a rotation with fixed axis and angle ϕ in the representation θ_ℓ is

$$\chi_\ell(\phi) = 1 + 2 \sum_{k=1}^{\ell} \cos(k\phi) . \qquad (13.105)$$

The symmetries of the cube are rotations through π, $2\pi/3$, and $\pi/2$ such that

$$\begin{aligned}
\chi_\ell(e) &= 2\ell + 1, \\
\chi_\ell(\mathscr{C}_2) &= (-1)^\ell, \\
\chi_\ell(\mathscr{C}_3) &= 1 - (\ell \bmod 3), \\
\chi_\ell(\mathscr{C}_2') &= (-1)^\ell, \\
\chi_\ell(\mathscr{C}_4) &= 1 + (\ell \bmod 2) - (\ell \bmod 4) .
\end{aligned} \qquad (13.106)$$

Now we are ready to determine the coefficients α_ℓ in the branching rules

$$\theta_\ell = \sum_\theta \alpha_\ell(\theta)\, \theta \quad \text{and} \quad \chi_\ell = \sum_\theta \alpha_\ell(\theta)\, \chi_\theta \qquad (13.107)$$

by using the orthonormality of the $\mathcal{O}_{24}$-characters:

$$\alpha_\ell(\theta) = \frac{1}{24} \sum_{g \in \mathcal{O}_{24}} \chi_\theta(g)\chi_\ell(g) .$$

Table 13.2 Angular momenta for the irreducible representations of the cubic group

	θ_0	θ_1	θ_2	θ_3	θ_4	θ_5	θ_6	θ_7	θ_8	θ_9	θ_{10}	θ_{11}	θ_{12}	θ_{13}	θ_{14}	θ_{15}
A_1	1	0	0	0	1	0	1	0	1	1	1	0	2	1	1	1
A_2	0	0	0	1	0	0	1	1	0	1	1	1	1	1	1	2
E	0	0	1	0	1	1	1	1	2	1	2	2	2	2	3	2
T_1	0	0	1	1	1	1	2	2	2	2	3	3	3	3	4	4
T_2	0	1	0	1	1	2	1	2	2	3	2	3	3	4	3	4

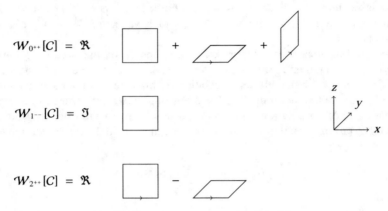

Fig. 13.13 Examples of operators with fixed quantum numbers J^{PC}

The characters are constant on a conjugacy class such that (13.106) yields

$$24\alpha_\ell(A_1) = 2\ell + 15 + 9(-1)^\ell + 6(\ell \bmod 2) - 8(\ell \bmod 3) - 6(\ell \bmod 4),$$
$$24\alpha_\ell(A_2) = 2\ell + 3 - 3(-1)^\ell - 6(\ell \bmod 2) - 8(\ell \bmod 3) + 6(\ell \bmod 4),$$
$$24\alpha_\ell(E) = 4\ell - 6 + 6(-1)^\ell + 8(\ell \bmod 3), \tag{13.108}$$
$$24\alpha_\ell(T_1) = 6\ell - 3 + 3(-1)^\ell - 6(\ell \bmod 2) + 6(\ell \bmod 4),$$
$$24\alpha_\ell(T_2) = 6\ell + 9 - 9(-1)^\ell + 6(\ell \bmod 2) - 6(\ell \bmod 4) .$$

Inserting these results back into (13.107) yields the branching rules given in Table 13.2; see also [47]. As expected θ_0 branches into A_1 and θ_1 branches into T_2. But already the five-dimensional representation θ_2 branches into two representations of the cubic group: $\theta_2 = E \oplus T_1$. In a lattice theory, we can project onto one of the five irreducible representations of the cubic group. In general the corresponding subspace contains states with different angular momenta. For example, according to Table 13.2, the angular momenta $\ell = 0, 4, 6, 8, 9, \ldots$ all contribute to the trivial representation A_1. This already illustrates the problem with filtering out representations with $\ell > 3$ on a cubic lattice. Figure 13.13 shows some simple combinations of Wilson loops with fixed quantum numbers J^{PC}.

Table 13.3 Glueball masses for different quantum numbers J^{PC} (taken from [48])

J^{PC}	0^{++}	2^{++}	0^{-+}	1^{+-}	2^{-+}	3^{+-}	3^{++}	1^{--}	2^{--}	3^{--}	2^{+-}	0^{+-}
m_G[MeV]	1710	2390	2560	2980	3940	3600	3670	3830	4010	4200	4230	4780

To extract glueball masses with sufficient accuracy from simulations, the overlap $\langle 1|\hat{O}|0\rangle$ should be as large as possible. We may increase the overlap by selecting a suitable linear combination of the operators with fixed quantum numbers in (13.104):

$$\mathscr{W}_{\theta PC} = \sum_{\mathscr{C}} \alpha(\mathscr{C})\,\mathscr{W}_{\theta PC}[\mathscr{C}] \quad \text{with} \quad \sum_{\mathscr{C}} |\alpha(\mathscr{C})|^2 = 1 \ . \tag{13.109}$$

The optimal weights $\alpha(\mathscr{C})$ used in this *smearing procedure* are extracted from Monte Carlo simulations. The masses of the lowest-lying glueballs, converted to physical units and corresponding to a Sommer parameter $r_0^{-1} = 410(20)$, are listed in Table 13.3 and are taken from [48]. The reader interested in details concerning the use of nonlocal smeared operators to improve the signal-to-noise ratio for small lattice spacings may consult [49] and the textbooks cited at the end of this chapter.

13.7   Gauge Theories at Finite Temperature

The partition function of a canonical ensemble at temperature T is given by

$$Z(\beta_T) = \operatorname{tr} e^{-\beta_T H}, \quad \beta_T = \frac{1}{k_B T} \tag{13.110}$$

We set the Boltzmann constant $k_B = 1$ and use the symbol β_T for the inverse temperature since β is reserved for the gauge coupling. In the path integral representation of the partition function and the thermal correlation functions, one integrates over fields which are periodic in the Euclidean time with period β_T. An exhaustive introduction to finite-temperature continuum gauge theories is contained in [50].

In the regularized lattice theory, $\beta_T = aN_0$. In order not to mess up finite-temperature and finite-volume effects, we always assume $N_0 \ll N_i$, $i = 1, 2, 3$, or equivalently that β_T is much smaller than the spatial extent of the system. In the continuum limit $a \to 0$, the inverse temperature β_T is kept fixed, and this implies $N_0, N_i \to \infty$. But when we perform the thermodynamic limit $N_i \to \infty$ at fixed lattice spacing a, then the number of lattice points in the time direction remains finite, and the lattice forms a cylinder as depicted in Fig. 13.14. From the partition function

$$Z(V, \beta_T) = \oint \prod_x d\phi_x \prod_\ell dU_\ell\, e^{-S_{\text{YMH}}(U,\phi)} , \tag{13.111}$$

Fig. 13.14 A gauge theory at finite temperature is discretized on a cylinder with circumferences $\beta_T = aN_0$. Even in the thermodynamic limit, there exist parallel transporters which wind around the periodic time direction—these are the Polyakov loops

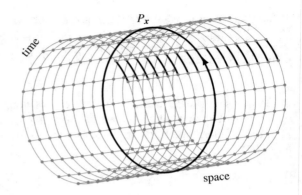

where one integrates over time-periodic fields,

$$\Phi_{x+N_0 e_0} = \phi_x, \quad U_{\langle x, y \rangle} = U_{\langle x+N_0 e_0, y+N_0 e_0 \rangle}, \tag{13.112}$$

one extracts the free energy density $f = -T \log Z/V$. The first derivatives of this density yield the internal energy density and pressure of the thermodynamic system:

$$\varepsilon = \frac{\partial \left(\beta_T f(V, \beta_T) \right)}{\partial \beta_T} \bigg|_V \quad \text{and} \quad p = -\frac{\partial \left(V f(V, \beta_T) \right)}{\partial V} \bigg|_{\beta_T}. \tag{13.113}$$

The specific heat c_V is given by the derivative of ε with respect to temperature at constant volume. In addition, using thermodynamic identities one can extract the speed of sound from ε and p. For extracting the pressure and energy density from lattice results extrapolated to the continuum limit, we refer to the nice review [51]. For sufficiently high temperatures, a pure gauge theory should behave as a gas of free gluons. This is actually not the case. For SU(3) one finds a deviation from the ideal gas behavior of (15–20)% even at temperatures as high as $T \approx 3T_c$, where T_c is the critical temperature above which gluons, confined to glueballs at low temperatures, become free and form a plasma [52].

We are not going further in this direction but instead discuss an interesting order parameter. At finite temperature there are additional (nonlocal) observables, namely, the thermal Wilson loops or Polyakov loops winding around the periodic time direction [53, 54]:

$$P_x = \operatorname{tr} \mathscr{P}(x), \quad \mathscr{P}(x) = \prod_{t=1}^{N_0} U_{(t,x),0}. \tag{13.114}$$

We already encountered these loop variables when we discussed the axial gauge on a periodic lattice; see (13.57). Polyakov loops are particular Wilson loops such that

$$\mathrm{e}^{-\beta_T \Delta F(x-y)} = \left\langle P_y^* P_x \right\rangle \tag{13.115}$$

defines the free energy ΔF which is needed to insert a static quark at x and a static antiquark at y into the thermodynamic system. In the confining phase, ΔF grows without limits when we try to separate the pair, whereas in the deconfined phase, we can separate the pair with a finite amount of energy ΔF. The cluster property

$$\left\langle P_y^* P_x \right\rangle \longrightarrow \langle P_y^* \rangle \langle P_x \rangle = |\langle P \rangle|^2 \quad \text{for} \quad |x - y| \to \infty$$

implies that

$$\langle P \rangle = \begin{cases} 0 & \text{confining phase} \\ P_0 \neq 0 & \text{deconfining phase .} \end{cases} \tag{13.116}$$

The expectation value of the Polyakov loop distinguishes between the phase with confined quarks and the phase with liberated quarks, similarly as the magnetization distinguishes between the ordered and disordered phases of spin systems.

13.7.1 Center Symmetry

In a pure gauge theory with nontrivial center, the Polyakov loop is an order parameter, similarly as the magnetization is an order parameter in the Ising model. The center of a group is the maximal subgroup consisting of elements which commute with all elements of the group. The center of SU(n) is the finite group

$$Z_n = \{z_p = z^P \mathbb{1} \,|\, z = e^{2\pi i/n}, \, p = 1, 2, \ldots, n\} \subset \text{SU(n)} . \tag{13.117}$$

If a gauge group contains a nontrivial center, then the corresponding gauge theory is invariant under global center transformation. A center transformation multiplies all time-oriented link variables in a fixed time slice with the same center element:

$$U_{(t,x),0} \to z_p U_{(t,x),0}, \quad z_p \in Z_n, \quad t \text{ fixed .} \tag{13.118}$$

For example, we could multiply all fat links in Fig. 13.14 with the same center element. A center transformation is just a nonperiodic gauge transformation which maps periodic fields into periodic fields. It follows that the lattice action is invariant since the traces of all parallel transporters around contactable loops (i.e., the plaquette variables) are invariant under center transformations; see Problem 13.3. But the Polyakov loops are not invariant since just one link variable in (13.114) picks up a center element such that

$$P_x \to z^P P_x . \tag{13.119}$$

Fig. 13.15 Histograms of the Polyakov loop for the SU(3) gauge theory at the critical temperature [56]. Below the critical temperature, the order parameter is zero, and above the critical temperature, it takes one of the three center values. The expectation value jumps at the critical temperature

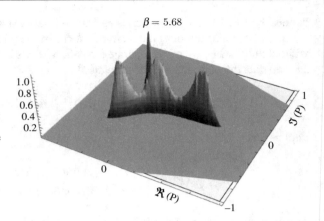

Hence, if the expectation value of P_x is nonzero, then the global center symmetry is broken. This means that in the confining phase, the center symmetry is realized, whereas in the deconfining phase, it is spontaneously broken.

Figure 13.15 shows the histogram of the Polyakov loop in SU(3) gauge theory just at the critical temperature. The order parameter jumps and this points to a first-order transition. The simulations reveal that the deconfinement transition in SU(n) gauge theories without matter is first order for $n \geq 3$. The transition becomes stronger with increasing n. The n-dependence of the critical temperature for $2 \leq n \leq 8$ in units of the string tension is well fitted by $T_c/\sqrt{\sigma} \approx 0.596 + 0.453/n^2$; see [55].

13.7.2 G₂ Gauge Theory

One may wonder whether a nontrivial center is essential for confinement since the Polyakov loop ceases to be an order parameter if the center is trivial. An interesting theory with trivial center is the Higgs model with exceptional gauge group G_2. This theory interpolates between pure G_2 gauge theory for $\kappa = 0$ and pure SU(3) gauge theory for $\kappa \to \infty$ [57]. The pure G_2 gauge theory shows a first-order confinement-deconfinement transition. The Polyakov loop jumps at the critical temperature and is very small but nonzero in the confining phase. Well below the critical temperature, it is difficult to measure its nonzero value in simulations. Hence, although P is not an order parameter in the strict sense, it is still a very useful quantity to spot the phase transition. Figure 13.16 taken from [58] shows the full phase diagram of the G_2 Higgs model in the (β, κ) plane on a $16^3 \times 6$ lattice. The Higgs field is in the seven-dimensional fundamental representation of G_2, and for $\beta = \infty$ the model collapses to a $O(7)$ nonlinear sigma model. This model shows a second-order transition from a $O(7)$-symmetric to a $O(7)$-broken phase, and this transition persists for finite β. At the same time, the first-order confinement-deconfinement transition at $\kappa = 0$ extends into the coupling constant plane until it meets the second-order line. The

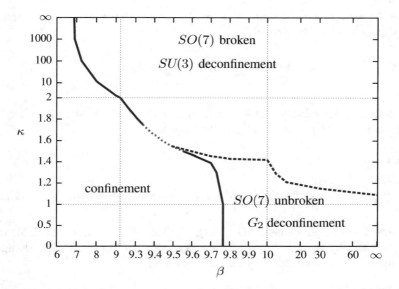

Fig. 13.16 Complete phase diagram of the G_2 Higgs model with scalar field in the fundamental representation in the (β, κ) plane on a $16^3 \times 6$ lattice [58]. The solid line indicates a first-order transition, the dashed line a second-order transition, and the dotted line a second-order transition or a crossover. Note that the scales are nonlinear

first-order line beginning at $\kappa = \infty$ almost meets the two other lines. There remains a small window with a crossover (or a weak transition) connecting the confining and deconfining phases.

13.8 Problems

13.1 (Elitzur Theorem for Z_2 Gauge Theory with Matter) Consider a Z_2 gauge theory coupled to a scalar field. The action with source term is

$$S = -\beta \sum_p U_p - \kappa \sum_{\langle x,y \rangle} \phi_x U_{\langle x,y \rangle} \phi_y + h \sum_x \phi_x + \sum_x V(\phi_x) \,,$$

where V is a Z_2 symmetric potential. Prove that

$$\lim_{h \downarrow 0} \langle \phi_x \rangle = 0$$

uniformly in Λ and the couplings.

13.2 (Metropolis Simulation of the Z_2 Gauge Model in Three Dimensions) Consider a three-dimensional Z_2 gauge theory with matter on a cubic lattice with

N^3 lattice sites (a related analysis of the four-dimensional theory is contained in [38]). The action is of the form

$$S = -\beta \sum_p U_p - \kappa \sum_{\langle x,y \rangle} s_x U_{xy} s_y \,,$$

where U_{xy} is Z_2-valued. The free energy density $f(\beta, \kappa) = -1/N^3 \log Z$ is proportional to the logarithm of the partition function.

(a) Derive the relations

$$\langle U_p \rangle = -\frac{1}{3} \frac{\partial}{\partial \beta} f(\beta, \kappa), \quad \text{and} \quad \langle s_x U_{\langle x,y \rangle} s_y \rangle = -\frac{1}{3} \frac{\partial}{\partial \kappa} f(\beta, \kappa) \,. \tag{13.120}$$

(b) Investigate the phase diagram in the (β, κ) plane with the help of the Metropolis algorithm by considering the expectation values (13.120). Before doing the simulation, you should study the limiting case $\beta = 0$.

13.3 (Center Transformations) Show that the center transformations introduced in Sect. 13.7 can be regarded as gauge transformations. Why do we admit these gauge transformations although they are not periodic in time. Show that parallel transporters around contractible loops are invariant under center transformations.

References

1. D. Ivanenko, G. Sardanashvily, The gauge treatment of gravity. Phys. Rept. **94**, 1 (1983)
2. F.W. Hehl, J.D. McCrea, E.W. Mielke, Y. Ne'eman, Metric affine gauge theory of gravity: field equations, Noether identities, world spinors, and breaking of dilation invariance. Phys. Rept. **258**, 1 (1995)
3. S. Pokorski, *Gauge Field Theories* (Cambridge University Press, Cambridge, 2000)
4. K. Huang, *Quarks, Leptons and Gauge Fields* (World Scientific, Singapore, 1992)
5. L. O'Raifeartaigh, *Group Structure of Gauge Theories* (Cambridge University Press, Cambridge, 1986)
6. A. Das, *Lectures on Quantum Field Theory* (World Scientific, Singapore, 2020)
7. F.J. Wegner, Duality in generalized Ising models and phase transitions without local order parameters. J. Math. Phys. **10**, 2259 (1971)
8. K.G. Wilson, Confinement of quarks. Phys. Rev. **D10**, 2445 (1974)
9. M. Creutz, L. Jacobs, C. Rebbi, Experiments with a gauge invariant Ising system. Phys. Rev. Lett. **42**, 1390 (1979)
10. M. Creutz, Confinement and the critical dimensionality of spacetime. Phys. Rev. Lett. **43**, 553 (1979)
11. M. Creutz, Monte Carlo simulations in lattice gauge theories. Phys. Rep. **95**, 201 (1983)
12. I. Montvay, G. Münster, *Quantum Fields on a Lattice* (Cambridge University Press, Cambridge, 2010)
13. H.J. Rothe, *Lattice Gauge Theories: An Introduction* (World Scientific, Singapore, 2012)
14. T. DeGrand, C. DeTar, *Lattice Methods for Quantum Chromodynamics* (World Scientific, Singapore, 2006)

15. C. Gattringer, C. Lang, *Quantum Chromodynamics on the Lattice*. Lect. Notes Phys., vol. 788 (Springer, Berlin, 2010)
16. L.P. Kadanoff, The application of renormalization group techniques to quarks and strings. Rev. Mod. Phys. **49**, 267 (1977)
17. J.B. Kogut, An introduction to lattice gauge theory and spin systems. Rev. Mod. Phys. **51**, 659 (1979)
18. J.B. Kogut, The lattice gauge theory approach to quantum chromodynamics. Rev. Mod. Phys. **55**, 775 (1983)
19. P. de Forcrand, O. Jahn, Comparison of SO(3) and SU(2) lattice gauge theory. Nucl. Phys. **B651**, 125 (2003)
20. R.L. Karp, F. Mansouri, J.S. Rho, Product integral formalism and non-Abelian Stokes theorem. J. Math. Phys. **40**, 6033 (1999)
21. R.L. Karp, F. Mansouri, J.S. Rho, Product integral representations of Wilson lines and Wilson loops, and non-Abelian Stokes theorem. Turk. J. Phys. **24**, 365 (2000)
22. R. Giles, Reconstruction of gauge potentials from Wilson loops. Phys. Rev. **D24**, 2160 (1981)
23. J. Fröhlich, G. Morchio, F. Strocchi, Higgs phenomenon without symmetry breaking order parameter. Nucl. Phys. **B190**, 553 (1981)
24. F. Hausdorff, Die symbolische Exponentialformel in der Gruppentheorie. Ber. Verh. Saechs. Akad. Wiss. Leipz. **58**, 19 (1906)
25. K. Wilson, in *Recent Developments of Gauge Theories*, ed. by G. 't Hooft et al. (Plenum, New York, 1980)
26. K. Symanzik, Continuum limit and improved action in lattice theories. 1. Principles and ϕ^4 theory. Nucl. Phys. **B226**, 187 (1983)
27. M. Luscher, P. Weisz, Computation of the action for on-shell improved lattice gauge theories at weak coupling. Phys. Lett. **B158**, 250 (1985)
28. K. Langfeld, Improved actions and asymptotic scaling in lattice Yang-Mills theory. Phys. Rev. **D76**, 094502 (2007)
29. J.M. Drouffe, J.B. Zuber, Strong coupling and mean field methods in lattice gauge theories. Phys. Rep. **102**, 1 (1983)
30. G. Arnold, B. Bunk, T. Lippert, K. Schilling, Compace QED under scrutiny: it's first order. Nucl. Phys. Proc. Suppl. **119**, 864 (2003)
31. K. Langfeld, B. Lucini, A. Rago, The density of states in gauge theories. Phys. Rev. Lett. **109**, 111601 (2012)
32. J. Carlsson, B. McKellar, SU(N) glueblall masses in 2+1 dimensions. Phys. Rev. **D68**, 074502 (2003)
33. S. Uhlmann, R. Meinel, A. Wipf, Ward identities for invariant group integrals. J. Phys. **A40**, 4367 (2007)
34. K. Osterwalder, E. Seiler, Gauge field theories on a lattice. Ann. Phys. **10**, 440 (1978)
35. E. Fradkin, S. Shenker, Phase diagrams of lattice gauge theories with Higgs fields. Phys. Rev. **D19**, 3682 (1979)
36. C. Bonati, G. Cossu, M. D'Elia, A. Di Giacomo, Phase diagram of the lattice SU(2) Higgs model. Nucl. Phys. **B828**, 390 (2010)
37. J. Greensite, *An Introduction to the Confinement Problem*. Lect. Notes Phys., vol. 972 (Springer, Berlin, 2020)
38. Y. Blum, P.K. Coyle, S. Elitzur, E. Rabinovici, S. Solomon, H. Rubinstein, Investigation of the critical behavior of the critical point of the Z2 gauge lattice. Nucl. Phys. **B535**, 731 (1998)
39. S. Elitzur, Impossibility of spontaneously breaking local symmetries. Phys. Rev. **D12** (1975) 3978
40. C. Itzikson, J.M. Drouffe, *Statistical Field Theory*, vol. I. Cambridge Monographs on Mathematical Physics (Cambridge University Press, Cambridge, 1991)
41. A. Maas, R. Sondenheimer, P. Törek, On the observable spectrum of theories with a Brout–Englert–Higgs effect. Ann. Phys. **402**, 18 (2019)
42. G.S. Bali, K. Schilling, C. Schlichter, Observing long color flux tubes in SU(2) lattice gauge theory. Phys. Rev. **D51**, 5165 (1995)

43. B. Wellegehausen, A. Wipf, C. Wozar, Casimir scaling and string breaking in G2 gluodynamics. Phys. Rev. **D83**, 016001 (2011)
44. E. Seiler, Upper bound on the color-confining potential. Phys. Rev. **D18**, 482 (1978)
45. C. Bachas, Convexity of the quarkonium potential. Phys. Rev. **D33**, 2723 (1986)
46. M. Lüscher, K. Symanzik, P. Weisz, Anomalies of the free loop wave equation in WKB approximation. Nucl. Phys. **B173**, 365 (1980)
47. M. Lax, *Symmetry Principles in Solid State and Molecular Physics* (Dover, New York, 2003)
48. Y. Chen et al., Glueball spectrum and matrix elements on anisotropic lattices. Phys. Rev. **D 73**, 014516 (2006)
49. M. Teper, An improved method for lattice glueball calculations. Phys. Lett. **183B**, 345 (1986)
50. J.I. Kapusta, C. Gale, *Finite-Temperature Field Theory: Principles and Applications* (Cambridge University Press, Cambridge, 2011)
51. F. Karsch, Lattice QCD at high temperature and density. Lect. Notes Phys. **583**, 209 (2002)
52. G. Boyd, J. Engels, F. Karsch, E. Laermann, C. Legeland, M. Lütgemeier, B. Petersson, Equation of state for the SU(3) gauge theory. Phys. Rev. Lett. **75**, 4169 (1995)
53. A.M. Polyakov, Quark confinement and topology of gauge groups. Nucl. Phys. **B120**, 429 (1977)
54. B. Svetitsky, L.G. Yaffe, Critical behavior at finite temperature confinement transitions. Nucl. Phys. **B210**, 423 (1982)
55. B. Lucini, M. Teper, U. Wenger, The high temperature phase transition in SU(n) gauge theories. JHEP **0401**, 061 (2004)
56. B. Wellegehausen, Effektive Polyakov-Loop Modelle für SU(n)- und G2-Eichtheorien (Effective Polyakov loop models for SU(n) and G2 gauge theories). Diploma Thesis, Jena (2008)
57. K. Holland, P. Minkowski, M. Pepe, U.J. Wiese, Exceptional confinement in G(2) gauge theory. Nucl. Phys. **B668**, 207 (2003)
58. B. Wellegehausen, A. Wipf, C. Wozar, Phase diagram of the lattice G2 Higgs Model. Phys. Rev. **D83**, 114502 (2011)

Two-Dimensional Lattice Gauge Theories and Group Integrals

14

In two dimensions a pure lattice gauge theory with the simple Wilson action can be solved analytically. With open boundary conditions and in the axial gauge, the partition function becomes a product of simple group integrals, and the area law behavior is exact for all values of the gauge coupling β. In this chapter we impose periodic boundary conditions in all directions, adequate for finite temperature and finite volume studies. On a torus the solution is a bit less trivial, and the exact solution can be used as a test bed for new Monte Carlo algorithms. First we study simple Abelian gauge models for which the calculation parallels our treatment of one-dimensional spin models. The second part deals with non-Abelian theories on the torus for which we use the character expansion and a recursion formula due to Migdal [1, 2] (for a review see [3])). We calculate the free energy and the potential energy between a static quark-antiquark pair. Two dimensional gauge theories confine a static $q\bar{q}$-pair, since in one space dimension the field energy cannot spread out in space. For weak couplings the string tension shows an exact Casimir scaling, similarly as in gauge theories in three and four dimensions. Towards the end of this chapter, we collect some facts on invariant group integration which are useful in strong-coupling expansions, mean field approximations, and exact solutions. Further material can be found in the textbook [4].

14.1 Abelian Gauge Theories on the Torus

For an Abelian theory, the calculation simplifies considerably if we choose an axial-type gauge for the link variables in the partition function

$$Z_V(\beta) = e^{-\beta V} \int \prod_\ell dU_\ell \prod_p e^{\beta \Re(U_p)}, \qquad U_p = \prod_{\ell \in \partial p} U_\ell . \qquad (14.1)$$

© The Author(s), under exclusive license to Springer Nature Switzerland AG 2021
A. Wipf, *Statistical Approach to Quantum Field Theory*, Lecture Notes
in Physics 992, https://doi.org/10.1007/978-3-030-83263-6_14

Fig. 14.1 If periodic boundary conditions are imposed, then only the variables on the marked links can be gauged to the identity

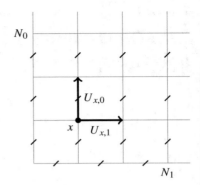

On page 348 we have argued that on a periodic lattice, the variables on the marked links in Fig. 14.1 can be transformed to the identity. As gauge-invariant variables, we may choose the plaquette variables and the *Polyakov loops* in both directions

$$P_0(x_1) = \prod_{n=1}^{N_0} U_{x+ne_0,0} \quad , \quad P_1(x_0) = \prod_{n=1}^{N_1} U_{x+ne_1,1} \,. \tag{14.2}$$

For periodic boundary conditions, not all plaquette variables are independent since $U_1 U_2 \ldots U_V = 1$. Only $V - 1$ plaquette variables are independent, and we choose these as new variables, together with a subset of Polyakov loops. To find this subset, we observe that the Polyakov loop at $x_1 + m$ is equal to the Polyakov loop at x_1 multiplied by the plaquette variables enclosed by the two loops. Thus, only one Polyakov loop in the time direction is independent. The same applies to Polyakov loops in the x_1 direction. Thus, there are $V + 1$ independent gauge-invariant variables. Together with the variables on the $V - 1$ marked links in Fig. 14.1, denoted by E', we obtain a complete set of $2V$ variables

$$\{U_p| \, p = 1, \ldots, V - 1\}, \quad P_0, \, P_1 \quad \text{and} \quad \{U_\ell| \, \ell \in E'\} \,. \tag{14.3}$$

Exploiting the invariance of the Haar measure, we can write

$$Z_V(\beta) = \mathrm{e}^{-\beta V} \int \prod_{p=1}^{V-1} \mathrm{d}U_p \prod_{p=1}^{V} \mathrm{e}^{\beta \Re U_p} \int \mathrm{d}P_0 \mathrm{d}P_1 \int \prod_{\ell \in E'} \mathrm{d}U_\ell \,, \tag{14.4}$$

wherein the plaquette variable U_V is given in terms of the other plaquette variables. The integrand does not depend on the link variables on E' and on the Polyakov loops, and with $1 = \int \mathrm{d}U_V \, \delta(1, U_1 U_2 \cdots U_V)$, we obtain

$$Z_V(\beta) = \mathrm{e}^{-\beta V} \prod_{p=1}^{V} \int \mathrm{d}U_p \, \mathrm{e}^{\beta \Re U_p} \, \delta\left(1, \prod U_p\right) \,. \tag{14.5}$$

For an Abelian group, every irreducible representation R_n is one-dimensional, and the formula (14.65) reads $\delta(1, U) = \sum_n R_n(U)$. In addition $R_n(UV) = R_n(U)R_n(V)$ such that the partition function for an Abelian gauge group factorizes

$$Z_V(\beta) = e^{-\beta V} \sum_n (Z_n)^V, \quad Z_n = \int dU \, e^{\beta \Re U} R_n(U), \qquad (14.6)$$

where the sum extends over all irreducible representation of the group. Expectation values of Wilson loops in a given representation R_{n_0} are calculated similarly. For an Abelian theory, Stokes theorem applies, and the Wilson loop is the product of plaquette variables

$$W[\mathscr{C}] = \prod_{\ell \in \mathscr{C}} R_{n_0}(U_\ell) = \prod_{p \in A} R_{n_0}(U_p), \qquad (14.7)$$

where A is an area with boundary $\mathscr{C}$. One obtains

$$\langle W[\mathscr{C}] \rangle = \frac{\sum_n Z_n^{V-A} Z_{n,n_0}^A}{\sum_n Z_n^V}, \quad Z_{n,n_0} = \int dU \, e^{\beta \Re U} R_n(U) R_{n_0}(U), \qquad (14.8)$$

such that the expectation value of the Wilson loop only depends on the charges of the static $q\bar{q}$ pair, the area of the loop, and the lattice size. Later we shall see that this formula implies confinement.

14.1.1 Z_2 Gauge Theory

There are two irreducible representations of $Z_2 = \{1, -1\}$. The trivial representation assigns 1 to every group element such that $Z_{\text{trivial}} = \cosh(\beta)$ and for the defining representation $Z_{\text{defining}} = \sinh(\beta)$. Thus, (14.6) yields

$$Z_V(\beta) = \left(\frac{1 + e^{-2\beta}}{2} \right)^V + \left(\frac{1 - e^{-2\beta}}{2} \right)^V. \qquad (14.9)$$

In the thermodynamic limit $V \to \infty$, we may neglect the last term and find the ground state energy density

$$e(\beta) = - \lim_{V \to \infty} \log \frac{Z_V(\beta)}{V} = - \log \frac{1 + e^{-2\beta}}{2}. \qquad (14.10)$$

In the weak-coupling limit $\beta \to \infty$, it converges to the constant value $\log(2)$. By contrast, in the strong-coupling limit, the energy density converges to $\beta = 1/g^2$. In between it is a monotonically decreasing function of g as depicted in Fig. 14.2.

Fig. 14.2 Ground state
energy density for the
two-dimensional Z_2-gauge
theory in the thermodynamic
limit $V \to \infty$ as function of
the coupling constant g. For
$g > 0$ it is an analytic
function of g

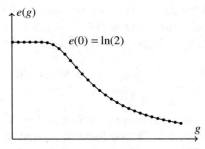

Fig. 14.3 Ground state
energy density of the
two-dimensional $U(1)$ gauge
theory in the infinite volume
limit

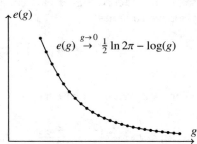

14.1.2 U(1) Gauge Theory

The continuous group $U(1)$ has infinitely many one-dimensional irreducible representation characterized by the integer charge,

$$R_n \left(e^{i\alpha} \right) = e^{in\alpha} , \quad n \in \mathbb{Z} , \tag{14.11}$$

such that Z_n in (14.6) is given by a modified Bessel function of order n,

$$Z_n(\beta) = \frac{1}{2\pi} \int_{-\pi}^{\pi} e^{\beta \cos\alpha + in\alpha} = I_n(\beta) . \tag{14.12}$$

Since $I_0(\beta) > I_1(\beta) > \ldots$, we obtain the following ground state energy density in the thermodynamic limit:

$$e(\beta) = \beta - \log I_0(\beta) . \tag{14.13}$$

In the weak-coupling regime, it diverges as $-\log(g)$, and for strong coupling it falls off as $1/g^2 - 1/4g^4 + \ldots$. Figure 14.3 shows the energy density as a function of the coupling g. In passing we note that if one is only interested in the thermodynamic limit, then one may neglect the constraint $U_1 \cdots U_V = 1$ imposed by the periodic boundary conditions. This already indicates that the thermodynamic limit $V \to \infty$ is insensitive to the boundary conditions. Actually this is true for two-dimensional gauge theories with general boundary conditions; see [5].

Equation (14.11) shows that $R_n R_{n_0} = R_{n+n_0}$ such that the formula for the Wilson loop (14.8) yields the following simple result in the thermodynamic limit:

$$\lim_{V \to \infty} \langle W[\mathscr{C}] \rangle = \left(\frac{I_{n_0}(\beta)}{I_0(\beta)} \right)^A . \tag{14.14}$$

We see that in all representations, the Wilson loops obey an area law. The string tension in the charge n_0 representation has the strong- and weak-coupling limits

$$\sigma_{n_0} = \log \frac{I_0(\beta)}{I_{n_0}(\beta)} \longrightarrow \begin{cases} n_0 \log(2/\beta) & \text{for } \beta \ll 1 \\ \log(1 + n_0^2/2\beta) & \text{for } \beta \gg 1 . \end{cases} \tag{14.15}$$

14.2   Non-Abelian Lattice Gauge Theories on the Torus

For free boundary conditions, it is easy to calculate the partition function and ground state energy density. For example, for the gauge group SU(2), one finds

$$e(\beta) = 2\beta - \log \left(\frac{I_1(2\beta)}{\beta} \right) . \tag{14.16}$$

The energy density shows a similar dependency on the gauge coupling as the energy density of the U(1) theory. In the strong-coupling regime $e \sim 1/g^2 - 1/8g^4 + \ldots$ and in the weak-coupling regime $e \sim 3 \log(g) + \frac{1}{2} \log(\pi/2)$.

For periodic boundary conditions in all directions, the models can still be solved, but the solution involves several steps. Since there is no simple Stoke theorem for non-Abelian theories, the constraint on the plaquette variables is not as simple as for Abelian theories. In place of Stokes theorem, we shall apply the gluing rule (14.63).

Gluing Loops and Migdal's Recursion Relation
Let us consider two loops $\mathscr{C}_x$, $\mathscr{C}'_x$ starting at x and sharing one edge as illustrated in Fig. 14.4. We denote the variable on the common edge by V. The parallel transporter along $\mathscr{C}_x$ is given by $U_{\mathscr{C}_x} = WV$, and the transporter along $\mathscr{C}'_x$ is $V^{-1}W'$. A class function obeys $f(\Omega U \Omega^{-1}) = f(U)$ and is a linear combination of the orthonormal characters,

$$f(U) = \sum_R c_R \chi_R(U) , \tag{14.17}$$

where the expansion coefficients are given by

$$c_R = (\chi_R, f) \equiv \int dU \, \bar{\chi}_R(U) f(U) . \tag{14.18}$$

Fig. 14.4 Gluing a class function on two loops which share a common link V

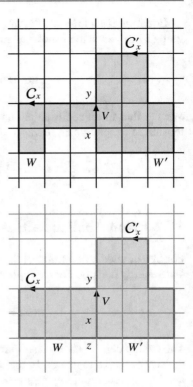

Fig. 14.5 Gluing together two loops at several connected links

This expansion allows for the integration over the common link variable V in the product function $f(U_{\mathscr{C}_x}) f(U_{\mathscr{C}'_x})$. By using the gluing rule (14.63), we find

$$\int dV f(U_{\mathscr{C}}) f(U_{\mathscr{C}'}) = \sum_R \frac{c_R^2}{d_R} \chi_R(WW') = \sum_R \frac{c_R^2}{d_R} \chi_R(U_{\mathscr{C} \circ \mathscr{C}'}) . \qquad (14.19)$$

If the two loops share several edges and if these edges are *connected*, then only one of the edges is glued. For example, if we glue together the loops $\mathscr{C}_x$ and $\mathscr{C}'_x$ along x, y in Fig. 14.5, the parallel transporter $U_{\langle x,z \rangle}$ in $\mathscr{C}_x$ and its inverse $U_{\langle z,x \rangle}$ in $\mathscr{C}'_x$ cancel in the class function $\chi_R(U_{\mathscr{C} \circ \mathscr{C}'})$. More generally, after gluing along one common edge, the link variables on the other common edges connected to the glued edge cancel on the right-hand side of (14.19). Hence, $U_{\mathscr{C} \circ \mathscr{C}'}$ in this formula describes the parallel transporter along the exterior boundary of the area enclosed by $\mathscr{C}$ and $\mathscr{C}'$. However, if the two loops $\mathscr{C}$ and $\mathscr{C}'$ share several unconnected edges, then only those variables cancel that are defined on common links connected to the glued edge. Now we can iterate the gluing process, for example, by gluing $\mathscr{C} \circ \mathscr{C}'$ and $\mathscr{C}''$ at a common edge V',

$$\int dV' dV f(U_{\mathscr{C}}) f(U_{\mathscr{C}'}) f(U_{\mathscr{C}''}) = \sum_R \frac{c_R^3}{d_R^2} \chi_R (U_{\mathscr{C} \circ \mathscr{C}' \circ \mathscr{C}''}) . \qquad (14.20)$$

Every surface A without holes is built from single plaquettes by gluing one plaquette after another along a connected set of edges, $A = p_1 \cup \cdots \cup p_n$. This way we obtain for a surface A without holes the *Migdal* recursion formula [1, 2]

$$\int \prod_{\ell=1}^{n-1} dV_\ell \prod f(U_{p_1}) \cdots f(U_{p_n}) = \sum_R d_R \left(\frac{c_R}{d_R} \right)^n \chi_R(U_{\partial A}) . \qquad (14.21)$$

14.2.1 Partition Function

Now we are ready to calculate the partition function with the help of the recursion formula applied to $f(U) = \exp(-\beta \operatorname{tr}(\mathbb{1} - \Re U_p))$. We assume A to be the union of all plaquettes such that the boundary ∂A contains the link variables V, V' and W depicted in Fig. 14.6. If we fix the link variables on the spatial boundary, then we obtain the partition function

$$Z(V, V') = \sum_R d_R \left(\frac{c_R}{d_R} \right)^V \int dW \, \chi_R(V^{-1} W^{-1} V' W) .$$

With the separation rule Eq. (14.64), the integral can be calculated,

$$Z(V, V') = \sum \left(\frac{c_R}{d_R} \right)^V \chi_R(V^{-1}) \chi_R(V') . \qquad (14.22)$$

Fig. 14.6 Calculating the partition function with the help of the gluing and separation rules

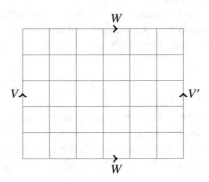

If in addition we choose periodic boundary conditions in the spatial direction, then we must set $V' = V$ and integrate over V,

$$Z_{\text{per}} = e^{-2\beta V} \sum_R \left(\frac{c_R(\beta)}{d_R} \right)^V , \quad c_R(\beta) = \int dU \, \bar{\chi}_R(U) \, e^{\beta \, \text{tr} \Re U} . \qquad (14.23)$$

The irreducible representations of SU(2) are characterized by their half-integer spin j. The spin-j representation has dimension $d_j = 2j + 1$, and its character is

$$\chi_j(U) = \frac{\sin(d_j \theta)}{\sin \theta} . \qquad (14.24)$$

With $\text{tr} \, \Re U = 2 \cos \theta$, the integral (14.23) yields a modified Bessel function:

$$c_j(\beta) = \frac{2}{\pi} \int_0^\pi d\theta \, (\sin \theta)^2 \frac{\sin(d_j \theta)}{\sin \theta} \, e^{2\beta \cos \theta} = -\frac{1}{\pi \beta} \int \sin(d_j \theta) \frac{d}{d\theta} e^{2\beta \cos \theta}$$

$$= \frac{d_j}{\pi \beta} \int \cos(d_j \theta) \, e^{2\beta \cos \theta)} = \frac{d_j}{\beta} I_{2j+1}(2\beta) .$$

Hence, we end up with the following exact formula for the partition function of the SU(2) gauge theory on the torus:

$$Z(\beta) = e^{-2\beta V} \left(\frac{1}{\beta} \right)^V \sum_{n=1,2,\dots}^\infty (I_n(2\beta))^V . \qquad (14.25)$$

In the thermodynamic limit, we recover the result (14.16) for the vacuum energy density.

14.2.2 Casimir Scaling of Polyakov Loops

At finite temperature we must impose periodic boundary conditions in the Euclidean time direction. In this section we compute the expectation value of correlators of the Polyakov loop $\mathscr{P}_x$ for static charges in a fixed representation R_0. The expectation value of one Polyakov loop is given by

$$\langle P \rangle = \frac{1}{Z} \int \prod_\ell dU_\ell \, e^{-S_{\text{gauge}}(U)} \chi_{R_0}(\mathscr{P}_x) . \qquad (14.26)$$

The straight Polyakov loop winding around the Euclidean time direction divides the lattice into two regions, denoted by A and A' in Fig. 14.7. We choose periodic boundary conditions in the spatial direction. First we glue together the plaquettes in each of the two regions. Decomposing the parallel transporters along ∂A and $\partial A'$ as

Fig. 14.7 Calculating the expectation value of the Polyakov loop with the cutting rule

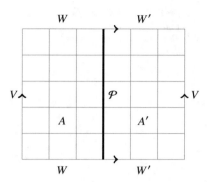

depicted in figure and using Migdal's recursion relation, we obtain for the integral in (14.26) the double sum

$$\sum_{R,R'} d_R \left(\frac{c_R}{d_R}\right)^A \chi_R \left(W V^{-1} W^{-1} \mathscr{P}\right) d_{R'} \left(\frac{c'_R}{d'_R}\right)^{A'} \chi_{R'} \left(\mathscr{P}^{-1} W'^{-1} V W'\right) \chi_{R_0}(\mathscr{P}) .$$

With the separation rule (14.64), the integration over W and W' can easily be done, and one obtains a double sum containing the factor $\chi_R(V^{-1}) \chi_{R'}(V)$. Owing to the orthogonality of the characters, the integration over V reduces the double sum to a single sum, and we end up with

$$\langle P \rangle = \frac{1}{Z} e^{-2\beta V} \sum_R \left(\frac{c_R}{d_R}\right)^V \int d\mathscr{P} \, \chi_{R_0}(\mathscr{P}) \chi_R(\mathscr{P}) \chi_R(\mathscr{P}^{-1}) .$$

We now focus again on the gauge group SU(2) with characters χ_j. The Clebsch-Gordan decomposition $\chi_{j_0} \chi_j = \chi_{j_0+j} + \cdots + \chi_{|j_0-j|}$ immediately yields

$$\int d\mathscr{P} \, \chi_{j_0}(\mathscr{P}) \chi_j(\mathscr{P}) \chi_j(\mathscr{P}^{-1}) = \begin{cases} 0 & \text{for half-integer } j_0 \\ 1 & \text{for integer } j_0 \text{ and } j \geq j_0/2 . \end{cases}$$

Thus, we obtain the interesting result that for static quarks transforming according to a half-integer spin representation, the Polyakov loop expectation value vanishes. This means that these particles are confined. Actually, a non-vanishing expectation value would violate the center symmetry. On the other hand, for static quarks transforming according to an integer spin representation of the gauge group, we obtain

$$\langle \chi_{j_0}(\mathscr{P}) \rangle = \frac{\sum_{n \geq 1} I^V_{j_0+n}(2\beta)}{\sum_{n \geq 1} I^V_n(2\beta)}, \quad j_0 \in \{0, 1, 2, \dots\} . \tag{14.27}$$

Fig. 14.8 Calculation of the two-point function of Polyakov loops via the gluing technique

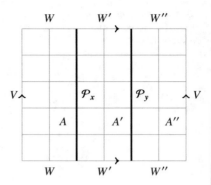

The terms with $n = 1$ dominate for large volumes V such that we find the asymptotic formula

$$\langle \chi_{j_0}(\mathscr{P}) \rangle \approx \left(\frac{I_{j_0+1}(2\beta)}{I_1(2\beta)} \right)^V , \quad j_0 \in \{0, 1, 2, \dots\} . \tag{14.28}$$

Since the ratio is less than one, the expectation value vanishes in the thermodynamic limit.

In order to compute the two-point function of the Polyakov loop corresponding to a static quark-antiquark pair in the representation R_0,

$$\langle \bar{\chi}_{R_0}(\mathscr{P}_x) \chi_{R_0}(\mathscr{P}_y) \rangle = \frac{1}{Z} \int \prod dU_\ell \, e^{-S_{\text{gauge}}(U)} \, \bar{\chi}_{R_0}(\mathscr{P}_x) \, \chi_{R_0}(\mathscr{P}_y) , \tag{14.29}$$

we divide the lattice into three domains A, A', A'' as depicted in Fig. 14.8. Again we impose periodic boundary conditions in all directions. First we glue the plaquettes in each of the three domains, and this yields a triple sum for the integral in (14.29). Then we glue together the variables W, W' and W'' in every domain in Fig. 14.8. After a further integration over V, the triple sum collapses to the double sum

$$e^{-2\beta V} \sum_{R,R'} \left(\frac{c_R}{d_R} \right)^{A+A'} \left(\frac{c_{R'}}{d_{R'}} \right)^{A'} \chi_R(\mathscr{P}_x) \bar{\chi}_{R_0}(\mathscr{P}_x) \bar{\chi}_{R'}(\mathscr{P}_x) \chi_{R'}(\mathscr{P}_y) \chi_{R_0}(\mathscr{P}_y) \bar{\chi}_R(\mathscr{P}_y) .$$

The final integration over $\mathscr{P}_y$ (or equivalently over $\mathscr{P}_x$) counts how often R appears in the tensor product $R_0 \otimes R'$. After the integration over the Polyakov loops, the absolute squares of these generalized Clebsch-Gordan coefficients appear on the right-hand side. In particular, for the gauge group SU(2) and static quarks in the fundamental representation, we find

$$\langle \text{tr} \, \mathscr{P}_x \, \text{tr} \, \mathscr{P}_y \rangle = \frac{e^{2\beta V}}{Z} \left(\frac{1}{\beta} \right)^V \sum_{n \geq 1} \left(I_n^{A+A''}(2\beta) I_{n+1}^{A'}(2\beta) + I_n^{A'}(2\beta) I_{n+1}^{A+A''}(2\beta) \right) , \tag{14.30}$$

with partition function given in (14.25). In the thermodynamic limit, both A and A'' tend to infinity, and only the term proportional to $I_1^{A+A''}$ contributes. By using the relation $V = A + A' + A''$, we end up with

$$\langle \text{tr } \mathscr{P}_x \text{ tr } \mathscr{P}_y \rangle = \left(\frac{I_2(2\beta)}{I_1(2\beta)} \right)^{A'} . \tag{14.31}$$

We see that the two-point function obeys a Wilson's area law. The free energy of the two static charges grows linearly with their separation,

$$f\left(|x - y| \right) = \sigma_{\frac{1}{2}} |x - y| \quad \text{with} \quad \sigma_{\frac{1}{2}} = -\log \left(\frac{I_2(2\beta)}{I_1(2\beta)} \right) . \tag{14.32}$$

Actually, one finds a similar behavior of static charges in higher representations of SU(2). The free energy rises linearly with the separation and defines a string tension

$$\sigma_j = -\log \left(\frac{I_{2j+1}(2\beta)}{I_1(2\beta)} \right) \longrightarrow \begin{cases} j(j+1)/\beta, & \beta \to \infty \\ 2j \log(1/\beta), & \beta \to 0. \end{cases} \tag{14.33}$$

Interestingly in the weak-coupling limit $\beta \to \infty$, the string tension σ_j for static $q\bar{q}$-pair in the representation j is proportional to the eigenvalue $j(j+1)$ of the Casimir operator. In two dimensions this *Casimir scaling* holds for higher groups as well. In general the string tension is proportional to the quadratic Casimir.

In higher dimensions there is no proof of Casimir scaling, although there are good reasons to believe that it holds. For example, for pure SU(2) and SU(3) gauge theories in three and four dimensions, there is conclusive numerical evidence for Casimir scaling from Monte Carlo simulations. In particular the simulations for SU(3) in four dimensions confirm Casimir scaling within 5% for separations up to 1 fm for static quarks with Casimir values (normalized by the Casimir of {3}) up to 7 [6]. Dynamical quarks can screen the static charges in which case we expect Casimir scaling only on intermediate scales. If the separation of the static $q\bar{q}$ pair becomes too large, then string breaking sets in, and the static potential flattens. String breaking has been observed in SU(2) gauge theory in [7, 8]. There are gauge theories where the gauge bosons can screen the static quarks as well. In such theories we expect string breaking even without matter. Indeed, for the exceptional G_2 gauge theory, Casimir scaling holds only at intermediate distances. At large separations one observes string breaking [9].

Pure gauge theories in two dimensions are analytic for $\beta > 0$. They behave similarly as one-dimensional spin chains, and in particular there is no phase transition. However, it was shown by Gross and Witten that, in two dimensions, the U(n) lattice gauge theory with Wilson action exhibits a third-order phase transition when n approaches infinity [10]. The phase transition is due to the presence of an infinite number of degrees of freedom in group space.

14.3   Invariant Measure and Irreducible Representations

Here we present some facts and formulas about invariant integration on group manifolds which are useful when dealing with lattice gauge theories. On every compact group, there exists a distinguished *Haar measure* which is positive, left-invariant, and right-invariant:

$$dU = d(\Omega U) = d(U\Omega), \quad \Omega \in G \,. \tag{14.34}$$

It was introduced by the Hungarian mathematician ALFRED HAAR back in 1933 and represents a generalization of the Lebesgue measure. When we normalize the invariant measure, i.e., $\int dU = 1$, then it is unique. The conditions (14.34) are equivalent to the left- and right-invariance of averages

$$\mathcal{M}(f) \equiv \int_G dU \, f(U) = \int_G dU \, f(\Omega U) = \int_G dU \, f(U\Omega) \tag{14.35}$$

for functions $f : G \to \mathbb{C}$. The map $\mathcal{M}$ is linear, positive, normalized, and invariant.
For a *finite group*, the average of a function is just its mean value,

$$\mathcal{M}(f) = \frac{1}{|G|} \sum_{U \in G} f(U) \,, \tag{14.36}$$

and the properties (14.35) are evident. Another simple example is the Abelian Lie group U(1) consisting of unimodular complex numbers parametrized by their phase according to $U = e^{i\alpha}$, $\alpha \in [-\pi, \pi)$. A function $f : U(1) \to \mathbb{C}$ is a 2π-periodic function of the real phase α, and its average is given by the integral

$$\mathcal{M}(f) = \frac{1}{2\pi} \int\limits_{-\pi}^{\pi} d\alpha \, f\left(e^{i\alpha}\right) \,. \tag{14.37}$$

For an Abelian group, left-multiplication and right-multiplication are identical. The invariance of the measure is easily proved,

$$\frac{1}{2\pi} \int\limits_{-\pi}^{\pi} d\alpha f\left(\Omega e^{i\alpha}\right) \overset{\Omega = e^{i\beta}}{=} \frac{1}{2\pi} \int\limits_{-\pi}^{\pi} d\alpha f\left(e^{i\beta + i\alpha}\right) = \frac{1}{2\pi} \int\limits_{-\pi}^{\pi} d\alpha f\left(e^{i\alpha}\right) \,.$$

Haar Measure of SU(2)
First we study the geometric meaning of left- and right-translations on the group.
To that aim we use the following bijective parametrization of the group elements:

$$
\alpha \longrightarrow U(\alpha) = \begin{pmatrix} \alpha_1 + i\alpha_2 & \alpha_3 + i\alpha_4 \\ -\alpha_3 + i\alpha_4 & \alpha_1 - i\alpha_2 \end{pmatrix} \quad \text{with} \quad \alpha = \begin{pmatrix} \alpha_1 \\ \alpha_2 \\ \alpha_3 \\ \alpha_4 \end{pmatrix} \in S^3 . \tag{14.38}
$$

By definition a Lie group is a differentiable manifold, and we see here explicitly that
the group SU(2) is identified with the sphere S^3. A short calculation reveals that the
left-translation $U \to \Omega U$ with $\Omega = U(\beta)$ is given by

$$
U(\beta)U(\alpha) = U\big(O(\beta)\alpha\big), \quad O(\beta)\alpha = \begin{pmatrix} \beta_1 & -\beta_2 & -\beta_3 & -\beta_4 \\ \beta_2 & \beta_1 & -\beta_4 & \beta_3 \\ \beta_3 & \beta_4 & \beta_1 & -\beta_2 \\ \beta_4 & -\beta_3 & \beta_2 & \beta_1 \end{pmatrix} \begin{pmatrix} \alpha_1 \\ \alpha_2 \\ \alpha_3 \\ \alpha_4 \end{pmatrix} . \tag{14.39}
$$

Since β has unit norm, the four-dimensional matrix $O(\beta)$ represents a rotation, i.e.,
$O^T O = \mathbb{1}$. Clearly the volume form on S^3, inherited from the embedding space $\mathbb{R}^4$,
is rotationally invariant. A similar argument shows that the induced volume form is
also invariant under right-translations, and this proves that

$$
dU = \delta\big(\alpha^2 - 1\big)\, d\alpha_1 d\alpha_2 d\alpha_3 d\alpha_4 \tag{14.40}
$$

is the unique Haar measure on the group SU(2). Alternatively one may introduce
"spherical coordinates" on the group manifold S^3, given by

$$
\begin{pmatrix} \alpha_1 \\ \alpha_2 \\ \alpha_3 \\ \alpha_4 \end{pmatrix} = \begin{pmatrix} \cos\theta \\ \sin\theta \cos\psi \\ \sin\theta \sin\psi \cos\varphi \\ \sin\theta \sin\psi \sin\varphi \end{pmatrix} , \tag{14.41}
$$

and this leads to the following parametrization of the group elements,

$$
U(\theta, \psi, \varphi) = \begin{pmatrix} \cos\theta + i\sin\theta \cos\psi & \sin\theta \sin\psi e^{i\varphi} \\ -\sin\theta \sin\psi e^{-i\varphi} & \cos\theta - i\sin\theta \cos\psi \end{pmatrix} . \tag{14.42}
$$

The angles are restricted to the intervals

$$
0 < \theta < \pi, \quad 0 < \psi < \pi \quad \text{and} \quad 0 < \varphi < 2\pi . \tag{14.43}
$$

For the spherical coordinates, the Haar measure reads

$$dU = \frac{1}{2\pi^2} (\sin\theta)^2 \sin\psi \, d\theta d\psi d\varphi \ . \tag{14.44}$$

Haar Measures for General Lie Groups

For other Lie groups, the Haar measure is constructed as follows: first one (locally) parametrizes the n-dimensional group with n parameters $\{\alpha^1, \ldots, \alpha^n\} = \alpha$. Then $dU U^{-1}$ is a linear combination of the differentials $d\alpha^a$, where the coefficients belong to the Lie algebra g, and the line element

$$ds^2 = -\text{tr}\left(dU U^{-1} \, dU U^{-1}\right) = \text{tr}\left(\frac{\partial U^{-1}}{\partial \alpha^a} \frac{\partial U}{\partial \alpha^b}\right) d\alpha^a d\alpha^b = g_{ab} \, d\alpha^a d\alpha^b \tag{14.45}$$

defines a left- and right-invariant metric on the group. Note that unitary groups have anti-hermitian $U^{-1}dU$ such that the minus sign in (14.45) yields a metric with positive signature. Now the Haar measure is proportional to the volume form associated with the invariant metric,

$$dU = \text{const} \sqrt{g} \, d\alpha^1 \cdots d\alpha^n, \quad g = \det(g_{ab}) \ . \tag{14.46}$$

The constant is fixed by the normalization of the measure. As an example consider the parametrization (14.42) of SU(2). The invariant line element is

$$ds^2 = d\theta^2 + \sin^2\theta d\psi^2 + \sin^2\theta \sin^2\psi d\phi^2 \ , \tag{14.47}$$

and the corresponding normalized volume form is the Haar measure (14.44). One may use the *exponential map* to parametrize the group: an element near the identity may be written as the exponential of an element T of the *Lie algebra*,

$$U = e^{iT} = e^{i(\alpha^1 T_1 + \cdots + \alpha^n T_n)} = U(\alpha) \ . \tag{14.48}$$

It is convenient to choose a trace-orthogonal basis of the Lie algebra,

$$\text{tr} \, T_a T_b = \kappa \, \delta_{ab} \ . \tag{14.49}$$

The parameters $\{\alpha^1, \ldots, \alpha^n\}$ are continuous local coordinates of the Lie group. Now we introduce the one-parameter group

$$U(t) = e^{itT} \quad \text{with} \quad U(0) = \mathbb{1}, \quad U(1) = U \ . \tag{14.50}$$

The n elements of the Lie algebra may be written as

$$L_a(t) = -i\frac{\partial U(t)}{\partial \alpha^a} U^{-1}(t) .$$

They satisfy the simple differential equations

$$\frac{\mathrm{d}L_a(t)}{\mathrm{d}t} = T_a + \mathrm{i}[T, L_a(t)] . \tag{14.51}$$

If we rewrite the L_a as linear combinations of the base elements T_a, i.e., $L_a = L_a{}^b T_b$, then the coefficient matrix $L = (L_a{}^b)$ fulfills the simple differential equation

$$\dot{L}(t) = \mathbb{1} + L(t)X, \quad X = (X_a{}^b), \quad X_a{}^b = f_{ac}{}^b \alpha^c . \tag{14.52}$$

The structure constants $f_{ab}{}^c$ of the Lie algebra satisfy the relation

$$[T_a, T_b] = \mathrm{i}f_{ab}{}^c T_c , \tag{14.53}$$

which means that they are real and antisymmetric in a and b (for compact groups). Hence, the matrix X is antisymmetric as well. The solution to the differential equation (14.52) for the matrix function $L(t)$ reads

$$L(t) = \int_0^t e^{(t-t')X} = \frac{e^{tX} - 1}{X} . \tag{14.54}$$

We now may calculate the invariant metric tensor as introduced in (14.45) by virtue of

$$g_{ab} = \mathrm{tr}\, L_a L_b|_{t=1} = \kappa\,(LL^T)_{ab}|_{t=1} , \tag{14.55}$$

where κ emerges from the normalization of the base elements. This leads us to the invariant volume form (up to a multiplicative factor)

$$\mathrm{d}V = \sqrt{g} \prod_a \mathrm{d}\alpha^a \propto (\det LL^T)_{t=1}^{1/2} \prod_a \mathrm{d}\alpha^a \tag{14.56}$$

which is proportional to the Haar measure. The matrix LL^T appearing in (14.56) has the form

$$LL^T|_{t=0} = -\frac{1}{X^2}\left(e^X - \mathbb{1}\right)\left(e^{-X} - \mathbb{1}\right) = -4X^{-2}\sinh^2(X/2)$$

$$= -\prod_{n \neq 0}\left(1 + X^2/(2\pi n)^2\right) , \tag{14.57}$$

where we have used the Weierstrass product representation of the sinh function. In particular, for the gauge group SU(2) with $T = \alpha^a \sigma_a$ we obtain

$$
X = \begin{pmatrix} 0 & -\alpha_3 & \alpha_2 \\ \alpha_3 & 0 & -\alpha_1 \\ -\alpha_2 & \alpha_1 & 0 \end{pmatrix} \quad \text{and} \quad \mathrm{d}V \propto \frac{8}{|\boldsymbol{\alpha}|^3} \sin^3\left(\frac{|\boldsymbol{\alpha}|}{2}\right) \mathrm{d}^3\alpha \ . \tag{14.58}
$$

14.3.1 The Peter-Weyl Theorem

The Haar measure is used to define an *invariant* scalar product for functions $G \to \mathbb{C}$,

$$
(f, g) \equiv \int_G \bar{f}(U)g(U)\,\mathrm{d}U \ . \tag{14.59}
$$

One obtains an orthonormal basis of the Hilbert space $L_2(G, \mathrm{d}U)$ of square integrable functions by considering all irreducible representations of the group G. We recall that a *representation* R of G is a homomorphism from the group to the group of invertible linear maps $V \to V$. This means that a representation preserves the *structure* of the group:

$$
R: G \longrightarrow L(V), \qquad R(U_1 U_2) = R(U_1)R(U_2), \qquad R(\mathbb{1}) = \mathbb{1} \ . \tag{14.60}
$$

For a fixed basis in V, we can identify a linear map with a matrix. Thus, a representation assigns to each group element U an invertible matrix such that the conditions (14.60) are satisfied. The dimension d_R of a representation R is given by the dimension of the vector space V. For example, the infinitely many representations of SU(2) are classified by the angular momentum j, and the dimension of the representation is $2j + 1$.

For a group with invariant measure every representation is equivalent to a unitary representation; hence, we may assume that the matrices R are unitary. A representation is called *irreducible* if the linear maps $\{R(U)|U \in G\}$ have no common *invariant* subspace in V, apart from the empty set and V itself. Let $\{R(U)\}$ denote the set of all irreducible representations. Then we have the following important theorem [11–14]:

Theorem 14.1 (Peter-Weyl Theorem) *The functions $\{R(U)^a{}_b\}$ define a complete orthogonal system of $L_2(\mathrm{d}U)$ with*

$$
\left(R^a{}_b, R'^c{}_d\right) \equiv \int \bar{R}^a{}_b(U)R'^c{}_d(U)\,\mathrm{d}U = \frac{\delta_{RR'}}{d_R}\,\delta^{ac}\delta_{bd} \ , \tag{14.61}
$$

where $d_R = \mathrm{tr}\, R(\mathbb{1})$ denotes the dimension of the representation R.

This theorem provides a generalization of the Fourier analysis of functions on the unit circle to functions on groups. A useful consequence of this theorem is

Lemma 14.1 *The characters $\chi_R(U) = \text{tr } R(U)$ of the irreducible representations form a orthogonal basis of the space of invariant functions $f(U) = f(\Omega U \Omega^{-1})$ in $L_2(dU)$. In particular, we have $(\chi_R, \chi_{R'}) = \delta_{RR'}$.*

This lemma can be used to decompose a reducible representation into its irreducible parts. In the strong-coupling expansions and the exact solutions of two-dimensional gauge theories, we shall need further useful identities which are the content of

Lemma 14.2 *The following identities hold:*

$$\text{orthogonality:} \quad \left(R^a_b, \chi_{R'}\right) = \frac{\delta_{RR'}}{d_R} \delta^a_b \,, \tag{14.62}$$

$$\text{gluing:} \quad \int d\Omega \, \chi_R(U\Omega^{-1})\chi_{R'}(\Omega V) = \frac{\delta_{RR'}}{d_R}\chi_R(UV) \,, \tag{14.63}$$

$$\text{separation:} \quad \int d\Omega \, \chi_R(\Omega U\Omega^{-1}V) = \frac{1}{d_R}\chi_R(U)\chi_R(V) \,, \tag{14.64}$$

$$\text{decomposition of } \mathbb{1}: \quad \sum_R d_R \, \chi_R(U) = \delta(\mathbb{1}, U) \,. \tag{14.65}$$

For example, the gluing property is proven quite easily:

$$\int d\Omega \, \chi_R(U\Omega^{-1})\chi_{R'}(\Omega V) = \sum_{a,b,c,d} R^a_b(U) \int d\Omega \, \bar{R}_a{}^b(\Omega)R'^c_d(\Omega)R'^d_c(V)$$

$$= \sum_{a,b,c,d} R^a_b(U) R'^d_c(V)\frac{\delta_{RR'}}{d_R}\delta^c_a\delta^b_d = \frac{\delta_{R,R'}}{d_R}\chi_R(UV) \,.$$

The decomposition of the identity follows from the orthogonality relation (14.61). Proofs and further relations are found in the rich literature on groups and representations, for example in [11–14].

14.4   Problems

14.1 (Solution of Z_2 Gauge Theories in Two Dimensions)   Calculate the partition function of the Z_2 gauge theory

$$Z_V(\beta) = \frac{1}{2^{|E|}} \sum_{\{U\}} e^{-\sum_p \beta(1-U_p)} = \frac{1}{2^{|E|}} \left(\frac{\cosh\beta}{e^\beta}\right)^V \sum_{\{U\}} \prod_p (1 + \tanh\beta U_p)$$

without gauge fixing. Thereby U_p denote a plaquette variable, and the link variables take the values $U_\ell \in \{-1, 1\}$. Expand the product, and use

$$\sum_U U^k = \begin{cases} 2 \; k \text{ odd} \\ 0 \; k \text{ even.} \end{cases}$$

The solution is very similar to that of the Ising spin chain.

14.2 (Peter-Weyl Theorem and Fourier Analysis) Apply the Peter-Weyl theorem to the group U(1), and show that it reduces to a well-known property of the Fourier series for periodic functions on an interval.

References

1. A.A. Migdal, Phase transitions in gauge and spin-lattice systems. JETP **69**, 1457 (1975)
2. K.R. Ito, Analytic study of the Migdal-Kadanoff recursion formula. Comm. Math. Phys. **93**, 379 (1984)
3. J.M. Drouffe, J.B. Zuber, Strong coupling and mean field methods in lattice gauge theories. Phys. Rep. **102**, 1–119 (1983) 1
4. H.J. Rothe, *Lattice Gauge Theories: An Introduction* (World Scientific, Singapore, 2012)
5. H.G. Dosch, V.V. Müller, Lattice gauge theory in two spacetime dimensions. Fort. der Physik **27**, 547 (1979)
6. G. Bali, Casimir scaling of SU(3) static potentials. Phys. Rev. **D62**, 114503 (2000)
7. O. Philipsen, H. Wittig, String breaking in non-Abelian gauge theories with fundamental matter fields. Phys. Rev. Lett. **81**, 4059 (1998)
8. M. Pepe, U.-J. Wiese, From decay to complete breaking: pulling the strings in SU(2) Yang-Mills theory. Phys. Rev. Lett. **102**, 191601 (2009)
9. B. Wellegehausen, A. Wipf, C. Wozar, Casimir scaling and string breaking in G2 gluodynamics. Phys. Rev. **D83**, 016001 (2011)
10. D.J. Gross, E. Witten, Possible third order phase transition in the large N lattice gauge theory. Phys. Rev. **D21**, 446 (1980)
11. T. Bröcker, T. tom Dieck, *Representations of Compact Lie Groups*. Graduate Texts in Physics (Springer, Berlin, 2010)
12. G. Segal, Lie groups, in *Lectures on Lie Groups and Lie Algebras* (Cambridge University Press, Cambridge, 1995)
13. C. Chevalley, *Theory of Lie Groups*. Dover Books on Mathematics (Dover, Mineola, 2018)
14. M. Hammermesh, *Group Theory and Its Application to Physical Problems* (Dover, New York, 2003)

Fermions on a Lattice

15

In the previous chapters we considered quantum field theories for bosons with spin 0 and spin 1 and discussed the regularization of these theories on a spacetime lattice. But all fundamental microscopic theories of nature contain both bosonic and fermionic fields. Hence, it remains to put fermions with spin 1/2 onto a lattice. Electrons, muons, or quarks are all described by a four-component spinor field $\psi(x) \in \mathbb{C}^4$. The corresponding quantum field $\hat{\psi}(x)$ creates and annihilates the particles together with their anti-particles which have identical mass but opposite charges. In this chapter we briefly recall the basic properties of a Dirac field in Euclidean space. By using anticommuting Grassmann variables, we formulate the path integral for Fermi fields. With the most naive approach, we encounter the species-doubling phenomenon—the fact that a naively discretized Dirac field leads to more excitations than expected. We discuss various proposals to discretize fermion fields; these include Wilson fermions, staggered fermions, and Ginsparg-Wilson fermions. Towards the end we shall comment on problems with formulating supersymmetric systems on a lattice. A discussion of fermions on a lattice is contained in several textbooks; see [1–7].

15.1 Dirac Equation

We assume that the reader has basic knowledge of the Dirac theory, and hence we will not present many details regarding particular properties of the Dirac equation and its solutions. In particular we only discuss the internal and spacetime symmetries in Euclidean space. We begin with the relativistic wave equation of a spinor field ψ, the *Dirac equation*, in Minkowski spacetime,

$$\left(i\slashed{\partial} - m\right)\psi(x) = 0, \quad \slashed{\partial} = \gamma^\mu \partial_\mu, \quad \{\gamma^\mu, \gamma^\nu\} = 2\eta^{\mu\nu}\mathbb{1}, \tag{15.1}$$

© The Author(s), under exclusive license to Springer Nature Switzerland AG 2021
A. Wipf, *Statistical Approach to Quantum Field Theory*, Lecture Notes
in Physics 992, https://doi.org/10.1007/978-3-030-83263-6_15

where $(\eta_{\mu\nu})$ denotes the metric tensor diag$(1, -1, -1, -1)$ and $\gamma^0, \ldots, \gamma^3$ are the four-dimensional gamma matrices that satisfy the relation in (15.1). In the following we shall consider Euclidean Fermi fields. The gamma matrices in Euclidean space γ_E^μ are related to those in Minkowski spacetime according to $\gamma_E^0 = \gamma^0$ and $\gamma_E^i = i\gamma^i$. They satisfy anticommutation rules with the Euclidean metric,

$$\{\gamma_E^\mu, \gamma_E^\nu\} = 2\delta^{\mu\nu}\mathbb{1} \ . \tag{15.2}$$

Since we are dealing with the Euclidean theory, we omit the index E in what follows. The Euclidean gamma matrices are hermitian

$$\gamma_\mu^\dagger = \gamma_\mu = \gamma^\mu \ , \tag{15.3}$$

and the Euclidean Dirac equation reads

$$D\psi(x) \equiv \left(\slashed{\partial} + m\right)\psi(x) = 0 \ , \tag{15.4}$$

with an anti-hermitian operator $\slashed{\partial}$. Lorentz transformations in Minkowski spacetime turn into rotations in four-dimensional Euclidean space, and a spinor field transforms under these rotations according to $\psi(x') = S\psi(x)$ with $S^\dagger S = \mathbb{1}$.[1] Here S is a spin transformation from the covering group of the rotation group SO(4). The bilinear $\bar{\psi}\psi$ is *not* invariant under "Lorentz transformations" in Euclidean space if we stick to the definition $\bar{\psi} = \psi^\dagger\gamma^0$, valid in Minkowski spacetime. In Euclidean space $\bar{\psi}$ should be considered as $\psi^\dagger$ when it comes to spin transformations. The hermitian matrix

$$\gamma_5 = \gamma^0\gamma^1\gamma^2\gamma^3 = \gamma_5^\dagger \tag{15.5}$$

anticommutes with all gamma matrices,

$$\{\gamma_5, \gamma^\mu\} = 0 \quad \text{and} \quad \gamma_5^2 = \mathbb{1} \ , \tag{15.6}$$

and has double degenerate eigenvalues ± 1. The Dirac operator D fulfills the relation

$$\gamma_5 D\gamma_5 = D^\dagger \tag{15.7}$$

which implies that all non-real eigenvalues appear in complex-conjugate pairs and that the determinant of D is real. This non-hermitian Dirac operator is also obtained by a careful derivation of the path-integral representation of the partition function.

---

[1] Recall that in Minkowski spacetime $S^\dagger = \gamma^0 S^{-1}\gamma^0$.

The Dirac equation (15.4) is the Euler-Lagrange equation derived from the action

$$S_F = \int d^4x \left(\frac{1}{2} \left(\bar{\psi}(x) \gamma^\mu \partial_\mu \psi(x) - \partial_\mu \bar{\psi}(x) \gamma^\mu \psi(x) \right) + m \bar{\psi}(x) \psi(x) \right) .$$

$$(15.8)$$

Up to a surface term, it can be written as

$$S_F = \int d^4x \, \bar{\psi}(x) D \psi(x) , \qquad (15.9)$$

and the latter form is most commonly used in the literature.

15.1.1 Coupling to Gauge Fields

Realistic theories contain several Dirac fields, and these fields combine to a field ψ with values in $V \otimes \mathbb{C}^4$. Here V is a vector space equipped with a G-invariant scalar product, i.e., $(\chi, \psi) = (\Omega \chi, \Omega \psi)$ for all $\Omega \in G$. Expanding χ and ψ in an orthonormal basis of V the scalar product reads $(\chi, \psi) = \sum_a \bar{\chi}_a \psi_a$, where each term is the Lorentz-invariant bilinear of the corresponding Dirac spinors. The linear transformation Ω only acts on the internal index a (and not on the spinor index) such that the Lagrangian density is invariant under global transformations,

$$\mathcal{L} = (\psi, D\psi) = (\psi', D\psi'), \quad \psi' = \Omega \psi, \quad \Omega \in G . \qquad (15.10)$$

Similarly as for a scalar field, we can promote the global symmetry to a local one by introducing a Lie-algebra valued gauge potential to define a covariant derivative. The resulting covariant Dirac operator reads

$$D = \slashed{D} + m, \quad \slashed{D} = \gamma^\mu D_\mu, \quad D_\mu = \partial_\mu - ig A_\mu , \qquad (15.11)$$

and the Lagrangian (15.9) with this operator is invariant under local gauge transformations $(\psi, A_\mu) \rightarrow (\psi', A'_\mu)$. If there is only one Dirac field, the gauge transformations are local $U(1)$ transformations,

$$\psi'(x) = e^{ig\lambda(x)} \psi(x), \quad \bar{\psi}'(x) = \bar{\psi}(x) e^{-ig\lambda(x)}, \quad A'_\mu(x) = A_\mu(x) + \partial_\mu \lambda(x) . \qquad (15.12)$$

In the chiral limit $m = 0$, the Lagrangian (15.9) for one Dirac field in addition is invariant under global chiral transformations,

$$\psi(x) \longrightarrow e^{\gamma_5 \alpha} \psi(x) \quad , \quad \bar{\psi}(x) \longrightarrow \bar{\psi}(x) e^{\alpha \gamma_5}, \qquad \alpha \in \mathbb{R} . \qquad (15.13)$$

In Euclidean space the chiral transformations form the *non-compact* group $\mathbb{R}_+$, whereas in Minkowski spacetime they form the compact group $U(1)$. The chiral symmetry group is enlarged for a multiplet of Dirac fields. In Chap. 17 we shall study in great detail the symmetries of Fermi systems in arbitrary dimensions.

15.2 Grassmann Variables

To motivate why we need anticommuting variables in the path-integral quantization of theories with fermions, we briefly return to bosonic fields. A scalar field entering the path integral is a real and commuting function,

$$[\phi(x), \phi(y)] = 0 \, . \tag{15.14}$$

This property may be regarded as the limiting case $\hbar \to 0$ of the commutation rules for the quantum field $\hat{\phi}$. A fermionic quantum field, on the other hand, must satisfy the equal-time anticommutation relations

$$\left\{ \hat{\psi}(t, \boldsymbol{x}), \hat{\psi}(t, \boldsymbol{y}) \right\} = 0, \quad \boldsymbol{x} \neq \boldsymbol{y} \, ,$$

so that the Pauli exclusion principle and Fermi–Dirac statistics are fulfilled. This serves as motivation for using an anticommuting field,

$$\{\psi(x), \psi(y)\} = 0, \quad \forall x, y \, , \tag{15.15}$$

in the fermionic path integral. We may view an anticommuting field as classical limit of a fermionic quantum field. Classical bosonic fields are built with commuting c-numbers, whereas "classical fermionic fields" are built with anticommuting Grassmann variables. We refer the reader interested in details to the literature [8]. The objects $\{\eta_i, \bar{\eta}_i\}$ form a complex *Grassmann algebra* if

$$\{\eta_i, \eta_j\} = \{\bar{\eta}_i, \bar{\eta}_j\} = \{\eta_i, \bar{\eta}_j\} = 0, \quad i, j = 1, 2, \ldots, n \tag{15.16}$$

holds. Note that the square of a Grassmann variable is zero. For the path integral, we need to integrate over functions of Grassmann variables. The integration is a linear map and is defined by the rules

$$\int \mathrm{d}\eta_i (a + b\eta_i) = b, \quad \int \mathrm{d}\bar{\eta}_i (a + b\bar{\eta}_i) = b \tag{15.17}$$

with arbitrary complex numbers a, b. The integration is invariant under a translation of the Grassmann variables.

15.2.1 Gaussian Integrals

Most path integrals for fermionic systems lead to Gaussian integrals of the form

$$Z = \int \mathscr{D}\bar{\eta}\mathscr{D}\eta \, e^{-\bar{\eta}A\eta}, \quad \bar{\eta}A\eta = \sum_{ij} \bar{\eta}_i A_{ij} \eta_j \,, \tag{15.18}$$

where the integration extends over all Grassmann variables

$$\mathscr{D}\bar{\eta}\mathscr{D}\eta = \prod d\bar{\eta}_i d\eta_i \,. \tag{15.19}$$

The Gaussian integral can be evaluated by an expansion of the exponential function. Due to the integration rules (15.17), the only non-vanishing contribution is given by

$$\frac{(-1)^n}{n!} \int \mathscr{D}\bar{\eta}\mathscr{D}\eta \, (\bar{\eta}A\eta)^n = (-1)^n \int \mathscr{D}\bar{\eta}\mathscr{D}\eta \sum_{i_1,\dots,i_n} \left(\bar{\eta}_1 A_{1i_1}\eta_{i_1}\right) \cdots \left(\bar{\eta}_n A_{ni_n}\eta_{i_n}\right)$$

$$= (-1)^n \int \mathscr{D}\bar{\eta}\mathscr{D}\eta \prod_i (\bar{\eta}_i \eta_i) \sum_{i_1,\dots,i_n} \varepsilon_{i_1\dots i_n} A_{1i_1} \cdots A_{ni_n}$$

$$= \int \prod_i (d\bar{\eta}_i \bar{\eta}_i \, d\eta_i \eta_i) \det A = \det A \,.$$

Thus, we end up with the simple formula

$$Z = \int \mathscr{D}\bar{\eta}\mathscr{D}\eta \, e^{-\bar{\eta}A\eta} = \det A \,. \tag{15.20}$$

Next we consider the slightly more general Gaussian integral

$$Z(\bar{\alpha}, \alpha) = \int \mathscr{D}\bar{\eta}\mathscr{D}\eta \, e^{-\bar{\eta}A\eta + \bar{\alpha}\eta + \bar{\eta}\alpha} \tag{15.21}$$

with Grassmann-valued sources $\alpha = (\alpha_1, \dots, \alpha_n)$ and $\bar{\alpha} = (\bar{\alpha}_1, \dots, \bar{\alpha}_n)$. We used the abbreviation $\bar{\eta}\alpha = \sum \bar{\eta}_i \alpha_i$. Shifting the integration variables $\bar{\eta}_i$, η_i according to

$$\eta \to \eta + A^{-1}\alpha \quad \text{and} \quad \bar{\eta} \to \bar{\eta} + \bar{\alpha}A^{-1}$$

and using the translational invariance of the integration (15.17), we obtain the following generalization of (15.20):

$$Z(\bar{\alpha}, \alpha) = \int \mathscr{D}\bar{\eta}\mathscr{D}\eta \, e^{-\bar{\eta}A\eta + \bar{\alpha}\eta + \bar{\eta}\alpha} = e^{-\bar{\alpha}A^{-1}\alpha} \det A \,. \tag{15.22}$$

Expanding both sides in powers of $\bar{\alpha}$, α, we find the useful formula

$$\langle \bar{\eta}_i \eta_j \rangle \equiv \frac{1}{Z} \int \mathscr{D}\bar{\eta}\mathscr{D}\eta \, e^{-\bar{\eta}A\eta} \, \bar{\eta}_i \eta_j = (A^{-1})_{ij} \, . \tag{15.23}$$

The integral is zero when the number of variables $\bar{\eta}$ and η do not match,

$$\langle \bar{\eta}_{i_1} \cdots \bar{\eta}_{i_n} \eta_{j_1} \cdots \eta_{j_m} \rangle = 0 \quad \text{if} \quad m \neq n \, . \tag{15.24}$$

In passing we note that path integrals for Majorana fermions lead to Gaussian integrals of the form

$$\int \mathscr{D}\eta \, e^{\frac{1}{2}\eta^t M \eta} = \text{Pf}(M) \quad \mathscr{D}\eta = \prod \text{d}\eta_a \, , \tag{15.25}$$

where $\eta = (\eta_1, \dots, \eta_{2N})$ denotes an even number of real Grassmann variables with anticommutation rules

$$\{\eta_a, \eta_b\} = 0 \, , \tag{15.26}$$

and M is a real and antisymmetric matrix. Up to a sign the Pfaffian $\text{Pf}(M)$ is the square root of $\det(M)$. Further properties of Grassmann integrals for real anticommuting parameters are discussed in problem 15.1.

15.2.2 Path Integral for Dirac Theory

After these algebraic preliminaries, we return to quantum field theory. An anticommuting field assigns several Grassmann variables to each spacetime point. For a Dirac field in four dimensions, these are the anticommuting variables $\{\psi_\alpha(x), \bar{\psi}_\alpha(x)\}$, where the spinor index α takes the values 1, 2, 3, 4. The "classical" Dirac field satisfies

$$\{\psi_\alpha(x), \psi_\beta(y)\} = \{\bar{\psi}_\alpha(x), \bar{\psi}_\beta(y)\} = \{\psi_\alpha(x), \bar{\psi}_\beta(y)\} = 0 \, . \tag{15.27}$$

The functional integration over the fermion field is the (formal) Grassmann integral

$$\int \mathscr{D}\psi\mathscr{D}\bar{\psi} \, \cdots \equiv \int \prod_x \prod_\alpha \text{d}\psi_\alpha(x) \, \text{d}\bar{\psi}_\alpha(x) \, \dots \, , \tag{15.28}$$

and the expectation value of an observable $\hat{A}$ is given by the functional integral

$$\langle 0|\hat{A}|0\rangle = \frac{1}{Z_F} \int \mathscr{D}\psi\mathscr{D}\bar{\psi} \, A(\bar{\psi}, \psi) \, e^{-S_F(\psi, \bar{\psi})} \, , \tag{15.29}$$

normalized by the partition function

$$Z_F = \int \mathscr{D}\psi \mathscr{D}\bar{\psi}\, e^{-S_F} \, . \tag{15.30}$$

The integrands contain the classical action S_F for the fermion field. Most physically relevant theories have a bilinear action

$$S_F = \int d^d x\, \mathscr{L}(\psi, \bar{\psi}), \quad \mathscr{L} = \bar{\psi}(x)D\psi(x) \, , \tag{15.31}$$

which contains the Dirac operator D. Exceptions are the Thirring, Gross-Neveu, and supergravity models which contain terms that are quartic in the Fermi fields. Four-Fermi theories are studied in great detail in Chap. 17.

Applying (15.20) we can calculate the partition function of a theory with bilinear action (15.31). Formally it is just the determinant of the Dirac operator,

$$Z_F = \int \mathscr{D}\psi \mathscr{D}\bar{\psi} \exp\left(-\int d^d x\, \bar{\psi}(x)D\psi(x)\right) = \det D \, . \tag{15.32}$$

Recall that the corresponding formula for a *complex* scalar fields reads

$$Z_B = \int \mathscr{D}\phi \mathscr{D}\bar{\phi} \exp\left(-\int d^d x\, \bar{\phi}(x)A\phi(x)\right) = \frac{1}{\det A} \, . \tag{15.33}$$

There are interesting field theories with an additional supersymmetry for which the contributions of the bosons and fermions to the partition function cancel. For these theories the bosonic operator A and the fermionic operator D are related. In a lattice regularization, the Dirac operator becomes a huge matrix, and one of the main difficulties in MC simulations is to handle this determinant and the inverse of D.

15.3 Fermion Fields on a Lattice

To obtain a lattice regularization for fermionic systems, we proceed similarly as for scalar field theories by substituting differentials by differences. In contrast to the bosonic fields, the most naive discretization is afflicted with the doubling problem, and we shall discuss how to deal with this problem. In many cases dimensionful parameters are measured in units defined by the lattice spacing a.

15.3.1 Lattice Derivative

The discretization of differential operators within fermionic systems is a subtle point since the field equations contain the first-order Dirac operator D. According to

(15.23) the two-point function (before averaging over the other fields) reads

$$G_F(x, y) = \langle \psi(x)\bar{\psi}(y) \rangle \equiv G_F(x - y) = \langle x \,|\, \frac{1}{D} \,|\, y \rangle, \qquad (15.34)$$

and the inverse of the Dirac operator depends on the discretization. We assume that the lattice Dirac operator is γ_5-symmetric

$$\gamma_5 D \gamma_5 = D^\dagger, \qquad (15.35)$$

such that its eigenvalues come in complex conjugated pairs $\{\lambda, \lambda^*\}$ and its determinant is real. To prove this proposition, we consider the characteristic polynomial of the matrix D on a finite lattice,

$$P(\lambda) \equiv \det(\lambda - D) = \det \gamma_5 (\lambda - D)\gamma_5 = \det\left(\lambda - D^\dagger\right) = P^*(\lambda^*). \qquad (15.36)$$

This means that if λ is a root of the characteristic polynomial, then λ^* is a root as well, and this proves the proposition. A Dirac operator $D = \slashed{\partial} + m + \mathcal{O}$ is γ_5-hermitian if the lattice derivatives are anti-hermitian and the operator $\mathcal{O}$ is hermitian with respect to the ℓ_2-scalar product on the space of lattice functions and also commutes with γ_5,

$$\gamma_5 D \gamma_5 = \gamma_5 \left(\slashed{\partial} + m + \mathcal{O}\right)\gamma_5 = -\slashed{\partial} + m + \mathcal{O} = \slashed{\partial}_\mu^\dagger + m + \mathcal{O}^\dagger = D^\dagger, \qquad (15.37)$$

where we used that the γ^μ anticommute with γ_5.

Forward and Backward Derivatives

The frequently used nearest-neighbor forward and backward derivatives

$$(\hat{\partial}_\mu f)(x) = f(x + e_\mu) - f(x), \qquad (\hat{\partial}'_\mu f)(x) = f(x) - f(x - e_\mu) \qquad (15.38)$$

are not anti-hermitian with respect to the scalar product $(f, g) = \sum_x \bar{f}(x)g(x)$. In fact for periodic boundary conditions, we have $\hat{\partial}_\mu^\dagger = -\hat{\partial}'_\mu$. Both derivatives define circulant matrices and thus commute with each other,

$$[\hat{\partial}_\mu, \hat{\partial}_\nu] = [\hat{\partial}'_\mu, \hat{\partial}'_\nu] = [\hat{\partial}_\mu, \hat{\partial}'_\nu] = 0. \qquad (15.39)$$

Plane waves on the periodic lattice

$$\varphi_p = \frac{1}{\sqrt{V}} e^{ipx}, \qquad p_\mu = \frac{2\pi}{N} n_\mu, \qquad n_\mu \in \mathbb{Z}_N \qquad (15.40)$$

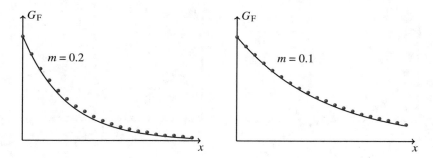

Fig. 15.1 The fermionic Green function for the forward derivative on the one-dimensional lattice with $N = 20$ lattice sites as well as its interpolating exponential function

are simultaneous eigenfunctions of the derivative operators,

$$\hat{\partial}_\mu \varphi_p = i\hat{p}_\mu e^{ip_\mu/2} \varphi_p, \quad \hat{\partial}'_\mu \varphi_p = i\hat{p}_\mu e^{-ip_\mu/2} \varphi_p, \quad \hat{p}_\mu = 2\sin\frac{p_\mu}{2} . \quad (15.41)$$

In particular the inverse of $\hat{\partial} + m$ on a one-dimensional lattice reads

$$\left\langle x \left| \frac{1}{\hat{\partial} + m} \right| y \right\rangle = \frac{1}{N} \sum_p \frac{e^{ip(x-y)}}{m + ie^{ip/2}\hat{p}} \xrightarrow{N \to \infty} \frac{1}{2\pi} \int_{-\pi}^{\pi} dp \, \frac{e^{ip(x-y)}}{m + ie^{ip/2}\hat{p}} . \quad (15.42)$$

This Green function is very well approximated by an exponential fit for small bare masses $m \lesssim 0.2$ or equivalently for large correlation lengths $\xi \gtrsim 5$. Figure 15.1 shows both the propagator and its exponential fit for the masses $m = 0.1$ and $m = 0.2$.

Antisymmetric Derivative

In place of the forward and backward derivatives, one may employ the *antisymmetric* discretization of ∂_μ, defined by

$$\mathring{\partial}_\mu = \tfrac{1}{2}\left(\hat{\partial}_\mu + \hat{\partial}'_\mu\right) \implies \left(\mathring{\partial}_\mu f\right)(x) = \frac{1}{2}\left(f(x + e_\mu) - f(x - e_\mu)\right) . \quad (15.43)$$

These commuting derivatives can be diagonalized simultaneously. The plane waves (15.40) are the eigenfunctions with eigenvalues

$$\mathring{\partial}_\mu \varphi_p(x) = i\mathring{p}_\mu \varphi_p(x), \quad \mathring{p}_\mu = \sin p_\mu , \quad (15.44)$$

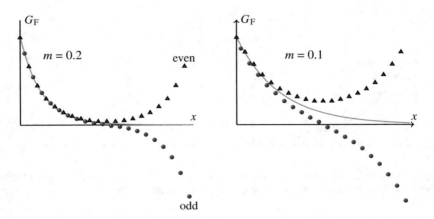

Fig. 15.2 The fermion Green function for the antisymmetric derivative on a one-dimensional lattice with $N = 40$

and we find the following Green function on the one-dimensional lattice:

$$\left\langle x \left| \frac{1}{\mathring{\partial} + m} \right| 0 \right\rangle = \frac{1}{N} \sum_p \frac{e^{ipx}}{m + i\mathring{p}} \xrightarrow{N \to \infty} \frac{1}{2\pi} \int_{-\pi}^{\pi} dp \, \frac{e^{ipx}}{m + i\mathring{p}} \, . \tag{15.45}$$

Figure 15.2 shows the Green function on the lattice with 40 sites. Note that the restriction of the propagator to even (odd) lattice sites defines an even (odd) lattice function. The functions on the two sublattices approach each other for small x, but they have opposite signs near $x = N$. This means that the lattice Green function with antisymmetric derivative oscillates with a large amplitude around the mean value 0 for $x \to N$. The figure also shows that the exponential function $\exp(-mx)$ fits the Green function rather well for $x \ll N$ and $5 \ll \xi \ll N/2$.

SLAC Derivative

Here we introduce yet another lattice derivative—the SLAC derivative [9, 10]—which can be used in fermionic systems without local gauge invariance.[2] This derivative yields the best values for the bound state energies of quantum mechanical Hamiltonians discretized on lattices of moderate sizes [13, 14]. In addition, simulations of supersymmetric Yukawa models show that in these models, observables approach their continuum values most rapidly when one employs the SLAC derivative [12, 15]. This lattice derivative has the interesting property of having exactly the same spectrum as the continuum derivative below the UV-cutoff. A particular simple choice of the SLAC derivative has the following matrix elements

---

[2] The problems with the SLAC derivative in gauge theories as discussed in [11] are absent in theories without local gauge invariance [12].

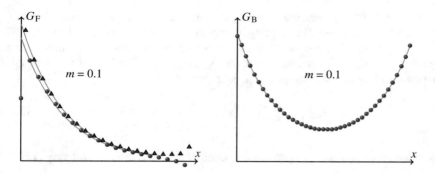

Fig. 15.3 The fermion and boson Green function on a chain with $N = 41$ sites, evaluated with the nonlocal SLAC derivative

in position space:

$$(\partial_{\mathrm{slac}})_{kk} = 0, \quad (\partial_{\mathrm{slac}})_{k \neq k'} = \frac{\pi}{N}(-)^{k-k'}\frac{1}{\sin(\pi t_{kk'})}, \quad t_{kk'} = \frac{k-k'}{N}. \quad (15.46)$$

Clearly, the SLAC derivative is antisymmetric. Further properties of this derivative are discussed in the Appendix to this chapter. Figure 15.3 shows the Green functions of the Dirac and Klein-Gordon operators

$$G_{\mathrm{F}}(x) = \langle x \,|\, \frac{1}{m + \partial_{\mathrm{slac}}} |0\rangle \quad \text{and} \quad G_{\mathrm{B}}(x) = \langle x \,|\, \frac{1}{m^2 - \partial_{\mathrm{slac}}^2} |0\rangle. \quad (15.47)$$

on a one-dimensional lattice. Near the origin the amplitude of the fermionic Green function overshoots since in momentum space the SLAC derivative jumps at the edge of the Brillouin zone. The overshooting is the well-known Gibbs phenomenon of Fourier transforms of discontinuous functions. The fits in Fig. 15.3 are normalized such that they match the propagator at $x = 2$ and $x = 3$, respectively. The interpolating function

$$G_{\mathrm{B}}(x) \sim \mathrm{const}\left(\mathrm{e}^{-mx} + \mathrm{e}^{-m(N-x)}\right)$$

approximates the bosonic Green function very well.

15.3.2 Naive Fermions on the Lattice

In this section we shall discuss various types of lattice fermions distinguished by the different discretization of the Dirac operator. Since the main focus is on the discretizations of $\partial\!\!\!/$, it is sufficient to consider free fermions. We do not specify the dimension of the Euclidean lattice. In d dimensions a Dirac spinor has $\Delta_f = 2^{[d/2]}$ components, where $[d/2]$ is the largest integer which is smaller or equal to $d/2$. For even d there exists a generalization of γ_5 which anticommutes with all γ^μ. If we use the forward (or backward) derivative $\hat{\partial}_\mu$ in the discretization of the continuum action (15.9) in the absence of gauge fields, we obtain

$$S_{\text{naive}} = \sum_x \bar{\psi}_x \hat{D} \psi_x, \quad \hat{D} = \gamma^\mu \hat{\partial}_\mu + m \ . \tag{15.48}$$

But in even dimensions, the operator $\hat{D}$ does not possess the γ_5-symmetry (15.7),

$$\gamma_5 \hat{D} \gamma_5 = -\hat{\slashed{\partial}} + m \neq D^\dagger \ ,$$

since the hermite conjugate of a forward derivative is the backward derivative. In addition, this implementation violates the cubic symmetry on a hypercubic lattice which is useful to recover the rotational $O(4)$ symmetry on large scales or in the continuum limit. Besides, the reflection hermiticity (the Euclidean counterpart of the hermiticity in Minkowski spacetime) is violated, and the theory in Minkowski spacetime lacks unitarity.

One may believe that the antisymmetric lattice derivative $\mathring{\partial}_\mu$ in (15.43) leads to an acceptable lattice Dirac operator since it is γ_5-hermitian. But this operator suffers from the *species doubling problem*. To see this more clearly, we calculate the eigenvalues, eigenfunctions, and Green function

$$G_{\text{F}}(x, y) = \langle \psi_x \bar{\psi}_y \rangle = \langle x | \frac{1}{\mathring{D}} | y \rangle, \quad \mathring{D} = \gamma^\mu \mathring{\partial}_\mu + m \tag{15.49}$$

with antisymmetric derivatives. First we observe that the lattice Laplacian $\mathring{\Delta}$ in

$$\mathring{D} \mathring{D}^\dagger = \left(\gamma^\mu \mathring{\partial}_\mu + m \right) \left(-\gamma^\mu \mathring{\partial}_\mu + m \right) = \left(-\mathring{\Delta} + m^2 \right) \mathbb{1} \tag{15.50}$$

only connects next-to-nearest neighbors on the lattice,

$$(\mathring{\Delta} f)(x) = \frac{1}{4} \sum_\mu \left(f(x + 2e_\mu) - 2f(x) + f(x - 2e_\mu) \right) \ . \tag{15.51}$$

The plane waves (15.40) are eigenvectors of $\mathring{\Delta}$ with eigenvalues

$$\mathring{p}^2, \quad \mathring{p}_\mu = \sin(p_\mu) \ , \tag{15.52}$$

and this spectrum causes the doubling problem. This becomes clear when we compare (15.51) with the standard lattice Laplacian

$$(\hat{\Delta} f)(x) = \left(\hat{\partial}'_\mu \hat{\partial}^\mu f \right)(x) = \sum_\mu \left(f(x + e_\mu) - 2f(x) + f(x - e_\mu) \right) \ , \tag{15.53}$$

Fig. 15.4 The spectra of two discretizations $\mathring{\Delta}$ and $\hat{\Delta}$ of the Laplace operator on a one-dimensional lattice. For a comparison we also plotted the dispersion relation of the continuum operator on the finite interval

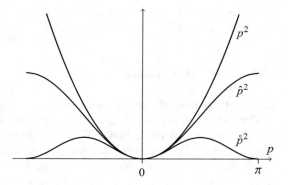

which connects nearest neighbors on the lattice and has the eigenvalues

$$\hat{p}^2, \qquad \hat{p}_\mu = 2\sin\left(\frac{p_\mu}{2}\right) . \tag{15.54}$$

Figure 15.4 shows the dispersion relations (15.52) and (15.54) together with the dispersion relation $p \to p^2$ of the continuum operator on the interval of "length" N. We see that for small p_μ, the three dispersion relations are identical and in particular that the constant function with zero momentum is an eigenfunction with eigenvalue zero (a zero mode) of all three operators. But for even N, the operator $\mathring{\Delta}$ has not only one but 2^d zero modes[3] with momenta in the first Brillouin zone,

$$p = (p_0, \ldots, p_{d-1}) \quad \text{and} \quad p_\mu \in \{0, \pi\} . \tag{15.55}$$

The Green function G_F of the naive lattice Dirac operator is

$$G_F(x, y) = \frac{1}{V}\sum_p G_F(p)\,e^{ip(x-y)}, \qquad G_F(p) = \frac{1}{i\gamma^\mu \mathring{p}_\mu + m} . \tag{15.56}$$

In the thermodynamic limit, the sum over the discrete momenta in (15.56) turns into a Riemann integral over the Brillouin zone,

$$G_F(x, y) \overset{N\to\infty}{\longrightarrow} \frac{1}{(2\pi)^d}\int_B d^4p\, G_F(p) . \tag{15.57}$$

For small momenta the propagator $G_F(p)$ reproduces the correct continuum propagator $G_F(p) \propto (i\gamma^\mu p_\mu + m)^{-1}$ and for massless fermions has a pole at $p = 0$. The problem is, however, that $G_F(p)$ has not just one but 2^d poles in the Brillouin

---

[3] Strictly speaking there is only one zero mode for odd N. But in the thermodynamic limit, the momenta on the edge of the Brillouin zone again give rise to zero modes.

zone. Thus, the most naive discretization of the continuum theory leads to a lattice discretization with 2^d fermionic species.

15.3.3 Wilson Fermions

K. Wilson has been aware of the doubling problem since the starting years of lattice field theories and suggested a modification of the action in order to get rid of the doublers in the continuum limit [16]. He added a particular momentum-dependent mass term—the Wilson term—to the naive action. This contribution gives a mass to the doublers. In the continuum limit, the doublers become infinitely heavy and thus unobservable. Unfortunately at the same time, the Wilson term breaks the chiral symmetry of massless theories explicitly. In more detail, on a lattice with lattice spacing a, the modified new action reads

$$S_{\rm w} = S_{\rm naive} - \frac{r}{2} \sum_x \bar{\psi}_x \, a\hat{\Delta}\psi_x = \sum_x \bar{\psi}_x D_{\rm w}\psi_x \,, \tag{15.58}$$

where the Wilson parameter r in the modified Dirac operator

$$D_{\rm w} = \mathring{D} - \frac{ar}{2}\hat{\Delta} \tag{15.59}$$

has values in the interval $(0, 1]$. For positive r the Wilson term proportional to r acts like a momentum-dependent mass such that in the massless limit, $D_{\rm w}$ does not anticommute with γ_5. This means that the chiral symmetry is explicitly *broken* by the Wilson term. In even dimensions $D_{\rm w}$ is γ_5-hermitian. Without gauge fields it has the eigenvalues

$$\lambda_p = \left(m + \frac{ar}{2}\,\hat{p}^2\right) \pm {\rm i}|\mathring{p}| \,, \tag{15.60}$$

where we recall the definitions of $\hat{p}_\mu$ and $\mathring{p}_\mu$,

$$\hat{p}_\mu = \frac{2}{a}\sin\left(\frac{ap_\mu}{2}\right), \quad \mathring{p}_\mu = \frac{1}{a}\sin(ap_\mu), \quad p_\mu \in \frac{2\pi}{a}\frac{n_\mu}{N} \,. \tag{15.61}$$

To localize the eigenvalues of $D_{\rm w}$ in the complex plane, it is convenient to use the parameters $t_\mu = -\cos(ap_\mu) \in [-1, 1]$. In the thermodynamic limit, the $\{p_\mu\}$ vary smoothly in the Brillouin zone such that the t_μ define a d-dimensional cube. The map $\{t_\mu\} \to \lambda(t)$ defined by (15.60) maps the edges of this cube into d ellipses with semi-major axes r and 1 and equidistant centers on the real axis,

$$\{(m + pr, 0)\}, \quad p = 1, 3, \ldots, 2d - 1 \,. \tag{15.62}$$

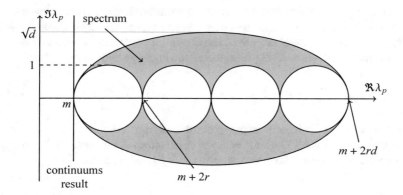

Fig. 15.5 The complex eigenvalues of the free massive Wilson operator D_{w} in the thermodynamic limit define the shaded region. Depicted is the spectrum for Wilson parameter $r = 1$ for which the inner ellipses become circles

Here we set $a = 1$ which means that we used a to set the scale. The ellipses are symmetrical to the real axis, touch each other on the real axis, and form the inner boundary of the spectrum of the Wilson operator, i.e., the set of eigenvalues of D_{w}. Next we observe that the momenta with equal coordinates $t_\mu = t$ are mapped to an ellipse with semi-major axes rd and $d^{1/2}$ and center at $(m + rd, 0)$. This ellipse encloses the smaller ellipses and forms the outer boundary of the spectrum.

Thus, the shaded area in Fig. 15.5 represents the spectrum of the *four*-dimensional Wilson operator in the thermodynamic limit. For $r \to 0$ we recover the spectrum of the naive Dirac operator $\overset{\circ}{D}$, i.e., the interval connecting $m - \mathrm{i} d^{1/2}$ and $m + \mathrm{i} d^{1/2}$. In the massless limit, the eigenvalues on the real axis at $0, 2r, 4r, \ldots$ become the annoying doublers of the naive Dirac operator when $r \to 0$. In one dimension there is only one inner ellipse, and this ellipse coincides with the outer one. Hence, all eigenvalues are located on the ellipse centered at $m + r$ with semi-major axes r and 1. The value $r = 1$ corresponds to the backward derivative, the value $r = -1$ to the forward derivative, and the value $r = 0$ to the antisymmetric derivative. In the latter case all eigenvalues are located on the interval $m + \mathrm{i}[-1, 1]$. In the limit $r \to 0$, the eigenvalues $m + 2r$ and m are degenerate, and we recover the species-doubling problem.

To investigate the naive continuum limit of the Wilson operator, we momentarily reinstall the lattice spacing a in the mass, eigenvalues, and momenta. With decreasing a the center of the outer ellipse at $(m + rd/a, 0)$ moves away from the origin, and at the same time the semi-major axes rd/a and $d^{1/2}/a$ become large. The same applies to the inner ellipses at

$$\left\{ \left(m + \frac{pr}{a}, 0 \right) \right\}, \quad p = 1, 3, \ldots, 2d - 1$$

with semi-major axes r/a and $1/a$. In the continuum limit, only the inner ellipse next to the imaginary axis and the outer ellipse remain, and they converge to the line $m \pm i|p|$ defining the spectrum of the continuum operator. More accurately, the dimensionful eigenvalues of D_w have the following expansion for small a:

$$\lambda_p = m + \frac{ar}{2}\,\mathring{p}^2 \pm i|\mathring{p}| = m \pm i|p| + \frac{1}{2}(ar)p^2 + O(a^2)\,. \tag{15.63}$$

Thus, for $a \rightarrow 0$ we recover the spectrum of the free Dirac operator in the continuum. For Wilson's choice $r = 1$, the eigenvalues contain errors of order a, to be compared with $O(a^2)$ for naive, staggered, or SLAC fermions.

15.3.4 Staggered Fermions

Staggered fermions are obtained from naive fermions by redistributing the spinor degrees of freedom across different lattice sites [17]. As a result, staggered fermions describe a theory with much less doublers as naive fermions. In addition they are relatively easy and fast to implement in simulations. To begin with, consider the naive fermion action for spinor field with $\Delta_f = 2^{[d/2]}$ components,

$$S_{\mathrm{naive}} = \sum_{x,\mu} \bar\psi_x \gamma^\mu (\mathring{\partial}_\mu \psi)_x + m \sum_x \bar\psi_x \psi_x\,. \tag{15.64}$$

Now we perform the site-dependent similarity transformation

$$\psi_x = T(x)\chi_x, \quad \bar\psi_x = \bar\chi_x T^\dagger(x), \quad T(x) = \gamma_0^{x_0} \cdots \gamma_{d-1}^{x_{d-1}}\,, \tag{15.65}$$

which diagonalizes the Dirac operator in spin space. With the help of

$$T^\dagger(x)\gamma^\mu T(x \pm e_\mu) = \Gamma_\mu(x)\mathbb{1}_{\Delta_f}, \quad \Gamma_\mu(x) = (-1)^{x_0+x_1\cdots+x_{\mu-1}}\,, \tag{15.66}$$

the action transforms into a sum of identical actions for the components χ_α of χ,

$$S_{\mathrm{naive}}[\bar\psi, \psi] = \sum_{\alpha=1}^{\Delta_f} S_s[\bar\chi_\alpha, \chi_\alpha], \quad S_s[\bar\chi, \chi] = \sum_x \chi_x^*(Q_s\chi)_x\,, \tag{15.67}$$

where the (normal) matrix Q_s acts on the one-component χ as follows[4]

$$(Q_s\chi)(x) = \sum_\mu \Gamma_\mu(x)(\mathring{\partial}_\mu\chi)_x + m\chi_x\,. \tag{15.68}$$

---

[4] To simplify the notation, we use the same symbol χ for the components of the field as for the original field with d_s components.

To decrease the number of doublers, we keep only *one* of the Δ_f identical terms, and this single term is just the action for staggered fermions. As a result we are dealing with only one fermionic degree of freedom per lattice site. The site-dependent phases $\Gamma_\mu(x)$ in Q_s are remnants of the Dirac structure. With this clever trick, one can reduce the 2^d-fold degeneracy by a factor Δ_f without losing the original chiral symmetry completely. The action for staggered fermions still admits an Abelian $U(1) \times \mathbb{R}_+$ symmetry. We show this by introducing the lattice function

$$\varepsilon(x) = (-1)^{x_0 + x_1 + \cdots + x_{d-1}} , \tag{15.69}$$

which is 1 on the even sublattice and -1 on the odd sublattice. In the massless limit, the action is a sum of nearest-neighbor interaction terms, and each term contains the product of the field on an even site and the field on an odd site. Thus, the reduced action retains the $U(1) \times \mathbb{R}_+$ part of the original symmetry:

$$\chi_x \to e^{ig\lambda + \alpha\varepsilon(x)} \chi_x \quad , \quad \chi_x^* \to \chi_x^* e^{-ig\lambda + \alpha\varepsilon(x)} . \tag{15.70}$$

Clearly the mass term $\chi_x^* \chi_x$ acquires a factor $e^{2\alpha\varepsilon(x)}$ and is not invariant under transformations with non-zero α. The symmetry (15.70) is enough to prevent the occurrence of mass counterterms in the renormalization process: $m_{\text{bare}} = 0$ implies $m_{\text{ren}} = 0$. The $\mathbb{R}_+$ symmetry becomes a flavor non-singlet axial symmetry in the continuum limit. Its possible spontaneous breaking produces a Goldstone boson for any value of the lattice spacing.

By keeping just one of the Δ_f identical contributions to the action in (15.67), we did not remove all of the 2^d doublers from the theory. In even dimensions we are still left with $2^{d/2}$ fermion species and in odd dimensions with $2^{(d+1)/2}$ species. For simplicity we assume now that d is even. From the one-component field χ, we can reconstruct $2^{d/2}$ flavors of Dirac spinors on 2^d disjoint sublattices $\{\Lambda_\varrho\}$ with lattice spacing $2a$. The sublattices are defined by the 2^d corners of a given elementary hypercube on the lattice, see Fig. 15.6, as follows: Let $\{f_\varrho\}$ be the lattice vectors pointing from a fixed corner x of the cube to all corners of the cube. In particular $f_0 = 0$. Then the field χ_ϱ on the sublattice Λ_ϱ is given by (we set $a = 1$)

$$\chi_{\varrho,x} = \chi_{2x + f_\varrho}, \quad \varrho = 0, \ldots, 2^d - 1 . \tag{15.71}$$

Finally from the one-component fields $\{\chi_\varrho\}$ on the coarse lattices, we may reconstruct $2^{d/2}$ flavors of Dirac fields ψ^f, usually called "tastes" to distinguish them from real flavors, with components ψ_α^f as follows:

$$\hat{\psi}_{\alpha,x}^f = \mathcal{N}_0 \sum_\varrho (T_\varrho)_{\alpha f} \chi_{\varrho,x}, \quad \alpha, f = 1, \ldots 2^{d/2} , \tag{15.72}$$

Fig. 15.6 Dirac spinor fields at x are constructed from the one-component field χ on the 2^d corners of the hypercubes of the lattice specified by the base point x and the vectors f_ϱ. The vector f_0 is the null vector

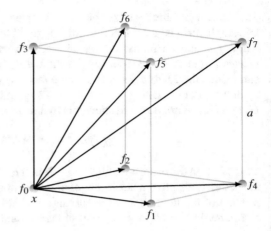

where the matrices T_ϱ are given by

$$T_\varrho = \gamma_0^{\varrho_0} \gamma_1^{\varrho_1} \cdots \gamma_{d-1}^{\varrho_{d-1}} \ . \tag{15.73}$$

The staggered fermion description has, however, some unpleasant features as well. Flavor (or taste) symmetry is explicitly broken and hoped to be recovered only in the continuum limit. In addition it is non-trivial to construct baryon operators with definite quantum numbers. Finally, in order to obtain a theory with a single physical flavor, one usually takes the fourth root of the fermionic determinant for staggered fermions. This is correct in the free theory, but we do not know for sure whether this nonlocal prescription makes sense non-perturbatively [18].

15.3.5 Nielsen–Ninomiya Theorem

The lattice fermions discussed so far are all afflicted with certain problems. Naive fermions and staggered fermions show fermion doubling, Wilson fermions break chiral symmetry explicitly, and staggered fermions break chiral symmetry partially. In this section we shall shed light on the doubling problem in a more general fashion with the help of the *Nielsen–Ninomiya no-go theorem* [19–21]. First, consider an arbitrary bilinear action

$$S = \sum_{x,y} \bar{\psi}_x M(x, y) \psi_y \tag{15.74}$$

for spin-1/2 lattice fermions. Because of translational invariance, the Dirac operator depends only on the difference $x - y$, i. e.,

$$M(x, y) = D(x - y) \ . \tag{15.75}$$

We may now ask the question, why the discretization of fermions without doublers and under conservation of chiral symmetry is as hard as it is? This question is answered by the no-go theorem which says that the phenomenon of fermion doubling occurs, provided that we only assume some general properties of the action as locality, hermiticity, and translational invariance. Thereby, one finds an equal number of left- and right-handed fermions. More precisely, the theorem states:

Theorem 15.1 (Nielsen–Ninomiya Theorem) *There exists no translational invariant Dirac operator that fulfills the following four properties:*

1. *locality:* $D(x - y) \lesssim e^{-\gamma|x-y|}$,
2. *continuum limit:* $\lim_{a\to 0} \tilde{D}(p) = \sum_\mu \gamma^\mu p_\mu$,
3. *no doublers:* $\tilde{D}(p)$ *is invertible if* $p \neq 0$,
4. *chirality:* $\{\gamma_5, D\} = 0$.

Locality implies the Fourier transform $\tilde{D}$ of D to be an analytic and periodic function of the momenta p_μ with period $2\pi/a$. The second and third assumptions guarantee the correct continuum limit of D. Reference [19–21] gives a proof of the theorem by homotopy theory. Readers interested in a proof based on elegant arguments from differential geometry should consult [22].

Proof We shall give a proof under the additional assumption [23]

$$\tilde{D}(p) = \sum \gamma^\mu \tilde{D}_\mu(p) \quad \text{with} \quad \tilde{D}_\mu(p) \in \mathbb{R} \tag{15.76}$$

in momentum space, where the *analytic* functions $\tilde{D}_\mu$ tend to p_μ for small momenta. Since the Brillouin zone shows the topology of a torus in d dimensions, $\tilde{D}_\mu$ defines a vector field $\tilde{D}$ on T^d. We now assign an index to every zero p_i of this vector field, whereby we assume the number of zeros to be finite. According to a theorem by HOPF and POINCARÉ, the sum of the indices of all zeros on a compact and oriented manifold is equal to the Euler characteristic of the manifold,

$$\sum_{\text{zeroes } p_i} \text{index}\left(\tilde{D}(p_i)\right) = \chi(\mathrm{T}^d) . \tag{15.77}$$

Thereby, the index of $\tilde{D}$ at a zero p_i is equal to the degree of the induced map (the winding number) from the boundary of a small ball centered at p_i into $\mathbb{R}^d - 0$. Figure 15.7 shows the vector field $\tilde{D}(p)$ of the naive Dirac operator in two dimensions in the first Brillouin zone. The vector field has index 1 at the zeros $p = (0, 0)$ and (π, π) and index -1 at the zeros $p = (0, \pi)$ and $(\pi, 0)$. The sum of the indices vanishes in accordance with the fact that the Euler characteristics of the torus vanishes, $\chi(\mathrm{T}^d) = 0$. The figure shows clearly the doublers at momenta $(\pi, 0)$, $(0, \pi)$, and (π, π).

Fig. 15.7 The vector field $\tilde{D}_\mu(p)$ of the naive Dirac operator in two dimensions. The vector field has four zeros in the Brillouin zone. Two zeros have winding number 1 and two zeros have winding number -1

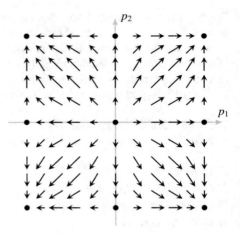

Similarly, on the d-dimensional torus, the zeros of $\tilde{D}$ occur in pairs with opposite indices. To see this we expand $\tilde{D}$ around a zero p_0 according to

$$\tilde{D}_\mu(p) = A_{\mu\nu}(p-p_0)^\nu + \ldots, \qquad A_{\mu\nu} = \frac{\partial \tilde{D}_\mu}{\partial p_\nu}\Big|_{p_0}.$$

The index of the vector field $\tilde{D}$ at its zero is equal to the sign of det A. For example, if A is diagonal at the zero $p_0 = 0$ then we have $\tilde{D}_\mu(p) = A_\mu p_\mu + O(p^2)$. If in four dimensions $A = \mathrm{diag}(1,1,1,1)$, the index is equal to 1 and we find

$$\bar{\psi}\gamma^\mu \tilde{D}_\mu(p)\psi \approx \bar{\psi}\gamma^\mu p_\mu \psi \tag{15.78}$$

in the vicinity of the zero. Recall that the field ψ transforms under a chiral transformation into $\exp(\alpha\gamma_5)\psi$. In contrast, if $A = \mathrm{diag}(-1,1,1,1)$, the vector field has index -1, and the Lagrangian density in the vicinity of the zero $p_0 = 0$ has the form

$$\bar{\psi}\gamma^\mu \tilde{D}_\mu(p)\psi \approx \bar{\psi}\gamma_5\gamma^0(\gamma^\mu p_\mu)\gamma^0\gamma_5\psi \equiv \bar{\chi}\not{p}\chi. \tag{15.79}$$

Clearly $\chi = \gamma^0\gamma_5\psi$ is to be interpreted as Dirac field of the doubler. Under chiral transformations it transforms according to $\exp(-\alpha\gamma_5)\chi$ such that the two fermion species have opposite chirality. Every pole of the massless propagator corresponds to a fermionic one-particle state. Thus, we conclude that a fermion of chirality $+1$ is always accompanied by a fermion of chirality -1.

15.4   Ginsparg-Wilson Relation and Overlap Fermions

The Nielsen–Ninomiya no-go theorem makes clear that under certain conditions, it is impossible to find a chirally invariant Dirac operator without doublers. Similarly as with many other no-go theorems, there is a way to bypass it. The solutions of the doubler problem are based on a paper published in 1982 by Ginsparg and Wilson [24]. They asked thereby the question "... how can a lattice theory serve to represent a continuum situation where the symmetry does not suffer explicit breaking?".

To begin with they considered a chirally invariant continuum theory which is mapped to a lattice theory via a block-spin transformation. In detail, the continuum field ϕ on $\mathbb{R}^d$ is related via a block-spin transformation to the lattice field ψ:

$$\psi_x = \int d^d y \, \alpha(x - y) \, \phi(y), \quad x \in \Lambda . \tag{15.80}$$

Thereby, the exact form of the weight function α is irrelevant. Then Ginsparg and Wilson analyzed the lattice action induced by the blocking transformation and how much it deviates from a chirally invariant action. They discovered that operators which solve the *Ginsparg-Wilson relation*

$$\gamma_5 D + D\gamma_5 = a D\gamma_5 D \tag{15.81}$$

yield optimal lattice operators. The lattice spacing a on the right-hand side ensures the proper continuum relation $\gamma_5 D + D\gamma_5 = 0$ on macroscopic scales or for sufficiently small a. By multiplying the relation (15.81) with $S = D^{-1}$ from both sides, we find

$$S\gamma_5 + \gamma_5 S = a\gamma_5 \quad \text{or} \quad S(x, y)\gamma_5 + \gamma_5 S(x, y) = a\gamma_5 \delta_{x,y} . \tag{15.82}$$

This means that chiral symmetry breaking as encoded in the propagator is *ultralocal*, i.e., we obtain a chirally invariant propagator for all finite distances $|x - y| > 0$. The property (15.82) is sufficient to preserve many relevant consequences of chiral symmetry as, e.g., the absence of an additive mass renormalization on the lattice.

The new developments have been triggered by the rediscovery of the Ginsparg-Wilson relation by P. Hasenfratz [25–27] and an explicit solution of the Ginsparg-Wilson relation by H. Neuberger [28, 29]. M. Lüscher noticed that the fermionic action

$$S_F = a^d \sum_{x,y} \bar{\psi}_x D(x - y) \, \psi_y \tag{15.83}$$

admits a continuous symmetry if the Dirac operator satisfies the Ginsparg-Wilson relation [30]. The symmetry is interpreted as lattice version of the chiral symmetry and reads

$$\psi \longrightarrow \psi_{(\alpha)} = e^{\alpha \gamma_5 (1 - aD/2)} \psi \quad \text{and} \quad \bar{\psi} \longrightarrow \bar{\psi}_{(\alpha)} = \bar{\psi} \, e^{\alpha (1 - aD/2) \gamma_5} \, . \quad (15.84)$$

It is not difficult to show that the bilinear $\bar{\psi} D \psi$ is invariant:

$$\frac{d}{d\alpha} \left(\bar{\psi}_{(\alpha)} D \psi_{(\alpha)} \right) = \bar{\psi}_{(\alpha)} \left\{ \left(1 - \tfrac{1}{2} aD \right) \gamma_5 D + D \gamma_5 \left(1 - \tfrac{1}{2} aD \right) \right\} \psi_{(\alpha)} \stackrel{(15.81)}{=} 0 \, .$$

However, in general the fermionic integration measure $\mathscr{D}\psi \, \mathscr{D}\bar{\psi}$ is not invariant under the deformed chiral transformations (15.84). This fact ensures the occurrence of the *axial anomaly* in presence of an external gauge field. Examples of lattice Dirac operators that satisfy the Ginsparg-Wilson relations are:

1. domain wall fermions [31–33],
2. overlap operators [28, 29, 34–36],
3. fixed point operators [25–27],
4. chirally improved operators [37–39].

In the following section, we shall discuss the overlap operators introduced by Neuberger and Narayanan.

15.4.1 Overlap Fermions

An elegant solution of the Ginsparg-Wilson relation—the overlap operator—has been constructed in the pioneering contributions [28, 29, 34–36]. The operator has the form

$$D_o = \frac{1}{a}(\mathbb{1} + V), \quad V = (D_w D_w^\dagger)^{-1/2} D_w \, , \quad (15.85)$$

where D_w is the Wilson operator (15.59) with negative m. Dividing D_w by its modulus yields a unitary operator V with its spectrum on the unit circle. Correspondingly, the overlap operator has its eigenvalues on a circle that touches the imaginary axis at the origin, as shown in Fig. 15.8. Let us show that the overlap operator solves the Ginsparg-Wilson relation (15.81). In terms of the unitary operator V, the left-hand side reads

$$D_o \gamma_5 + \gamma_5 D_o = \frac{2}{a} \gamma_5 + \frac{1}{a} \{ \gamma_5, V \} \, . \quad (15.86)$$

Fig. 15.8 The eigenvalues of the overlap operator are located on a Ginsparg-Wilson circle that touches the imaginary axis at the origin

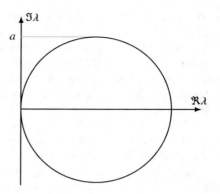

To rewrite the right-hand side, we use the following properties of the Wilson operator

$$[D_{\rm w}, D_{w}^{\dagger}] = 0, \quad \gamma_5 D_{\rm w} = D_{w}^{\dagger}\gamma_5, \quad [D_{\rm w} D_{\rm w}^{\dagger}, \gamma_5] = 0 , \tag{15.87}$$

which imply $V\gamma_5 V = \gamma_5$. Now the right-hand side of the relation reads

$$a D_o \gamma_5 D_o = \frac{1}{a}(\gamma_5 + \{\gamma_5, V\} + V\gamma_5 V) = \frac{2}{a}\gamma_5 + \frac{1}{a}\{\gamma_5, V\} , \tag{15.88}$$

and it is equal to the left-hand side in (15.86).

15.4.2 Locality

One disadvantage of the overlap operator is the appearance of the inverse of the modulus $(D_{\rm w} D_{\rm w}^{\dagger})^{1/2}$ of the Wilson operator. It is not evident that D_o is a local operator. In general, we distinguish between *ultralocal* operators, where $D(x - y)$ vanishes exactly if $|x - y| > \ell$, and *local* operators, where $D(x - y)$ decreases exponentially as a function of the distance $|x - y|$ on the lattice. In the continuum limit local operators become *exact* local operators. By considering the spectral representation of the operator D_o, we can analyze its locality behavior. Since

$$D_{\rm w} D_{\rm w}^{\dagger} = -\mathring{\Delta} + \left(m - \frac{ar}{2}\Delta\right)^2 , \tag{15.89}$$

the overlap operator in Fourier space may be written as

$$a\tilde{D}_o(p) = 1 + \left\{i\gamma^{\mu}\mathring{p}_{\mu} + m + \frac{ar}{2}\hat{p}^2\right\} \left\{\mathring{p}^2 + \left(m + \frac{ar}{2}\hat{p}^2\right)^2\right\}^{-1/2} . \tag{15.90}$$

The operator satisfies the first three conditions stated in the Nielsen–Ninomiya theorem. In particular, $\tilde{D}_o(p)$ is analytical, and thus the operator $D_o(x - y)$ in position space vanishes exponentially with increasing distances $|x - y|$. One assumption of the Nielsen–Ninomiya theorem is not satisfied, since the chiral symmetry is realized differently. The Ginsparg-Wilson relation allows for a very soft breaking of the usual chiral symmetry such that the propagating states are effectively chiral and all physical consequences of chiral symmetry are preserved.

To summarize: The overlap operator by Neuberger and Narayanan is local, shows no fermion doubling, and preserves (a deformed) chiral symmetry. However, since the inversion of $D_\mathrm{w} D_\mathrm{w}^\dagger$ in presence of gauge fields is quite time-consuming, the overlap operator may not always be the best choice in simulations. In addition, if $m = 0$ there may appear zero modes for non-vanishing gauge fields. Indeed, Ginsparg-Wilson operators obey an exact Atiyah-Singer index theorem on the lattice, and thus there are zero modes for topologically non-trivial background fields [40].

15.5   Yukawa Models on the Lattice

Yukawa models can be used to describe the strong nuclear force between nucleons mediated by pions. The Yukawa interaction between fermions and scalar or pseudo-scalar particles is also used in the Standard Model of particle physics to describe the coupling between the Higgs field and massless quark and lepton fields. Through spontaneous symmetry breaking, these fermions acquire a mass proportional to the vacuum expectation value of the Higgs field. A Yukawa interaction has the form

$$y\bar{\psi}\phi\psi \quad \text{or} \quad iy\bar{\psi}\gamma_5\phi\psi \ . \tag{15.91}$$

If the fermions and (pseudo)scalars transform according to a non-trivial representation of some internal symmetry group G, then the trilinears must be invariant. The full Euclidean action of the Yukawa model (with scalar field) reads

$$S = \int \mathrm{d}^d x \, \mathscr{L}(\phi, \psi), \quad \mathscr{L} = \frac{1}{2}(\nabla\phi)^2 + V(\phi) + \bar{\psi}(\partial\!\!\!/ + m)\psi + y\bar{\psi}\phi\psi \ , \tag{15.92}$$

where the potential V is G-invariant $V(\Omega\phi) = V(\phi)$. Suppose that the classical potential V has a minimum at a constant field $\phi_0 \neq 0$ which is not invariant under the symmetry transformations. Then the classical vacuum configuration breaks the internal symmetry. Expanding the action about ϕ_0 in powers of $\chi = \phi - \phi_0$, we see that the Yukawa interaction yields a term $y\phi_0\bar{\psi}\psi$, which is just a mass term for the fermions with fermion mass $y\phi_0$. The field χ is known as a Higgs field.

The same spontaneous symmetry breaking occurs in the quantized theory if the scalar field acquires a non-zero vacuum expectation value, similar to a non-zero magnetization in spin models. The lattice formulation enables us to study Yukawa theories at strong Yukawa coupling y. On a lattice the expectation value of an observable is given by

$$\langle O(\phi, \psi, \bar{\psi}) \rangle = \frac{1}{Z} \int \mathscr{D}\phi \mathscr{D}\psi \mathscr{D}\bar{\psi} \; O(\phi, \psi, \bar{\psi}) \, e^{-S[\phi, \psi, \bar{\psi}]} \,, \tag{15.93}$$

where S is some lattice version of the continuum action (15.92), e.g.,

$$S = \frac{1}{2} \sum_{\langle x, y \rangle} (\phi_x - \phi_y)^2 + \sum_x V(\phi_x) + \sum_{x,y} \bar{\psi}_x D(x - y)\psi_y + y \sum_y \bar{\psi}_x \phi_x \psi_x \,. \tag{15.94}$$

In the normalizing partition function, we can integrate over the fermionic degrees of freedom and obtain the determinant of $D + y\phi$,

$$Z = \int \mathscr{D}\phi \mathscr{D}\psi \mathscr{D}\bar{\psi} \, e^{-S[\phi, \psi, \bar{\psi}]} = \int \mathscr{D}\phi \, \det(D + y\phi) \, e^{-S_B(\phi)} \tag{15.95}$$

where S_B denotes the bosonic part of the action, i.e., the ψ-independent part of S. For a real scalar field, the operator $D + y\phi$ is γ_5-hermitian and hence has real determinant. But the sign of the determinant may depend on the scalar field and may give rise to the sign problem, in particular for a strong Yukawa coupling y.

15.5.1 Higgs Sector of Standard Model

The Higgs sector of the Standard Model of particle physics represents a $4d$ Yukawa model. A careful analysis of this sector reveals that it defines a trivial theory. Triviality refers to the behavior of the renormalized quartic coupling constant λ_r of the scalar field in dependence on the cutoff parameter Λ of the regularized theory. In a renormalizable theory, the cutoff parameter can be sent to infinity while holding a finite (and small) number of physical quantities constant, making the predictions for all physical observables eventually independent of the introduced auxiliary parameter Λ, as desired. In a trivial theory, however, the renormalized coupling constants must vanish as function of the cutoff parameter in the limit $\Lambda \to \infty$, leading to a free, non-interacting theory when we try to remove the cutoff. Thus, the Higgs sector of the Standard Model can only be considered as an effective theory connected with a non-removable cutoff parameter, which can be interpreted as the maximal scale up to which the underlying effective theory can be trusted. As a consequence the renormalized quartic coupling constant at a given cutoff Λ is bounded from above according to $\lambda_r(\Lambda) \leq \lambda_{\mathrm{up},r}(\Lambda)$. This bound translates into an *upper bound* $m_{\mathrm{up},H}(\Lambda)$ on the Higgs boson mass, and this bound decreases with increasing cutoff [41]. Early attempts to simulate the Higgs-Yukawa sector of the Standard Model on a lattice had problems with removing the fermion doublers from the spectrum while maintaining chiral symmetry. With the overlap construction for the free Dirac operator to simulate the chirally invariant Higgs-Yukawa model, reliable upper and lower bounds on the Higgs boson mass as a function of the cutoff parameter have been obtained in [42]. The authors concluded that for a Higgs boson mass of about 125 GeV, the Standard Model can be valid up to very large cutoff scales.

As it turned out, the value $m_H = 125.10 \pm 0.14$ GeV measured at the ATLAS and CMS experiments at the Large Hadron Collider [43] is close to the lower bound. Thus, an elementary and weakly coupled Higgs boson allows one to extend the validity of the Standard Model up to very high energy, maybe as high as the Planck scale. But the maximum ultraviolet scale depends on the level of approximation involved, as has been demonstrated within the FRG method by admitting higher-dimensional operators in the flow equation [44].

15.5.2 Supersymmetric Yukawa Models

Supersymmetry is an important ingredient of modern high-energy physics beyond the standard model. Since boson masses are protected by supersymmetry in such theories with chiral fermions, it helps to reduce the hierarchy and fine-tuning problems drastically. However, as low-energy physics is manifestly not supersymmetric, this symmetry has to be broken at some energy scale. In the case of supersymmetric field theories, a lattice formulation is hampered by the fact that the supersymmetry algebra closes on the generator of infinitesimal translations which do not exist on a discretized spacetime. A related fact is that lattice derivatives do not satisfy the Leibniz rule implying that supersymmetric actions will in general not be invariant under lattice supersymmetries. In generic lattice formulations, there are no discrete remnants of supersymmetry transformations on the lattice; in such theories, supersymmetry in the continuum limit can only be recovered by appropriately fine-tuning the bare couplings of all supersymmetry-breaking counterterms.

Particular simple supersymmetric Yukawa theories—Wess–Zumino models in two dimensions—have been the subject of intensive analytic and numerical investigations [45–48]. The models with two supersymmetries contain one Dirac spinor and two real scalar fields. The simpler model with just one supersymmetry contains one real Majorana spinor field and one real scalar field and has the action

$$S = \int \mathrm{d}^2 x \, \frac{1}{2} \left((\partial_\mu \phi)^2 + \bar{\psi} D \psi + \mathcal{P}(\phi)^2 \right), \quad D = \not{\partial} + \mathcal{P}'(\phi) , \tag{15.96}$$

with superpotential $\mathcal{P}$. It is invariant under the supersymmetry transformations

$$\delta \phi = \bar{\epsilon} \psi, \quad \delta \psi = \left(\not{\partial} \phi - \mathcal{P}(\phi) \right) \epsilon . \tag{15.97}$$

The constant and anticommuting Majorana spinor ε parametrizes the infinitesimal supersymmetry transformation. With an even superpotential, the bosonic part of the action is invariant under a reflection of the scalar field $\phi \rightarrow -\phi$. Since

$$\not{\partial} + \mathcal{P}'(-\phi) = \not{\partial} - \mathcal{P}'(\phi) = -\gamma_5 \left(\not{\partial} + \mathcal{P}'(\phi) \right) \gamma_5 , \tag{15.98}$$

the fermion determinant $\det(D)$ is also invariant under a reflection of the scalar field. In a series of papers, various discretizations of models with one or two supersymmetries have been studied and compared [12, 15, 49]. For the models based on the SLAC derivative (introduced on p. 404) for both bosonic and fermionic fields, the fermionic operator is γ_5-hermitian, and the internal continuum symmetries are realized on the lattice. When one analyzes supersymmetric Ward identities and the particle spectrum, one sees that the models with SLAC derivative are superior to other discretizations and yield accurate results already on moderately sized lattices.

Let us consider the model with one supersymmetry and even superpotential

$$\mathscr{P}(\phi) = \frac{\mu_0^2}{\sqrt{2\lambda}} + \sqrt{\frac{\lambda}{2}}\phi^2 \implies V(\phi) \equiv \frac{\mathscr{P}(\phi)^2}{2} = \frac{\mu_0^2}{2}\phi^2 + \frac{\lambda}{4}\phi^4 + \text{const}. \quad (15.99)$$

It has vanishing Witten index and hence (at least) two degenerate ground states. One expects that for fixed λ and $\mu_0^2 \ll 0$, the system cannot tunnel between the two ground states so that supersymmetry is unbroken. On the other hand, for $\mu_0^2 > 0$ both ground state energies are lifted above zero, and supersymmetry is broken [50, 51]. Two comments should be made at this point. Firstly the analysis of the divergent diagrams in Fig. 15.9 shows that a logarithmic renormalization of the bare mass parameter is necessary to cancel divergent contributions. The renormalization procedure amounts to a normal ordering of interaction terms with respect to a mass parameter in the symmetric phase; for details see [49]. The dimensionless renormalized coupling $f = \lambda/\mu^2$ distinguishes between the Z_2-symmetric phase and broken phase or the phase without supersymmetry and the phase with supersymmetry. Figure 15.10 shows the continuum-extrapolated expectation values of the superpotential. For strong couplings f, the expectation value vanishes such that supersymmetry is realized and the Z_2-symmetry is broken. On the other hand, for weak couplings we are in the Z_2-symmetric phase, and supersymmetry is dynamically broken. Secondly, the fermionic integral for a theory with Majorana fermions yields the Pfaffian in place of the determinant. Up to a sign the Pfaffian of an antisymmetric matrix is the square root of its determinant. Further properties of the Pfaffian are discussed in problem 15.1.

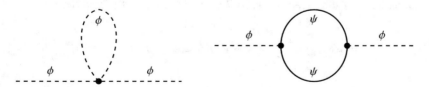

Fig. 15.9 The divergent Feynman diagrams of the Wess–Zumino model in the Z_2 symmetric phase

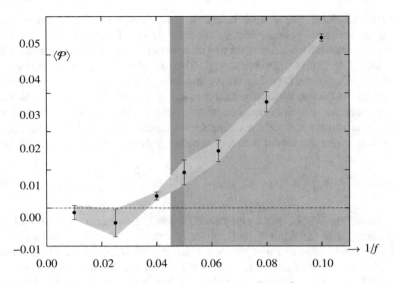

Fig. 15.10 Dimensionless prepotential $\mathcal{P}$ over inverse coupling f^{-1}. The shaded region on the right is the Z_2 symmetric phase where SUSY is broken, while the small shaded region in the middle gives the error bound for the Z_2 phase transition

15.6   Coupling to Lattice Gauge Fields

As on p. 397, we consider a Dirac field with values in $V \otimes \mathbb{C}^{\Delta_f}$. A global symmetry transformation $\psi \rightarrow \Omega \psi$ acts in the vector (flavor) space V such that the Ω-invariance of the scalar product implies the invariance of the lattice action

$$S_F = \sum_x \bar{\psi}_x (D\,\psi)_x \ . \tag{15.100}$$

Replacing lattice derivatives by covariant lattice derivatives, one obtains an action which is invariant under (local) gauge transformations $\psi_x \rightarrow \Omega_x \psi_x$. For Wilson fermions with Wilson parameter r, the Dirac operator has the form (we set $a = 1$)

$$(D_{\mathrm{w}})_{xy} = (m + rd)\,\delta_{xy} - \frac{1}{2} \sum_{\mu=0}^{d-1} \left((1 + \gamma^\mu) U_{y,-\mu} \delta_{x,y-e_\mu} + (1 - \gamma^\mu) U_{y,\mu} \delta_{x,y+e_\mu} \right) \ . \tag{15.101}$$

A parametrization introduced by Wilson follows from rescaling

$$\psi \rightarrow \frac{1}{\sqrt{m+rd}}\,\psi \quad , \quad \bar{\psi} \rightarrow \bar{\psi}\,\frac{1}{\sqrt{m+rd}}$$

and yields the following form for the gauge-invariant action:

$$S_{\mathrm{W}} = \sum_x \bar{\psi}_x \psi_x - \kappa \sum_{x,\mu} \left(\bar{\psi}_{x-e_\mu} (r + \gamma^\mu) U_{x,-\mu} \psi_x + \bar{\psi}_{x+e_\mu} (r - \gamma^\mu) U_{x,\mu} \psi_x \right) ,$$

(15.102)

where $\kappa = (2m + 2rd)^{-1}$ is called *hopping parameter*. In quantum chromodynamics (QCD) m is a diagonal matrix in flavor space, and r is usually chosen flavor independent, mostly $r = 1$. The action for Wilson fermions breaks the chiral symmetry explicitly. Unfortunately, in lattice QCD this leads to a variety of complications. In particular, recovering chiral symmetry in the continuum limit requires unnatural fine-tuning of the bare fermion mass or, equivalently, the hopping parameter.

Gauge theories in particle physics contain both fermions and gauge bosons, and one is confronted with lattice path integrals of the form

$$Z = \int \prod_\ell \mathrm{d}U_\ell \prod_x \mathrm{d}\psi_x \mathrm{d}\bar{\psi}_x \, e^{-S_{\mathrm{gauge}}(U) - S_{\mathrm{F}}(\psi, \bar{\psi})}$$

(15.103)

$$= \int \prod_\ell \mathrm{d}U_\ell \, \det(D[U]) \, e^{-S_{\mathrm{gauge}}(U)}$$

(15.104)

$$= \int \prod_\ell \mathrm{d}U_\ell \, \mathrm{sign}(\det D) \, (\det M)^{1/2} \, e^{-S_{\mathrm{gauge}}(U)} ,$$

(15.105)

where $M = D^\dagger D$ is hermitian and non-negative such that $\det M \geq 0$. For a γ_5-hermitian Dirac operator, $\det(D)$ is real and $\mathrm{sign}(\det D) \in \{1, -1\}$. In more general situations and in particular at finite baryon number densities, the sign function can be a complex phase, leading to the notorious fermion sign problem [52]. Standard Monte Carlo techniques are only applicable if the measure in the path integral is positive. Therefore, we assume that $\det(D)$ either has a fixed sign or else simulate with $(\det M)^{1/2} \geq 0$ and treat $\mathrm{sign}(\det D)$ as an insertion into the path integral. In the latter situation where $\det D$ changes sign, the reweighing with $\mathrm{sign}(\det D)$ is a delicate issue and, in particular for strong coupling, may fail.

In simulations it is too expensive to calculate the determinant of D or $M = D^\dagger D$ for every gauge field configuration—the computational cost for $\det(M)$ grows with the third power of the lattice volume. Thus, one often introduces N_{PF} complex pseudofermion fields [53] and rewrites the partition function as

$$(\det M)^{1/2} = \int \prod_p \mathscr{D}\phi_p^\dagger \mathscr{D}\phi_p \, e^{-S_{\mathrm{PF}}}, \quad S_{\mathrm{PF}} = \sum_{p=1}^{N_{\mathrm{PF}}} \left(\phi_p, M^{-q} \phi_p \right) ,$$

(15.106)

where $q\,N_{PF} = 1/2$. If det D is positive, then the resulting path integral

$$Z = \int \prod_\ell dU_\ell \mathscr{D}\phi \mathscr{D}\phi^* \, e^{-S_{gauge}(U) - S_{PF}(U,\phi,\phi^\dagger)} \tag{15.107}$$

can then be estimated with a HMC algorithm with force given by the gradient of the nonlocal action $S_{gauge} + S_{PF}$. In the rational HMC dynamics, M^{-q} is replaced by a rational approximation according to

$$M^{-q} \approx \alpha_0 + \sum_{r=1}^{N_R} \frac{\alpha_r}{M + \beta_r} \, , \tag{15.108}$$

where the number of terms N_R depends on the required accuracy of the rational approximation and the spectral range of M. The coefficients α and β can be calculated with the Remez algorithm [54]. The force terms in the rHMC dynamics contain the inverse of the matrices $M + \beta_r$ acting on a vector, and this mapping can be approximated with the help of a (multi-mass) conjugate gradient solver. The efficiency of the rHMC algorithm crucially depends on the lowest eigenvalues, i.e., the condition number $\lambda_{max}/\lambda_{min}$ of the hermitian operator M used in the rational approximation.

Observables of interest in particle physics are, for example, hadron masses, decay widths, weak matrix elements, or form factors, and nowadays these quantities can be estimated by lattice simulations on high-performance computer clusters. Although this is a very important field of research in particle physics, we refrain from studying this any further but refer the interested reader to the nice textbook of GATTRINGER and C. LANG [4]. Instead we turn to gauge theories under extreme conditions.

15.7   Finite Temperature and Density

Finite temperatures are introduced via the boundary conditions in the Euclidean time directions: bosonic fields must be periodic, and fermionic fields must be antiperiodic in Euclidean time with period $\beta = 1/k_b T$. The antiperiodic boundary conditions for fermions originate from the anticommutation relations for fermion field operators. Using this fact and the cyclicity of the trace, we see that the thermal Green function for the Euclidean field operator $\hat{\psi}(\tau, x) = \exp(\tau \hat{H})\hat{\psi}(0, x)\exp(-\tau \hat{H})$ satisfies

$$G_\beta(\tau, x; 0, y) = \frac{1}{Z}\mathrm{tr}\left(e^{-\beta \hat{H}} T \hat{\psi}(\tau, x)\hat{\psi}(0, y)\right)$$

$$= -\frac{1}{Z}\mathrm{tr}\left(e^{-\beta \hat{H}} T \hat{\psi}(\tau, x)\hat{\psi}(\beta, y)\right) = -G_\beta(\tau, x, \beta, y) \, ,$$

where T is the time ordering. We conclude that fermion fields must be antiperiodic in Euclidean time inside the path integral, cp. problem 15.4 on p. 429.

Dynamical fermions in a fundamental representation of the gauge group break the center symmetry of the pure gauge theory (see the discussion on p. 371) explicitly, and this affects the Polyakov loop. In full QCD one finds that the expectation value of the Polyakov loop settles on the real positive axis. In the confined low-temperature phase, it settles near the origin and in the high-temperature deconfined phase near the trace of the center element $\mathbb{1}$. The Polyakov loop and Wilson loop are no longer true order parameters.

In our theoretical world, we may change the mass parameter m entering the Dirac operator. If we let the parameter approach infinity, the quarks decouple, and we are in the so-called quenched situation without dynamical quarks. In the quenched approximation, the SU(3) theory shows a first-order phase transition from the confined to the deconfined phase. Lowering the quark masses, the latent heat decreases, and the transition becomes weaker until the transition turns into a crossover. At physical quark masses, the confined and deconfined regions in parameter space are analytically connected [55]. The lattice computation now agrees on the position of the crossover temperature for physical quark masses ($m_\pi \approx$ 140 MeV) at $T_c \approx 170$ MeV [56,57]. There is clear evidence that the susceptibilities do not diverge.

Finite Baryonic Density

In extreme situations, such as heavy ion collisions or ultradense matter in neutron stars, the baryon number density may exceed the density of atomic nuclei. In these extreme situations we must switch to the partition function of the grand canonical ensemble with quark chemical potential μ multiplying the quark number operator $\hat{N}_q$,

$$Z(\beta, \mu) = \text{tr} \left(e^{-\beta(\hat{H} - \mu \hat{N}_q)} \right) . \tag{15.109}$$

Sometimes the baryon number operator $\hat{N}_B = \hat{N}_q/3$ and baryon chemical potential $\mu_B = 3\mu$ are used instead. The quark number density is the zero component of the conserved Noether vector current $j^\mu = \bar{\psi} \gamma^\mu \psi$ connected to the U(1) invariance of the theory. The corresponding Noether charge

$$\hat{N}_q = \int d^3 x \, \hat{\bar{\psi}} \gamma^0 \hat{\psi} \tag{15.110}$$

commutes with the Hamiltonian. We can treat the combination $\hat{H} - \mu \hat{N}_q$ as effective Hamiltonian and directly write down the path integral by simply adding

$$-\mu Q = \mu \int d^d x \, \bar{\psi} \gamma^0 \psi \tag{15.111}$$

to the Euclidean action. Thus, at finite temperature and finite chemical potential, the fermionic part of the Euclidean continuum action takes the form

$$S_{\mathrm{F}} = \int_0^\beta \mathrm{d}\tau \int \mathrm{d}^{d-1}x \, \bar\psi \left(\slashed{D} + m + \mu\gamma^0 \right) \psi \; . \tag{15.112}$$

Note that the additional term containing μ is obtained by shifting an Abelian gauge potential according to $A_0 \to A_0 - \mathrm{i}\mu$. This partly explains why the partition function of an Abelian gauge theory in a finite box and subject to periodic boundary conditions in the spatial directions does not depend on the chemical potential [58]. On the lattice one is tempted to add a term $\mu \sum_x \bar\psi_x \gamma^0 \psi_x$ to the lattice action. This simplistic ansatz runs into problems with unphysical divergences of the energy density, though. Closer inspection of the continuum situation clarifies the problem. Determining the Noether current for the lattice action gives the current expressed by nearest-neighbor terms. An easy way to find the correct expression is to recall that the elementary parallel transporter changes under the shift $A_0 \to A_0 - \mathrm{i}\mu$ according to

$$U_{x,\mu} \approx \mathrm{e}^{\mathrm{i}a A_\mu(x)} \longrightarrow \mathrm{e}^\mu U_{x,\mu} \; . \tag{15.113}$$

Thus, the Wilson operator at bare fermion mass m and real chemical potential μ reads

$$\begin{aligned} D_{\mathrm{w},xy}(\mu) = {} & (m+rd)\delta_{xy} \\ & - \frac{1}{2}\sum_\nu \left((r+\gamma^\nu)\,\mathrm{e}^{-\mu\delta_{0,\nu}} U_{y,-\nu}\delta_{x,y-e_\nu} \right. \\ & \left. + (r-\gamma^\nu)\,\mathrm{e}^{\mu\delta_{\nu,0}} U_{y,\nu}\delta_{x,y+e_\nu} \right) \; . \end{aligned} \tag{15.114}$$

and similarly for staggered, twisted mass, or overlap fermions. The Euclidean action

$$S = S_{\mathrm{gauge}} + \sum_{x,y} \bar\psi_x D_{xy}(\mu)\,\psi_y \tag{15.115}$$

enters the grand partition function for the theory with N_f flavors of Dirac fermions at finite volume, temperature, and chemical potential

$$Z(V,T,\mu) = \oint \prod_\ell \mathrm{d}U_\ell \left(\det_{\mathrm{ap}} D(\mu) \right)^f \mathrm{e}^{-S_{\mathrm{gauge}}(U)} \; . \tag{15.116}$$

The index at the determinant indicates that it must be calculated with respect to antiperiodic boundary conditions in the imaginary time direction. Unfortunately the operators $D(\mu)$ have complex determinants for $\mu \neq 0$, and it is impossible to apply stochastic methods directly to estimate the thermodynamic potential or thermal correlation functions at finite baryon density. Dealing with such complex determinants in simulations is one of the most urgent problems in gauge theories at finite densities. Various ways have been suggested to circumvent the problem:

- One generates ensembles with $\mu = 0$ and treats the μ-dependent part of $\exp(-S)$ as insertion. This reweighing requires the exact evaluation of $\det D$ and works well on small lattices [59].
- One performs a Taylor expansion around $\mu = 0$ where the expansion coefficients are estimated with the ensemble at $\mu = 0$. This works well for small μ [60].
- For imaginary chemical potential, the determinant becomes real, and one can estimate $Z(V, T, i\mu)$. The result is analytically continued to real chemical potential. The analytic continuation of an approximation may be a bad approximation to the analytic continuation, and indeed the method only works well for small μ [61].

For larger chemical potentials, the only results on the phase diagram at finite temperature and finite density are obtained from functional methods or model calculations that crucially rely on truncations or model building [62,63]. A different strategy is to investigate QCD-like theories without a sign problem, having as much features in common with QCD as possible. An example of such a theory is two-color QCD with a real determinant at finite density. The phase diagram of the two-flavor model as a function of temperature and net baryon density has been investigated in [64,65]. Unfortunately the theory has no fermionic baryons and thus cannot support a "neutron star." A gauge theory without sign problem and with fermionic baryons is G_2-QCD which can be broken to real-life QCD with a scalar field in the seven-dimensional fundamental representation [66,67]. All irreducible representation of this exceptional groups can be chosen real, and as a consequence the fermionic determinant at finite baryon density is real and non-negative [68]. The theory can be simulated at finite temperature and finite baryon density, and the resulting phase diagram looks similar to the expected phase diagram for QCD, depicted in Fig. 15.11.

15.8 Problems

15.1 (Pfaffian) Let $\eta_1, \ldots, \eta_{2N}$ be an even number of anticommuting real Grassmann variables, $\{\eta_a, \eta_b\} = 0$. Prove that the Gaussian integral over such variables yields the Pfaffian,

$$\int d\psi_1 \cdots d\psi_{2N} \, e^{\frac{1}{2}\psi^t M \psi} = \frac{1}{2^N N!} \varepsilon_{a_1 b_1 \ldots a_N b_N} M_{a_1 b_1} \cdots M_{a_N b_N} = \mathrm{Pf}(M) .$$

$$(15.117)$$

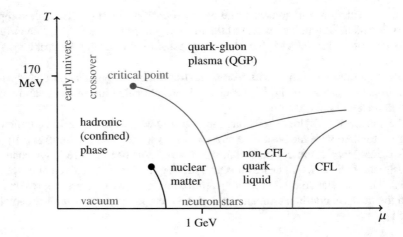

Fig. 15.11 Conjectured form of the phase diagram of QCD at finite temperature and baryon density. At high temperature we expect a deconfined quark-gluon plasma. At low temperatures and low densities, we are in the confined hadronic phase. If we increase the density at low temperature, then we move from the vacuum sector into the phase with nuclear matter. Increasing the density even further, we encounter a quark phase. At ultrahigh densities we expect to find a color-flavor-locked (CFL) phase with color-superconducting quark matter. There is a crossover between the confined and deconfined phases at low densities

By doubling the degrees of freedom, prove the important identity

$$\det M = (\mathrm{Pf}(M))^2 \ .$$

Transform the Grassmann variables in (15.117) according to $\eta \rightarrow R\eta$, and show

$$\mathrm{Pf}\left(R^t M R\right) = \det(R)\,\mathrm{Pf}(M) \ .$$

Prove that for an antisymmetric matrix M of dimensional $2N$ we have

$$\mathrm{Pf}(M^t) = (-1)^N \mathrm{Pf}(M) \ .$$

Show, by using the relation between the Pfaffian and determinant, that

$$\delta \log \det(M) = \mathrm{tr}\left(M^{-1}\delta M\right) \Longrightarrow \delta \log \mathrm{Pf}(M) = \frac{1}{2}\mathrm{tr}\left(M^{-1}\delta M\right) \ .$$

Let us assume that the antisymmetric M is a tensor product of a symmetric matrix S and an antisymmetric matrix A. By transforming both matrices into their normal forms, prove that

$$\mathrm{Pf}\,M = \mathrm{Pf}(S \otimes A) = (\det S)^{\dim A} \cdot (\mathrm{Pf}A)^{\dim S} \ .$$

15.2 (Staggered Fermions) Prove the identity (15.66) which was used to diagonalize the naive Wilson operator to obtain the operator for staggered fermions.

15.3 (Supersymmetric Actions) Show that the action (15.96) is left invariant
by the supersymmetry transformation (15.97), which contains a constant and
anticommuting Majorana parameter ε.

15.4 (Fermions at Finite Temperature) We have seen that in the path integral for
fermions at finite temperature T, one integrates over anticommuting fields which are
antiperiodic in Euclidean time, $\psi(\tau + \beta_T, x) = -\psi(\tau, x)$. First show that formally
the path integral for free fermions with $\mathscr{L}_F = \bar{\psi}(\partial\!\!\!/ + m)\psi$ can be rewritten as

$$Z_\beta = \oint \mathscr{D}\psi \mathscr{D}\bar{\psi}\, e^{-S_F} \propto \det_{\mathrm{ap}}(\partial\!\!\!/ + m)$$

$$= \det_{\mathrm{ap}}^{1/2}\left[\left(-\Delta + m^2\right) \mathbb{1}_4\right] = \det_{\mathrm{ap}}^2\left(-\Delta + m^2\right),$$

where the determinants are calculated on the space of antiperiodic functions. You
may exploit that $\partial\!\!\!/ + m$ is γ_5-hermitian. Now calculate $\det_{\mathrm{ap}}(-\Delta + m^2)$ on this
space (e.g., with the zeta-function regularization used in Sect. 5.2 to calculate
the determinant with respect to periodic boundary conditions), and show that the
resulting free energy density is that for free fermions in a box.

Appendix: The SLAC Derivative

We introduce the SLAC derivative on a one-dimensional periodic lattice with
equidistant sampling points

$$x_k = x_0 + k, \quad k = 1, \ldots, N . \tag{15.118}$$

The set of lattice functions $x_k \to \psi_k \in \mathbb{C}$, equipped with the scalar product

$$(\phi, \psi) = \sum_{k=1}^{N} \bar{\phi}_k \psi_k , \tag{15.119}$$

defines a Hilbert space. If ψ is normalized to one, we may interpret $|\psi_k|^2$ as the
probability of finding the particle described by the wave function ψ at site x_k. Then
the expectation value of the position operator is given by

$$\langle \hat{x} \rangle_\psi = \langle \bar{\psi}|\hat{x}|\psi \rangle = \sum x_k |\psi_k|^2 \equiv \sum_{kk'} \bar{\psi}_k x_{kk'} \psi_{k'} . \tag{15.120}$$

As expected, the position operator $\hat{x}$ is diagonal in real space such that its matrix elements vanish if $k \neq k'$. To introduce the SLAC derivative, we switch to momentum space with wave functions $\tilde{\psi}(p_\ell) \equiv \tilde{\psi}_\ell$ given by

$$\tilde{\psi}_\ell = \frac{1}{\sqrt{N}} \sum_{k=1}^{N} e^{-ip_\ell x_k} \psi_k, \quad \ell = 1, \ldots, N . \tag{15.121}$$

The inverse Fourier transformation reads

$$\psi_k = \frac{1}{\sqrt{N}} \sum_{\ell=1}^{N} e^{ip_\ell x_k} \tilde{\psi}_\ell, \quad k = 1, \ldots, N . \tag{15.122}$$

We choose the $\{p_\ell\}$ symmetric with respect to the origin,

$$p_\ell = \frac{2\pi}{N} \left(\ell - \frac{N+1}{2} \right) , \tag{15.123}$$

and with this choice the number of sites must be odd to obtain periodic wave functions, and it must be even to obtain antiperiodic wave functions.

Now we seek a lattice momentum operator $\hat{p}$ which is diagonal in momentum space and has eigenvalues p_ℓ. This means that below the cutoff, it has exactly the same eigenvalues as the continuum operator on the interval. Similarly as in the continuum, we interpret $|\tilde{\psi}_\ell|^2$ as probability for finding the eigenvalue p_ℓ of $\hat{p}$. Then the mean value of $f(\hat{p})$ is

$$\langle f(\hat{p}) \rangle_\psi = \sum_\ell f(p_\ell) |\tilde{\psi}_\ell|^2 = \frac{1}{N} \sum_\ell \sum_{kk'} e^{ip_\ell(x_k - x_{k'})} f(p_\ell) \bar{\psi}_k \psi_{k'}$$

$$= \sum_{kk'} \bar{\psi}_k f(p)_{kk'} \psi_{k'}, \quad f(p)_{kk'} = \frac{1}{N} \sum_\ell e^{ip_\ell(x_k - x_{k'})} f(p_\ell) . \tag{15.124}$$

Of course the operator $f(\hat{p})$ is non-diagonal in position space, and to find its matrix elements $f(p)_{kk'}$, we define the generating function

$$Z(x) = \frac{1}{N} \sum_{\ell=1}^{N} e^{iNp_\ell x} = \frac{\sin(\pi N x)}{N \sin(\pi x)} . \tag{15.125}$$

The matrix elements are obtained by differentiation,

$$f(p)_{kk'} = f\left(\frac{1}{iN} \frac{d}{dx} \right) Z(x) \Big|_{x=t_{kk'}}, \quad t_{kk'} = \frac{k - k'}{N} . \tag{15.126}$$

In particular we find

$$p_{kk} = 0, \quad p_{k \neq k'} = \frac{\pi}{iN}(-)^{k-k'} \frac{1}{\sin(\pi t_{kk'})}, \tag{15.127}$$

and these matrix elements define the SLAC derivative $\partial_{\text{slac}} = i\hat{p}$ in position space.

References

1. J. Smit, *Introduction to Quantum Field Theories on a Lattice* (Cambridge University Press, Cambridge, 2002)
2. I. Montvay, G. Münster, *Quantum Fields on a Lattice* (Cambridge University Press, Cambridge, 2010)
3. H.J. Rothe, *Lattice Gauge Theories: An Introduction* (World Scientific, Singapore, 2012)
4. C. Gattringer, C. Lang, *Quantum Chromodynamics on the Lattice*. Lecture Notes in Physics, vol. 788 (Springer, Berlin, 2012)
5. E. Seiler, *Gauge Theories as a Problem of Constructive Quantum Field Theory and Statistical Mechanics*. Lecture Notes in Physics, vol. 159 (Springer, Berlin, 2014)
6. V. Mitrjushkin, G. Schierholz (eds.), *Lattice Fermions and Structure of the Vacuum* (Kluwer Academic, Dordrecht, 2000)
7. S. Chandrasekharan, U.-J. Wiese, An introduction to chiral symmetry on the lattice. Prog. Part. Nucl. Phys. **53**, 373 (2004)
8. G. Roepstorff, *Path Integral Approach to Quantum Physics* (Springer, Berlin, 1996)
9. S.D. Drell, M. Weinstein, S. Yankielowicz, Variational approach to strong coupling field theory. 1. ϕ^4 Theory. Phys. Rev. **D14**, 487 (1976)
10. S.D. Drell, M. Weinstein, S. Yankielowicz, Strong coupling field theories: 2. fermions and gauge fields on a lattice. Phys. Rev. **D14**, 1627 (1976)
11. L.H. Karsten, J. Smit, The vacuum polarization with SLAC lattice fermions. Phys. Lett. **B85**, 100 (1979)
12. G. Bergner, T. Kaestner, S. Uhlmann, A. Wipf, Low-dimensional supersymmetric lattice models. Ann. Phys. **323**, 946 (2008)
13. A. Kirchberg, D. Laenge, A. Wipf, From the Dirac operator to Wess-Zumino models on spatial lattices. Ann. Phys. **316**, 357 (2005)
14. J. Förster, A. Saenz, U. Wolff, Matrix algorithm for solving Schrödinger equations with position-dependent mass or complex optical potentials. Phys. Rev. **E86**, 016701 (2012)
15. T. Kaestner, G. Bergner, S. Uhlmann, A. Wipf, C. Wozar, Two-dimensional Wess-Zumino models at intermediate couplings. Phys. Rev. **D78**, 095001 (2008)
16. K. Wilson, *New Phenomena in Subnuclear Physics* (Plenum, New York, 1977)
17. L. Susskind, Lattice fermions. Phys. Rev. **D16**, 3031 (1977)
18. M. Creutz, Chiral anomalies and rooted staggered fermions. Phys. Lett. **B649**, 230 (2007)
19. H. Nielsen, M. Ninomiya, Absence of neutrinos on a lattice (I). Proof by homotopy theory. Nucl. Phys. **B185**, 20 (1981)
20. H. Nielsen, M. Ninomiya, Absence of neutrinos on a lattice (II). Intuitive topological proof. Nucl. Phys. **B193**, 173 (1981)
21. L.H. Karsten, J. Smit, Lattice fermions: species doubling, chiral invariance and the triangle anomaly. Nucl. Phys. **B183**, 103 (1981)
22. D. Friedan, A proof of the Nielsen Ninomiya theorem. Commun. Math. Phys. **85**, 481 (1982)
23. C. Itzykson, J.M. Drouffe, *Statistical Field Theory I* (Cambridge University Press, Cambridge, 1991)
24. P.H. Ginsparg, K.G. Wilson, A remnant of chiral symmetry on the lattice. Phys. Rev. **D25**, 2649 (1982)

25. P. Hasenfratz, Lattice QCD without tuning, mixing and current renormalization. Nucl. Phys. **B525**, 401 (1998)
26. P. Hasenfratz, Prospects for perfect actions. Nucl. Phys. Suppl. **63**, 53 (1998)
27. P. Hasenfratz, S. Hauswirth, T. Jorg, F. Niedermayer, K. Holland, Testing the fixed point QCD action and the construction of chiral currents. Nucl. Phys. **B643**, 280 (2002)
28. H. Neuberger, Exactly massless quarks on the lattice. Phys. Lett. **B417**, 141 (1998)
29. H. Neuberger, More about exactly massless quarks on the lattice. Phys. Lett. **B427**, 353 (1998)
30. M. Lüscher, Exact chiral symmetry on the lattice and the Ginsparg-Wilson relation. Phys. Lett. **B428**, 342 (1998)
31. D. Kaplan, A method for simulating chiral fermions on the lattice. Phys. Lett. **B288**, 342 (1992)
32. Y. Shamir, Chiral fermion from lattice boundaries. Nucl. Phys. **B406**, 90 (1993)
33. V. Furman, Y. Shamir, Axial symmetries in lattice QCD with Kaplan fermions. Nucl. Phys. **B439**, 54 (1995)
34. S.A. Frolov, A.A. Slavnov, An invariant regularization of the standard model. Phys. Lett. **B309**, 344 (1993)
35. R. Narayanan, H. Neuberger, Infinitely many regulator fields for chiral fermions. Phys. Lett. **B302**, 62 (1993)
36. R. Narayanan, H. Neuberger, Chiral determinants as an overlap of two Vacua. Nucl. Phys. **B412**, 574 (1994)
37. C. Gattringer, I. Hip, New approximate solutions of the Ginsparg-Wilson equation: tests in 2D. Phys. Lett. **B480**, 112 (2000)
38. C. Gattringer, A new approach to Ginsparg-Wilson fermions. Phys. Rev. **D63**, 114501 (2001)
39. C. Gattringer, et al., Quenched spectroscopy with fixed point and chirally improved fermions. Nucl. Phys. **B677**, 3 (2004)
40. P. Hasenfratz, V. Laliena, F. Niedermayer, The index theorem in QCD with a finite cutoff. Phys. Lett. **427**, 125 (1998)
41. N. Cabibbo, L. Maiani, G. Parisi, R. Petronzio, Bounds on the Fermions and Higgs boson masses in grand unified theories. Nucl. Phys. **B158**, 295 (1979)
42. P. Gerhold, K. Jansen, Upper Higgs boson mass bounds from a chirally invariant lattice Higgs-Yukawa model. J. High Energy Phys. **04**, 094 (2010); Lower Higgs boson mass bounds from a chirally invariant lattice Higgs-Yukawa model with overlap fermions. J. High Energy Phys. **07**, 025 (2009)
43. P.A. Zyla, et al., Review of particle physics (particle data group). Prog. Theor. Exp. Phys. **8**, 083C01 (2020)
44. A. Eichhorn, H. Gies, J. Jaeckel, T. Plehn, M.M. Scherer, The Higgs mass and the scale of new physics. J. High Energy Phys. **04**, 022 (2015)
45. S. Elitzur, E. Rabinovici, A. Schwimmer, Supersymmetric models on the lattice. Phys. Lett. **B119**, 165 (1982)
46. M. Beccaria, C. Rampino, World-line path integral study of supersymmetry breaking in the Wess-Zumino model. Phys. Rev. **D67**, 127701 (2003)
47. S. Catterall, S. Karamov, Exact lattice supersymmetry: the two-dimensional $N = 2$ Wess-Zumino model. Phys. Rev. **D65**, 09450 (2002)
48. S. Catterall, S. Karamov, A lattice study of the two-dimensional Wess-Zumino model. Phys. Rev. **D68**, 014503 (2003)
49. C. Wozar, A. Wipf, Supersymmetry breaking in low dimensional models. Ann. Phys. **327**, 774 (2012)
50. J. Bartels, J.B. Bronzan, Supersymmetry on a lattice. Phys. Rev. **D28**, 818 (1983)
51. F. Synatschke, G. Gies, A. Wipf, Phase diagram and fixed point structure of two dimensional N=1 Wess-Zumino models. Phys. Rev. **80**, 085007 (2009)
52. M. Troyer, U.J. Wiese, Computational complexity and fundamental limitations to fermionic quantum Monte Carlo simulations. Phys. Rev. Lett. **94**, 170201 (2005)
53. D.H. Weingarten, D.N. Petcher, Monte Carlo integration for lattice gauge theories with fermions. Phys. Lett. **B99**, 333 (1981)

54. W. Frazer, A survey of methods of computing minimax and near-minimax polynomial approximations for functions of a single independent variable. J. ACM **12**, 295 (1965)
55. F. Brown, et al., On the existence of a phase transition for QCD with three light quarks. Phys. Rev. Lett. **65**, 2491 (1990)
56. Y. Aoki, G. Endrödi, Z. Fodor, S. Katz, K. Szabó, The order of the quantum chromodynamics transition predicted by the standard model of particle physics. Nature **443**, 675 (2006)
57. A. Bazavov, et al., Equation of state and QCD transition at finite temperature. Phys. Rev. **D80**, 014504 (2009)
58. I. Sachs, A. Wipf, Generalized thirring models. Ann. Phys. **249**, 380 (1996)
59. Z. Fodor, S.D. Katz, Lattice determination of the critical point of QCD at finite T and μ. J. High Energy Phys. **0203**, 014 (2002)
60. C.R. Allton, et al., The QCD thermal phase transition in the presence of a small chemical potential. Phys. Rev. **D66**, 074507 (2002)
61. P. de Forcrand, O. Philipsen, The QCD phase diagram for small densities from imaginary chemical potential. Nucl. Phys. **B642**, 290 (2002)
62. R.D. Pisarski, Quark gluon plasma as a condensate of SU(3) Wilson lines. Phys. Rev. **D62**, 111501 (2000)
63. T.K. Herbst, J.M. Pawlowski, B.J. Schaefer, The phase structure of the Polyakov–quark-meson model beyond mean field. Phys. Lett. **B696**, 58 (2011)
64. J.B. Kogut, M.A. Stephanov, D. Toublan, J.J.M. Verbaarschot, A. Zhitnitsky, QCD-like theories at finite baryon density. Nucl. Phys. **B582**, 477 (2000)
65. S. Hands, S. Kim, J.I. Skullerud, A quarkyonic phase in dense two color matter? Phys. Rev. **D81**, 091502 (2010)
66. K. Holland, P. Minkowski, M. Pepe, U.J. Wiese, Exceptional confinement in G(2) gauge theory. Nucl. Phys. **B668**, 207 (2003)
67. B. Wellegehausen, C. Wozar, A. Wipf, Phase diagram of the lattice G(2) Higgs model. Phys. Rev. **D83**, 114502 (2011)
68. A. Maas, L. von Smekal, B. Wellegehausen, A. Wipf, The phase diagram of a gauge theory with fermionic baryons. Phys. Rev. **D86**, 111901 (2012)

Finite Temperature Schwinger Model

16

The study of exactly soluble field theories has always received a good deal of attention in the hope that they might shed some light on more realistic theories. One such model is the Schwinger model or QED_2—quantum electrodynamics in two spacetime dimensions [1]. It serves as an important tool in illustrating various (related) field theoretical concepts such as mass generation, dynamical symmetry breaking, charge shielding, or fermion trapping [2].

At zero temperature the model with massless fermions has been solved some time ago by using operator methods [3] and the path integral formulation [4]. Some properties of the model (e.g., the non-trivial vacuum structure) are more transparent in the operator approach. Others, for example, the role of the chiral anomaly, are better seen in the path integral approach. Solutions of QED_2 on finite Euclidean spacetime have clarified various global aspects of the theory. For example, Jayewardena studied the system on a *two*-dimensional sphere and emphasized the role of the fermionic zero modes [5]. Massless QED_2 has also been solved on a two-dimensional disk with self-adjoint local bag boundary conditions for the Fermi field [6, 7].

We have seen earlier in Sects. 5.2 and 15.7 that expectation values in the canonical or grand canonical ensembles are given by the Euclidean path integral over Bose (Fermi) fields which are periodic (antiperiodic) in imaginary time with period $\beta = 1/T$. In order to avoid infrared divergences, which sometimes plague the analysis of two-dimensional gauge theories, we quantize the system in a finite spatial box, such that the Euclidean Dirac operator has discrete eigenvalues. In other words, we assume that spacetime is a finite cylinder $[0, \beta] \times [0, L]$ with finite volume $V = \beta \cdot L$. The finite-temperature periodicity conditions in time direction are supplemented by suitable boundary conditions in the spatial direction. For example, one may impose parity-breaking bag boundary conditions. The corresponding solution has been presented in [8] and yields an interesting interpretation of the θ-parameter in QED_2. In this chapter we impose periodic boundary conditions in the spatial direction. This means that Euclidean spacetime is a 2-torus as depicted in

© The Author(s), under exclusive license to Springer Nature Switzerland AG 2021
A. Wipf, *Statistical Approach to Quantum Field Theory*, Lecture Notes
in Physics 992, https://doi.org/10.1007/978-3-030-83263-6_16

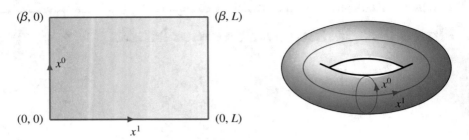

Fig. 16.1 Gauge invariant fields are periodic in spatial direction with period L and temporal direction with period β, such that we may view Euclidean spacetime as 2-torus $\mathcal{T}$

Fig. 16.1. Solutions of the Schwinger model on the torus were presented for Dirac-Kähler fermions in [9] and for Dirac fermions in [10]. Here I follow the latter work obtained in collaboration with Ivo Sachs.

The main results will be analytic expressions for the effective actions and zero modes in all topological sectors and the temperature and volume dependence of the chiral condensate. We shall see that the torus calculation nicely splits the chiral anomaly responsible for the non-vanishing condensate into its high- and low-energy parts. We will obtain explicit results for (thermal) Wilson loop correlators and in particular their temperature dependence as well as the gauge invariant fermionic two-point functions. Towards the end of the chapter, we shall comment on generalizations of the presented results and the relevance of the exact solution for lattice simulations and more recent development in quantum field theory.

A correct lattice regularization of a gauge theory with dynamical fermions should reproduce perturbation theory for smooth gauge fields and should reproduce instanton physics correctly. In particular the second feature and the chiral properties are relevant for the physics of the Schwinger model. This explains why the model has played and still plays a prominent role in the formulation of chiral lattice gauge theories [11]. It serves as a test bed for checking new formulations and algorithms to deal with dynamical fermions on the lattice. Variants of the Schwinger model have been used as a benchmark to test methods against the fermion sign problem in numerical Monte Carlo simulations. For example, the tensor network methods [12] (in particular the matrix product states ansatz) have been applied to the Schwinger model in [13].

16.1   The Massless Schwinger Model

We begin with introducing the Schwinger model or QED_2 in 2 Euclidean dimensions with massless Dirac fermions. We shall emphasize the global features involved when considering the system on a torus. The classical action of the model is

$$S[A, \bar{\psi}, \psi] = \int d^2x \, \mathcal{L}(A, \bar{\psi}, \psi), \qquad \mathcal{L} = \frac{E^2}{2} - \bar{\psi} i \slashed{D} \psi \qquad (16.1)$$

and contains the field strength and Euclidean Dirac operator acting on spinor fields,

$$E = F_{01} = \partial_0 A_1 - \partial_1 A_0 \quad \text{and} \quad \slashed{D} = \gamma^\mu \left(\partial_\mu - ieA_\mu \right). \tag{16.2}$$

We need not pick a particular representation for the γ^μ but should keep in mind that in Euclidean spacetime with signature $(+, +)$, they are hermitian. The Lagrangian is invariant under global $U(1)_A$ chiral transformations,

$$\psi(x) \mapsto e^{\alpha \gamma_*} \psi(x) \quad \text{and} \quad \bar{\psi}(x) \mapsto \bar{\psi}(x) e^{\alpha \gamma_*}, \quad \gamma_* = -i\gamma^0 \gamma^1. \tag{16.3}$$

It is well known that this symmetry does not exist on the quantum level due to the chiral anomaly. The anomalous breaking of $U(1)_A$ is driven by tunneling between topologically different gauge sectors as described by gauge field configurations supporting fermionic zero modes—so-called instantons [14]. Since instanton-induced amplitudes are suppressed at high temperature, we expect that the $U(1)_A$ is almost restored at sufficiently high temperature. We shall see explicitly that this is the case in the Schwinger model: at low temperature instanton configurations induce a chiral condensate $\langle \bar{\psi} \psi \rangle$ which vanishes exponentially fast at high temperature $T \gg m_\gamma$, where m_γ is the anomaly-induced "photon mass." Actually, in two dimensions the electric charge e has the mass dimension, and the model is super-renormalizable. The coupling e plays a role similar to Λ_{QCD}, and the anomaly-induced mass m_γ in QED_2, which is proportional to e, mimics the mass of η' in QCD.

The generating functional for the Green functions is

$$Z[J, \eta, \bar{\eta}] = C^{-1} \int \mathscr{D}A \mathscr{D}\bar{\psi} \mathscr{D}\psi \; e^{-S[A, \bar{\psi}, \psi] - S_{gf}[A] + (J, A) + (\bar{\eta}, \psi) + (\bar{\psi}, \eta)}, \tag{16.4}$$

where the c-number vector source $J^\mu(x)$ and Grassmann-valued spinorial sources $\bar{\eta}(x)$ and $\eta(x)$ couple to the gauge potential and spinor fields via

$$(J, A) = \int d^2x \, J^\mu(x) A_\mu(x), \quad (\bar{\eta}, \psi) = \int d^2x \, \bar{\eta}(x) \psi(x), \tag{16.5}$$

and similarly for $(\bar{\psi}, \eta)$. Below we shall use a decomposition of the gauge potential (corresponding to a particular gauge fixing) for which the Faddeev-Popov determinant is independent of A such that no ghosts are required in the path integral. The normalization constant C is chosen such that $Z[0, 0, 0] = 1$.

In a first step, we evaluate the fermionic path integral

$$Z[A; \eta, \bar{\eta}] = \int \mathscr{D}\bar{\psi} \mathscr{D}\psi \; e^{\int \bar{\psi} i \slashed{D} \psi + (\bar{\eta}, \psi) + (\bar{\psi}, \eta)} \tag{16.6}$$

and thereby treat the gauge field as an external field. On the torus the hermitian Dirac operator $i\slashed{D}$ has discrete real eigenvalues and may possess normalizable zero modes. Let us assume that there are ν zero modes $\psi_1, \ldots, \psi_\nu$. In addition there exist an infinite number of excited eigenmodes $\psi_{\nu+1}, \psi_{\nu+2}, \ldots$ with non-zero real eigenvalues.

The number of zero modes can be determined as follows: First we note that since the hermitian γ_* in (16.3) anticommutes with Dirac operator, the excited modes come in pairs with opposite real eigenvalues,

$$\mathrm{i}\slashed{D}\psi = \lambda\psi \implies \mathrm{i}\slashed{D}(\gamma_*\psi) = -\lambda(\gamma_*\psi)\,, \tag{16.7}$$

such that for $\lambda \neq 0$ the eigenfunctions ψ and $\gamma_*\psi$ are orthogonal fields in the Hilbert space $L_2(\mathcal{T}) \times \mathbb{C}^2$, which means $(\psi, \gamma_*\psi) = 0$. In passing we note that similar statements hold for non-Abelian Euclidean gauge theories on even dimensional compact spaces without boundary. Even dimensional since only in dimension $d = 2, 4, 6, \ldots$ there exists a generalization γ_* of γ_5. We assume a compact space without boundary since then the spectrum of $\mathrm{i}\slashed{D}$ is discrete and left-right symmetric.[1] The pairing of excited eigenmodes implies that their contribution to the (supersymmetric) partition function vanishes, also called Witten supersymmetry index [15]

$$\mathrm{tr}\left(\gamma_* \mathrm{e}^{t\slashed{D}^2}\right) = \sum_{p=1}^{\nu}(\psi_p, \gamma_*\psi_p) + \sum_{p=\nu+1}^{\infty} \mathrm{e}^{-t\lambda_p^2}(\psi_p, \gamma_*\psi_p) = \sum_{p=1}^{\nu}(\psi_p, \gamma_*\psi_p)\,. \tag{16.8}$$

There is no pairing of zero modes. Indeed, on the kernel of the Dirac operator, γ_* and $\mathrm{i}\slashed{D}$ commute, and we may assume that a (normalized) zero mode is either right- or left-handed, which means $\gamma_*\psi = \psi$ or $\gamma_*\psi = -\psi$. Let us assume that there are n_+ right-handed and n_- left-handed zero modes. Then the last sum in (16.8) is just the difference between the number of right- and left-handed zero modes, which is the index of the Dirac operator.

The result (16.8) implies that the trace is independent of the proper time t such that we may as well employ the small-t expansion of the heat kernel of $(\mathrm{i}\slashed{D})^2$ (for more details you may consult the review [16])

$$\langle x|\mathrm{e}^{t\slashed{D}^2}|x\rangle \sim \frac{1}{4\pi t}\left(1 + \gamma_*eEt + O(t^2)\right). \tag{16.9}$$

Thus, to calculate the trace (16.8) we multiply (16.9) with γ_*, take the trace over Dirac indices, and integrate over spacetime. This way one obtains the index theorem for the massless Dirac operator in 2 Euclidean dimensions (more on index theorems

---

[1] Local boundary conditions are in conflict with chiral symmetry. For example, if ψ obeys hermitian bag-boundary conditions, then $\gamma_*\psi$ does not fulfill these conditions; see the analysis in [7].

can be found in [17])

$$n_+ - n_- = \frac{e}{2\pi} \int d^2x \, F_{01} \equiv \frac{1}{2\pi} \Phi \, . \tag{16.10}$$

It relates the index of the Dirac operator to the flux of F_{01}. This formula shows that for a non-vanishing Φ, there are fermionic zero modes and that Φ is quantized in integer multiples of 2π. Actually, other versions of the index theorem arise in various physical systems. Generally speaking, an index theorem relates global properties of a gauge (or gravitational) field and of spacetime to the number of zero modes of an elliptic differential operator acting on this spacetime. Most important in this context is the Atiyah-Singer index theorem [18]. One of the earliest and most important applications is 't Hooft's resolution of the U(1) problem [19]. The theorem contributed to our understanding of anomalies [20] and in particular how instantons—these are non-Abelian gauge field configuration with a nonzero flux—can lead to the breaking of global symmetries. For a lucid explanation, the reader may consult Sidney Coleman's Erice lectures "The uses of instantons" [21] or the exhaustive text [22].

To determine $n_\pm$ separately, we decompose the gauge potential eA_μ into a global instanton potential eA_μ^{inst} with constant field strength Φ/V and a local fluctuation $e\delta A_\mu = -\varepsilon_{\mu\nu}\partial_\nu\varphi$ with periodic and dimensionless φ about the instanton,

$$eA_\mu = eA_\mu^{\text{inst}} - \varepsilon_{\mu\nu}\partial_\nu\varphi, \quad eE = \frac{\Phi}{V} + \Delta\varphi \, . \tag{16.11}$$

In two dimensions we may use the identity $\gamma_\mu\gamma_* = -i\varepsilon_{\mu\nu}\gamma_\nu$ to prove that

$$\slashed{D}_A = e^{\gamma_*\varphi} \, \slashed{D}_{A^{\text{inst}}} \, e^{\gamma_*\varphi} \, . \tag{16.12}$$

It follows that the number of fermionic zero modes is independent of the periodic field φ and hence is the same for A and A^{inst}. For the simple instanton potential, the second-order operator

$$\slashed{D}_{\text{inst}}^2 = D_{\text{inst}}^2 + \gamma_* \frac{\Phi}{V} \tag{16.13}$$

is non-positive, and we see at once that all zero modes are either right- or left-handed for non-vanishing Φ. Only in the sector with $\Phi = 0$ it can happen that there are both right- and left-handed zero modes. What we have shown then is that for $\Phi \neq 0$, either n_+ or n_- is zero such that there are exactly $\nu = |\Phi|/2\pi$ zero modes. If the flux is positive, they are all right-handed, else they are left-handed. The property that either n_+ or n_- must be zero is sometimes called vanishing theorem.

After having counted the number of zero modes, we proceed by expanding the Dirac field in an adapted *orthonormal* eigenbase as

$$\psi(x) = \sum_{p=1}^{\nu} \alpha_p \psi_p(x) + \sum_{p=\nu+1}^{\infty} \beta_p \psi_p(x) \tag{16.14}$$

and similarly $\bar{\psi}$, so that the source terms in (16.5) split into a sum containing the zero modes and a sum containing the excited modes,

$$(\bar{\eta}, \psi) = \sum_{p \leq \nu}(\bar{\eta}, \psi_p)\alpha_p + \sum_{p > \nu}(\bar{\eta}, \psi_p)\beta_p \,, \quad (\bar{\psi}, \eta) = \sum_{p \leq \nu}\bar{\alpha}_p(\bar{\psi}_p, \eta) + \sum_{p > \nu}\bar{\beta}_p(\bar{\psi}_p, \eta) \,.$$

Inserting this decomposition back into (16.6) and using that for an orthonormal base $\mathscr{D}\bar{\psi}\mathscr{D}\psi = \mathscr{D}\bar{\alpha}\mathscr{D}\alpha\mathscr{D}\bar{\beta}\mathscr{D}\beta$, the Grassmannian integral over the α's can easily be done since the action does not depend on them. This way one finds for the zero-mode contribution to (16.6)

$$\int \mathscr{D}\bar{\alpha}\mathscr{D}\alpha \prod_{p=1}^{\nu} e^{(\bar{\eta}, \psi_p)\alpha_p}\, e^{\bar{\alpha}_p(\bar{\psi}_p, \eta)} = \prod_{p=1}^{\nu}(\bar{\eta}, \psi_p)(\bar{\psi}_p, \eta) \,. \tag{16.15}$$

The remaining Gaussian integral over the Grassmann variables β is performed by shifting the β's (and similarly the $\bar{\beta}$'s) according to

$$\beta_p \longrightarrow \beta_p - \frac{1}{\lambda_p}(\bar{\psi}_p, \eta), \quad p > \nu \,.$$

After this shift the β integration yields

$$\int \mathscr{D}\bar{\beta}\mathscr{D}\beta \prod_{p=\nu+1}^{\infty} e^{\lambda_p \bar{\beta}_p \beta_p + (\bar{\eta}, \psi_p)\beta_p + \bar{\beta}_p(\bar{\psi}_p, \eta)} = e^{-\int \bar{\eta}(x)G_e(A;x,y)\eta(y)} \cdot \det'(i\slashed{D}) \,,$$

$$\tag{16.16}$$

where $\det'$ denotes the determinant with the zero eigenvalues omitted and

$$G_e(A; x, y) = \sum_{p=\nu+1}^{\infty} \frac{\psi_p(x)\psi_p^{\dagger}(y)}{\lambda_p} \tag{16.17}$$

is the Green function on the space orthogonal to the zero modes. It obeys the differential equation

$$i\slashed{D}\, G_e(A; x, y) = \delta(x - y) - P(x, y), \quad \text{where} \quad P(x, y) = \sum_{p=1}^{\nu} \psi_p(x)\psi_p^{\dagger}(y) \tag{16.18}$$

projects onto the subspace spanned by the zero modes. Due to the pairing property of the excited modes, this Green function anticommutes with γ_*:

$$\gamma_* G_e(A; x, y)\gamma_* = -G_e(A; x, y) \,. \tag{16.19}$$

Inserting the expressions (16.15) and (16.16) into (16.6), we end up with

$$Z[A; \bar{\eta}, \eta] = \prod_{p=1}^{\nu} (\bar{\eta}, \psi_p)(\bar{\psi}_p, \eta) \ e^{-\int \bar{\eta}(x) \, G_e(A;x,y) \, \eta(y)} \, \det'(i\slashed{D}) \,. \tag{16.20}$$

To determine $Z(A; \bar{\eta}, \eta)$ for a fixed potential, we need the ν zero modes of $i\slashed{D}$ besides its determinant and Green function with zero modes omitted. The (not normalized) fermionic $2n$-point functions for a given gauge field are obtained by differentiation with respect to the external sources

$$\int \mathscr{D}\bar{\psi}\mathscr{D}\psi \ \psi_{\alpha_1}(x_1)\bar{\psi}_{\beta_1}(y_1)...\psi_{\alpha_n}(x_n)\bar{\psi}_{\beta_n}(y_n) \, e^{\int \bar{\psi} \, i\slashed{D}\psi}$$

$$= \frac{\delta^{2n}}{\delta\eta^{\beta_n}(y_n)\delta\bar{\eta}^{\alpha_n}(x_n)...\delta\eta^{\beta_1}(y_1)\delta\bar{\eta}^{\alpha_1}(x_1)} \, Z[A; \bar{\eta}, \eta]\Big|_{\bar{\eta}=\eta=0} \,. \tag{16.21}$$

Using for $Z[A; \bar{\eta}, \eta]$ the result (16.20) we immediately read off that:

1. The fermionic partition function

$$Z[A; 0, 0] = \det(i\slashed{D}) \tag{16.22}$$

   which enters the normalization constant C in (16.4) is only non-zero for gauge fields with vanishing flux.
2. The 2-point functions are only non-zero if $\Phi = 0$ or $\Phi = \pm 2\pi$. From (16.19) it follows that the expectation values

$$\langle \bar{\psi} P_\pm \psi \rangle, \quad \text{where} \quad P_\pm = \frac{1}{2}(1 \pm \gamma_*) \tag{16.23}$$

   project onto the right- and left-handed spinors and are non-vanishing only for $\Phi = \pm 2\pi$, when the Dirac operator has one zero mode ψ_1 of chirality ± 1, and then

$$\int \mathscr{D}\bar{\psi}\mathscr{D}\psi \ e^{\int \bar{\psi} \, i\slashed{D}\psi} \ \bar{\psi}(x) P_\pm \psi(x) = -\text{tr}\left(\psi_1^\dagger(x) P_\pm \psi_1(x)\right) \det'(i\slashed{D}) \,. \tag{16.24}$$

3. Similar results hold for the higher $2n$-point functions. For the 4-point functions, only gauge potentials with fluxes $\Phi \in \{0, \pm 2\pi, \pm 4\pi\}$ contribute. More gener-

ally, for expectation values of the operators with chiral charge $\Delta n = n_+ - n_-$

$$O(\psi) = (\bar{\psi} P_+ \psi)^{n+} (\bar{\psi} P_- \psi)^{n-}, \qquad O(e^{\alpha \gamma_*} \psi) = e^{\Delta n \, \alpha} O(\psi), \qquad (16.25)$$

only potentials with flux $\Phi = 2\pi \Delta n$ yield a non-vanishing result [23, 24]. For example, only gauge potentials with flux $\Phi = \pm 2\pi n$ contribute in the path integral for $(\bar{\psi} P_\pm \psi)^n$, and then only the zero modes and effective actions— but not the excited Green functions—are required to calculate the expectation values. This observation will simplify the calculation of the chiral condensate considerably.

16.2   Effective Action: Anomaly-Induced Local Part

We shall integrate the chiral anomaly to determine the φ-dependence of $\det' i \slashed{D}$ for a general gauge potential (16.11). In doing so we shall apply the deformation technique discussed in [25]. We present this method in some detail since it can be applied to find anomaly-induced terms in a variety of field theories. For example, one can solve the massless Thirring model with this technique [26]. Or one may integrate the conformal anomaly of quantum fields conformally coupled to gravity to calculate the anomaly-induced effective Polyakov action in $1 + 1$ dimensions [27] or the anomaly-induced effective Riegert action in $1 + 3$ dimensions [28].

To solve the Schwinger model, we consider the one-parameter family of Dirac operators

$$\slashed{D}_\varepsilon = e^{\gamma_* \varphi \varepsilon} \slashed{D}_{\text{inst}} e^{\gamma_* \varphi \varepsilon} \quad \text{with} \quad \frac{\mathrm{d}}{\mathrm{d}\varepsilon} \slashed{D}_\varepsilon = \{\gamma_* \varphi, \slashed{D}_\varepsilon\}, \qquad (16.26)$$

which, according to (16.12), interpolates between $\slashed{D}_{\text{inst}}$ and $\slashed{D}$, and calculate the variation of the zeta-function regularized determinants (introduced on page 90)

$$\log \det' i \slashed{D}_\varepsilon = \frac{1}{2} \log \det'\left(- \slashed{D}_\varepsilon^2\right) = -\frac{1}{2} \frac{\mathrm{d}}{\mathrm{d}s} \zeta_{\slashed{D}_\varepsilon^2}(s) \, |_{s=0} \,. \qquad (16.27)$$

Recall that the prime means that we leave out the zero modes. They do not enter the definition of the zeta-function, which is the analytic continuation of

$$\zeta_{\slashed{D}_\varepsilon^2}(s) = \sum_{p=\nu+1}^{\infty} \mu_p^{-s}, \qquad \mu_p = \lambda_p^2, \qquad (16.28)$$

to $\Re(s) \leq 1$. The μ_p are the ε-dependent positive eigenvalues of the square of $i \slashed{D}_\varepsilon$. Using (16.26) in the Feynman-Hellman formula

$$\frac{\mathrm{d}}{\mathrm{d}\varepsilon} \lambda_p = \left(\psi_p, i \frac{\mathrm{d}\slashed{D}_\varepsilon}{\mathrm{d}\varepsilon} \psi_p\right) = 2\lambda_p (\psi_p, \gamma_* \varphi \, \psi_p), \qquad (16.29)$$

the variation of the zeta-function can be written as

$$\frac{d}{d\varepsilon} \zeta_{\not{D}_\varepsilon^2}(s) = -s \sum_{p>v} \mu_p^{-s-1} \frac{d\mu_p}{d\varepsilon} = -4s \sum_{p>v} \mu_p^{-s} (\psi_p, \gamma_* \varphi \, \psi_p). \tag{16.30}$$

This can be further rewritten as a Mellin transform

$$\frac{d}{d\varepsilon} \zeta_{\not{D}_\varepsilon^2}(s) = -\frac{4s}{\Gamma(s)} \int_0^\infty dt \, t^{s-1} \sum_{p>v} e^{-t\mu_p} (\psi_p, \gamma_* \varphi \, \psi_p)$$

$$= -\frac{4s}{\Gamma(s)} \int dt \, t^{s-1} \mathrm{tr}'\left(e^{t\not{D}_\varepsilon^2} \gamma_* \varphi\right), \tag{16.31}$$

where the prime denotes the trace in the subspace spanned by the excited eigenmodes. Inserting the asymptotic expansion (16.9) and taking care of the prime by subtracting the zero-mode contribution, we find

$$\frac{d}{d\varepsilon} \frac{d}{ds} \zeta_{\not{D}_\varepsilon^2}(s) \,|_{s=0} = -\frac{2\varepsilon}{\pi} \int d^2x \, e E \varphi + 4 \int d^2x \, \mathrm{tr}\left(P_\varepsilon(x,x) \gamma_* \varphi(x)\right). \tag{16.32}$$

The projection density $P_\varepsilon(x,x)$ is the projection kernel in (16.18) on the diagonal, but of the deformed, operator. Observe that the zero modes of $\not{D}_\varepsilon$ and $\not{D}_{\mathrm{inst}}$ are related as

$$\psi_p^{(\varepsilon)} = e^{\mp\varepsilon\varphi} \psi_{\mathrm{inst},p}, \qquad p = 1, \ldots, v \tag{16.33}$$

with negative and positive sign in the exponent for right- and left-handed zero modes. These modes are in general not orthonormal, and the normmatrix

$$N_{pr}(\varepsilon) = (\psi_p^{(\varepsilon)}, \psi_r^{(\varepsilon)}) \tag{16.34}$$

is not the identity, so that the projection density reads

$$P_\varepsilon(x,x) = \sum_{pr} \psi_p^{(\varepsilon)}(x) N_{pr}^{-1}(\varepsilon) \, (\psi_r^{(\varepsilon)})^\dagger(x). \tag{16.35}$$

Now it follows from (16.33) and (16.34) that

$$\frac{d}{d\varepsilon} \log \det\left(N(\varepsilon)\right) = -2 \int d^2x \, \mathrm{tr}\left(P_\varepsilon(x,x) \gamma_* \varphi(x)\right), \tag{16.36}$$

which shows that the last term in (16.32) is a ε-derivative. Now we can easily integrate the anomaly equation (16.32) over ε from 0 to 1. This way we end up with

$$\det'(i\not{D}) = \det\left(\frac{N}{N_{\mathrm{inst}}}\right) \det'(i\not{D}_{\mathrm{inst}}) \, \exp\left(\frac{e}{2\pi} \int E\varphi\right). \tag{16.37}$$

This useful result which we obtained by integrating the anomaly presents an almost factorization of the determinants into an A^{inst}-dependent global and a φ-dependent anomaly-induced local factor. But the factorization is not complete since $\det N = \det \mathcal{N}(1)$ still contains a coupling between the instanton potential and the fluctuation via (16.33). Furthermore, since E in (16.11) depends on both Φ and φ, there is an apparent coupling in the last factor in (16.37). However, we shall see later that not all functions φ are permitted and that for the allowed ones this factor does not depend on Φ. Note that the anomaly induces a mass term for the photon—this is the last factor—and the induced mass $m_\gamma = e/\sqrt{\pi}$ is the same as on the infinite plane. It is the analog of the η'-mass in QCD. We shall further elaborate on this mass in Problem 16.1.

16.3 Effective Actions: Global Part

With the (almost) factorization of the zero modes in (16.33) and effective action in (16.37) into local and global contributions, the problem of finding the 2-point functions (16.24) reduces to the problem of computing the zero modes and effective actions for the instanton potentials A^{inst}. The index theorem tells us that on a torus $\mathcal{T}$, the flux

$$\Phi = \int_{\mathcal{T}} e F_{01}\, \mathrm{d}x_0 \mathrm{d}x_1 = \oint_{\partial \mathcal{T}} e A_\mu \mathrm{d}x^\mu \qquad (16.38)$$

is a multiple of 2π. If Φ is not zero, then the gauge potential cannot be periodic, since for a periodic A_μ the last integral would vanish. We can only impose the weaker and physically better motivated condition that gauge invariant fields are periodic. This means that the gauge potential must be periodic up to "gauge transformations".[2]

For the instanton potential with constant field strength, we choose a particular simple representative in each topological sector characterized by the flux Φ,

$$e A_0^{\text{inst}} = -\frac{\Phi}{V}x^1 + \frac{2\pi}{\beta}h_0 + \partial_0\lambda \quad \text{and} \quad e A_1^{\text{inst}} = \frac{2\pi}{L}h_1 + \partial_1\lambda . \qquad (16.39)$$

Below we shall prove that every gauge potential has the decomposition (16.11) with periodic φ and instanton potential (16.39) depending on a periodic gauge function λ, constant toron field h_μ, and quantized flux Φ. The instanton potential and full potential are both periodic in time and modulo a gauge transformation periodic in space,

$$A_\mu(x^0 + \beta, x^1) = A_\mu(x^0, x^1), \quad A_\mu(x^0, x^1 + L) = A_\mu(x^0, x^1) + \frac{1}{e}\partial_\mu\alpha(x) , \qquad (16.40)$$

---

[2] More precisely: A_μ defines a connection on a non-trivial U(1) bundle over the two-torus.

with transition function

$$\alpha(x) = -\frac{\Phi}{\beta}x^0 \,.$$
(16.41)

Similarly, spinor fields are antiperiodic in time and modulo a gauge transformation periodic in space,

$$\psi(x^0 + \beta, x^1) = -\psi(x^0, x^1)\,, \quad \psi(x^0, x^1 + L) = e^{i\alpha(x)}\psi(x^0, x^1)\,.$$
(16.42)

Written as one form, the decomposition (16.11) with instanton potential (16.39) takes the form

$$eA = \delta\varphi + eA^{\text{inst}} = \delta\varphi + d\lambda + h - \frac{\Phi}{V}x^1 dx^0 \,.$$

In the topologically trivial sector $\Phi = 0$, it is the Hodge decomposition into co-exact, exact, and harmonic parts. The gauge-invariant eigenvalues of the Dirac operator do not depend on the gauge function λ. The harmonic part enters the non-integrable phase factors

$$W[\mathscr{C}] = e^{ie\oint_{\mathscr{C}} A} \implies W[\mathscr{C}] = e^{2\pi i(h_0 n^0 + h_1 n^1)} \, W[\mathscr{C}] \big|_{h=0}$$
(16.43)

for loops $\mathscr{C}$ winding n^0-times around the torus in the temporal direction and n^1-times in the spatial direction.

16.3.1 Topologically Trivial Sector

We begin with calculating the fermion determinant in the topologically trivial sector with vanishing flux. In this sector $(i\slashed{D}_{\text{inst}})^2 = (\nabla - h)^2 \mathbb{1}$ has the double degenerate eigenvalues

$$\mu_m = \left(\frac{2\pi}{\beta}\right)^2 (m_0 - a_0)^2 + \left(\frac{2\pi}{L}\right)^2 (m_1 - a_1)^2\,, \quad \begin{pmatrix} a_0 \\ a_1 \end{pmatrix} = \begin{pmatrix} \frac{1}{2} + h_0 \\ h_1 \end{pmatrix}\,.$$
(16.44)

The corresponding zeta-function $\zeta(s) = \sum \mu_m^{-s}$ is the well-studied Epstein zeta-functions [29]. With the help of the Poisson resummation formula (see Problem 16.2), its derivative at $s = 0$ can be expressed in terms of Jacobi theta-functions as [30, 31]

$$\frac{d}{ds}\zeta(s)|_{s=0} = -2\log\left| \frac{1}{\eta(i\tau)} \Theta \begin{bmatrix} \frac{1}{2} + a_0 \\ \frac{1}{2} - a_1 \end{bmatrix}(0, i\tau) \right|\,, \quad \text{where} \quad \tau = \frac{L}{\beta}$$
(16.45)

is the ratio of the two circumferences of the torus and

$$\eta(i\tau) = q^{1/24} \prod_{n>0}(1 - q^n) \quad \text{with} \quad q = e^{-2\pi\tau} \tag{16.46}$$

is Dedekind eta-function. We have adopted the conventions in [32] for the theta-functions:

$$\Theta\begin{bmatrix} a \\ b \end{bmatrix}(z, i\tau) = \sum_{Z} e^{-\pi\tau(n+a)^2 + 2\pi i(n+a)(z+b)} . \tag{16.47}$$

Taking the degeneracy of the eigenvalues into account, the zeta-function regularized determinant (16.27) in the sector with zero flux reads

$$\det(i\slashed{D}_{\text{inst}}) = \left| \frac{1}{\eta(i\tau)} \Theta\begin{bmatrix} \frac{1}{2} + a_0 \\ \frac{1}{2} - a_1 \end{bmatrix}(0, i\tau) \right|^2 . \tag{16.48}$$

The determinant does not change under gauge transformations with windings, under which a_0 and a_1 are shifted by an integer. This symmetry and further properties of the instanton determinant are encoded in the way the theta-function changes under modular transformations [32].

16.3.2 Topologically Non-trivial Sectors

In (16.7) we have argued that the excited eigenmodes of $i\slashed{D}$ come in orthogonal pairs $\{\psi, \gamma_*\psi\}$ with opposite eigenvalues. Since γ_* commutes with the squared operator $(i\slashed{D})^2$, the chiral projections $\{P_+\psi, P_-\psi\}$ are eigenmodes of $(i\slashed{D})^2$ with equal eigenvalue but opposite chirality. Besides these excited eigenmodes, there exist $\nu = |\Phi|/2\pi$ right- or left-handed zero modes. Since in the two chiral sectors $(i\slashed{D}_{\text{inst}})^2$ in (16.13) are the same, up to a constant,

$$(i\slashed{D}_{\text{inst}})^2|_{\gamma_*=-1} = (i\slashed{D}_{\text{inst}})^2|_{\gamma_*=1} + \frac{2\Phi}{V} , \tag{16.49}$$

we conclude that for $\Phi > 0$, the ν zero modes have chirality $\gamma_* = +1$. These zero modes are at the same time excited modes with eigenvalues $2\Phi/V$ and chirality -1. Due to pairing there must exist ν excited modes with the same eigenvalues but chirality $+1$ and so on. Thus, $(i\slashed{D}_{\text{inst}})^2$ possesses the following discrete spectrum:

$$\mu_n = \begin{cases} 0 & \text{degeneracy} = |\Phi|/2\pi , \\ 2n|\Phi|/V & \text{degeneracy} = |\Phi|/\pi . \end{cases} \tag{16.50}$$

Contrary to the situation in the zero-flux sector, the spectrum does not depend on the harmonic part h of the potential. The zeta-function is now proportional to the ordinary Riemann zeta-function, and one obtains

$$\det'(i D_{\text{inst}}) = \left(\frac{\pi V}{|\Phi|}\right)^{|\Phi|/4\pi} \tag{16.51}$$

for the primed determinant in the topologically non-trivial sectors.

Inserting (16.48) and (16.51) into (16.37), we end up with the following formulas for the determinants in the sectors with v zero modes [10]:

$$\det(i D) = \left|\frac{1}{\eta(i\tau)}\Theta\right|^2 \exp\left(\frac{1}{4\pi}\int \varphi\Delta\varphi\right), \qquad v = 0, \tag{16.52}$$

$$\det'(i D) = \left(\frac{V}{2v}\right)^{v/2} \det\frac{N}{N_{\text{inst}}} \exp\left(\frac{1}{4\pi}\int \varphi\Delta\varphi + \frac{v}{V}\int\varphi\right), \quad v \neq 0, \tag{16.53}$$

where Θ is the theta-function in (16.48). With the help of (16.11), we expressed E in terms of φ and Φ and made the dependence on the electric charge explicit.

16.4 Computing the Zero Modes

It suffices to calculate the zero modes for the instanton potential, since D can be deformed into D_{inst}; see (16.33). When calculating expectation values of gauge invariant operators, the integrals over λ cancel in the numerator and denominator. Thus, we may set $\lambda = 0$ such that A^{inst} becomes time independent and the zero modes are proportional to $\exp(i\omega_p x^0)$ with Matsubara frequency ω_p. If we further eliminate the harmonic part of A^{inst}, we are led to the ansatz

$$\chi_p(x) = e^{i\omega_p x^0} e^{2\pi i h_1 x^1/L} \xi_p(x^1), \qquad \omega_p = (2p-1)\frac{\pi}{\beta}. \tag{16.54}$$

Inserting this ansatz into the zero-mode equation $D_{\text{inst}}^2\chi_p = 0$ yields

$$\left(\frac{d^2}{dy_p^2} - \frac{\Phi^2}{V^2}y_p^2 + \frac{|\Phi|}{V}\right)\xi_p = 0, \quad \text{with} \quad y_p = x^1 + \frac{L}{v}(p - a_0). \tag{16.55}$$

It follows that ξ_p is the ground state wave function of a harmonic oscillator

$$\xi_p(x^1) = \xi(y_p), \qquad \xi(y) = \exp\left(-\frac{|\Phi|}{2V}y^2\right), \qquad p \in \mathbb{Z}. \tag{16.56}$$

The functions $\chi_p(x)$ do not obey the boundary conditions (16.42) in the spatial direction, but the correct eigenmodes can be constructed as superpositions of them. To find these superpositions, we note that

$$\chi_p(x^0, x^1+L) = e^{-i\Phi x^0/\beta} e^{2i\pi h_1} \chi_{p+v}(x^0, x^1), \tag{16.57}$$

which implies that the sums

$$\psi_{\text{inst},p}(x) = \left(\frac{2v}{\beta^2 V}\right)^{1/4} \sum_{n\in\mathbb{Z}} e^{2i\pi nh_1} \chi_{p+nv}(x), \qquad p = 1, \ldots, v \qquad (16.58)$$

satisfy the boundary conditions and are the v zero modes of the Dirac operator with instanton potential. Actually, these eigenmodes define an orthonormal basis of the zero-mode subspace. They can be written in terms of Jacobi theta-functions (we drop a constant phase) as follows [10]:

$$\psi_{\text{inst},p} = \left(\frac{2v}{\beta^2 V}\right)^{1/4} e^{2\pi i[h_0 x^0/\beta - vx^0 x^1/V]} \; \Theta \begin{bmatrix} x^1/L + (p - a_0)/v \\ vx^0/\beta + a_1 \end{bmatrix} (0, iv\tau),$$

$$(16.59)$$

where τ has been introduced in (16.45). The norm matrix $\mathcal{N}_{\text{inst}}$ of this orthonormal basis is the identity matrix. The v unnormalized zero modes which enter the norm-matrix $\mathcal{N}$ can now be obtained from the zero modes (16.59) by the transformation (16.33) with $\varepsilon = 1$.

For the two-point functions (16.24), only the sectors with $\Phi = \pm 2\pi$ contribute, and since then the normalized zero mode is just $\psi/\sqrt{N}$, and the normalization constant cancels with $\mathcal{N}$ in (16.53), such that

$$\int \mathcal{D}\bar{\psi}\mathcal{D}\psi \; e^{\int \bar{\psi}i\not{D}\psi} \; \bar{\psi}(x)P_{\pm}\psi(x) = -\frac{|\Theta_{\pm}|^2}{\beta} e^{\mp 2\varphi} \; \exp\left(\frac{1}{4\pi}\int \varphi\Delta\varphi + \frac{v}{V}\int \varphi\right),$$

$$(16.60)$$

where $\Theta_{\pm}$ are the theta-functions in (16.59) with $p = 1$ and $v = \pm 1$:

$$\Theta_{\pm} = \Theta \begin{bmatrix} x^1/L \mp a_0) \\ x^0/\beta \pm a_1 \end{bmatrix} (0, i\tau). \qquad (16.61)$$

We used transformation properties of the θ-functions to obtain this canonical form.

16.5 Chiral Condensate at Finite T and L

After having calculated the fermionic path integral, we are left with the functional integrals over the A-field. Below we shall evaluate the two-point functions

$$\langle \bar{\psi} P_{\pm}\psi \rangle = \frac{\int \mathcal{D}A \, e^{-\frac{1}{2}\int E^2} \int \mathcal{D}\bar{\psi}\mathcal{D}\psi \; e^{\int \bar{\psi}i\not{D}\psi} \; \bar{\psi}(x)P_{\pm}\psi(x)}{\int \mathcal{D}A \, e^{-\frac{1}{2}\int E^2} \int \mathcal{D}\bar{\psi}\mathcal{D}\psi \; e^{\int \bar{\psi}i\not{D}\psi}}, \qquad (16.62)$$

where E is given in terms of φ and Φ in (16.11). Since the integrands in the numerator and denominator depend on $\{\varphi, h_\mu, \Phi\}$, it is natural to change integration variables from A_μ to these dimensionless fields.

Clearly there is a one-to-one mapping $E \leftrightarrow \{\Delta\varphi, \Phi\}$ with periodic φ. This implies a one-to-one mapping $E \leftrightarrow \{\varphi, \Phi\}$ if we demand that φ is orthogonal to the kernel of Δ, which means that it integrates to zero. Second, for given $\{\varphi, \Phi\}$ the gauge invariant Wilson loops $W[\mathscr{C}]$ in (16.43) are in one-to-one correspondence to the phases $\exp(2i\pi h_\mu)$ and thus to the component h_μ of the toron field modulo 1. Finally, from $e\partial_\mu A^\mu = \Delta\lambda$ we see that there is also a one-to-one correspondence between the divergence of A and periodic gauge functions λ which integrate to zero. To summarize: we have shown that the transformation

$$A_\mu \longrightarrow \{\Phi, \varphi, h_\mu, \lambda\} \tag{16.63}$$

defined in (16.11) and (16.39) with periodic fields φ and λ and quantized flux Φ is one to one if

$$\int_\mathcal{T} \varphi = \int_\mathcal{T} \lambda = 0 \quad \text{and} \quad 0 \le h_\mu < 1. \tag{16.64}$$

To actually calculate the bosonic path integral, we need the Jacobian of the transformation (16.63). For this purpose we expand all periodic fields in eigenmodes of $-\Delta$, for example,

$$\varphi(x) = \sum_{k \ne 0} \varphi_k e_k(x) \quad \text{with} \quad e_k(x) = \frac{1}{\sqrt{V}} e^{ik \cdot x}, \quad k_0 \in \frac{2\pi}{\beta}\mathbb{Z}, \quad k_1 \in \frac{2\pi}{L}\mathbb{Z}, \tag{16.65}$$

where the wave vector $k = 0$ is absent since φ must integrate to zero. In terms of the coefficients, the transformations (16.11) and (16.39) read in the zero-flux sector

$$\begin{pmatrix} eA_{0,k} \\ eA_{1,k} \end{pmatrix} = i \begin{pmatrix} -k_1 & k_0 \\ k_0 & k_1 \end{pmatrix} \begin{pmatrix} \varphi_k \\ \lambda_k \end{pmatrix} \quad (k \ne 0)$$

$$\begin{pmatrix} eA_{0,0} \\ eA_{1,0} \end{pmatrix} = 2\pi \begin{pmatrix} \sqrt{\tau} & 0 \\ 0 & 1/\sqrt{\tau} \end{pmatrix} \begin{pmatrix} h_0 \\ h_1 \end{pmatrix}.$$

The Jacobian of this transformation is just $J = (2\pi)^2 \det'(-\Delta)$ and thus independent of the dynamical fields. In the sectors with non-zero flux, we can write $eA_\mu = -\Phi/V \cdot x^1 \delta_{\mu 0} + e\delta A_\mu$, and the above transformation applies then to $e\delta A$, so that

$$\int \mathscr{D}A_\mu = \sum_\nu \mathscr{D}\delta A_\mu = J \sum_\nu \int_0^1 dh_0 dh_1 \int \prod_{k \ne 0} d\varphi_k d\lambda_k. \tag{16.66}$$

In expectation values the same field-independent Jacobian appears in the numerator and denominator and thus cancels. In addition, for gauge invariant operators, the λ integrals in expectation values cancel as well (the condition $\lambda = 0$ corresponds to the ghost free Lorentz gauge). Inserting (16.53) and (16.60) into (16.62), we find

$$\langle \bar{\psi} P_{\pm} \psi \rangle = -\frac{1}{\beta} \, e^{-2\pi^2/e^2 V} \, \frac{\int d^2 h \, |\Theta_{\pm}|^2 \int \mathcal{D}\varphi \, e^{-\Gamma[\varphi] \mp 2e\varphi(x)}}{\int d^2 h \, |\frac{1}{\eta} \Theta|^2 \int \mathcal{D}\varphi \, e^{-\Gamma[\varphi]}} , \tag{16.67}$$

with anomaly-induced effective action

$$\Gamma[\varphi] = \frac{1}{2e^2} \int_{\mathcal{T}} d^2 x \, \varphi \, (\Delta^2 - m_\gamma^2 \Delta) \, \varphi . \tag{16.68}$$

The theta-functions $\Theta_{\pm}$ and Θ are given in (16.61) and (16.48). Note that the last term in the exponent in (16.60) is absent since φ must integrate to zero. Finally, using

$$\int d^2 h \, |\Theta_{\pm}|^2 = \int d^2 h \, |\Theta|^2 = \sqrt{\frac{1}{2\tau}} \tag{16.69}$$

and performing the Gaussian functional integral yields

$$\langle \bar{\psi} P_{\pm} \psi \rangle = -\frac{|\eta(i\tau)|^2}{\beta} \, e^{-2\pi^2/e^2 V} \, e^{2e^2 K_e(0)} , \tag{16.70}$$

with Green function corresponding to the effective action Γ,

$$K_e(x - y) = \langle x | \frac{1}{\Delta^2 - m_\gamma^2 \Delta} | y \rangle . \tag{16.71}$$

The index e indicated that the constant zero mode is omitted in the spectral decomposition of the Green function. Using the eigenmodes (16.65) of the Laplacian, we obtain the series

$$K_e(x - y) = \frac{1}{m_\gamma^2} \sum_{k \neq 0} \left(\frac{1}{k^2} - \frac{1}{k^2 + m_\gamma^2} \right) e_k(x) e_k^*(y) , \tag{16.72}$$

where k^2 with wave vectors (16.66) are the eigenvalues of $-\Delta$. For $x = y$ the product $e_k(x) e_k^*(y)$ is just $1/V$, and we may use the identity derived in Problem 16.3 to sum over k_1. Setting $k_0 = 2\pi n/\beta$, we end up with the convergent series

$$m_\gamma^2 K_e(0) = \frac{1}{m_\gamma^2 V} - \frac{\coth(L m_\gamma /2)}{2\beta m_\gamma} + \frac{\tau}{12} + \frac{1}{2\pi} \sum_{n>0} \left(\frac{\coth(n\pi\tau)}{n} - \frac{\coth(\pi\tau\nu_n)}{\nu_n} \right) , \tag{16.73}$$

where $v_n^2 = n^2 + (\beta m_\gamma/2\pi)^2$. The first sum on the right-hand side can be related to Dedekind η-function as follows:

$$\sum_{n>0} \frac{1 - \coth(\pi\tau n)}{2n} = -\sum_{n,r>0} \frac{e^{-2\pi\tau nr}}{n} = \sum_r \log\left(1 - e^{-2\pi\tau r}\right) = \frac{\pi\tau}{12} + \log\eta(i\tau).$$

It cancels against the third term on the right-hand side in (16.73) and the factor $|\eta^2(i\tau)|$ in (16.70). Also note that the V-dependent exponential factor in (16.70) cancels against the first term on the right-hand side in (16.73) so that finally

$$\langle \bar\psi P_\pm \psi\rangle = -\frac{1}{\beta} \exp\left(-\frac{\pi}{\beta m_\gamma}\coth\left(\frac{1}{2}Lm_\gamma\right)\right) \cdot e^{F(\beta m_\gamma)} e^{-2H(\beta m_\gamma,\tau)}, \qquad (16.74)$$

where we defined the functions

$$F(a) = \sum_{n>0}\left(\frac{1}{n} - \frac{1}{\sqrt{n^2 + (a/2\pi)^2}}\right), \qquad (16.75)$$

$$H(a,\tau) = \sum_{n>0} \frac{1}{\sqrt{n^2 + (a/2\pi)^2}} \cdot \frac{1}{e^{\tau\sqrt{(2\pi n)^2 + a^2}} - 1}. \qquad (16.76)$$

This is the exact form of the condensate in the right- and left-handed sectors we have been aiming at. The result simplifies considerably if we let $L \to \infty$, in which case $H(.,.) \to 0$ and $\coth(.) \to 1$ so that

$$\langle\bar\psi\psi\rangle = -\frac{2}{\beta}\exp\left(-\frac{\pi}{\beta m_\gamma}\right) e^{F(\beta m_\gamma)}, \qquad (16.77)$$

where we used that $P_+ + P_- = 1$. The formula (16.74) for the temperature-dependent chiral condensate in a finite box and the corresponding formula (16.77) after the limit $L \to \infty$ have been taken, are the main results of this chapter, and are taken from [10]. The $L \to \infty$ condensate is depicted in Fig. 16.2.

To find an alternative representation to (16.77) which immediately yields the low-temperature limit, we regularize the individual sums in (16.75) according to

$$\sum \frac{1}{n^{1+s}} \sim \frac{1}{s} + \gamma + O(s)$$

$$\sum \frac{1}{(n^2 + a^2)^{\frac{1}{2}+s/2}} \sim \frac{1}{s} - \frac{1}{2a} - \log\frac{a}{2} + 2\sum_{\ell=1}^{\infty} K_0(2\pi\ell a) + O(s), \qquad (16.78)$$

Fig. 16.2 Temperature
dependence of the chiral
condensate in the limit
$L \to \infty$

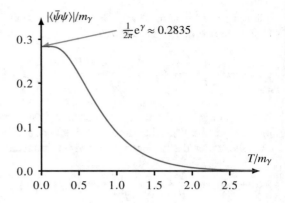

where $\gamma = 0.57721\ldots$ is the Euler-Mascheroni constant. The difference of the two
sums is finite as s tends to zero and inserting it into (16.75) and (16.77) yields

$$\langle \bar{\psi}\psi \rangle = -\frac{m_\gamma}{2\pi} e^\gamma\, e^{2I(\beta m_\gamma)}, \quad \text{where} \quad I(a) = \int\limits_0^\infty \frac{1}{1 - e^{a\cosh(t)}}\, dt\,. \tag{16.79}$$

Up to a sign the last integral is the last sum in (16.78). This form of the condensate
on the infinite cylinder has been obtained within the Hamiltonian approach to QED$_2$
in [33]. Since the integral I tends to zero for low temperatures, we find

$$\langle \bar{\psi}\psi \rangle \longrightarrow -\frac{m_\gamma}{2\pi}\, e^\gamma \qquad \text{for} \quad T/m_\gamma \to 0\,, \tag{16.80}$$

and this is the old result for the zero temperature Schwinger model [4].

To study the high-temperature limit, we observe that $F(a)$ tends to zero as $a \ll 1$
or if the temperature is big compared to the induced photon mass. In this region

$$\langle \bar{\psi}\psi \rangle \longrightarrow -2T\, e^{-\pi T/m_\gamma} \qquad \text{for} \quad T/m_\gamma \to \infty\,. \tag{16.81}$$

We see that the chiral condensate falls off exponentially for $T \gg m_\gamma$. For
intermediate temperatures $0 < T \sim m_\gamma$, we evaluate the sum (16.75) in the
exponent of (16.77) numerically. The result is depicted in Fig. 16.2. Although the
condensate behaves similar to an order parameter in a system which shows a second-
order phase transition, it does not vanish at any finite temperature. In a strict sense,
the chiral symmetry remains broken at all finite temperatures, similarly as in one-
flavor QCD.

Zero Temperature and Finite Volume
Returning to (16.70) and noting that

$$\frac{|\eta(i\tau)|^2}{\beta} = \frac{\eta(i\tau)\eta(i/\tau)}{\sqrt{V}} \qquad (\tau = L/\beta) \qquad (16.82)$$

is invariant under an exchange of β with L, and that the same holds true for $K_e(0)$, we may view (16.77) as dependence of the zero-temperature condensate on the spatial size L of the system. Thus, Fig. 16.2 can be interpreted as the change of $\langle\bar{\psi}\psi\rangle(L)$ due to finite (spatial) size effects. In particular, if the compactified spatial size shrinks to zero, the chiral condensate behaves as

$$\langle\bar{\psi}\psi\rangle(L) \longrightarrow -\frac{2}{L}\,e^{-\pi/Lm_\gamma} \qquad \text{for} \quad L \to 0 \qquad (16.83)$$

and thus vanishes exponentially (and non-analytically) for small spatial volume.

θ-Angle
In the decomposition (16.66) of the functional integral over the gauge fields, we have integrated over the different $U(1)$-principal bundles labeled by the index of $i\slashed{D}$ or by the flux of F_{01}. Thereby, we assumed that the relative normalization between contributions with different ν is one. To recover the θ-vacuum structure [3], we must allow for a relative phase factor $\exp(i\theta)$ between the topological sectors ν and $(\nu + 1)$—similarly as in the path integral for a particle on a punctured plane. Then the condensates $\langle\bar{\psi}P_\pm\psi\rangle$ pick up the phases $\exp(\pm i\theta)$, and the right-hand sides of (16.77) and (16.79–16.81) are multiplied by $\cos(\theta)$. The same happens if one adds the explicit θ-term $\propto i\theta F_{01}$ to the classical Lagrangian (16.1). Adding such a θ-term is more interesting in the massive Schwinger model, and the physical interpretation of such a term has been extensively discussed in the pioneering work by S. Coleman in [34].

16.6 Wilson Loops, Field Strength, and 2-Point Function

In this section we determine expectation values of Wilson loops and correlations of the field strength. For such gauge invariant bosonic operators $O[A]$ only gauge fields with zero flux contribute to the path integral

$$\langle O[A]\rangle = \sqrt{2\tau}\,\frac{\int \mathscr{D}\varphi\,e^{-\Gamma[\varphi]}\int d^2h\,|\Theta|^2\,O[\varphi, h]}{\int \mathscr{D}\varphi\,e^{-\Gamma[\varphi]}}. \qquad (16.84)$$

Here Γ is the effective action (16.68) which is quadratic in φ, Θ is the theta-function in (16.48), and we used the second formula in (16.69). Hence, up to an integration over the harmonics h_μ, we remain with Gaussian integrals over the periodic scalar field φ.

16.6.1 Correlation Functions of the Field Strength

The n-point functions of the field strength are particularly simple, since $E(x)$ does not depend on the toron field h_μ and the integration over these compact variables just cancels the factor $\sqrt{2\tau}$ in (16.84). Since $\Gamma[\varphi]$ is even in φ, all correlation functions with an odd number of the field strength $E = \Delta\varphi$ vanish. The $2n$-point functions are

$$\langle E(x_1) \cdots E(x_{2n}) \rangle = \Delta_1 \cdots \Delta_{2n} \langle \varphi(x_1) \cdots \varphi(x_{2n}) \rangle \,.$$

The correlation functions of φ are calculated with the quadratic effective action Γ. Applying Wick's theorem they are given as products of the 2-point function

$$\langle \varphi(x)\varphi(y) \rangle = K_e(x - y) \,. \tag{16.85}$$

The Green function K_e corresponding to Γ has been introduced in (16.71). Since $-\Delta\varphi_m = k^2\varphi_m$, we immediately get

$$\langle E(x)E(y) \rangle = \sum_{k \neq 0} e_k^\dagger(x)e_k(y)\frac{k^2}{k^2 + m_\gamma^2} = -\Delta_x G_e(x - y, m_\gamma) \,, \tag{16.86}$$

where G_e is the two-point function of a free scalar field with mass m_γ, minus the contribution of the constant mode $e_{k=0}$. Now we use the completeness of the orthonormal modes e_m and obtain

$$\langle E(x)E(y) \rangle = \delta(x - y) - m_\gamma^2 G(x - y; m_\gamma) \,, \tag{16.87}$$

where now G is the 2-point function with constant mode included. Using the result (16.145) in Problem 16.3, the propagator can be rewritten as

$$G(\xi; m_\gamma) = \frac{1}{2\beta}\sum_{n \in \mathbb{Z}}\frac{\cosh\left(\frac{1}{2}M_n(L - 2\xi_1)\right)}{M_n \sinh\left(\frac{1}{2}M_n L\right)}\,\mathrm{e}^{2\pi i n\xi_0/\beta} \,, \quad 0 \leq \xi_1 \leq L \,, \tag{16.88}$$

with mass parameter M_n depending on the Matsubara frequency,

$$M_n^2 = m_\gamma^2 + \omega_n^2 \,, \quad \omega_n = \frac{2\pi n}{\beta} \,. \tag{16.89}$$

In the limit $L \to \infty$, we obtain the Green function on the infinite cylinder

$$G(\xi; m_\gamma) = \frac{\mathrm{e}^{-m_\gamma|\xi_1|}}{2\beta m_\gamma} + \frac{1}{\beta}\sum_{n>0}\frac{\mathrm{e}^{-M_n|\xi_1|}}{M_n}\cos(\omega_n\xi_0) \,. \tag{16.90}$$

This shows explicitly that the 2-point function of the field strength falls off exponentially.

16.6.2 Wilson Loops and Charge Screening

We consider contractible loops as we did in Sects. 13.6 and 13.6.2. A contractible loop $\mathscr{C}$ does not wind around the torus and is the oriented boundary of a region $\mathscr{D}$ on the torus, $\mathscr{C} = \partial \mathscr{D}$. Only gauge fields with zero flux contribute to expectation values of Wilson loops

$$W[\mathscr{C}] = e^{ie' \oint_{\mathscr{C}} A} = e^{ie' \int_{\mathscr{D}} E} \,, \tag{16.91}$$

such that these expectation values are given by

$$\langle W[\mathscr{C}] \rangle = \frac{\int \mathscr{D}\varphi \, e^{-\Gamma[\varphi]} \, \exp\left(\frac{ie'}{e} \int_{\mathscr{D}} \Delta\varphi \right)}{\int \mathscr{D}\varphi \, e^{-\Gamma[\varphi]}} \,. \tag{16.92}$$

Note that we allowed for a static test charge e' which may be different to the charge of the dynamical fermions. If e' is an integer multiple of e, we call it integer charge, else we call it fractional charge. With Wick's theorem and (16.87), the resulting Gaussian integral yields

$$\langle W[\mathscr{C}] \rangle = \exp\left(-\frac{e'^2}{2} \int_{\mathscr{D} \times \mathscr{D}} \langle E(x) E(y) \rangle \, d^2x d^2y \right)$$

$$= \exp\left(-\frac{e'^2}{2} \left[A_{\mathscr{C}} - m_\gamma^2 \int_{\mathscr{D} \times \mathscr{D}} G(x - y, m_\gamma) \right] \right) \tag{16.93}$$

$$= \exp\left(\frac{e'^2}{2} \int_{\mathscr{D} \times \mathscr{D}} \Delta_x G(x - y; m_\gamma) \right), \tag{16.94}$$

where $A_{\mathscr{C}}$ is the area enclosed by the loop, and in the last step we used that G is the massive Green function. Choosing for $\mathscr{D}$ the rectangle $\mathscr{D} = [0, T] \times [0, R]$, we can extract the static potential

$$V(R) = -\lim_{T \to \infty} \frac{1}{T} \log\langle W[R, T] \rangle \,, \tag{16.95}$$

as introduced in Sect. 13.6.1. The explicit calculation of the exponent in (16.94) for a rectangular region is performed in Problem 16.4 and gives

$$-\frac{\log \langle W[R, T]\rangle}{T} = \frac{2e'^2}{T} \int_0^R d\xi_1 \left(G(0, \xi_1; m_\gamma) - G(T, \xi_1; m_\gamma)\right)(R - \xi_1)$$

$$+\frac{2e'^2}{T} \int_0^T d\xi_0 \left(G(\xi_0, 0; m_\gamma) - G(\xi_0, R; m_\gamma)\right)(T - \xi_0). \quad (16.96)$$

To study the interaction between two static electric charges at a large separation R, we consider the force between these charges,

$$F(R) = -\frac{d}{dR} V(R) = 2e'^2 \lim_{T \gg m_\gamma} \frac{1}{T} \int_0^R d\xi_1 \, G(T, \xi_1; m_\gamma)$$

$$+2e'^2 \frac{\partial}{\partial R} \lim_{T \gg m_\gamma} \int_0^T d\xi_0 \, G(\xi_0, R; m_\gamma) \left(1 - \frac{\xi_0}{T}\right). \quad (16.97)$$

For temperatures small compared to the induced photon mass m_γ and loops with edges long compared to the corresponding Compton length, both contributions on the right-hand side in (16.97) are exponentially small, since the massive propagator falls off exponentially as $\exp(-m_\gamma|\xi|)$. For example, the propagator on $\mathbb{R}^2$ is $G(\xi, m_\gamma) = K_0(m_\gamma|\xi|)/2\pi$. We conclude that at low temperature $T \ll m_\gamma$, the interaction between static electric charges is exponentially small. Any test charge e'—be it integer or fractional—is screened by the fluctuating fermion field, and hence there is no confinement of test charges. The screening of fractional charges is a particular feature of the massless Schwinger model.

Indeed, the screening disappears, and confinement is restored as soon as the dynamical fermions acquire a non-zero mass and the charge e' of the static test charge is fractional. Actually for small masses and $e' \ll e$, the string tension is proportional to the product of the fermion mass and condensate [2, 35].

16.6.3 Polyakov Loops (Thermal Wilson Loops)

Polyakov loops and their equilibrium expectation values have been considered earlier in this book. They serve as an order parameter for the breaking of a center symmetry, and in Sect. 13.7 we used these variables to discuss confinement in gauge theories at finite temperature. In Sect. 14.2.2 we showed that two-dimensional pure gauge theories (without matter fields) are confining and that expectation values of Polyakov loops exhibit Casimir scaling of the string tension. Since we have just seen that massless fermions in $1 + 1$ dimensions screen all test charges, we expect a qualitatively different behavior of Polyakov loops with massless fermions as compared to the quenched system without fermions.

The Polyakov loop variable of a static charge e is

$$\mathscr{P}_u = \exp\left(ie \int_0^\beta A_0(x^0, u)\, dx^0\right) \tag{16.98}$$

and depends on the harmonics because it winds once around the periodic time direction. Since $e \int A_0 dx^0 = 2\pi h_0 - \partial_1 \int \varphi\, dx^0$, the integration over the harmonics h_μ for a product of q Polyakov loops yields a factor

$$\int d^2 h\, |\theta|^2\, e^{2\pi i q h_0} = \sqrt{\frac{1}{2\tau}} e^{-\pi q^2/2\tau}, \tag{16.99}$$

where the θ-function originates from the determinant in the zero-flux sector (16.52). The remaining φ-integration is Gaussian

$$\langle \mathscr{P}_{u_1} \cdots \mathscr{P}_{u_q} \rangle = e^{-\pi q^2/2\tau} \frac{\int \mathscr{D}\varphi\, e^{-\Gamma[\varphi]+(j,\varphi)}}{\int \mathscr{D}\varphi\, e^{-\Gamma[\varphi]}}, \tag{16.100}$$

with static source localized at the positions $u_1, \ldots, u_q$ of the charges

$$j(x) = i \sum_i \delta'(x_1 - u_i). \tag{16.101}$$

The Gaussian path integral with subsequent partial integration in the integration over space leads to

$$\langle \mathscr{P}_{u_1} \cdots \mathscr{P}_{u_q} \rangle = \exp\left(-\frac{\beta}{2} \sum_{i,j=1}^q V(u_i - u_j)\right). \tag{16.102}$$

We absorbed τ-dependent exponential factor in (16.100) in the function

$$V(u) = \frac{\pi}{L} - \frac{e^2}{\beta} \int_0^\beta dx^0 dy^0 \Delta K_e(x^0 - y^0, u) = e^2 \int d\xi^0\, G(\xi^0, u) \tag{16.103}$$

and used the fact that we may replace the second derivative of the periodic K_e with respect to the spatial coordinate by the Laplacian of K_e under the integral over x^0 and y^0. The term π/τ just converts the zero-mode subtracted Green function G_e into the full Green function G in the last integral. With the explicit form of this Green function in (16.88), we obtain

$$V(u) = \frac{\pi m_\gamma}{2} \frac{\cosh \frac{1}{2} m_\gamma (L - 2|u|)}{\sinh(\frac{1}{2} m_\gamma L)} \quad \text{with} \quad V(0) = \frac{\pi m_\gamma}{2} \coth \frac{m_\gamma L}{2}. \tag{16.104}$$

Hence, the explicit expression for the two-point functions is

$$\langle \mathscr{P}_u \mathscr{P}_0 \rangle = \langle \mathscr{P}_0 \rangle^2 \exp\left(-\beta V(u)\right), \quad \text{with} \quad \langle \mathscr{P}_0 \rangle = \exp\left(-\frac{1}{2}\beta V(0)\right).$$
(16.105)

The normalized higher n-point functions

$$\mathscr{P}(u_1 \ldots, u_q) \equiv \frac{\langle \mathscr{P}_{u_1} \cdots \mathscr{P}_{u_q} \rangle}{\langle \mathscr{P}_0 \rangle^q}$$
(16.106)

are then just products of the normalized two-point function

$$\mathscr{P}(u_1, \ldots, u_q) = \prod_{i<j} \mathscr{P}(u_i, u_j).$$
(16.107)

According to the discussion in Sect. 13.7, the object

$$- T \log \mathscr{P}(u_1, \ldots, u_q) = \frac{1}{2} \sum_{i \neq q} V(u_i - u_j)$$
(16.108)

is interpreted as (zero energy subtracted) free energy of q static charges at positions $u_1, \ldots, u_q$. The result (16.107) means that the free energy of q charges is just the sum of the free energies of the individual pairs of charges.

Let us compare the result with the potential extracted from a rectangular Wilson loop with maximal temporal length $T = \beta$. For such loops the ξ^1-integral in (16.96) vanishes, since G is periodic in ξ_0 with period β, and the integral over ξ_0 results in

$$- \frac{\log\langle W[R, \beta] \rangle}{T} = V(0) - V(R),$$
(16.109)

with potential V given in (16.104). Thus, the potentials calculated with two Polyakov loops and a rectangular Wilson loop with maximal temporal extend coincide, up to the zero-point energy and the sign. This must be the case, since the gauge potentials are periodic in time, see (16.40), such that the contributions of the two horizontal lines marked with a double stroke in Fig. 16.3 to $W[\mathscr{C}]$ cancel. However, the two Polyakov loops describe equally charged sources, whereas the Wilson loop describes sources with opposite charges. This is why the interaction extracted from two Polyakov loops is repulsive and the interaction from a rectangular Wilson loop attractive.

On the infinite cylinder $L = \infty$, the normalized correlation function for 2 Polyakov loops has the simple form

$$\mathscr{P}(r, 0) = \exp\left(-\frac{\pi}{2}\beta m_\gamma \, e^{-m_\gamma r}\right).$$
(16.110)

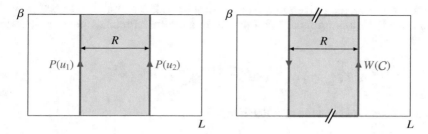

Fig. 16.3 The static potential extracted from two Polyakov loops (left panel) is, up to a sign, the same as that extracted from a Wilson loop with maximal extend in temporal direction (right panel)

The (zero energy subtracted) free energy between two static charges falls off exponentially, and we find a repulsive Yukawa force

$$F = -\frac{\mathrm{d}V(r)}{\mathrm{d}r} = \frac{\mathrm{e}^2}{2}\,\mathrm{e}^{-m_\gamma r} \tag{16.111}$$

between two static charges. Again we see that the classical Coulomb force between two static charges, which in $1+1$ dimensions is linearly rising and hence confining, is screened by massless dynamical fermions. The screening length is just the Compton wave length of the induced "photon mass."

16.6.4 Gauge-Invariant Fermionic Two-Point Functions

Fermion bilinears $\bar\psi(x)\Gamma\psi(y)$ are not gauge- invariant since the Dirac field at different positions transforms with a different phase under local gauge transformations. Gauge invariance imposes the introduction of a phase factor in order to maintain gauge invariance. In particular the gauge-invariant chiral two-point functions

$$S_\pm(x,y) = \langle \bar\psi(x)\,\mathrm{e}^{ie\int_y^x A_\mu \mathrm{d}\xi^\mu}\,P_\pm\,\psi(y)\rangle \tag{16.112}$$

contain as phase factor the Wilson line connecting the arguments x and y. They depend explicitly on the path $\mathscr{C}$ connecting y with x and can be calculated in a similar fashion as the chiral condensates (16.67) which they must reproduce for coinciding points [36]. Only gauge field with flux 2π contributes to S_+ and using the factorization of the zero mode and explicit expression of the fermion determinant in the sectors with no and with one zero-mode yields

$$S_+(x,y) = -\sqrt{\frac{V}{2}}\,\mathrm{e}^{-\frac{2\pi}{m_\gamma^2 V}}\,\frac{\int \mathrm{d}^2h\,\psi_{\mathrm{inst}}^\dagger(x)\,\psi_{\mathrm{inst}}(y)\int \mathscr{D}\varphi\,\mathrm{e}^{-\Gamma[\varphi]-\varphi(x)-\varphi(y)+ie\int_y^x A}}{\int \mathrm{d}^2h\,|\frac{1}{\eta}\Theta|^2\int \mathscr{D}\varphi\,\mathrm{e}^{-\Gamma[\varphi]}}, \tag{16.113}$$

with zero-mode ψ_{inst} given in (16.59) with $\nu = 1$. For simplicity we take $p = 0$ (and not 1) in this formula. The integration over the harmonics in the numerator yields

$$\int d^2h \, \psi_{\text{inst}}^\dagger(x) \, \psi_{\text{inst}}(y) \, e^{ie \int_y^x A^{\text{inst}}} = \frac{1}{V} e^{-\frac{\pi}{2V}(x-y)^2} . \tag{16.114}$$

The left-hand side contains the line integral

$$ie \int_y^x A = ie \int_0^1 A_\mu(x(\tau)) \, \dot{x}^\mu(\tau) \, d\tau, \tag{16.115}$$

and to calculate this line integral, we must pick a path from y to x. For a potential with non-zero field strength, this integral depends on the chosen path. We choose a straight line connecting the two endpoints,

$$x^\mu(\tau) = \tau x^\mu + (1-\tau)y \quad \text{with} \quad x(0) = y, \quad x(1) = x. \tag{16.116}$$

Let us note that the harmonic gauge potential alone would give an additional factor in (16.114) which breaks translation invariance.[3] The additional factor cancels against the contribution of the part $A_0 = -2\pi x_1/V$ in the instanton potential. As a result one ends up with the simple translation invariant result (16.114).

It remains to integrate over φ in the numerator and denominator of (16.113). The integration in the numerator is Gaussian with effective action $\Gamma[\varphi]$ and source

$$j(z) = ie\varepsilon_{\mu\nu}\frac{\partial}{\partial z^\nu} \int d\tau \, \delta(z - x(\tau))\dot{x}^\mu(\tau) - \delta(z - x) - \delta(z - y), \tag{16.117}$$

such that the Gaussian φ-integration leads to

$$\frac{\int \mathcal{D}\varphi \, e^{-\Gamma[\varphi]+(j,\varphi)}}{\int \mathcal{D}\varphi \, e^{-\Gamma[\varphi]}} = e^{Q(\xi)}, \quad Q(\xi) = \frac{e^2}{2} \int d^2z \, d^2z' \, j(z)K_e(z - z')j(z'), \tag{16.118}$$

where we already anticipated that the result depends on the difference $\xi = x - y$ only. The calculation of this double integral with Green function (16.72) is Problem 16.7. One finds the following series expansion:

$$Q(\xi) = e^2 K_e(0) + e^2 K_e(\xi) \tag{16.119}$$

$$-\frac{e^2}{V} \sum_{k \neq 0} \left(\frac{k_0\xi_1 - k_1\xi_0}{k \cdot \xi}\right)^2 \frac{1}{k^2(k^2 + m_\gamma^2)} (1 - \cos(k \cdot \xi)).$$

---

[3] The explicit calculation of this factor is done in Problem 16.6.

With the series expansion (16.72), the sum of the ξ-dependent contributions simplify,

$$Q(\xi) = 2e^2 K_e(0) - \frac{\pi}{V} \sum_{k \neq 0} \frac{\xi^2}{(k \cdot \xi)^2} \frac{m_\gamma^2}{k^2 + m_\gamma^2} \left(1 - \cos(k \cdot \xi)\right). \qquad (16.120)$$

Inserting this result together with (16.114) into (16.113) and comparing with the expression for the chiral condensate in (16.70), we conclude that

$$S_+(\xi) = \langle \bar{\psi}(0) P_+ \psi(0) \rangle \, e^{-R(\xi)} = S_+(0) \, e^{-R(\xi)}, \quad \xi = x - y, \qquad (16.121)$$

where the dependence on $\xi = x - y$ is encoded in the function

$$R(\xi) = \frac{\pi}{V} \sum_k \frac{\xi^2}{(k \cdot \xi)^2} \frac{m_\gamma^2}{k^2 + m_\gamma^2} \left(1 - \cos(k \cdot \xi)\right). \qquad (16.122)$$

The last sum includes $k = 0$. The contribution with $k = 0$ produces the term $\xi^2/2V$ coming from the integration over the harmonics; see (16.114).

Let us consider the equal time correlator in more detail and thus assume $\xi^0 = 0$. Then the series simplifies considerably. Using the two formulas in problem 16.3, we can sum over k_1 and obtain for the dependence on $r = |x_1 - y_1|$ the series representation

$$R(r) = \frac{\pi m_\gamma r}{4} \coth \frac{\beta m_\gamma}{2} + \frac{\pi m_\gamma^2}{2\beta} \sum_n \frac{1}{M_n^3} \left(\frac{\cosh \frac{1}{2} M_n (L - 2r)}{\sinh \frac{1}{2} M_n L} - \coth \frac{M_n L}{2} \right),$$
$$(16.123)$$

where the n-dependent mass M_n depends on the Matsubara frequency ω_n and has been defined in (16.89). If the separation r exceeds $L/2$, then our parallel transporter is not anymore along the shortest geodesic from x to y. Then one has go the "other way around" the torus. This means that (16.122) is the correct exponent for $r \leq L/2$.

In the thermodynamic limit $m_\gamma L \to \infty$, the hypergeometric functions simplify,

$$R(r) \to \frac{\pi m_\gamma r}{4} \coth \frac{\beta m_\gamma}{2} + \frac{\pi m_\gamma^2}{2\beta} \sum_n \frac{1}{M_n^3} \left(e^{-M_n r} - 1 \right). \qquad (16.124)$$

For widely separated points $m_\gamma r \gg 1$, the sum converges to an r-independent constant, and the two-point function falls off exponentially

$$S_+(r) = S_+(0) \, e^{-R(r)} \sim \text{const} \cdot e^{-m_\beta r}, \quad m_\beta = \frac{\pi m_\gamma}{4} \coth \frac{\beta m_\gamma}{2}. \qquad (16.125)$$

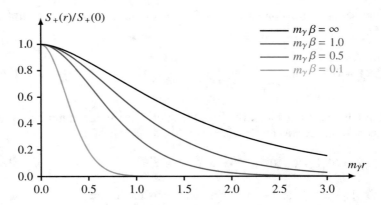

Fig. 16.4 The gauge invariant two-point function S_+ in (16.121) at equal time $\xi_0 = 0$ and $r = |\xi|$, for several values of the inverse temperature measured in units defined by the induced mass

For low temperature the effective mass m_β converges to the zero-temperature value $\pi m_\gamma / 4$, but it grows when the system is heated. For arbitrary separations r of the (equal-time) arguments of the two-point function, we calculated the sum (16.124). The r-dependence for four values of the inverse temperature is depicted in Fig. 16.4. Again we see how the effective mass increases with increasing temperature.

16.7   Massive Multi-Flavor Schwinger Model on the Lattice

A mass term for the fermions breaks chiral symmetry explicitly, and this is the main reason why the massive and massless models behave differently on a qualitative level. For example, a fractional static charge is not screened by the dynamical fermions in the massive model—only an integer charge is screened. Actually, in what follows we admit N_f fermion flavors with equal mass m. There are many works devoted to the massless and massive multi-flavor models. The reader may consult the older work [37] and the more recent one [38] to learn more about the physics of the models with N_f species of fermions.

Let us point out some major differences between the multi-flavor systems without and with mass. The path integral with N_f *massless* fermion fields just produces the result (16.20) for the massless one-flavor model to the power N_f. Thus, in the sectors with minimal non-vanishing fluxes $\pm 2\pi$, the fermion partition function contains a factor

$$\prod_{a=1}^{N_f} (\bar{\eta}_a, \psi_1)(\bar{\psi}_1, \eta_a) , \tag{16.126}$$

since there exists one zero-mode ψ_1 for every flavor. Thus, there are N_f sources η_a and N_f sources $\bar{\eta}_a$ in the $\Phi = \pm 2\pi$ sectors, and this proliferation leads to a vanishing condensate for $N_f > 1$. This is different to the system with N_f *massive* fermion fields, when the Dirac operator

$$i(\slashed{D} + m) \otimes \mathbb{1}_{N_f} \tag{16.127}$$

has no zero modes and the fermionic partition function

$$Z[A; \bar{\eta}, \eta] = \exp\left(-\int d^2x\, d^2y \sum_{a=1}^{N_f} \bar{\eta}_a(x)\, G(A; x, y)\, \eta_a(y) \right) \det^{N_f}(i\slashed{D} + im) \tag{16.128}$$

contains no product of zero modes as in (16.126). Instead, the Green function $G(A; x, y)$ of $i\slashed{D} + im$ does not anticommute with γ_*, and a chiral condensate can form. For $m \neq 0$ we cannot just integrate the chiral anomaly to solve the model analytically, since the gauge potential cannot be removed by a local chiral rotation.

There is no *spontaneous breaking* of a continuous global symmetry with associated Goldstone bosons in two dimensions.[4] Indeed, the breaking of the axial $U_A(1)$ in the Schwinger model is triggered by the chiral anomaly, similar to the *anomalous breaking* of the $U_A(1)$ symmetry in QCD. As a result the symmetry can't get restored in the one-flavor model at finite temperature. We expect a similar behavior in the massive one-flavor model with explicit breaking of the chiral symmetry $SU(N_f)$ and explicit and anomalous breaking of the axial $U_A(1)$. For the one-flavor model, the chiral limit $m \to 0$ should be smooth since there exists a mass gap even for the limiting model. Indeed it has been argued that the small-mass perturbation theory is IR-finite [39].

In the multi-flavor model, a condensate would break the continuous chiral $SU(N_f)$ symmetry of the massless model spontaneously. According to the Mermin-Wagner theorem, this is not possible in *two* dimensions. In the massive model, the symmetry is broken explicitly, and a condensate can form. But since a naive small-mass perturbation theory is plagued with IR-divergences, the dependence of the condensate on m could be non-analytic. This already indicates that for several flavors, the limit $m \to 0$ could be more subtle. Let us present some selected results for the massive one- and multi-flavor models in more detail.

---

[4] if the global symmetry is non-compact then there exists a variant of the theorem.

One-Flavor Model

The massive model with $N_f = 1$ has been studied in the IR-finite mass perturbation expansion in [39]. Within this expansion one can determine the leading corrections to the condensate,

$$\frac{\langle \bar{\psi} \psi \rangle_m}{\langle \bar{\psi} \psi \rangle_0} \approx 1 + 0.41225 \cdot \frac{m}{m_\gamma} \frac{\langle \bar{\psi} \psi \rangle_0}{m_\gamma} + O(m^2) \,, \tag{16.129}$$

and to the susceptibility. A resummation of the perturbation series for correlation functions with the help of Dyson-Schwinger equations reveals the spectrum of stable and unstable bound states from the two-point function. For $\theta \neq 0$ there are two stable particles: the Schwinger boson with mass μ relating to the mass m_γ of the model with massless fundamental fermions,

$$\frac{\mu^2}{m_\gamma^2} \approx 1 + 3.5621 \cdot \frac{m}{m_\gamma} - 2.0833 \cdot \frac{m^2}{m_\gamma^2} + \dots \,, \tag{16.130}$$

and a second bound state with mass μ' given by

$$\left(\frac{\mu'}{2\mu} \right)^2 \approx 1 - 7.83 \cdot \frac{m^2}{m_\gamma^2} + O \left(\frac{m^3}{m_\gamma^3} \right) . \tag{16.131}$$

For $\theta = 0$ there is an additional mixed bound state. Unstable higher bound states may be formed out of an arbitrary number of the stable particles.

Multi-Flavor Model

In several works the massive Schwinger model with N_f flavors of Dirac fermions has been studied for light fermions with $m \ll e$ [34, 37, 40]. One finds in mass perturbation theory, partly based on bosonization techniques, that the particle spectrum involves a massive Schwinger boson (the analog of η') with mass

$$m_\gamma \approx e \sqrt{\frac{N_f}{\pi}} + O(m) \tag{16.132}$$

and "quasi-Goldstone particles" with mass

$$\frac{m_G}{e} \approx C_{N_f} \left(\frac{m}{e} \right)^{N_f / N_f^+} \,, \quad C_2 \approx 2.008, \, \dots \tag{16.133}$$

where we defined $N_f^+ = N_f + 1$. From a semi-classical analysis, one finds the same formula for large masses and $N_f = 2$, but with $C_2 \approx 2.163$ [41]. A chiral condensate forms for all values of N_f, and for light fermions one finds

$$\langle \bar{\psi} \psi \rangle \sim D_{N_f} \, m \left(\frac{e}{m} \right)^{2/N_f^+} \,, \quad D_2 \approx 0.388, \, \dots \, . \tag{16.134}$$

Actually with $D_1 = e^{\gamma}/2\pi$, this result holds for one flavor as well since then the exponent is 1 and the mass m cancels such that one obtains the condensate (16.80). In passing we note that identical results are obtained when one solves the Schwinger model with chirality breaking bag-boundary conditions, in which case the fermion mass is replaced by the (inverse) size of the box enclosing the system [7].

A relevant parameter in the massive multi-flavor model as well as in QCD is the dimensionless Leutwyler-Smilga parameter [42]

$$x = m\langle\bar{\psi}\psi\rangle V , \tag{16.135}$$

which decides whether the topological charge of a gauge field is physically relevant ($x \ll 1$) or irrelevant ($x \gg 1$). For example, for $x \ll 1$ the partition function is completely dominated by configurations from the topologically trivial sector. For large values of x, configurations with non-zero topological charge contribute as well. Arguments based on anomaly matching and global symmetries predict a simple Gaussian distribution of the topological charge ν (actually it is further assumed that the Compton wavelength of the pseudo-Goldstone bosons is larger than the box size)

$$p_\nu \sim e^{-\nu^2/2\langle\nu^2\rangle}, \quad \langle\nu^2\rangle = \frac{x}{N_f}, \quad x \gg 1 . \tag{16.136}$$

This prediction has been compared with results of lattice simulations; see the discussion below. The Leutwyler-Smilga parameter x also indicates whether the spacetime volume V is small or large or whether the spontaneous breaking of chiral symmetry (in QCD) is mild or strong.

θ-Angle

Based on a weak coupling $e/m \ll 1$ and strong coupling $e/m \gg 1$ analysis Coleman conjectured that the one-flavor model shows a phase transition at $\theta = \pi$ and some finite value of m [34]. At weak coupling the $\mathbb{Z}_2$ symmetry of the model at $\theta = \pi$ is spontaneously broken, and at strong coupling the $\mathbb{Z}_2$ symmetry is realized in the vacuum. In [39] the θ-dependence of the condensate, particle masses, and vacuum energy have been calculated in mass perturbation expansion. Coleman conjecture has meanwhile been confirmed by lattice simulations with staggered fermions and the standard Wilson form for the pure gauge action in [43]. As an order parameter for the $\mathbb{Z}_2$ symmetry (a function of), the topological charge density was used. The explicit and accurate result

$$\frac{m_{\text{crit}}}{e} = 0.3335(2) \tag{16.137}$$

for the value of the critical mass at $\theta = \pi$, where a second-order phase transition is seen, has been extracted with the density matrix renormalization group approach [44]. The critical mass m_{crit} marks a critical endpoint, since for all $m > m_{\text{crit}}$ the model shows first-order transitions and for $m < m_{crit}$ there is no phase transition. There is good evidence that at the critical point, the systems fall into the universality class of the Ising model.

16.7.1 Lattice Simulations

In [45] the authors studied the scalar condensate $\langle \bar{\psi} \psi \rangle$ and the topological susceptibility

$$\chi_{top} = \frac{1}{2\pi} \int d^2 x \, \langle E(x) E(0) \rangle \tag{16.138}$$

of the massive model with $N_f = 0, 1,$ and 2 flavors, mainly to confront staggered and overlap fermions with analytical results. For example, for 1 and 2 light flavors the results (16.134) read

$$N_f = 1 : \quad \frac{\langle \bar{\psi} \psi \rangle}{m_\gamma} \sim \frac{e^\gamma}{2\pi}, \qquad N_f = 2 : \quad \frac{\langle \bar{\psi} \psi \rangle}{m_\gamma} \sim 0.451 \left(\frac{m}{m_\gamma} \right)^{1/3} \tag{16.139}$$

and have been compared with results obtained with various simulations.[5]

Staggered Fermions
The main results in [45] are that, at *finite lattice spacing* and in a finite box, simulations with staggered fermions lead to a qualitatively different dependence of the condensate on the fermion mass m as analytically known. But with only a few smearing steps, staggered fermions do much better: the mass at which the simulation results deviate from the analytic result is dramatically lowered. Similar statements hold for the topological susceptibility. The staggered artifacts get drastically reduced by smearing, resulting in a good agreement with analytic results down to very small quark masses. In a subsequent work, the same authors demonstrated that the condensate of the massless 1-flavor model can be reproduced with rooted staggered fermions if the limit $m \to 0$ is taken *after* letting $a \to 0$ [46] and that the order of the two limits is essential. For all observable considered in the work (most involve only sea quarks), staggered fermions give the correct result in the continuum limit, provided one takes the chiral and continuum limits in the correct order.

Overlap Fermions
The Schwinger model played a pivotal role in construction of lattice chiral gauge theories which avoid the Nielsen-Ninomiya no-go theorem discussed in Sect. 15.3.5 and maintain the important features of chiral fermions. In the pioneering work on

---

[5] A slightly different constant 0.4329 in place of 0.451 is given in [37].

overlap fermions [11], it was demonstrated that

- the lattice overlap correctly reproduces the fermion determinant and two-point function in the topologically trivial sector,
- properties of the fermion zero modes (shape, quantum numbers) in the non-trivial sectors are correctly reproduced by the overlap construction.

Hasenfratz, Laliena, and Niedermayer [47] and Lüscher [48] explicitly showed that any lattice Dirac fermion operator satisfying the Ginsparg-Wilson relation must obey the Index theorem exactly. Since the overlap operator explicitly solves the Ginsparg-Wilson relation, it must obey the index theorem. However, on a finite lattice it is not clear a priori how the zero modes and the index theorem are recovered. But it turns out that already on a very small 6×6 lattice, the exact zero modes with definite chirality are reproduced to a very high accuracy, and the index theorem and vanishing theorem are satisfied exactly [11, 49]. Also, it was demonstrated that overlap fermions successfully reproduce the analytic results for the condensate and topological susceptibility [45, 46].

Other Fermion Species
The massive $N_f = 2$ lattice Schwinger model with Wilson, hypercube, fully twisted mass, and overlap fermion action and its approach to the continuum has been the subject of study in [50]. For that purpose the scaling behavior of the pseudo-scalar mass and the condensate (in units of e) have been determined for several values of the dimensionless lattice constant ea. The results for the above fermion species were compared with the continuum prediction. Since Wilson and hypercube fermions are afflicted with an additive mass renormalization, the critical fermion mass m_{crit} was computed by the vanishing of the PCAC fermion mass. The simulation results show that for all fermion species, m_G/e scales linear in a^2.

Tensor Networks
Tensor network (TN) methods are based on the Hamiltonian formulation of a quantum many-body system [12]. Thereby, one exploits the variational character-ization of the ground state(s) (or excited states or equilibrium states) to find a good approximation for this state. The success of the method rests on the observation that one can characterize a suitable and relatively small subspace in the huge Hilbert space $\mathscr{H}$ of the many-body system to find a good approximation to the true ground state. The method can be applied to quantum field theories discretized on a (finite) spatial lattice, since these are particular quantum many-body systems with many degrees of freedom.

To find suitable and sufficiently small subspaces of $\mathscr{H}$, one uses the fact that for many (gapped) systems the physically interesting states are less entangled than they could be. More accurately, their entanglement entropy $S(\varrho_A)$ of a subregion A does not scale extensively with the size of the subregion as one might expect, but only with its boundary area $|\partial A|$. This so-called area law can be proven for a class of quantum systems, which include gapped free bosonic and fermionic models. These findings can be seen as an evidence that gapped many-body systems generically have the property that the ground state satisfies an area law. When applying the TN method, one often approximates the ground state or low-lying states by matrix product states since (at least in lower dimensions) these states obey an area law.

The main reason for the interest in the TN approach in gauge theories is its ability, shown so far in several condensed matter models, to deal with theories which exhibit the notorious sign problem in MC simulations. When applied to gauge theories in lower dimensions (see [51] for a recent review), one often reformulates the lattice gauge theory as spin model. In fact, since the Schwinger model is arguably the most simple relativistic theory of gauge-matter interaction, it has been extensively used to probe the TN method as an alternative to conventional Monte Carlo-based lattice simulations. Actually, most approaches are based on the lattice Hamiltonian for staggered fermions with a compact U(1) gauge field [52] and use open boundary conditions. The TN methods allow for a very good precision, much better than standard Monte Carlo techniques, due to the possibility of reaching much smaller lattice spacings.

For example, in [53] the one-flavor finite-temperature Schwinger model for massless and massive fermions has been investigated in great detail. These studies involve additional extrapolations besides the ubiquitous continuum limit and infinite volume extrapolation. In particular one must choose a sufficiently large bond dimension (which determines the number of variational parameters) to minimize the truncation effects. In addition, at finite temperature the matrix product states are replaced by matrix product operators. For zero fermion mass, the TN approach is in excellent agreement with the analytic results (16.79). For finite fermion masses, the agreement is (at present) only on a qualitative level.

16.8   Problems

16.1 (Induced Photon Mass)   The only dynamical field in the massless Schwinger model is the "transverse photon" with mass m_γ. This means that the gauge boson acquires a mass without breaking gauge invariance. Here we discuss the dynamical generation of a gauge invariant mass in more detail. We have seen in the main text that the path integral over the fermion yields the anomaly-induced effective action

$$\Gamma[\varphi] = \frac{1}{2e^2} \int_{\mathcal{T}} d^2x \, \varphi \, (\Delta^2 - m_\gamma^2 \Delta) \, \varphi \,.$$

Express this action in terms of A_μ in (16.11) in the zero-flux sector and for $h_\mu = 0$. How come that a gauge boson can become massive without breaking gauge invariance?

Hint: the transverse part of the potential is

$$A_\mu^T = A_\mu - \frac{1}{\Delta} \partial_\mu A^\mu \,.$$

Show that the divergence $\partial_\mu A^\mu$ is just $\Delta\lambda$. Since the spacetime average of λ is zero, we can determine the transverse part without ambiguities. Expressing Γ in terms of A_μ^T, you should obtain

$$\Gamma = \frac{1}{2} \int_{\mathcal{T}} d^2 x \, A_\mu^T (-\Delta + m_\gamma^2) \, A_\mu^T \,, \qquad (16.140)$$

and this is just the action of a massive field.

16.2 (Poisson Resummation) Let $f : \mathbb{R}^d \mapsto \mathbb{C}$ be a square integrable function and

$$\tilde{f}(y) = \int_{\mathbb{R}^d} d^d x \, e^{2\pi i y \cdot x} f(x)$$

its Fourier transform. Prove the Poisson resummation formula

$$\sum_{m \in \mathbb{Z}^d} \tilde{f}(m) = \sum_{n \in \mathbb{Z}^d} f(n) \,. \qquad (16.141)$$

The Fourier transformation maps Gaussian functions into Gaussian functions and for

$$f(y) = e^{-\pi(y-a)G^{-1}(y-a) - 2\pi i x \cdot y} \qquad (16.142)$$

one finds the interesting relation

$$\sum_{n \in \mathbb{Z}^d} e^{-\pi(n-a) G^{-1}(n-a)} = \sqrt{\det G} \sum_{m \in \mathbb{Z}^d} e^{-\pi m G m - 2\pi i m \cdot a} \,. \qquad (16.143)$$

Apply the result to the Jacobi theta-function defined in (16.47).

Hint: the functions $\exp(2\pi i m \cdot x)$ form an orthonormal basis on the space of periodic function on $[0, 1]^d$ such that

$$\sum_{m \in \mathbb{Z}^d} e^{2\pi i m \cdot x} = \sum_{n \in \mathbb{Z}^d} \delta(x - n) \,.$$

Multiply this relation with a test function $f(x)$ and integrate.

16.3 (Some Useful Series) Proof the following identities

$$S_1(\phi) = \sum_{m \in \mathbb{Z}} \frac{e^{2\pi i m \phi}}{a^2 + m^2} = \frac{\pi}{a} \frac{\cosh \pi a(1 - 2\phi)}{\sinh \pi a} \quad (0 \le \phi \le 1) \qquad (16.144)$$

$$S_2(\phi) = \sum_{m \ne 0} \frac{1 - e^{2\pi i m \phi}}{(a^2 + m^2)\, m^2} = \frac{\pi}{a^3} \left(\frac{\cosh \pi a(1 - 2\phi)}{\sinh \pi a} - \coth a\pi \right)$$

$$+ \frac{2\pi^2}{a^2} \phi(1 - \phi) . \qquad (16.145)$$

The first sum is over all integers and the second over all integers but 0. Both sums define periodic functions of ϕ. They are equal to the expressions on the right-hand sides for ϕ in the unit interval. For ϕ outside $[0, 1]$, we must take ϕ modulo 1. If you have problems with the derivation, you may consult the book by Oberhettinger [54].

16.4 (Double Integral Over Translation Invariant Functions) The solution of this problem is used in the main body of the text when calculating the static potential. Let us consider a double integral over x and y,

$$I_1 = \int_0^L \int_0^L dx dy \, f(x, y) ,$$

with $f(x, y)$ only depending on the difference $\xi = x - y \in \mathbb{R}$. Show that

$$I_1 = 2 \int_0^L d\xi \, \big(f(\xi) + f(-\xi) \big) (L - \xi) .$$

Now let us assume that G is a translation invariant Green function in *two* dimensions, $G(x, y) = G(\xi)$ with $\xi = x - y$ in $\mathbb{R}^2$. We also assume that G only depends on the modulus of the components of ξ (which is true for the Green functions of interest). Consider the integral

$$I_2 = - \int_{\mathscr{D}} d^2 x \int_{\mathscr{D}} d^2 y \, \Delta_x G(x - y) ,$$

where x and y are both integrated over the rectangular region $\mathscr{D} = [0, T] \times [0, R]$. Use the above result to show that this *four*-dimensional integral reduces to the sum of two *one*-dimensional integrals,

$$I_2 = 4 \int_0^R d\xi_1 \, \big(G(0, \xi_1) - G(T, \xi_1) \big) (R - \xi_1)$$

$$+ 4 \int_0^T d\xi_0 \, \big(G(\xi_0, 0) - G(\xi_0, R) \big) (T - \xi_0) . \qquad (16.146)$$

16.5 (Explicit Integral of Green Function on the Cylinder) We consider the integral (16.93) with Green function on the infinite cylinder (16.90). Show, for example, by an explicit integration, that for $\mathscr{D} = [0, T] \times [0, R]$ we have

$$\int_{\mathscr{D}} d^2x \int_{\mathscr{D}} d^2y \, G(x - y) = \frac{2}{\beta} \sum_{n=-\infty}^{\infty} \frac{1}{M_n^3 \omega_n^2} \left(1 - \cos(\omega_n T)\right)\left(e^{-M_n R} + M_n R - 1\right),$$

where $M_n^2 = m_\gamma^2 + \omega_n^2$ depends on the Matsubara frequency ω_n. For low temperatures the Riemann sum becomes a Riemann integral, and we obtain

$$\int_{\mathscr{D} \times \mathscr{D}} G(x - y) = \frac{2}{\pi} \int_0^\infty \frac{d\omega}{\omega^2 M^3(\omega)} \left(1 - \cos(\omega T)\right)\left(e^{-RM(\omega)} + M(\omega)R - 1\right)$$

with $M^2(\omega) = m_\gamma^2 + \omega^2$. For $m_\gamma R \gg 1$ the term $M(\omega)R$ between the last brackets gives the dominant contribution. Show that

$$\frac{2m_\gamma^2}{\pi} \int_0^\infty \frac{d\omega}{\omega^2 M^3(\omega)} \left(1 - \cos(\omega T)\right) M(\omega) R = TR - \frac{R}{m_\gamma}.$$

This result is used to show that the string tension vanishes.

16.6 (Integration Over the Harmonics) We consider the integral (16.114) over the harmonic field

$$I = \int_0^1 d^2 h_\mu \, \psi_{\text{inst}}^\dagger(x) \, \psi_{\text{inst}}(y) \, e^{2\pi i \int_y^x (h_0 dx^0/\beta + h_1 dx^1/L)}$$

with normalized zero mode ψ_{inst} given by the series in (16.58) with $\nu = 1$ and $p = 0$ (which gives the same result as $p = 1$). Convince yourself that, after integration over h_1, the double sum and double integral collapses to the following single sum and integral over $a_0 = h_0 + \frac{1}{2}$,

$$I = \left(\frac{2}{\beta^2 V}\right)^{1/2} \sum_n \int_{1/2}^{3/2} da_0 \, e^{2\pi i (a_0 - n)(x^0 - y^0)/\beta} \, \xi\left(x^1 - (a_0 - n)L\right) \xi\left(y^1 - (a_0 - n)L\right)$$

containing the Gaussian wave function ξ in (16.56). Show that the sum and integral can be combined to one integral over the real line,

$$I = \frac{\sqrt{2}}{V^{3/2}} \int_{\mathbb{R}} du \, e^{2\pi i \xi_0 u / V} \, \xi(x^1 - u) \, \xi(y^1 - u) = \frac{1}{V} \exp\left(-\frac{\pi}{2V}\xi^2 + \frac{i\pi}{V}\xi_0(x_1 + y_1)\right),$$

where $\xi = x - y$. Note that I is not translation invariant.

16.7 (Gauge Invariant Two-Point Function) In the main text, we studied the gauge-invariant 2-point function for fermions and encountered the integral

$$J = \int d^2z \, d^2z' \int d\tau \, d\tau' \dot{x}^\mu(\tau) \dot{x}^\alpha(\tau') \varepsilon_{\mu\nu} \varepsilon_{\alpha\beta} \delta(z - x(\tau)) \delta(z' - x(\tau')) \frac{\partial^2}{\partial z^\nu \partial z^\beta} K_e(z - z') \,,$$

where $x(\tau)$ is the straight path (16.116) with tangential vector $\dot{x} = \xi = x - y$ and K_e the Green function (16.72). Show that

$$J = \varepsilon_{\mu\nu} \varepsilon_{\alpha\beta} \int_0^1 d\tau \int_0^1 d\tau' \, \xi^\mu \xi^\alpha \partial_\nu \partial_\beta K(\tau \xi - \tau' \xi) \,,$$

where $\partial_\beta K(z)$ is the derivative of K with respect to the argument z^β. Show that the integral has the double-series expansion

$$J = -\frac{2}{V} \sum_{k \neq 0} \left(\frac{k_1 \xi_0 - k_0 \tau \xi_1}{k \cdot \xi} \right)^2 \frac{1}{k^2(k^2 + m_\gamma^2)} \big(1 - \cos(k \cdot \xi) \big) \tag{16.147}$$

with wave vectors k the components of which are $k_0 = \frac{2\pi}{\beta} m_0$ and $k_1 = \frac{2\pi}{L} m_1$ with integers m_0 and m_1.

References

1. J. Schwinger, Gauge invariance and mass 2. Phys. Rev. **128**, 2425 (1962)
2. S. Coleman, R. Jackiw, L. Susskind, Charge shielding and quark confinement in the massive Schwinger model. Annals Phys. **93**, 267 (1975); L.S. Brown, Gauge invariance and mass in a two-dimensional model. Nuovo Cimento **29**, 617 (1963)
3. J.H. Lowenstein, J.A. Swieca, Quantum electrodynamics in two-dimensions. Ann. Phys. **68**, 172 (1971); A.Z. Capri, R. Ferrari, Schwinger model, chiral symmetry, anomaly and θ vacua. Nuovo Cimento **A62**, 273 (1981)
4. N.K. Nielsen, B. Schroer, Topological fluctuations and breaking of chiral symmetry in gauge theories involving massless fermions. Nucl. Phys. **B120**, 62 (1977); M. Hortacsu, K.D. Rothe, B. Schroer, Generalized QED in two-dimensions and functional determinants. Phys. Rev. **D20**, 3293 (1979)
5. C. Jayewardena, Schwinger model on S(2). Helv. Phys. Acta **61**, 636 (1988)
6. J. Balog, P. Hrasko, The fermion boundary condition and the θ-angle in QED$_2$. Nucl. Phys. **B245**, 118 (1984)
7. S. Durr, A. Wipf, Gauge theories in a bag. Nucl. Phys. **B443**, 201 (1995)
8. S. Durr, A. Wipf, Finite temperature Schwinger model with chirality breaking boundary conditions. Annals Phys. **255**, 333 (1997)
9. H. Joos, The geometric Schwinger model on the torus I. Helv. Phys. Acta **63**, 670 (1990); H. Joos, S.I. Azakov, The geometric Schwinger model on the torus II. Helv. Phys. Acta **67**, 723 (1994)
10. I. Sachs, A. Wipf, Finite temperature Schwinger model. Helv. Phys. Acta **65**, 652 (1992)
11. R. Narayanan, H. Neuberger, A construction of lattice chiral gauge theories. Nucl. Phys. **B443**, 305 (1995)

12. R. Orus, A practical introduction to tensor networks: matrix product states and projected entangled pair states. Annals Phys. **349**, 117 (2014); J.C. Bridgeman, C.T. Chubb, Handwaving and interpretive dance: an introductory course on tensor networks. J. Phys. A: Math. Theor. **50**, 223001 (2017); S.J. Ran, E. Tirrito, C. Peng, X. Chen, L. Tagliacozzo, G. Su, M. Lewenstein, *Tensor Network Contractions: Methods and Applications to Quantum Many-Body Systems*. Lecture Notes in Physics vol. 964 (Springer, Berlin, 2020)

13. M.C. Bañuls, K. Cichy, K. Jansen, J.I. Cirac, The mass spectrum of the Schwinger model with matrix product states. JHEP **11**, 158 (2013)

14. G. t'Hooft, Computation of the quantum effects due to a four-dimensional pseudoparticle. Phys. Rev. **D14**, 3432 (1976)

15. E. Witten, Constraints on supersymmetry breaking. Nucl. Phys. **B202**, 253 (1982)

16. D.V. Vassilevich, Heat kernel expansion: user's manual. Phys. Rept. **388**, 279 (2003)

17. P.B. Gilkey, *The Index Theorem and the Heat Equation*. (Publish or Perish, Boston, 1974)

18. M.F. Atiyah, I.M. Singer, The index of elliptic operators on compact manifolds. Bull. Am. Math. Soc. **69**, 422 (1963); The index of elliptic operators. Ann. Math. **87**, 484 (1968); ibid. 546

19. G. t'Hooft, Symmetry breaking through Bell-Jackiw anomalies. Phys. Rev. Lett. **37**, 8 (1976)

20. J.S. Bell, R. Jackiw, A PCAC puzzle $\pi^0 \to YY$ in the σ model. Phys. Lett. **59B**, 85 (1969); S.L. Adler, Axial vector vertex in spinor electrodynamics. Phys. Rev. **117**, 2426 (1969)

21. S. Coleman, The use of instantons, in *The Whys of Subnuclear Physics*, vol. 15, ed. by A. Zichichi (Springer, Boston, 1979)

22. R.A. Bertlmann, in *Anomalies in Quantum Field Theory*. International Series of Monographs on Physics, vol. 91 (Clarendon Press, Oxford, 2000)

23. A. Fayyazuddin, T.H. Hansson, M.A. Nowak, J.J.M. Verbaarshot, I. Zahed, Finite temperature correlators in the Schwinger model. Nucl. Phys. **B425**, 553 (1994)

24. S. Azakov, The Schwinger model on the torus. Fortsch. Phys. **45**, 589 (1997)

25. S. Blau, M. Visser, A. Wipf, Determinants, Dirac operators, and one loop physics. Int. J. Mod. Phys. **6**, 1467 (1989); Determinants of conformal wave operators in four dimensions. Phys. Lett. **209B**, 2019 (1988)

26. I. Sachs, A. Wipf, Generalized Thirring models, Ann. Phys. **249**, 380 (1996)

27. A.M. Polyakov, Quantum Gravity in two dimensions. Mod. Phys. Lett. **A2**, 893 (1987)

28. R.J. Riegert, A nonlocal action for the trace anomaly. Phys. Lett. **134B**, 56 (1984); E.S. Fradkin, A.A. Tseytlin, Conformal anomaly in Weyl theory and anomaly free superconformal theories. Phys. Lett. **134B**, 187 (1984); S.D. Odintsov, I.L. Shapiro, Perturbative approach to induced quantum gravity. Class. Quant. Grav. **8**, L57 (1991); I. Antoniadis, P.O. Mazur, E. Mottola, Conformal symmetry and central charges in 4 dimensions. Nucl. Phys. **B388**, 627 (1992)

29. P. Epstein, Zur Theorie allgemeiner Zetafunktionen I, II. Math. Ann. **56**, 615 (1903); **63**, 205 (1907)

30. L. Alvarez-Gaumé, G. Moore, C. Vafa, Theta functions, modular invariance, and strings. Commun. Math. Phys. **106**, 1–40 (1986)

31. S. Blau, M. Visser, A. Wipf, Analytical results for the effective action. Int. J. Mod. Phys. **A6**, 5408 (1992)

32. D. Mumford, *Tata Lectures on Theta* (Birkhäuser, Boston, 1983)

33. N. Manton, The Schwinger model and its axial anomaly. Ann. Phys. **159**, 220 (1985); J.E. Hetrick, Y. Hosotani, QED on a circle. Phys. Rev. **D38**, 2621 (1988)

34. S. Coleman, More about the massive Schwinger model. Ann. Phys. **101**, 239 (1976)

35. D.J. Gross, I.R. Klebanov, A.V. Maatytsin, A.V. Smilga, Screening versus confinement in (1+1) dimensions. Nucl. Phys. **B461**, 109 (1996)

36. J.V. Steele, J.J.M. Verbaarschot, I. Zahed, The Invariant fermion correlator in the Schwinger model on the torus. Phys. Rev. **D51**, 5915 (1995)

37. J.E. Hetrick, Y. Hosotani, S. Iso, The massive multi-flavor Schwinger model. Phys. Lett. **B350**, 92 (1995)

38. T. Misumi, Y. Tanizaki, M. Ünsal, Fractional θ angle, 't Hooft anomaly, and quantum instantons in charge-q multi-flavor Schwinger model. JHEP **07**, 018 (2019)

39. C. Adam, Massive Schwinger model within mass perturbation theory. Ann. Phys. **259**, 1 (1997)
40. A.V. Smilga, On the fermion condensate in Schwinger model, Phys. Lett. **278**, 371 (1992); Critical amplitudes in two-dimensional theories. Phys. Rev. **D55**, 443 (1996)
41. C. Gattringer, preprint hep-th/9503137
42. H. Leutwyler, A.V. Smilga, Spectrum of Dirac operator and role of winding number in QCD. Phys. Rev. **D46**, 5607 (1992)
43. V. Azcoiti, E. Follana, E. Royo-Amondarain, G. Di Carlo, A. Vaquero Avilés-Casco, Massive Schwinger model at finite θ. Phys. Rev. **D97**, 014507 (2018)
44. T.M.R. Byrnes, P. Sriganesh, R.J. Bursill, C.J. Hamer, Density matrix renormalization group approach to the massive Schwinger model. Phys. Rev. **D66** (2002) 013002
45. S. Durr, C. Hoelbling, Staggered versus overlap fermions: a study in the Schwinger model with $N(f) = 0, 1, 2$. Phys. Rev. **D69**, 034503 (2004)
46. S. Durr, C. Hoelbling, Scaling tests with dynamical overlap and rooted staggered fermions. Phys. Rev. **D71**, 054501 (2005)
47. P. Hasenfratz, V. Laliena, F. Niedermayer, The index theorem in QCD with a finite cutoff. Phys. Lett. **B427**, 125 (1998)
48. M. Lüscher, Exact chiral symmetry on the lattice and the Ginsparg-Wilson relation. Phys. Lett. **B428**, 342 (1998)
49. T.W. Chiu, The Spectrum and topological charge of exactly massless fermions on the lattice. Phys. Rev. **D58**, 074511 (1998)
50. N. Christian, K. Jansen, K. Nagai, B. Pollakowski, Scaling test of fermion actions in the Schwinger model. Nucl. Phys. **B739**, 60 (2006)
51. M.C. Bañuls, R. Blatt, J. Catani, A. Celi, J.I. Cirac, Simulating Lattice gauge theories with quantum technologies. Eur. Phys. J. **D74**, 165 (2020)
52. T. Banks, L. Susskind, J.B. Kogut, Strong coupling calculations of lattice gauge theories: (1+1) dimensional exercises. Phys. Rev. **D13**, 1043 (1976)
53. M.C. Bañuls, K, Cichy, K. Jansen, H. Saito, Chiral condensate in the Schwinger model with matrix product operators. Phys. Rev. **D93**, 094512 (2016)
54. F. Oberhettinger, *Fourier Expansions* (Academic, New York and London, 1973)

Interacting Fermions   17

In this chapter we study interacting four-Fermi theories in two and three spacetime dimension. Their Lagrangian density contains—besides the ubiquitous Dirac term $\bar{\psi}(\mathrm{i}\partial\!\!\!/ + m)\psi$—a Lorentz invariant interaction term with four powers of the Fermi field,

$$\mathscr{L}_I \propto (\bar{\psi}M\psi)(\bar{\psi}M\psi)\,. \tag{17.1}$$

Different types of four-Fermi theories have been introduced and studied by Thirring [1], Gross and Neveu [2], and by Nambu and Jona-Lasinio [3]. These are interacting relativistic theories with application in particle physics, condensed matter physics, and mathematical physics.

We shall investigate interacting Fermi systems with a small number N_f of flavors and in the limit $N_f \to \infty$, when the saddle point approximation becomes exact. The fermions may be in thermal contact with a heat bath, may exchange particles with the bath, and may be subject to a magnetic field B. Particular emphasis is on global symmetries and their breaking; the dependency on the external control parameters T, μ, and B; and—more technically—the sign problem in the fermionic path integral. On the way we calculate fermion determinants depending on these control parameters and for constant and spatially modulated condensate fields.

First we collect some relevant facts about relativistic interacting continuum models. Since our main focus is on the thermodynamics, we shall consider the *Euclidean theories*. Via the Hubbard–Stratonovich transformation, we shall introduce auxiliary fields to eliminate the four-Fermi terms. These auxiliary fields couple to the Dirac field, and one is led to calculating fermion determinants for Dirac operators containing scalar, pseudo-scalar, or vector fields. We shall embark on a general discussion of the sign problem in the continuum model and discuss the regularization and renormalization of the fermion determinant both with a momentum cutoff and with the ζ-function method. We shall solve the models in the limit where the number of flavors N_f tends to infinity and in particular study the

© The Author(s), under exclusive license to Springer Nature Switzerland AG 2021   475
A. Wipf, *Statistical Approach to Quantum Field Theory*, Lecture Notes
in Physics 992, https://doi.org/10.1007/978-3-030-83263-6_17

condensate $\langle \bar{\psi} \psi \rangle$ as a function of the temperature, chemical potential, and magnetic field. A subclass of models considered shows a non-trivial phase diagram with symmetric and broken phases. In even dimensions chiral symmetry can be broken, and in odd dimensions parity-breaking is possible.

At the end of this chapter, we shall present known results about inhomogeneous phases of two-dimensional four-Fermi theories in the large N_f-limit. This shows explicitly that for low T and high μ, translation invariance can be broken in the $N_f \to \infty$ limit. Recent Monte Carlo simulations of the model revealed an inhomogeneous phase even for a finite number of flavors [4].

Many momentum integrals of rotation invariant functions are proportional to the volume of the unit sphere in $\mathbb{R}^d$ divided by an appropriate power of 2π,

$$\mathscr{C}_d = \frac{|S^{d-1}|}{(2\pi)^d} = \frac{\pi^{\frac{d+1}{2}} \Gamma(d)}{\Gamma(\frac{d+1}{2})}, \quad \mathscr{C}_1 = \frac{1}{\pi}, \quad \mathscr{C}_2 = \frac{1}{2\pi}. \tag{17.2}$$

When computing fermion determinants and thermodynamic potentials, one often encounters the following ratio of two Γ-functions,

$$r_\alpha(s) = \frac{\Gamma(s - \alpha/2)}{\Gamma(s)} \tag{17.3}$$

with simple zeros at $s \in -\mathbb{N}_0$ and simple poles at $s \in \alpha/2 - \mathbb{N}_0$. One also obtains Gaussian sums over Matsubara frequencies which can be rewritten with the help of the resummation formula discussed in Problem 16.2,

$$K_\beta(t) = \sum_{n \in \mathbb{Z}} e^{-t(\omega_n - i\mu)^2} = \frac{\beta}{\sqrt{4\pi t}} \sum_{n \in \mathbb{Z}} (-1)^n e^{-\beta^2 n^2/4t - n\beta\mu}, \quad \omega_n = \frac{2\pi}{\beta}(n + \tfrac{1}{2}). \tag{17.4}$$

Imposing periodic boundary conditions for free fermions on a wire of length L, one meets momentum sums

$$K_L(t) = \sum_{m \in \mathbb{Z}} e^{-t(2\pi m/L)^2} = \frac{L}{\sqrt{4\pi t}} \sum_{m \in \mathbb{Z}} e^{-L^2 m^2/4t}. \tag{17.5}$$

The corresponding sum for free fermions interacting with a constant vector field in a box of linear extent L and volume $V_s = L^{d-1}$ reads

$$K_L^{d-1} = \sum_{\boldsymbol{m} \in \mathbb{Z}^{d-1}} e^{-t(\boldsymbol{p_m} + \boldsymbol{v})^2} = \frac{V_s}{(4\pi t)^{\frac{1}{2}(d-1)}} \sum_{\boldsymbol{m} \in \mathbb{Z}^{d-1}} e^{-L^2 m^2/4t + iL\boldsymbol{m}\cdot\boldsymbol{v}}, \quad \boldsymbol{p_m} = \frac{2\pi}{L}\boldsymbol{m}. \tag{17.6}$$

All Gaussian sums are proportional to particular Jacobi theta-functions. For low temperature and large box sizes, these functions converge to

$$\lim_{\beta\to\infty}\lim_{\mu\to 0} K_\beta(t) = \frac{\beta}{\sqrt{4\pi t}} \quad\text{and}\quad \lim_{L\to\infty} K_L(t) = \frac{L}{\sqrt{4\pi t}}. \tag{17.7}$$

These results will be useful when we study interacting fermions in the large N_{f} limit.

17.1 Symmetries of Fermi Systems

We assume that the Fermi fields transform according to an irreducible representation of the Euclidean spin group.[1] In even dimensions there is one irreducible representation, and in odd dimensions there are two (γ^μ and $-\gamma^\mu$ are inequivalent). In Euclidean spacetime we can and will choose hermitian γ^μ-matrices as in (15.3). An irreducible spinor has

$$d_s = 2^{[d/2]} \tag{17.8}$$

components. Here $[x]$ denotes the largest integer less or equal to x. Thus in 2 and 3 dimensions the γ^μ are 2×2 matrices and in 4 and 5 dimensions 4×4 matrices. Furthermore, there exists a hermitian generalization of γ_5 in (15.5), denoted by γ_*, which commutes or anticommutes with all γ^μ,

$$\gamma_* = -\mathrm{i}^{[d/2]}\gamma^0\cdots\gamma^{d-1} = \gamma_*^\dagger, \quad \gamma_*\gamma^\mu + (-1)^d\gamma^\mu\gamma_* = 0, \quad \gamma_*^2 = \mathbb{1}. \tag{17.9}$$

In odd dimensions γ_* commutes with all γ^μ and is $\mathbb{1}$ in one irreducible representation and $-\mathbb{1}$ in the other irreducible representation.

In this section we study the global symmetries of the theory with N_{f} free fermions with identical mass described by the Lagrangian density

$$\mathscr{L}_0 = \sum_{a=1}^{N_{\mathrm{f}}} \bar\psi_a(\slashed{\partial} + m)\psi_a \equiv \bar\psi(\slashed{\partial} + m)\psi, \quad \psi = \begin{pmatrix}\psi_1 \\ \vdots \\ \psi_{N_{\mathrm{f}}}\end{pmatrix}, \tag{17.10}$$

where in the last expression $\slashed{\partial}$ and m act in the same way on all N_{f} flavors (the purists would write $\mathbb{1}_{N_{\mathrm{f}}} \otimes (\slashed{\partial} + m)$). Besides the (Euclidean) spacetime symmetries these are chiral rotations and charge conjugation.

Four-Fermi theories have a Lagrangian

$$\mathscr{L} = \mathscr{L}_0 + \mathscr{L}_I \tag{17.11}$$

---

[1] We emphasize this point since in applications in condensed matter theory often reducible *four*-component Fermi fields in 3 spacetime dimensions are considered.

with four-Fermi interaction

$$\mathscr{L}_I = -\frac{1}{2N_f} \sum_p g_p^2 \left(\bar{\psi} M_p \psi\right)^2, \qquad \bar{\psi} M_p \psi = \sum_{a,b} \bar{\psi}_a (M_p)_{ab} \psi_b \,. \tag{17.12}$$

The terms in the sum over p are characterized by the transformation properties of the bilinears $\bar{\psi} M_p \psi$. They should be spacetime tensors such that Lorentz invariance is manifest. The following bilinears are used in the systems considered below:

$$S(x) = \bar{\psi}(x)\psi(x) \,,$$

$$P(x) = i\bar{\psi}(x)\gamma_*\psi(x) \quad \text{and} \quad P_p(x) = i\bar{\psi}(x)\gamma_* T_p \psi(x) \,, \tag{17.13}$$

$$J^\mu(x) = \bar{\psi}(x)\gamma^\mu\psi(x) \,.$$

In even dimensions S is a scalar, and in odd dimensions $S = \pm iP$ is a pseudo-scalar. The matrices T_p in the definition of the P_p are (trace-orthogonal) generators of the flavor group $SU(N_f)$. For 2 flavors we may choose $T_p = \sigma_p$.

$U_V(N_f)$ Vector Flavor-Symmetry (Isospin Symmetry)

The unitary flavor rotations mix the different flavors

$$\psi \mapsto U\psi, \quad \bar{\psi} \to \bar{\psi} U^\dagger, \quad U \in U(N_f) \,. \tag{17.14}$$

They do not act on the spin indices and leave the Lagrangian $\mathscr{L}_0$ in (17.10) invariant. More generally, all completely antisymmetric tensor fields

$$T^{\mu_1\mu_1\cdots\mu_n} = \sum_{a=1}^{N_f} \bar{\psi}_a \gamma^{\mu_1\mu_2\cdots\mu_n} \psi_a, \quad \gamma^{\mu_1\mu_2\cdots\mu_n} = \gamma^{[\mu_1}\gamma^{\mu_2}\cdots\gamma^{\mu_n]} \tag{17.15}$$

are invariant under these rotations. In particular the fields S, P, and J^μ in (17.13) are flavor singlets. The pseudo-scalar fields P_p transform with the adjoint representation

$$P_p \mapsto \sum R_{pq} P_q, \quad \text{where} \quad (R_{pq}) = \text{tr}\left(T_q^\dagger U^\dagger T_p U\right) \in SO(N_f^2 - 1), \tag{17.16}$$

such that the four-Fermi term $\sum P_p P_p$ is a singlet as well.

The Abelian subgroup $U_V(1)$ transforms all flavors in the same way,

$$\psi \mapsto e^{i\alpha}\psi, \quad \bar{\psi} \mapsto \bar{\psi} e^{-i\alpha} \,. \tag{17.17}$$

It remains a symmetry in case the different flavors have different masses, and the associated Noether charge is just the conserved fermion number. The global $U_V(1)$

symmetry can be promoted to a local gauge symmetry by introducing an Abelian gauge potential.

$U(N_f) \times U(N_f)$ Chiral Symmetry

In *even dimensions* we can define chiral (left- and right-handed) fermions,

$$\psi_\pm = P_\pm \psi = \frac{1}{2}(\mathbb{1} \pm \gamma_*)\,\psi \,, \tag{17.18}$$

and rotate these independently among each other,

$$\psi_\pm \mapsto U_\pm \psi_\pm, \quad \bar{\psi}_\pm \mapsto \bar{\psi}_\pm U_\mp^{-1} \quad \text{with} \quad U_\pm \in U(N_f)\,. \tag{17.19}$$

Given the transformations of $\psi_\pm$, one obtains the transformations of $\bar{\psi}_\pm$ by the requirement, that the Lagrangian $\mathscr{L}_0$ with $m = 0$ is left invariant. The chiral symmetry is an extension of the vector-flavor symmetry (17.14). Only tensor fields $T^{\mu_1\mu_2\cdots}$ with an odd number of indices are invariant under all chiral transformations. If the Lagrangian contains a mass term (with equal mass for all flavors), only the diagonal subgroup consisting of the $U(N_f)$ vector flavor rotations with $U_- = U_+$ forms a symmetry.

In contrast, the *axial transformations* with $U_- = U_+^{-1}$ transform right- and left-handed chiral fermions in the opposite way. Taken together the vector and axial transformations generate the full chiral symmetry. The axial transformations contain the Abelian group of flavor-blind $U_A(1)$ axial transformations

$$\psi \mapsto e^{i\alpha\gamma_*}\psi, \quad \bar{\psi} \mapsto \bar{\psi}e^{i\alpha\gamma_*}\,. \tag{17.20}$$

For each flavor the bilinears S and P in (17.13) form an SO(2)-vector under this $U_A(1)$,

$$\begin{pmatrix} S \\ P \end{pmatrix} \mapsto \begin{pmatrix} \cos 2\alpha & \sin 2\alpha \\ -\sin 2\alpha & \cos 2\alpha \end{pmatrix} \begin{pmatrix} S \\ P \end{pmatrix}\,. \tag{17.21}$$

This means that the particular four-Fermi term

$$S^2 + P^2 \tag{17.22}$$

is invariant not only under $U_V(N_f)$ transformations but also under $U_A(1)$ axial rotations. It is not invariant under general axial transformations.

$\mathbb{Z}_2$ Charge Conjugation

The charge conjugation maps every particle into its anti-particle. In Euclidean spacetime the conjugation acts on a spinor field as

$$\psi \longmapsto \psi_c = \mathscr{C}\psi^*\,, \tag{17.23}$$

with charge conjugation matrix $\mathscr{C}$. This matrix conjugates each γ_μ into its transpose or negative transpose,

$$\gamma_\mu^T = \eta_c \,\mathscr{C}^{-1}\gamma_\mu\mathscr{C}, \quad \eta_c = \pm 1 \implies \mathscr{C}^{-1}\gamma_*^T\mathscr{C} = \eta_c'\gamma_*, \tag{17.24}$$

where $\eta_c' = (-1)^{d/2}$ in even dimensions and 1 in odd dimensions (where γ_* is proportional to the identity). In even dimensions there exists one $\mathscr{C}$ with $\eta_c = 1$ and another one with $\eta_c = -1$. In $1 + 4n$ dimensions, there exists a $\mathscr{C}$ with $\eta_c = 1$ and in $3 + 4n$ dimensions with $\eta_c = -1$; see Problem 17.1.

$\mathbb{Z}_2$ Parity Symmetry
    The parity operation is an improper Lorentz transformation and a symmetry of physical systems where we can neglect the weak interaction. But we shall see later that it may be spontaneously broken in odd spacetime dimensions. As parity transformation we choose the reflection at the hyperplane orthogonal to the last coordinate,

$$x \mapsto x' = Px, \quad P = (P^\mu_\nu) = \mathrm{diag}(1, 1, \dots, 1, -1). \tag{17.25}$$

Scalar fields preserve their sign, and pseudo-scalar fields reverse their sign under this operation. It is possible to define parity transformations for ψ and $\bar\psi$ such that the massless Lagrangian $\mathscr{L}_0$ is a scalar field and the action is Lorentz-invariant. Later we shall argue that in every dimension there exists a parity matrix $\mathscr{P}$ with

$$\mathscr{P}^{-1}\gamma^\mu\mathscr{P} = \alpha P^\mu_\nu\gamma_\nu \implies \mathscr{P}^{-1}\gamma_*\mathscr{P} = -\alpha^d\,\gamma_*, \quad \alpha^2 = 1. \tag{17.26}$$

The Lagrangian $\mathscr{L}_0$ with massless fermions is a scalar if the fields transform as

$$\psi(x) \mapsto \psi_P(x') = \mathscr{P}\psi(x), \quad \bar\psi(x) \mapsto \bar\psi_P(x') = \alpha\,\bar\psi(x)\mathscr{P}^{-1}. \tag{17.27}$$

In *even dimensions* there is the parity matrix $\mathscr{P} = i\gamma_*\gamma_{d-1}$ with $\alpha = 1$. Then the bilinears $S(x)$ and $P(x)$ in (17.13) are scalar and pseudo-scalar fields, and a mass term $m\bar\psi\psi$ in the action is parity invariant. In *odd dimensions* there is only a parity matrix with $\alpha = -1$, for example, $\mathscr{P} = \gamma_{d-1}$; see Problem 17.2. Now the bilinear $S(x)$ is a pseudo-scalar, and a mass term breaks parity explicitly.

In multi-flavor models, there is more freedom to construct a parity matrix. For example, in odd-dimensional models with an even number of flavors, one can again define a matrix $\mathscr{P}$ fulfilling (17.26) with $\alpha = 1$. For example, by a dimensional torus reduction from 4 to 3 dimensions, the number of flavors doubles, and one can define a $\mathscr{P}$ with $\alpha = 1$. The fermions of the reduced model transform according to a *four*-dimensional reducible transformation of the spin group, and their symmetries have been discussed in great detail in [5, 6].

17.2 Four-Fermi Theories

Four-Fermi terms come with coupling constants with mass dimension $2 - d$. Thus, in 2 dimensions they are renormalizable in perturbation theory. In addition they asymptotically free or conformal. But above 2 dimensions, they cease to be renormalizable in perturbation theory. However, they are renormalizable above 2 and below 4 dimensions in the large N_f expansion [7–9]. In these dimensions they provide a simple realization of the concept of asymptotic safety [10], a notion introduced some time ago by S. Weinberg in the context of quantum gravity [11]. Four-Fermi theories in 2 and 3 dimensions are of relevance in solid-state physics, where they describe low-energy electronic properties of materials like graphene [12, 13] and high-temperature superconductors [14] or have been applied to the description of optical lattices [15]. In the following we shall focus on relativistic four-Fermi theories and in particular the Thirring, Gross–Neveu, and chiral Gross–Neveu models.

17.2.1 Thirring Model

It has been introduced by Walter Thirring back in 1958 [1] and describes a vector-vector four-Fermi interaction

$$\mathscr{L}_{\text{Th}} = \mathscr{L}_0 + \frac{g^2}{2N_f}(\bar{\psi}\gamma^{\mu}\psi)^2 \,. \tag{17.28}$$

The two-dimensional massless model is an exactly solvable conformal field theory for any number of flavors. It can be solved with the deformation method introduced and applied in Chap. 16, and the n-point correlation functions are known explicitly [16, 17]. The massive model can be solved with the Bethe ansatz (cp. the tutorial [18] for beginning graduate students)— which allows one to calculate the mass spectrum and scattering matrix elements.

The model in odd dimensions has an $U(N_f)$ vector flavor symmetry, and the massless model in even dimensions admits the full $U(N_f) \times U(N_f)$ chiral symmetry.

17.2.2 (Chiral) Gross–Neveu Model

The Gross–Neveu model is a field theory for Dirac fermions interacting via a scalar-scalar four-Fermi term

$$\mathscr{L}_{\text{GN}} = \mathscr{L}_0 - \frac{g^2}{2N_f}(\bar{\psi}\psi)^2 \,. \tag{17.29}$$

In 1974 the model has been introduced by David Gross and André Neveu as a *two*-dimensional asymptotically free toy model for quantum chromodynamics (QCD)

[2]. It admits an $U(N_f)$ vector flavor symmetry[2] and in even dimensions and for massless fermions admits an additional discrete $\mathbb{Z}_2$ chiral symmetry

$$\psi \mapsto i\gamma_*\psi, \quad \bar{\psi} \mapsto i\bar{\psi}\gamma_* \quad \text{with} \quad \bar{\psi}\psi \mapsto -\bar{\psi}\psi. \tag{17.30}$$

Gross and Neveu studied the system in the limit $N_f \to \infty$ and demonstrated that it is asymptotically free. Even for vanishing fermion mass, a non-zero chiral condensate $\langle\bar{\psi}\psi\rangle$ forms and breaks chiral symmetry spontaneously. The broken phase is gapped, and the masses are non-analytic functions of the dimensionless coupling constant. In 3 dimensions a condensate breaks parity symmetry spontaneously.

The *chiral Gross–Neveu model* is a natural $U_A(1)$-invariant extension of the GN model and can be formulated in even dimensions. Its Lagrangian

$$\mathscr{L}_{\text{cGN}} = \mathscr{L}_0 + \frac{g_v^2}{2N_f}(\bar{\psi}\gamma^\mu\psi)^2 - \frac{g_s^2}{2N_f}\left((\bar{\psi}\psi)^2 + (\bar{\psi}i\gamma_*\psi)^2\right), \tag{17.31}$$

contains the $U_A(1)$-invariant Thirring coupling and the $U_A(1)$-invariant interaction discussed in (17.22). Hence, it is invariant under $U_V(N_f)$ vector and $U_A(1)$ axial transformations. A chiral condensate would break this continuous symmetry, and — according to the Mermin-Wagner theorem—should not form in 2 dimensions. But a condensate may form in the limit $N_f \to \infty$, where the theorem can be bypassed.

Unfortunately in the literature one finds different names for the same four-Fermi system. Moreover, one finds the same name for different models. For example, the $U_A(1)$ invariant theory without Thirring coupling is also called chiral GN model or *two*-dimensional Nambu-Jona-Lasinio model.

17.2.3 Nambu-Jona-Lasinio Model

Already in 1961 the NJL model has been introduced by Giovanni Jona-Lasinio and Yoichiro Nambu [3] and independently by Valentin Vaks and Anatoly Larkin [19] to highlight the chiral properties of the strong interaction at low energies. It was introduced as an effective four-Fermi theory of nucleons and mesons with chiral symmetry, paralleling the construction of Cooper pairs from electrons in the BCS theory of superconductivity. It is formulated for two flavors in an even dimensional spacetime. The Lagrangian

$$\mathscr{L}_{\text{NJL}} = \mathscr{L}_0 - \frac{g^2}{4}\left((\bar{\psi}\psi)^2 + (\bar{\psi}i\gamma_*\boldsymbol{\sigma}\psi)^2\right), \tag{17.32}$$

---

[2] The $U(N_f)$ symmetry can be extended to an $O(2N_f)$ symmetry. This is made explicit by rewriting the Dirac spinor in terms of its Majorana components.

where the Pauli matrices in $\boldsymbol{\sigma} = (\sigma_1, \sigma_2, \sigma_3)$ act on the flavor indices, is invariant under $U_V(2)$ vector flavor transformations. For $m = 0$ it is also invariant under $U_A(1)$ axial transformations. Less obvious is the invariance under axial flavor transformations, under which the chiral fermion fields transform as

$$\psi_+ \mapsto U\psi_+, \quad \psi_- \mapsto U^{-1}\psi_-, \quad \bar{\psi}_+ \mapsto \bar{\psi}_+ U, \quad \bar{\psi}_- \mapsto \bar{\psi}_- U^{-1}, \qquad (17.33)$$

with $U \in SU(2)$. A short calculation reveals that the scalar bilinear and the pseudo-scalar isovector bilinears entering the Lagrangian transform according to

$$\begin{pmatrix} \bar{\psi}\psi \\ i\bar{\psi}\gamma_*\boldsymbol{\sigma}\psi \end{pmatrix} \mapsto \begin{pmatrix} \cos 2\theta & \sin 2\theta \, \boldsymbol{e}^T \\ -\sin 2\theta \, \boldsymbol{e} & \mathbb{1} + (\cos 2\theta - 1) \, \boldsymbol{e}\boldsymbol{e}^T \end{pmatrix} \begin{pmatrix} \bar{\psi}\psi \\ i\bar{\psi}\gamma_*\boldsymbol{\sigma}\psi \end{pmatrix}, \qquad (17.34)$$

where the angle θ and the unit vector $\boldsymbol{e}$ parametrize the $SU(2)$ matrix

$$U = \exp(i\theta \, \boldsymbol{e} \cdot \boldsymbol{\sigma}). \qquad (17.35)$$

The four-dimensional real matrix in (17.34) is a proper rotation in $\mathbb{R}^4$ which proves that $\mathscr{L}_{\mathrm{NJL}}$ is invariant under axial transformations (17.33). We conclude that the massless 2-flavor NJL model admits the maximal possible $U(2) \times U(2)$ chiral symmetry. More general two-flavor models with maximal chiral symmetry were studied in [20].

17.2.4 Relations Between Different Four-Fermi Theories

It may happen that two seemingly different looking four-Fermi theories are equivalent on account of a Fierz identity [5]. Since irreducible spinors in two and three dimensions have only *two* components, there is a more direct way to relate different one-flavor models. Indeed, for a *two*-component anticommuting ψ an arbitrary four-Fermi term $(\bar{\psi}A\psi)^2$ at a fixed spacetime point must be proportional to $\bar{\psi}_1\psi_1\bar{\psi}_2\psi_2$. For one flavor

$$(\bar{\psi}A\psi)^2 = \det A \, (\bar{\psi}\psi)^2, \quad d = 2, 3. \qquad (17.36)$$

1. *One-flavor models in two dimensions:* The hermitian 2×2 matrices γ^μ square to one, have vanishing trace and thus determinant -1. We conclude

$$\mathscr{L}_{\mathrm{Th}} = \mathscr{L}_0 - g^2(\bar{\psi}\psi)^2, \quad N_f = 1, \, d = 2. \qquad (17.37)$$

This means that the Thirring model is equivalent to the GN model with rescaled coupling[3] Thus, the massless GN model is invariant not only under $U_V(1) \times \mathbb{Z}_2$ but also under the larger symmetry group $U_V(1) \times U_A(1)$.

2. *One-flavor models in three dimensions:* Here one has the relation

$$\mathscr{L}_{Th} = \mathscr{L}_0 - \frac{3g^2}{2}(\bar{\psi}\psi)^2, \quad N_f = 1, \ d = 3, \tag{17.38}$$

such that the Thirring model is equivalent to the GN model with rescaled g.

3. Comparable simple equivalences do not exist for multi-flavor models or models in higher dimensions. An arbitrary Fierz identity in 3 and 2 dimensions follows from the identity discussed in Problem 17.4. For example, the Lagrangian of the *three*-dimensional massless N_f-flavor Thirring model can be rewritten as

$$\mathscr{L}_{Th} = \mathscr{L}_{GN} - \frac{g^2}{N_f} \sum_{a,b} (\bar{\psi}^a \psi^b)(\bar{\psi}^b \psi^a), \quad d = 3. \tag{17.39}$$

This means that the Thirring coupling is converted into a Gross–Neveu plus a tensor-tensor coupling.

17.2.5 Hubbard–Stratonovich Transformation

It is possible to eliminate the four-Fermi terms in the Lagrangian

$$\mathscr{L} = \mathscr{L}_0 - \frac{1}{2N_f} \sum_p g_p^2 (\bar{\psi} M_p \psi)^2, \tag{17.40}$$

by a Hubbard–Stratonovich transformation [21]. Thereby, one introduces an auxiliary tensor field $\phi_p(x)$ for each four-Fermi term. Indeed, the theory with Lagrangian (17.40) is equivalent to the model with Lagrangian

$$\mathscr{L} = \mathscr{L}_0 + \sum_p \phi_p(\bar{\psi}_p M_p \psi) + \frac{N_f}{2} \sum_p \frac{\phi_p^2}{g_p^2}. \tag{17.41}$$

The equivalence of (17.40) and (17.41) on the classical level is proven with the help of the field equation for the auxiliary fields $\phi_p(x)$,

$$\phi_p = -\frac{g_p^2}{N_f} \bar{\psi} M_p \psi. \tag{17.42}$$

---

[3] Actually, all one-flavor four-Fermi theories in two and three dimensions are equivalent.

The equivalence also holds for the quantized system, since the auxiliary fields are non-dynamical, enter the Lagrangian at most quadratically, and thus can be integrated over in the path integral. The auxiliary field ϕ_p must have the same tensor structure as $\bar{\psi} M_p \psi$, such that after contraction of tensor indices the terms linear in the auxiliary fields in (17.41) are Lorentz scalars.

The chiral GN model with Lagrangian (17.31) is bosonized with a vector field v_μ, scalar field σ, and pseudo-scalar field π,

$$\mathscr{L} = \bar{\psi} \mathscr{D} \psi + N_f \mathscr{V}_B, \quad \mathscr{V}_B = \frac{1}{2g_v^2} v_\mu v^\mu + \frac{1}{2g_s^2} (\sigma^2 + \pi^2). \tag{17.43}$$

The Dirac operator contains the auxiliary fields,

$$\mathscr{D} = \gamma^\mu (\partial_\mu + i v_\mu) + m + \sigma + i \gamma_* \pi. \tag{17.44}$$

For the Thirring model, $g_s = 0$ and the fields σ and π are absent. For the GN model, the vector field v_μ and pseudo-scalar field π are absent. Similarly, the two-flavor NJL model (17.32) is bosonized with 1 scalar and 3 pseudo-scalar fields.

Symmetries of Massive Versus Massless Systems

In Table 17.1 we list some properties of massive and massless four-Fermi models considered in this chapter. Therein we used the abbreviations in (17.13) for the fermion bilinears. In even dimensions the scalar field S and pseudo-scalar field P are independent. In odd dimensions $S = \pm i P$ is a pseudo-scalar. Depending on the particular form of the interaction and dimension of spacetime, the models admit different global symmetries. The symmetry groups of the massless systems are listed in the third to last column. A mass term or a condensate $\langle \bar{\psi} \psi \rangle$ breaks this symmetry to a smaller symmetry group listed in the second to last column. The last column lists the symmetries which are broken by a mass term or condensate.

17.3 Dirac Operator

With a Hubbard–Stratonovich transformation, one linearizes the non-linear field equations for the Fermi field. The transformed Lagrangian is quadratic in the Fermi fields and contains a coupling of the fermions to an auxiliary vector field, scalar field, and pseudo-scalar field. Thus, one is led to consider an Euclidean Dirac operator of the form

$$\mathscr{D} = \gamma^\mu D_\mu + \sigma + i \pi \gamma_* + \mu \gamma^0, \quad D_\mu = \partial_\mu + i v_\mu \tag{17.45}$$

In odd dimensions π is missing. We are interested in Fermi systems in thermal equilibrium at finite temperature and fermion density and thus included a chemical potential μ for the fermion charge; see the discussion in Sect. 15.7.

Table 17.1 The interaction terms of the Thirring, GN, cGN, and NJL models, the auxiliary fields after bosonization, the symmetries of the massless and massive models. The last column lists the broken symmetries

Model	Interaction	Aux. field	Dim	Symmetry $m = 0$	Symmetry $m \neq 0$	Breaking
Thirring	$\lambda J_\mu J^\mu$	v_μ	Even	$U(N_f) \times U(N_f)$	$U_V(N_f)$	Axial
			Odd	$U(N_f) \times \mathbb{Z}_2$	$U(N_f)$	Parity
GN	$-\lambda S^2$	σ	Even	$U(N_f) \times \mathbb{Z}_2$	$U(N_f)$	$\mathbb{Z}_2$ chirality
			Odd	$U(N_f) \times \mathbb{Z}_2$	$U(N_f)$	Parity
Chiral GN	$\lambda_v J_\mu J^\mu - \lambda_s (S^2 + P^2)$	σ, π, v_μ	Even	$U(N_f) \times U_A(1)$	$U(N_f)$	Axial $U_A(1)$
$N_f = 2$ NJL	$-\lambda(S^2 + \boldsymbol{P}^2)$	$\sigma, \boldsymbol{\pi}$	Even	$U(2) \times U(2)$	$U(2)$	Axial

The grand canonical potential is given by the Euclidean path integral over Fermi and auxiliary bosonic fields, and the temperature dependence comes from the periodicity conditions on the fields in imaginary time:

ψ antiperiodic: $\quad \psi(\tau + \beta, \boldsymbol{x}) = -\psi(\tau, \boldsymbol{x})$

σ, π periodic: $\quad \sigma(\tau + \beta, \boldsymbol{x}) = \sigma(\tau, \boldsymbol{x}), \ \pi(\tau + \beta, \boldsymbol{x}) = \pi(\tau, \boldsymbol{x})$. (17.46)

Since the action is quadratic in the Fermi fields, the Gaussian path integral over these fields produces the fermion determinant $\det \mathscr{D}$ of the operator (17.45). Since $\mathscr{D}$ is flavor-blind, it suffices to study its spectral properties for one flavor of irreducible fermions. This will be done for arbitrary spacetime-dependent auxiliary background fields. Actually, since $\mathscr{D}$ is neither hermitian nor anti-hermitian, there may be a sign problem, and we shall discuss this problem in some detail. Depending on the dimension, we can make stronger or weaker statements about $\det \mathscr{D}$.

17.3.1 Spectral Properties and Furry Theorem

Above we have studied the global symmetries of Fermi systems. Here we will study the implications for the spectrum of the Euclidean Dirac operator $\mathscr{D}$. This includes a generalization of Furry theorem [22] which is relevant when one calculates radiation corrections in quantum electrodynamics (QED).

Charge Conjugation

Let us assume that ψ is an eigenfunction of the Dirac operator (17.45) with arbitrary auxiliary background fields and chemical potential, $\mathscr{D}\psi = \lambda\psi$. To see which equation the charge-conjugated spinor field ψ_c in (17.23) obeys, we take the complex conjugate of the equation $\mathscr{D}^*\psi^* = \lambda^*\psi^*$. The γ^μ are hermitian such that their complex conjugate is equal to their transpose. Now we act with the charge conjugation matrix $\mathscr{C}$ and make use of (17.24) to arrive at the mode equation for ψ_c. With the abbreviation $\eta_c'' = -\eta_c\eta_c'$, one finds

$$\mathscr{D}(\mu, v_\mu, \sigma, \pi)\psi = \lambda\psi \implies \mathscr{D}(\mu, -v_\mu, \eta_c\sigma, \eta_c''\pi)\psi_c = \eta_c\lambda^*\psi_c. \qquad (17.47)$$

Hence, under a charge conjugation the eigenvalue λ of $\mathscr{D}$ transforms into $\pm\lambda^*$. Since $\mathscr{D}$ possesses an even number of eigenvalues[4] its determinant is therefore mapped into its complex conjugate under charge conjugation,

$$\det \mathscr{D}(\mu, v_\mu, \sigma, \pi) = \det{}^* \mathscr{D}(\mu, -v_\mu, \eta_c\sigma, \eta_c''\pi). \qquad (17.48)$$

---

[4] Since ψ has an even number of components this is certainly true on a finite lattice.

For example, in two and three dimensions, we have

$$d = 2: \quad \det{}^* \mathscr{D}(\mu, v_\mu, \sigma, \pi) = \det \mathscr{D}(\mu, -v_\mu, \sigma, \pi) \tag{17.49}$$

$$\det{}^* \mathscr{D}(\mu, v_\mu, \sigma, \pi) = \det \mathscr{D}(\mu, -v_\mu, -\sigma, -\pi),$$

$$d = 3: \quad \det{}^* \mathscr{D}(\mu, v_\mu, \sigma) = \det \mathscr{D}(\mu, -v_\mu, -\sigma). \tag{17.50}$$

The two relations in $d = 2$ dimensions imply that the determinant is invariant under a simultaneous sign change of σ and π. This property holds in all even dimensions and implies certain properties of Feynman diagrams with closed fermion loops only and a fixed number of external boson lines. These diagrams appear in a perturbative expansion of the determinant in powers of the auxiliary fields. A similar property holds for the fermion determinant in QED: Expanding $\log \det(\not{D} + m)$ in powers of the photon field only even powers of the photon field appear. There is no scattering of three real photons in QED. This is because of charge conjugation symmetry, as shown by Furry [22].

Hermitian Conjugation We assume that the derivative operators ∂_μ are anti-hermitian. This applies, for example, to the partial derivatives in the continuum, the SLAC derivatives on a lattice, and naive lattice derivatives, but not to Wilson fermions. Then the hermitian conjugate of the Dirac operator is

$$\mathscr{D}^\dagger(\mu, v_\mu, \sigma, \pi) = -\mathscr{D}(-\mu, v_\mu, -\sigma, \pi). \tag{17.51}$$

We conclude that in any dimension, the determinant has the property

$$\det{}^* \mathscr{D}(\mu, v_\mu, \sigma, \pi) = \det \mathscr{D}(-\mu, v_\mu, -\sigma, \pi). \tag{17.52}$$

Together with (17.48) one can prove further properties of $\det \mathscr{D}$. In addition, when one considers particular subsectors of the general $four$-Fermi model, for example, the Thirring, GN, or NJL model, then stronger statements hold true.

Chiral Transformations In even dimensions we can conjugate the Dirac operator (17.45) with γ_* and find

$$\det \mathscr{D}(\mu, v_\mu, \sigma, \pi) = \det \mathscr{D}(\mu, v_\mu, -\sigma, -\pi). \tag{17.53}$$

Thus, we reproduce the result (17.50) in even dimensions.

Limiting Cases
    Finally we collect and combine the results (17.48) and (17.52) for the cGN model and see what they imply for the Thirring and Gross–Neveu model.

1. *Thirring model:* The determinant is real for vanishing μ or v_μ,

$$\det \mathscr{D}(\mu, v_\mu) = \det{}^* \mathscr{D}(\mu, -v_\mu) = \det \mathscr{D}(-\mu, -v_\mu) = \det{}^* \mathscr{D}(-\mu, v_\mu).$$
(17.54)

2. *GN-model:* Under a sign change of μ or/and σ, the determinant goes into itself or its complex conjugate,

$$d = 2n: \quad \det \mathscr{D}(\mu, \sigma) = \det \mathscr{D}(\mu, -\sigma) = \det \mathscr{D}(-\mu, \sigma) \in \mathbb{R}$$

$$d = 1 + 4n: \quad \det \mathscr{D}(\mu, \sigma) = \det \mathscr{D}(-\mu, -\sigma) \in \mathbb{R}$$
(17.55)

$$d = 3 + 4n: \quad \det \mathscr{D}(\mu, \sigma) = \det{}^* \mathscr{D}(\mu, -\sigma) = \det \mathscr{D}(-\mu, \sigma).$$

3. *cGN-model without Thirring interaction:* One finds

$$d = 4n: \quad \det \mathscr{D}(\mu, \sigma, \pi) = \det \mathscr{D}(-\mu, \sigma, \pi)$$
(17.56)

$$= \det{}^* \mathscr{D}(\mu, -\sigma, \pi) = \det{}^* \mathscr{D}(\mu, \sigma, -\pi)$$

$$d = 2 + 4n: \quad \det \mathscr{D}(\mu, \sigma, \pi) = \det \mathscr{D}(\mu, -\sigma, -\pi) = \det \mathscr{D}(-\mu, -\sigma, \pi) \in \mathbb{R}.$$

Eigenvalues for Particular Background Fields
To actually calculate the determinant of $\mathscr{D}(\mu, v_\mu, \sigma, \pi)$, we observe that

$$\gamma^0 \mathscr{D} = (D_0 + \mu) + h$$
(17.57)

contains the hermitian Dirac Hamiltonian[5]

$$h = i\alpha^i D_i + \gamma^0 \sigma + i\gamma^0 \gamma_* \pi, \quad \alpha_i = -i\gamma^0 \gamma^i$$
(17.58)

with squares to the positive and hermitian operator

$$h^2 = -\boldsymbol{D}^2 - \frac{i}{4}[\gamma^i, \gamma^j] F_{ij} + \sigma^2 + \pi^2 - \gamma^\mu \partial_\mu (\sigma + i\gamma_* \pi),$$
(17.59)

where F_{ij} is the field strength of v_i. Recall that in odd dimensions π is missing.

The two operators $\mathscr{D}$ and $\gamma^0 \mathscr{D}$ should have the same determinant. Since the latter has a direct relation to the Dirac Hamiltonian and thus to the calculation in the canonical approach, we shall study $\det(\gamma^0 \mathscr{D})$ in what follows. Let us assume that D_0 commutes with h. For static background fields and a constant v_0, this condition is

---

[5] Although γ^μ are the matrices in Euclidean space, h is Hamiltonian in Minkowski spacetime.

fulfilled. Let us further assume that the eigenvalues of h come in pairs with opposite signs,

$$h\phi_m = \varepsilon_m \phi_m \quad \text{and} \quad h\tilde{\phi}_m = -\varepsilon_m \tilde{\phi}_m, \quad \varepsilon_m > 0. \tag{17.60}$$

This pairing property is in general not true. But it is true for the GN model or for the general model with constant background fields.

For static backgrounds we can factorize the time dependence of the eigenfunctions

$$\psi_{n,m} = e^{i\omega_n x_0} \phi_m(\boldsymbol{x}), \quad \omega_n = \frac{2\pi}{\beta}(n + \tfrac{1}{2}), \tag{17.61}$$

where we imposed antiperiodic boundary conditions in the imaginary time direction. These are eigenfunctions of the operator (17.57),

$$(\gamma^0 \mathscr{D}) \psi_{nm} = (i\omega_n + \mu_v + \varepsilon_m) \psi_{mn}, \quad \mu_v = \mu + iv_0. \tag{17.62}$$

Collecting the eigenvalues with $\pm\varepsilon_m$ in the determinant (and neglecting an overall sign), we get

$$(\det \mathscr{D})^2 = \prod_{n,m} \left((\omega_n - i\mu_v)^2 + \varepsilon_m^2\right) = \det H, \quad H = -(D_0 + \mu)^2 + h^2. \tag{17.63}$$

Here the product is over all ε_m—and not only over the positive ones— hence we get the square of the determinant. Collecting the terms with $\pm\omega_n$, we see that the determinant is real, although the individual eigenvalues are not. Even more, there is no sign problem since the squared determinant has a fixed sign.

From (17.63) follows that the ζ-function regularization yields

$$\log \det \mathscr{D} = -\frac{1}{2}\frac{d}{ds}\Big|_{s=0} \zeta_H(s). \tag{17.64}$$

Since H in (17.63) is the sum of two commuting operators, its partition function factorizes

$$\operatorname{tr} e^{-tH} = \sum_n e^{-t(\omega_n - i\mu_v)^2} \sum_m e^{-t\varepsilon_m} = K_\beta(t) \operatorname{tr} e^{-th^2}, \tag{17.65}$$

where K_β is the function (17.4) with $\mu \to \mu_v$. We conclude that ζ_H has the form

$$\zeta_H(s) = \frac{1}{\Gamma(s)} \int_0^\infty ds\, t^{s-1} K_\beta(t) \operatorname{tr} e^{-th^2}. \tag{17.66}$$

Thus, with the above assumptions we are left with calculating the partition function of h^2 and computing the s-derivative of the Mellin transform (17.66).

In the large N_f-limit, we would expect that the relevant auxiliary fields are constant (this will be discussed subsequently), in which case the eigenvalues of h^2 are

$$\varepsilon_m^2 = (p_m + v)^2 + \varrho^2, \quad p_m = \frac{2\pi}{L}m, \quad \varrho^2 = \sigma^2 + \pi^2. \tag{17.67}$$

The degeneracy of the eigenvalue ε_m^2 is d_s, such that the ζ-function takes the form

$$\zeta_H(s) = \frac{d_s}{\Gamma(s)} \int_0^\infty ds\, t^{s-1} K_\beta(t)\, K_L^{d-1}(t), \tag{17.68}$$

where the trace K_L^{d-1} is given in (17.6).

17.4   Large N_f Limit of Gross–Neveu Models

For arbitrary auxiliary fields, we can calculate $\det \mathscr{D}$ explicitly for the *two*-dimensional massless multi-flavor Thirring models [17] and the *two*-dimensional massless one-flavor GN model which is equivalent to the Thirring model. For GN models with more flavors and four-Fermi theories in higher dimensions, this is not possible anymore. But in the 't Hooft limit when the number of flavors N_f tends to infinity, the path integral over the auxiliary fields is localized at fields which minimize the "effective action." Assuming translation invariance, a minimizing configuration is homogeneous. Then we can compute the effective potential, thermodynamic potential, and chiral condensate at finite temperature and fermion density.

We shall explicitly calculate the partition function of the massless GN model at finite temperature and density in 2 and 3 spacetime dimensions. The calculation for the massive model is not much different, and we shall comment on the corresponding results for other four-Fermi theories towards the end of the chapter. We begin with the action after the Hubbard–Stratonovich transformation,

$$S = \int d^d x\, \mathscr{L}, \quad \mathscr{L} = \frac{N_f}{2g^2} \sigma^2 + \bar{\psi} \mathscr{D} \psi, \quad \mathscr{D} = \slashed{\partial} + \sigma + \mu \gamma^0. \tag{17.69}$$

The Dirac operator is flavor blind such that the Gaussian integral over the N_f Fermi fields in the partition function

$$Z = \int \mathscr{D}\sigma\, \mathscr{D}\psi\, \mathscr{D}\bar{\psi}\, e^{-S} \tag{17.70}$$

yields the determinant of the Dirac operator $\mathscr{D}$. This implies that the partition function takes the form

$$Z = \int \mathscr{D}\sigma \, e^{-N_f S_{\text{eff}}[\sigma]}, \quad S_{\text{eff}} = \frac{1}{2g^2} \int d^d x \, \sigma^2 - \log \det \mathscr{D}, \tag{17.71}$$

where now $\mathscr{D}$ is the operator acting on one flavor.[6]

Chiral Condensate

We have argued that a non-vanishing condensate $\langle \bar{\psi}\psi \rangle$ breaks the discrete $\mathbb{Z}_2$ axial symmetry in even dimensions and the $\mathbb{Z}_2$ parity symmetry in odd dimensions. Thermal expectation values of moments of the condensate field are related to moments of the auxiliary field by Ward identities. The most simple Ward identity follows from the fact that the integral of a derivative is zero (on a finite spacetime lattice, this is a rigorous statement). Thus, the following integral vanishes

$$0 = \frac{1}{Z} \int \mathscr{D}\sigma \, \mathscr{D}\psi \, \mathscr{D}\bar{\psi} \, \frac{\delta}{\delta\sigma(x)} e^{-S} = -\left\langle \frac{\delta S}{\delta\sigma(x)} \right\rangle. \tag{17.72}$$

This way one finds

$$\langle \bar{\psi}(x)\psi(x) \rangle = -\frac{N_f}{g^2} \langle \sigma(x) \rangle, \quad \bar{\psi}\psi = \sum_a \bar{\psi}_a \psi_a. \tag{17.73}$$

We see that the expectation value of the auxiliary scalar field serves as order parameter for chiral symmetry or parity. A further Ward identity, derived in Problem 17.5, relates the fluctuations of $\bar{\psi}\psi$ and σ:

$$\langle (\bar{\psi}\psi)(x)(\bar{\psi}\psi)(y) \rangle = \left(\frac{N_f}{g^2} \right)^2 \langle \sigma_x \sigma_y \rangle - \frac{N_f}{g^2} \delta(x-y). \tag{17.74}$$

Saddle Point Approximation and Gap Equation

In the large N_f limit, the absolute minimum of S_{eff} dominates the path integral over σ in (17.71), and one concludes that the free energy per flavor is

$$F = -\frac{1}{\beta N_f} \log Z = \frac{1}{\beta} \min_\sigma S_{\text{eff}}[\sigma], \quad N_f \to \infty. \tag{17.75}$$

In other words, the functional integral is localized at fields that minimize the effective action. A minimizing field solves the Euler–Lagrange equation for S_{eff}. To derive this so-called self-consistent *gap equation*, we vary the (regularized) effective action and thereby use $\delta \log \det \mathscr{D} = \text{tr} \, \mathscr{D}^{-1} \delta \mathscr{D}$. We evaluate the trace in position

---

[6] We use the same symbol $\mathscr{D}$ for the Dirac operator acting on 1 flavor and on N_f flavors.

space since the variation $\delta \mathscr{D} = \delta \sigma$ is diagonal in this representation,

$$\delta S_{\text{eff}}[\sigma] = \frac{1}{g^2} \int d^d x \, \sigma(x) \, \delta\sigma(x) - \int d^d x \, \text{tr}_{\text{D}} \langle x | \frac{1}{\mathscr{D}} | x \rangle \, \delta\sigma(x) \,. \quad (17.76)$$

The trace under the integral refers to the spin degrees of freedom. The minimizing background fields obey the self-consistency equations

$$\sigma(x) = g^2 \, \text{tr}_{\text{D}} \langle x | \frac{1}{\mathscr{D}} | x \rangle \,. \quad (17.77)$$

Since a thermal equilibrium state should be translation invariant, it seems natural to assume that the minimizing field is homogeneous. And for homogeneous fields, the determinant can be calculated, and the gap equation can be solved. Thereby, one needs to regularize the theory in the ultraviolet, and this will be done below. It has been realized, however, that for high-density systems, this homogeneity assumption may fail [23–26]. We shall come back to the breaking of translation invariance.

17.5   Effective Potential of GN Models at Finite T and μ

For a constant σ, the effective action (17.75) is equal to the effective potential, multiplied by the spacetime volume $V = \beta V_s$,

$$S_{\text{eff}}(\sigma) = V U_{\text{eff}}(\sigma) \,. \quad (17.78)$$

The potential is the sum of the contribution at zero temperature and density marked with a superscript 0 (it includes the classical term) plus a correction term marked with a superscript 1,

$$U_{\text{eff}} = U_{\text{eff}}^{(0)} + U_{\text{eff}}^{(1)}$$

$$U_{\text{eff}}^{(0)} = \lim_{\mu,T \to 0} U_{\text{eff}} = \frac{\sigma^2}{2g^2} - \frac{1}{V} \log \det^{(0)} \mathscr{D}, \quad U_{\text{eff}}^{(1)} = -\frac{1}{V} \log \det^{(1)} \mathscr{D}, \quad (17.79)$$

The UV-convergent term $U_{\text{eff}}^{(1)}$ vanishes when $\mu, T \to 0$. According to (17.75) the minimum of U_{eff} is the free energy density.

In the large N_{f} limit, the potential need not be and will not be convex in contrast to the exact effective potential defined as Legendre transform of the Schwinger function (see Sect. 5.3). It is the constraint effective potential [27] for a scalar theory with action $N_{\text{f}} S_{\text{eff}}$, where one disregards all inhomogeneous fluctuations $\sigma(x)$ contributing to the functional integral.

17.5.1 Effective Potentials for Homogeneous Background Fields

The infinite product $\mathscr{D}$-eigenvalues is divergent and must be regularized. Various regularizations are feasible: one may introduce a covariant cutoff for $p_\mu p^\mu$ or for the spatial momentum squared p^2. Or one may use the dimensional or Pauli–Villars regularization. Here we apply the ζ-function regularization which has been introduced in (5.22) and extensively used in the solution of the Schwinger model in Sect. 16.2. We will compare with alternative regularizations in a following section.

For a constant auxiliary scalar field σ, the eigenvalues of the Dirac Hamiltonian h in (17.58) come in pairs with opposite signs

$$\varepsilon_m = \pm\left(p_m^2 + \sigma^2\right)^{1/2} \tag{17.80}$$

and we may use the result (17.68), i.e.,

$$\zeta_H(s) = \frac{d_s}{\Gamma(s)} \int_0^\infty dt\, t^{s-1} K_\beta(t)\, K_L^{d-1}(t), \tag{17.81}$$

where v_μ and π are set to zero in the K-functions (17.4) and (17.6). The derivative of ζ_H determines the determinant according to (17.64) and thus enters the expression for the effective potential,

$$U_{\text{eff}}(\sigma) = \frac{\sigma^2}{2g^2} + \frac{1}{2V}\frac{d}{ds}\bigg|_{s=0} \zeta_H(s), \quad V = \beta V_s. \tag{17.82}$$

Zero-Temperature Contribution
We insert for K_β in (17.81) the alternating series in (17.4). In the zero temperature and density limit, only the term with $n = 0$ of this series contributes and

$$\zeta_H^{(0)}(s) = \lim_{T,\mu\to 0} \zeta_H(s) = \frac{\beta}{\sqrt{4\pi}}\frac{d_s}{\Gamma(s)} \int_0^\infty dt\, t^{s-1-\frac{1}{2}} K_L^{d-1}(t). \tag{17.83}$$

Next we insert for K_L^{d-1} the second series representation in (17.6) (with $v = 0$) and end up with the Mellin transform of a sum over $m \in \mathbb{Z}^{d-1}$,

$$\zeta_H^{(0)}(s) = \frac{V}{(4\pi)^{d/2}}\frac{d_s}{\Gamma(s)} \int_0^\infty dt\, t^{s-1-\frac{d}{2}} e^{-t\sigma^2}\left(1 + \sum_{m\neq 0} e^{-L^2 m^2/4t}\right). \tag{17.84}$$

This representation nicely separates the infinite-volume contributions—this is the term with $m = 0$— from the finite size corrections. The integrals of the individual terms can be expressed in terms of modified Bessel functions; see page 92. Using this result leads to a decomposition into an infinite volume contribution $\zeta_H^{(0,\infty)}$ plus a finite volume correction $\zeta_H^{(0,L)}$,

$$\zeta_H^{(0,\infty)}(s) = \frac{V}{(4\pi)^{d/2}}\, d_s r_d(s)\, \sigma^{d-2s}, \tag{17.85}$$

$$\zeta_H^{(0,L)}(s) = \frac{V}{(4\pi)^{d/2}}\frac{2d_s}{\Gamma(s)}\sigma^{d-2s}\sum_{m\neq 0}\left(\frac{2}{|m|\,\sigma L}\right)^{d/2-s} K_{s-d/2}\big(|m|\sigma L\big), \tag{17.86}$$

where $r_\alpha(s)$ has been introduced in (17.3) and w.l.o.g we assumed that $\sigma \geq 0$.

Odd Dimensions

For odd d the ratio of Gamma functions $r_d(s)$ is proportional to s for small s, and the formula (17.82) leads to

$$U_{\text{eff}}^{(0)} = \frac{\sigma^2}{2g^2} - \frac{1}{2}\left(\frac{-1}{\pi}\right)^{[d/2]}\frac{\sigma^d}{d!!} + \left(\frac{\sigma}{2\pi L}\right)^{d/2} d_s \sum_{m\neq 0}\frac{K_{d/2}\left(|m|\sigma L\right)}{|m|^{d/2}}. \tag{17.87}$$

In the derivation we assumed $\sigma \geq 0$. The effective potential is even in σ such that the term σ^d actually reads $|\sigma|^d$ and the potential is non-differentiable at the origin. This is why the model in the limit $N_f \to \infty$ has vastly different critical exponents as expected from Landau's theory. The latter assumes that the free energy is analytic in the order parameter and thus can be expanded in a power series in σ^2 for a $\mathbb{Z}_2$-symmetric system (see Sect. 7.6).

In $5+4n$ dimensions the potential is unbounded from below since the coefficient of the leading term σ^d becomes negative. This instability in higher dimensions is not related to the particular scheme we used.

Even Dimensions

For even d the ratio $r_d(s)$ and its derivative are both non-zero at $s = 0$, and the latter is proportional to the harmonic number $H_{d/2}$,

$$r_d(0) = \frac{(-1)^{d/2}}{(d/2)!} \quad \text{and} \quad r'_d(0) = H_{d/2}\, r_d(0), \qquad H_p = \sum_{k=1}^{p}\frac{1}{k}. \tag{17.88}$$

As a result the s-derivative of σ^{d-2s} in (17.85) produces a term $\propto \log \sigma^2$, which comes together with an undetermined scale factor.[7] Taking this into account leads to

$$U_{\mathrm{eff}}^{(0)} = \frac{\sigma^2}{2g^2} - \frac{1}{2}\left(\frac{-1}{\pi}\right)^{d/2}\frac{\sigma^d}{d!!}\left(\log\frac{\sigma^2}{\mu_0^2} - H_{d/2}\right) + \left(\frac{\sigma}{\pi L}\right)^{d/2}\sum_{m\neq 0}\frac{K_{d/2}(|m|\sigma L)}{|m|^{d/2}}.$$

(17.89)

Again the potential is not analytic at the origin, since it contains a term $\propto \sigma^d \log \sigma^2$. In $4 + 4n$ dimensions it is unbounded from below.

Finite-Temperature Contribution

To compute the finite temperature and density correction $\zeta_H^{(1)} = \zeta_H - \zeta_H^{(0)}$, we insert the alternating sum (17.4) without the $n = 0$ term into the Mellin transformation (17.81). The individual terms of the double sum can be integrated with (5.31)[8] and give

$$\zeta_H^{(1)}(s) = \frac{\beta}{\sqrt{\pi}}\frac{2d_s}{\Gamma(s)}\sum_{n>0,\,p}(-1)^n\left(\frac{\beta n}{2\varepsilon_p}\right)^{s-1/2}\cosh(n\beta\mu)\,K_{s-1/2}(\beta n\varepsilon_p),$$

(17.90)

with positive one-particle energies ε_p. Since $1/\Gamma(s)$ vanishes at the origin, the derivative of this ζ-function at $s = 0$ is easily worked out: it is the same expression but without the Gamma function and evaluated at $s = 0$. The function $\sqrt{x}\,K_{1/2}(x)$ is proportional to e^{-x} such that the sum over n is given by two geometric series which are identified as Taylor expansions of the logarithmic function. This yields the following finite temperature and density contribution to the effective potential,

$$U_{\mathrm{eff}}^{(1)} = -\frac{d_s}{2\beta V_s}\sum_p\left(\log\left(1 + \mathrm{e}^{-\beta(\varepsilon_p+\mu)}\right) + (\mu \to -\mu)\right).$$

(17.91)

In the thermodynamic limit, the momentum sum turns into a momentum integral and

$$U_{\mathrm{eff}}^{(1)} = -\frac{d_s}{2\beta(2\pi)^{d-1}}\int \mathrm{d}p\left(\log\left(1 + \mathrm{e}^{-\beta(\varepsilon_p+\mu)}\right) + (\mu \to -\mu)\right).$$

(17.92)

We have recovered the well-known results for a relativistic gas of fermions with mass σ at finite temperature and fermion density.

---

[7] For dimensional reasons one should consider $\det(\mathcal{D}/\mu_0)$ where μ_0 is just this scale factor.

[8] The series converges for $|\mu| \leq \sigma$.

17.5.2 The Effective Potentials in Two and Three Dimensions

The ζ-function regularization yields a particularly renormalized effective potential. Still we may perform a finite renormalization by imposing some "physical" condition. Here we assume that the system is in the broken phase at zero temperature, zero fermion density, and infinite box size. Then the minimum of $U_{\text{eff}}^{(0,\infty)}$ is at some prescribed $\sigma_0 \neq 0$. This way we can trade the coupling g and (in two dimensions) the scale parameter μ_0 for the minimum σ_0. Recall that in the large N_{f} limit σ_0 is equal to the expectation value of σ. According to the Ward identity (17.73), it is then proportional to the condensate $\langle \bar{\psi}\psi \rangle$ at vanishing T and μ and infinite L.

Two Dimensions

The renormalization condition relates the minimizing field to the scaling parameter according to $\sigma_0 = \exp(-\pi/g^2)\mu_0$, such that for $\sigma \geq 0$

$$U_{\text{eff}} = U_{\text{eff}}^{(0)} + U_{\text{eff}}^{(1)}, \quad U_{\text{eff}}^{(0)}(\sigma) = \frac{\sigma^2}{4\pi}\left(\log\frac{\sigma^2}{\sigma_0^2} - 1\right) + \frac{2\sigma}{\pi L}\sum_{m>0}\frac{K_1(m\sigma L)}{m}.$$

$$(17.93)$$

In the thermodynamic limit $L \to \infty$, the sum over the Bessel functions is zero and

$$U_{\text{eff}}^{(1)} = -\frac{T}{\pi}\int_0^\infty \mathrm{d}p \, \log\left(1 + \mathrm{e}^{-\beta(\varepsilon_p+\mu)}\right) + (\mu \to -\mu). \quad (17.94)$$

Figure 17.1 shows the potential for vanishing μ. We used units defined by σ_0. The system is in the symmetric phase at high temperature (dotted line) and in the broken phase at low temperature (full line). The dashed line shows the potential for the

Fig. 17.1 Effective potential in 2 dimensions for $\mu = 0$ and three values of T. U_{eff}, σ, μ, and T are measured in units defined by σ_0. $T = 0.567$ is the critical temperature for zero density

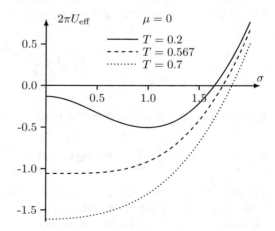

Fig. 17.2 Effective potential in $d = 2$ at zero temperature for four values of μ. U_{eff}, σ, and μ are measured in units defined by σ_0

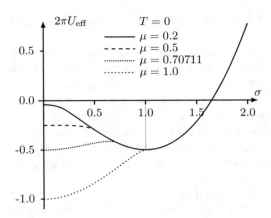

critical temperature. At zero temperature the effective potential simplifies to

$$2\pi U_{\text{eff}}(\sigma) \xrightarrow{T\to 0} \begin{cases} \sigma^2 \log(\sqrt{\mu^2 - \sigma^2} + \mu) - \frac{1}{2}\sigma^2 - \mu\sqrt{\mu^2 - \sigma^2}, & |\sigma| < \mu \\ \sigma^2 \log(|\sigma|/\sigma_0) - \frac{1}{2}\sigma^2, & |\sigma| \geq \mu, \end{cases}$$
(17.95)

where we assumed a non-negative μ (recall that the potential is even in μ). Figure 17.2 shows the zero-temperature potential for various values of μ. Note that the minimum jumps from σ_0 to 0 when μ reaches the critical value $\mu_c = \sigma_0/\sqrt{2}$. The system shows a density-driven first-order transition.

Three Dimensions

We insert the explicit form of the modified Bessel function $K_{3/2}$ and trade the renormalized coupling g for the minimum σ_0 of the effective potential $U^{(0,\infty)}$ at infinite volume, $g^2 = -2\pi/\sigma_0$. Then the potential takes the form

$$U_{\text{eff}}(\sigma) = \frac{\sigma^2}{12\pi}(2\sigma - 3\sigma_0) + \frac{1}{2\pi L^3} \sum_{m \neq 0} e^{-|m||\sigma|L} \left(\frac{\sigma L}{m^2} + \frac{1}{|m|^3}\right) + U_{\text{eff}}^{(1)}(\sigma),$$
(17.96)

where in the thermodynamic limit the sum over m vanishes and

$$U_{\text{eff}}^{(1)}(\sigma) = -\frac{T}{2\pi}\int_0^\infty dp\, p\, \log\left(1 + e^{-\beta(\varepsilon_p + \mu)}\right) + (\mu \to -\mu).$$
(17.97)

Figure 17.3 shows the effective potential for vanishing μ. The system is in the symmetric phase at high temperature (dotted line) and in the broken phase at low temperature (full line). The dashed line is the potential for the critical temperature.

Fig. 17.3 Effective potential in $d = 3$ for $\mu = 0$ and 3 values of T. U_{eff}, σ, μ, and T are measured in units defined by σ_0. For $\mu = 0$ the critical temperature is $T_c = 1/(2 \log 2) \approx 0.72135$

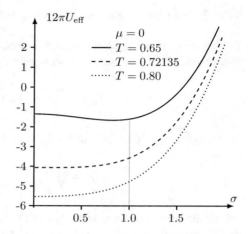

Fig. 17.4 Effective potential in $d = 3$ at zero temperature for three values of μ. U_{eff}, σ, and μ are measured in units defined by σ_0

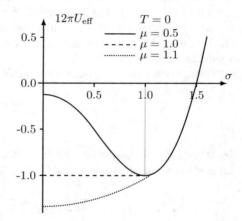

The potential takes the following simple form for $T \to 0$:

$$12\pi\, U_{\text{eff}}(\sigma) \xrightarrow{T \to 0} \begin{cases} 3(\mu - \sigma_0)\sigma^2 - \mu^3 & \text{for} \quad |\sigma| < \mu \\ (2|\sigma| - 3\sigma_0)\sigma^2 & \text{for} \quad |\sigma| \geq \mu. \end{cases} \tag{17.98}$$

It is depicted for three different fermion densities in Fig. 17.4. For the critical value $\mu_c = \sigma_0$, the potential is flat for $0 \leq \sigma \leq \sigma_0$ and shows a density-driven transition at the border between the second and first order. For small temperature the minimum varies rapidly from $\sigma \approx \sigma_0$ to $\sigma = 0$ when μ increases only slightly. Nevertheless, these are second-order transitions.

17.5.3 Gap Equations and Phase Diagrams

For a homogeneous auxiliary field, the right-hand side of the gap equation (17.77) is x-independent. Then the integration over x just produces the spacetime volume V

$$\frac{\sigma}{g_\Lambda^2} = \frac{1}{V} \, \mathrm{tr}\left(\frac{1}{\mathscr{D}}\right). \tag{17.99}$$

With the known eigenvalues of $\mathscr{D}$, the trace is readily calculated,

$$\frac{\sigma}{g_\Lambda^2} = \frac{d_s}{V} \sum_{n,p} \frac{\sigma}{(\omega_n - i\mu)^2 + \varepsilon_p^2}, \tag{17.100}$$

with σ-dependent one-particle energies ε_p. As expected, the sum over the momenta is UV-divergent and requires some regularization. But independent of the regularization, there is always the trivial solution $\sigma = 0$. In what follows we are interested in non-trivial solutions of the regularized and later renormalized gap equation. We shall employ the non-covariant sharp momentum cutoff and thus sum only over spatial momenta with $|p| < \Lambda$.

The sum over the Matsubara frequencies is evaluated with the help of

$$\sum_{n \in \mathbb{Z}} \frac{1}{(n + x)(n + y)} = -\pi \, \frac{\cot(\pi x) - \cot(\pi y)}{x - y}. \tag{17.101}$$

In addition, in the thermodynamic limit, the Riemann sum over the spatial momenta turns into a Riemann integral, and the regularized gap equation reads

$$\frac{\sigma}{g_\Lambda^2} = \frac{d_s \mathscr{C}_{d-1}}{4} \sigma \int_0^\Lambda \mathrm{d}p \, \frac{p^{d-2}}{\varepsilon_p} \left(\tanh \frac{\beta(\varepsilon_p + \mu)}{2} + \tanh \frac{\beta(\varepsilon_p - \mu)}{2} \right), \tag{17.102}$$

where p denotes the length of p. The factor $\mathscr{C}_{d-1}$ is proportional to the volume of the unit sphere in $\mathbb{R}^{d-1}$; see Eq. (17.2). At this point we renormalize the GN model by assuming that for vanishing T and μ (and in the thermodynamic limit), the gap equation has a prescribed solution $\sigma_0 \neq 0$. This is exactly the renormalization condition used in the ζ-function regularization. Since for vanishing T and μ the tanh-functions in (17.102) are 1, a non-trivial solution satisfies

$$\frac{1}{g_\Lambda^2} = \frac{d_s \mathscr{C}_{d-1}}{2} \int_0^\Lambda \mathrm{d}p \, \frac{p^{d-2}}{\varepsilon_p(\sigma_0)}. \tag{17.103}$$

This equation determines the cutoff dependence of the bare coupling.

Now we use the relation (17.103) to eliminate the bare coupling in the regularized gap equation (17.102). The resulting momentum integral is finite below 4 dimensions, and we can send the cutoff Λ to infinity. This yields the *renormalized gap equation*

$$0 = \sigma \int_0^\infty dp \, p^{d-2} \left(\frac{1}{\varepsilon_p} \tanh \frac{\beta(\varepsilon_p + \mu)}{2} + \frac{1}{\varepsilon_p} \tanh \frac{\beta(\varepsilon_p - \mu)}{2} - \frac{2}{\varepsilon_p(\sigma_0)} \right).$$

$$(17.104)$$

One can check that the right-hand side is just the σ-derivative of the ζ-function regularized potential in (17.87), (17.89), and (17.92). The $\mathbb{Z}_2$ symmetry is spontaneously broken if the Gap equation has a non-trivial solution σ with $U_{\mathrm{eff}}(\sigma) < U_{\mathrm{eff}}(0)$.

Two Dimensions

The gap equation for a non-zero extremum of the effective potential reads

$$\frac{1}{2} \log \frac{\sigma^2}{\sigma_0^2} + \int_0^\infty \frac{dp}{\varepsilon_p(\sigma)} \left(\frac{1}{1 + e^{\beta(\varepsilon_p + \mu)}} + \frac{1}{1 + e^{\beta(\varepsilon_p - \mu)}} \right) = 0, \qquad (17.105)$$

and it can be solved analytically at zero temperature only. Then the integral vanishes for $\mu < \sigma$ which implies $\sigma = \sigma_0$. We see that for all $\mu \leq \sigma_0$, there exists the solution σ_0. In addition, for $\sigma < \mu$ the integral does not vanish, and another nontrivial solution $\sigma(\mu)$ emerges, given by

$$\left(\frac{\sigma(\mu)}{\sigma_0} \right)^2 = \frac{2\mu}{\sigma_0} - 1. \qquad (17.106)$$

It is a local maximum of U_{eff} which exists for μ in the interval $\frac{1}{2}\sigma_0$ and σ_0.

Thus, we have the following situation at *zero temperature:* For $0 \leq \mu \leq \frac{1}{2}$ (in units defined by σ_0), the origin is a local maximum and $\sigma = 1$ the absolute minimum of the effective potential. When μ exceeds the value $\frac{1}{2}$, then the origin turns into a local minimum, and at the same time a local maximum develops at $\sigma(\mu)$, given by (17.106). This means that at this threshold value of μ, a metastable symmetric state develops. If the chemical potential reaches the value $\mu_c = 1/\sqrt{2} \approx 0.707$, then the symmetric phase becomes stable and the broken phase metastable. This defines the critical value of the chemical potential. If we further increase the chemical potential, then at $\mu = 1$ the local minimum at $\sigma = 1$ disappears—it merges with the local maximum at $\sigma(\mu)$. We see that the system shows a density-driven first-order phase transition. Overdense and underdense metastable states also exist for higher temperatures up to the Lifshitz point at $(\mu, T) \approx (0.608, 0.318)$, which is marked by a dot in Fig. 17.5. At the dashed line on the left, the metastable symmetric phase appears. It becomes stable at the solid line where the broken phase becomes metastable. At the dashed line on the right, the broken phase disappears. At the solid line, the order parameter jumps from 0 to a σ_{min} at a density characterized

Fig. 17.5 The phase
transition line for the GN
model in two dimensions.
The temperature and
chemical potentials are
plotted in units defined by the
scale σ_0. The dots mark the
parameter sets for which the
effective potential is plotted
in Figs. 17.1 and 17.2

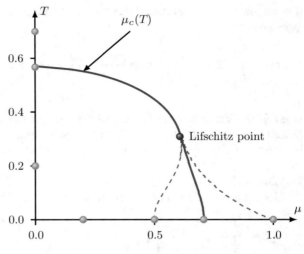

Fig. 17.6 The order
parameter in the
homogeneous equilibrium
states as function of the
control parameters μ and T.
For small T and $\mu = \mu_c(T)$,
the order parameter jumps

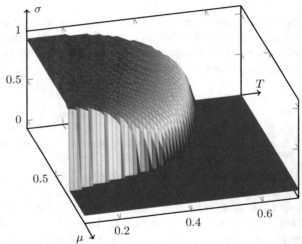

by $\mu_c(T)$. This jump is clearly seen in Fig. 17.6, which shows the order parameter
in the stable equilibrium state as function of μ and T.

The transitions at low density and high temperature— to the left of the Lifshitz
point—are of second order. This means that at the Lifshitz point, the second-order
line emanating from the T-axis at $T_c = e^\gamma/\pi \approx 0.567$ turns into a first-order
transition hitting the μ-axis at $\mu_c = 1/\sqrt{2} \approx 0.707$. This large-$N_f$ phase diagram
has been calculated in [28]. Later we shall argue, that it must be revised. Instead
of a first-order transition from one homogeneous into another homogeneous state,
the two-dimensional system shows second-order transitions from a homogeneously
broken into an inhomogeneous equilibrium phase and from this inhomogeneous
phase into the symmetric phase.

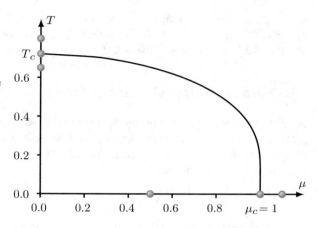

Fig. 17.7 The phase transition line for the GN model in three dimensions. The temperature and chemical potentials are plotted in units defined by the scale σ_0. The dots mark the parameter sets for which the effective potential is plotted in Figs. 17.3 and 17.4

Three Dimensions

With the help of $\varepsilon_p \mathrm{d}\varepsilon_p = p\,\mathrm{d}p$, the integral (17.104) can be calculated in three dimensions. One finds that a non-trivial solution solves the transcendental equation

$$\beta\sigma_0 = \beta\sigma + \log\left(1 + e^{-\beta(\sigma+\mu)}\right) + \log\left(1 + e^{-\beta(\sigma-\mu)}\right). \tag{17.107}$$

Clearly, for vanishing T and μ, we recover our renormalization condition $\sigma = \sigma_0$. With increasing temperature and chemical potential, the value of the non-trivial solution decreases. It vanishes for

$$\cosh\left(\frac{\mu}{2T}\right) = \frac{1}{2}e^{\sigma_0/2T}, \tag{17.108}$$

and this relation between T and μ/T defines the phase transition line separating the symmetric and broken phases which is plotted in Fig. 17.7. This line intersects the μ and T axis at the points

$$\mu_c = \sigma_0 \text{ for } T = 0, \qquad T_c = \frac{\sigma_0}{2\log 2} \text{ for } \mu = 0. \tag{17.109}$$

The shape of the effective potential at the *six* marked points in the phase diagram are depicted in Figs. 17.3 and 17.4.

17.6   Effective Potential of cGN Model with Momentum Cutoff

For some readers the elegant ζ-function method may look a bit formal and arbitrary. It may be reassuring to see that a more conventional regularization of the fermion determinant yields the same renormalized results. In this section we sketch the calculation with a non-covariant sharp cutoff for the spatial momenta. The calculation is a bit more involved than the ζ-function regularization but yields additional information about the cutoff dependence of the bare parameters.

17.6.1 Non-covariant Momentum Cutoff

Instead of merely reproducing the previous results, we now couple the fermions to a constant scalar, pseudo-scalar, and vector field and thus calculate the regularized determinant of the Dirac operator $\mathcal{D}$ in (17.45). Again the one-particle energies come in pairs with opposite signs

$$\pm \varepsilon_{\boldsymbol{p}}, \quad \text{where} \quad \varepsilon_{\boldsymbol{p}}^2 = (\boldsymbol{p} + \boldsymbol{v})^2 + \varrho^2, \tag{17.110}$$

such that the determinant of $\mathcal{D}$ is related to that of H as in (17.63). The one-particle energy $\varepsilon_{\boldsymbol{p}}$ depends on $\boldsymbol{p}$ and not only its modulus p. It depends only on the SO(2)-invariant combination $\varrho^2 = \sigma^2 + \pi^2$ on the fields σ and π, and the minimally coupled constant vector field v_μ enters via the combination $\boldsymbol{p} + \boldsymbol{v}$. For vanishing vector field, the energy $\varepsilon_{\boldsymbol{p}}$ reduces to the energy ε_p,

$$\lim_{\boldsymbol{v} \to 0} \varepsilon_{\boldsymbol{p}} = \varepsilon_p = \left(p^2 + \varrho^2\right)^{1/2}, \tag{17.111}$$

and the two energies are distinguished by their indices. If the product over the eigenvalues was convergent and the momenta continuous, then the determinant would not depend on the vector field at all. Hence, already from the start we could anticipate that in the large N_f-limit, the Thirring model is dynamically less interesting than the GN model.

Our point of departure is the divergent product representation (17.63). The product of the factors with $\pm \omega_n$ can be reshuffled as follows:

$$\prod_n \left((\omega_n - i\mu_v)^2 + \varepsilon_{\boldsymbol{p}}^2\right) = F(\mu_v)F(-\mu_v), \quad F(\mu_v) = \prod_{n \geq 0} \frac{\omega_n^2 + (\varepsilon_{\boldsymbol{p}} + \mu_v)^2}{\omega_n^2}. \tag{17.112}$$

The product on the left is even in μ_v (and in ϱ). In thermal field theory, one often encounters infinite products like $F(\mu)$, many of which can be evaluated with the following useful theorem [29, 30]:

Theorem 17.1 *Let $n \geq 1$ be an integer, and let $\alpha_1, \ldots, \alpha_n$ and $\beta_1, \ldots, \beta_n$ be non-zero complex numbers, none of which is a negative integer. If $\sum \alpha_i = \sum \beta_i$, then*

$$\prod_{k \geq 0} \frac{(k + \alpha_1) \cdots (k + \alpha_n)}{(k + \beta_1) \cdots (k + \beta_n)} = \frac{\Gamma(\beta_1) \cdots \Gamma(\beta_n)}{\Gamma(\alpha_1) \cdots \Gamma(\alpha_n)}. \tag{17.113}$$

Otherwise, the infinite product in (17.113) diverges.

To prove the convergence of the product, one considers its logarithm which for large k has the leading contribution $\sum_i (\alpha_i - \beta_i) \sum_k 1/k$. Clearly the coefficient of the divergent sum over k must vanish to get a finite answer. This is just the condition in the theorem.

We conclude that the regulated product over the Matsubara frequencies (17.112) converges. We apply the theorem to this product and find

$$F(\mu_v) = \prod_{n>0} \frac{(n + \tfrac{1}{2} + iz)(n + \tfrac{1}{2} - iz)}{(n + \tfrac{1}{2})(n + \tfrac{1}{2})} = \cosh(\pi z), \quad \pi z = \frac{\beta}{2}(\varepsilon_p + \mu_v).$$

(17.114)

This leaves us with a divergent infinite product over the spatial momenta. We regularize it by taking the finite product over momenta with $p \leq \Lambda$, where in this section p denotes the length of the spatial momentum. Then we arrive at the following regularized expression for logarithm of the determinant,

$$\log \det \mathscr{D} = \frac{d_s}{2} \sum_{p \leq \Lambda} \left(\log \cosh \frac{\beta(\varepsilon_p + \mu_v)}{2} + \log \cosh \frac{\beta(\varepsilon_p - \mu_v)}{2} \right).$$ (17.115)

Finally, by using $\cosh(\pi z) = \pi z + \log(1 + e^{-2\pi z}) - \log 2$ and dropping an irrelevant constant, the regularized potential can be written as sum of two terms,

$$U_{\text{eff}} = U_{\text{eff}}^{(0,\Lambda)} + U_{\text{eff}}^{(1)}, \quad U_{\text{eff}}^{(0,\Lambda)} = \mathscr{V}_{\text{B}} + \varepsilon_{\text{cas}}^{(\Lambda)} = \mathscr{V}_{\text{B}} - \frac{d_s}{2V_s} \sum_{p \leq \Lambda} \varepsilon_p.$$ (17.116)

The first term contains the classical potential $\mathscr{V}_{\text{B}}$ in (17.43) with bare parameters $g_{\text{b,v}}$ and $g_{\text{b,s}}$ and the regularized zero-point energy $\varepsilon_{\text{cas}}^{(\Lambda)}$. The zero-point energy depends on the spatial components of the vector potential and on ϱ. The second contribution $U_{\text{eff}}^{(1)}$ comprises the finite density and temperature corrections. In the *convergent* sum defining these corrections, we can send the cutoff to infinity and recover the result (17.91) with μ replaced by the shifted μ_v and σ replaced by ϱ.

We analyze the zero-point or Casimir energy density in the limit $L \to \infty$, where it is given by the momentum integral

$$\varepsilon_{\text{cas}}^{(\Lambda)} = -\frac{d_s}{2(2\pi)^{d-1}} \int_{p \leq \Lambda} d\mathbf{p}\, \varepsilon_p.$$ (17.117)

In a first step we extract the v-dependence of the integral. Actually it depends only on the modulus of v since the cutoff is rotational invariant in momentum space. Since ε_p depends on v and p only via their sum, the momentum integral would not depend on v at all if no cutoff was necessary. Because of this dependence on the sum, the coefficients in the small-v expansion of ε_p are all total p-derivatives,

$$\varepsilon_p = \varepsilon_p + v \cdot \nabla_p \varepsilon_p + \frac{1}{2}(v \cdot \nabla_p)^2 \varepsilon_p + O(v^3). \tag{17.118}$$

Recall that ε_p on the right-hand side is the one-particle energy without v; see Eq. (17.111). The n'th order term in this expansion falls off as p^{1-n} for large momenta. By power counting one infers that in two and three dimensions, only the terms of order zero and two contribute to the integral defining the Casimir energy[9]

$$\varepsilon_{\text{cas}}^{(\Lambda)} = -\frac{d_s \mathscr{C}_{d-1}}{2} \int_0^\Lambda \mathrm{d}p\, p^{d-2} \varepsilon_p - \frac{d_s \mathscr{C}_{d-1}}{4} \int_0^\Lambda \mathrm{d}p\, p^{d-2}(v \cdot \nabla_p)^2 \varepsilon_p, \tag{17.119}$$

with $\mathscr{C}_{d-1}$ defined in (17.2). One finds that the second-order term is finite in two dimensions and linearly divergent in three dimensions. Up to terms which vanish for large cutoffs Λ, it is independent of ϱ,

$$d = 2: \quad \varepsilon_{\text{cas}}^{(\Lambda)} - \varepsilon_{\text{cas}}^{(\Lambda)}(v = 0) = -\frac{\Lambda v^2}{2\pi \varepsilon_\Lambda(\varrho)} \sim -\frac{v^2}{2\pi}, \tag{17.120}$$

$$d = 3: \quad \varepsilon_{\text{cas}}^{(\Lambda)} - \varepsilon_{\text{cas}}^{(\Lambda)}(v = 0) = -\frac{\Lambda^2 v^2}{8\pi \varepsilon_\Lambda(\varrho)} \sim -\frac{v^2}{8\pi}\Lambda. \tag{17.121}$$

Actually, with the ζ-function regularization the fermion determinant does not generate terms quadratic in the vector field.

We continue with the more difficult task of calculating the ϱ-dependent contribution, that is, the first integral in (17.119). To handle the UV divergences, we isolate the leading terms in the series expansion of the ε_p in inverse powers of p,

$$\varepsilon_p = \varepsilon_p^{\text{sub}} + \delta_p, \quad \text{with} \quad \delta_p = p \sum_{n=0}^{[d/2]} c_n \left(\frac{\varrho^2}{p^2}\right)^n, \quad c_n = \binom{\frac{1}{2}}{n}. \tag{17.122}$$

The integral with the subtracted one-particle energy $\varepsilon_p^{\text{sub}}$ exists since the polynomial δ_p in ϱ contains the leading terms in the expansion of ε_p in inverse momenta. At this point we need to distinguish between odd and even dimensions.

In odd dimensions the momentum integrals exist in the infrared, and we conclude

$$\int_0^\Lambda \mathrm{d}p\, p^{d-2} \varepsilon_p = \int_0^\Lambda \mathrm{d}p\, p^{d-2} \varepsilon_p^{\text{sub}} - \Lambda^d \sum_{0 \leq n < d/2} \frac{c_n}{d - 2n} \left(\frac{\varrho}{\Lambda}\right)^{2n}. \tag{17.123}$$

By construction the integral with $\varepsilon_p^{\text{sub}}$ exists in the UV as well, and its cutoff can be removed. It can be calculated in closed form.

---

[9] The momentum integrals of terms with odd powers of v vanish because of rotational symmetry.

In even dimensions the last term in the sum representing δ_p in (17.122) leads to a divergent momentum integral in the infrared—the integral is logarithmically divergent on small and large scales. Since the original integral with ε_p is well-behaved for small momenta, we split

$$\int_0^\Lambda dp\, p^{d-2}\varepsilon_p = \int_0^{\mu_0} dp\, p^{d-2}\varepsilon_p + \int_{\mu_0}^\Lambda dp\, p^{d-2}\varepsilon_p^{\mathrm{sub}} + \int_{\mu_0}^\Lambda dp\, p^{d-2}\delta_p\,.$$

(17.124)

The sum of the integrals on the right-hand side does not depend on the just introduced momentum scale μ_0, but the scale is needed to separate the finite and logarithmically divergent contributions.

17.6.2 Renormalization

As we did earlier, we choose the bare parameters such that the minimum of $U_{\mathrm{eff}}^{(0,\Lambda)}$ with respect to ϱ is at a fixed ϱ_0 and the term quadratic in v has coefficient $1/2g_{\mathrm{v}}^2$. The second renormalization condition seems a bit artificial. But in the large N_{f} limit, the vector field decouples from the interesting dynamics, and the renormalization prescription in the Thirring sector is not of utmost interest. With these conditions we find the following relations between the bare and renormalized parameters,

$$d = 2:\quad \frac{2\pi}{g_{\mathrm{b,s}}^2} = \log\left(\frac{2\Lambda}{\varrho_0}\right)^2,\quad \frac{1}{g_{\mathrm{b,v}}^2} = \frac{1}{g_{\mathrm{v}}^2} + \frac{1}{\pi}\,,\qquad (17.125)$$

$$d = 3:\quad \frac{2\pi}{g_{\mathrm{b,s}}^2} = \Lambda - \varrho_0,\quad \frac{1}{g_{\mathrm{b,v}}^2} = \frac{1}{g_{\mathrm{v}}^2} + \frac{\Lambda}{4\pi}\,.\qquad (17.126)$$

Note that in two dimensions the vector couplings are related by a finite renormalization. This is not just an artifact of the large N_{f}-approximation. It is known that the exactly soluble two-dimensional (massless) Thirring model is a conformal field theory with vanishing β-function for any number of flavors [16, 17].

When we remove the cutoff (and thereby discard divergent constant terms which are absorbed in a redefinition of the vacuum energy), we recover the zero temperature and density *renormalized potentials* in two and three dimensions[10]

$$d = 2:\quad U_{\mathrm{eff}}^{(0)} = \frac{v^2}{2g_{\mathrm{v}}^2} + \frac{\varrho^2}{4\pi}\left(\log\frac{\varrho^2}{\varrho_0^2} - 1\right),\qquad (17.127)$$

$$d = 3:\quad U_{\mathrm{eff}}^{(0)} = \frac{v^2}{2g_{\mathrm{v}}^2} + \frac{\varrho^2}{12\pi}\left(2\varrho - 3\varrho_0\right).\qquad (17.128)$$

---

[10] We skip the v_0-contribution, which in a non-covariant regularization needs separate treatment.

Comparing with the expressions (17.93) and (17.96) obtained within the ζ-function regularization, we see that we reproduced identical results (of course, after setting $v = 0$ and identifying ϱ with σ). We still would have to add the finite temperature and density correction $U_{\text{eff}}^{(1)}$ depending on ϱ and v_0 to map out the phase diagram in the space of control parameters. But for constant σ, π fields, and vanishing vector field, we fall back to the Gross-Neveu model and its phase diagram. The only non-trivial dependence on the vector fields comes via the shift $\mu \rightarrow \mu_v = \mu + iv_0$. The spatial components v decouple from the other degrees of freedom and cannot change the previously obtained results for the GN model.

In four dimensions the situation is different. The cutoff potential without v_μ-field,

$$U_{\text{eff}}^{(0,\Lambda)} = \frac{\varrho^2}{2g_{\text{b,s}}^2} - \frac{\varrho^4}{16\pi^2}\left(\log\frac{\varrho^2}{4\Lambda^2} + \frac{1}{2}\right) - \frac{\varrho^4}{32\pi^2} - \frac{\Lambda^2\varrho^2}{4\pi^2} - \frac{\Lambda^4}{4\pi^2}, \qquad (17.129)$$

has a divergent term $\propto \varrho^4$ which cannot be removed by merely redefining the vacuum energy and bare coupling $g_{\text{b,s}}$. This already signals that the GN model is not renormalizable in the large N_{f} expansion, in contrast to the models in lower dimensions.

17.7 Magnetic Catalysis in the GN Model

In this section we study the impact of an external magnetic field on the GN model at finite temperature and finite fermion density. We assume a homogeneous B-field and would like to answer the question how the condensate field depends on the control parameters T, μ, and B. For example, does a magnetic field inhibit or amplify the breaking of chiral symmetry in even dimensions or of parity symmetry in odd dimensions? In an infinite spatial volume, these questions have been addressed in [31–34].

The gauge potential of a constant magnetic field depends linearly on the coordinates and apparently breaks translation invariance. We may still expect homogeneity when we consider gauge-invariant observables. Below we shall present some surprising and unexpected results, especially in lower dimensions and concerning the non-commutativity of certain limits in the space of control parameters.

In three or more dimensions,[11] the relevant Dirac operator

$$\mathscr{D} = \gamma^v D_v + \sigma + \mu\gamma^0 \qquad (17.130)$$

---

[11] In Problem 17.8 we shall consider the simpler case $d = 2$.

contains a homogeneous scalar field, chemical potential, and potential A_ν entering the covariant derivatives D_ν. We shall see that the problem factorizes into a non-trivial two-dimensional problem in the plane determined by the constant field strength $F_{\mu\nu}$ and a trivial problem in the directions orthogonal to this plane. The dynamics in the plane gives rise to the well-known Landau levels. To be specific we assume that the only non-trivial component of the field strength tensor is F_{12}, such that the non-trivial dynamics takes place in the 1-2 plane. As gauge potential we choose

$$A_1 = -Bx_2 \quad \text{and} \quad A_\nu = 0, \ \nu \neq 1, \tag{17.131}$$

with

$$F_{12} = -\partial_2 A_1 = B. \tag{17.132}$$

The eigenvalues are even functions of the magnetic field, and thus we may always assume $B \geq 0$.

Spectrum of the Dirac Operator

To calculate the eigenvalues of the Dirac operator, we split the operator according to

$$\mathscr{D} = \mathscr{Q}_\parallel + \mathscr{Q}_\perp + \sigma \tag{17.133}$$

with *anticommuting* operators

$$\mathscr{Q}_\parallel = \gamma^1 D_1 + \gamma^2 D_2 \quad \text{and} \quad \mathscr{Q}_\perp = \gamma^a \partial_a + \mu\gamma^0, \quad a = 0, 3, \dots \tag{17.134}$$

For a vanishing chemical potential both operators are anti-hermitian. The involutory hermitian matrix $\gamma_{12} = -i\gamma_1\gamma_2$ anticommutes with $\mathscr{Q}_\parallel$ and commutes with $\mathscr{Q}_\perp$ such that

$$[\mathscr{Q}_\parallel, \gamma_{12}\mathscr{Q}_\perp] = 0. \tag{17.135}$$

It follows that $\mathscr{Q}_\parallel$ and $\gamma_{12}\mathscr{Q}_\perp$ can be diagonalized at the same time,

$$\mathscr{Q}_\parallel \chi_{mn} = i\lambda_m \chi_{mn} \quad \text{and} \quad \gamma_{12}\mathscr{Q}_\perp \chi_{mn} = i\mu_n \chi_{mn}. \tag{17.136}$$

The matrix γ_{12} has similar properties and plays the same role as γ_* in the Schwinger model. Due to its (anti)commutation relations with the $\mathscr{Q}$-operators, the transformed mode $\tilde{\chi}_{mn} = \gamma_{12}\chi_{mn}$ is a simultaneous eigenfunction as well,

$$\mathscr{Q}_\parallel \tilde{\chi}_{mn} = -i\lambda_m \tilde{\chi}_{mn} \quad \text{and} \quad \gamma_{12}\mathscr{Q}_\perp \tilde{\chi}_{mn} = i\mu_n \tilde{\chi}_{mn}. \tag{17.137}$$

Without chemical potential λ_m and μ_n are both real. Clearly, the two-dimensional subspace spanned by the orthogonal modes χ_{mn}, $\tilde{\chi}_{mn}$ is invariant under the action of the Dirac operator

$$\mathscr{D} \begin{pmatrix} \chi_{mn} \\ \tilde{\chi}_{mn} \end{pmatrix} = \begin{pmatrix} \sigma + \mathrm{i}\lambda_m & \mathrm{i}\mu_n \\ \mathrm{i}\mu_n & \sigma - \mathrm{i}\lambda_m \end{pmatrix} \begin{pmatrix} \chi_{mn} \\ \tilde{\chi}_{mn} \end{pmatrix} . \tag{17.138}$$

We conclude that the determinant of $\mathscr{D}$ is given by the regularized infinite product

$$(\det \mathscr{D})^2 = \prod_{mn} \left(\lambda_m^2 + \mu_n^2 + \sigma^2 \right) . \tag{17.139}$$

Since $\gamma_{12}\mathscr{Q}_\perp$ and $\mathscr{Q}_\perp$ have the same square, we finally observe that

$$(\det \mathscr{D})^2 = \det H, \quad \text{with} \quad H = -\mathscr{Q}_\parallel^2 - \mathscr{Q}_\perp^2 + \sigma^2 . \tag{17.140}$$

The three operators adding up to H commute with one another.

Flux Quantization and Landau Levels

The effective Hamiltonian in (17.140) splits into a non-trivial part in the plane defined by the field strength and a trivial part in the remaining directions. In the last chapter, we calculated the energies and eigenfunctions of the two-dimensional operator $\mathscr{Q}_\parallel$ with constant magnetic field, subject to periodic boundary conditions. Strictly speaking, the spinor fields are periodic up to a gauge transformation given by non-trivial transition functions; see Sect. 16.3. We have seen that the index theorem implies that the magnetic flux through the xy-plane with area $L_1 L_2$ is quantized,

$$\Phi = B L_1 L_2 = 2\pi k, \quad \text{with} \quad k \in \mathbb{Z}, \tag{17.141}$$

see Eq. (16.10). In Sect. 16.3.2 we have calculated the eigenvalues of $\mathscr{Q}_\parallel$. Since this operator anticommutes with γ_{12}, its non-zero eigenvalues come in pairs with opposite signs. More precisely,

$$\lambda_m = \pm\sqrt{mB}, \quad m \in \mathbb{N}_0 \quad \text{degeneracy} \quad \frac{d_s}{2}k = \frac{d_s \Phi}{4\pi} . \tag{17.142}$$

We assumed a positive B in which case all zero modes have $\gamma_{12} = -1$. For a negative B, they have $\gamma_{12} = 1$, and B and Φ are replaced by their absolute values in (17.142).

Heat Kernel and ζ-Function We consider the ζ-function of the second-order operator H defined in (17.140),

$$\zeta_H(s) = \frac{1}{\Gamma(s)} \int_0^\infty \mathrm{d}t \, t^{s-1} \mathrm{tr} \, \mathrm{e}^{-tH} , \tag{17.143}$$

and make use of the fact that the three terms in the decomposition (17.140) commute, such that

$$\zeta_H(s) = \frac{1}{\Gamma(s)} \int_0^\infty dt \, t^{s-1} e^{-t\sigma^2} \, \mathrm{tr} \, e^{t\mathcal{D}_\parallel^2} \, \mathrm{tr} \, e^{t\mathcal{D}_\perp^2} \,. \tag{17.144}$$

Our gauge potential (17.130) depends only on the coordinate x_2 such that all momenta with the exception of p_1 and p_2 can be diagonalized.[12] It follows that $-\mathcal{D}_\perp^2$ has eigenvalues

$$(\omega_n - i\mu)^2 + \boldsymbol{p}_\perp^2 \,, \tag{17.145}$$

with Matsubara frequencies ω_n and quantized momenta in the $3, 4, \ldots$-directions.

The last trace in (17.144) involves a sum over these frequencies, and the dependence on temperature and chemical potential stems from this trace. It is given by

$$\mathrm{tr} \, e^{t\mathcal{D}_\perp^2} = K_\beta(t) \, K_L^{d-3}(t) \tag{17.146}$$

with the K-functions defined in (17.4) and (17.5). The dependence on the magnetic field originates from the first trace involving $\mathcal{D}_\parallel$ in (17.144). Without this trace we got the ζ-function of the GN model in *two* dimensions less. This is the reason why one expects a *dimensional reduction* when the higher Landau levels are not occupied. With the known eigenvalues of $\mathcal{D}_\parallel$ and their degeneracy, we obtain

$$\mathrm{tr} \, e^{t\mathcal{D}_\parallel^2} = \frac{d_s \Phi}{4\pi} \left(1 + 2 \sum_{m \in \mathbb{N}} e^{-2mtB} \right) = \frac{d_s \Phi}{4\pi} \coth(tB), \quad B > 0 \,. \tag{17.147}$$

For a large area of the 2-torus with coordinates x_1 and x_2, the Φ-quantization allows for densely packed values of B. Thus, for sufficiently large areas, we may safely ignore the quantization of the magnetic field.

ζ-Function of H

Inserting these results into (17.144) yields the ζ-function for the second-order operator H,

$$\zeta_H(s) = \frac{d_s \Phi}{4\pi} \frac{1}{\Gamma(s)} \int_0^\infty dt \, t^{s-1} K_\beta K_L^{d-3} \coth(tB) \,. \tag{17.148}$$

---

[12] With our choice of A_ν, the eigenmodes are superpositions of functions with different p_1.

Inserting for K_β the alternating series in (17.4), we observe that ζ_H splits into a zero temperature and density part plus a finite-temperature correction,

$$\zeta_H(s) = \zeta_H^{(0)} + \zeta_H^{(1)}, \tag{17.149}$$

given by the Mellin transformations,

$$\zeta_H^{(0)} = \frac{\beta\phi_d}{\Gamma(s)} \int_0^\infty dt\, K_L^{d-3} t^{s-\frac{3}{2}} e^{-t\sigma^2} \coth(tB), \quad \phi_d = \frac{d_s\Phi}{(4\pi)^{3/2}} \tag{17.150}$$

$$\zeta_H^{(1)} = \frac{\beta\phi_d}{\Gamma(s)} \int_0^\infty dt\, K_L^{d-3} t^{s-\frac{3}{2}} e^{-t\sigma^2} \sum_{n\neq 0}(-1)^n e^{-n^2\beta^2/4t - n\mu\beta} \coth(tB). \tag{17.151}$$

This result will be the starting point when we investigate the magnetic catalysis in the large-N_f GN model.

17.7.1 The Effective Potential in Three Dimensions

In three spacetime dimensions $d_s = 2$, and the first ζ-function reads

$$\zeta_H^{(0)}(s) = \frac{\beta\phi_3}{\Gamma(s)} \int_0^\infty dt\, t^{s-\frac{3}{2}} e^{-t\sigma^2} \coth(tB), \quad \phi_3 = \frac{\Phi}{4\pi^{3/2}}. \tag{17.152}$$

This integral defines the function ζ_H for $\Re(s) > 3/2$. To find its analytic continuation, we use

$$\coth(tB) = \frac{2}{1-e^{-2tB}} - 1 = \sum_{\ell\geq 0} g_\ell\, e^{-2\ell tB}, \quad g_0 = 1, \ g_{\ell\geq 1} = 2, \tag{17.153}$$

to express the integral in terms of the Hurwitz zeta-function. For $\Re(s) > 1$ this function is defined by the integral or series (see 3.551 in [35])

$$\zeta(s,a) = \frac{1}{\Gamma(s)} \int_0^\infty \frac{t^{s-1}e^{-at}}{1-e^{-t}} dt = \sum_{n=0}^\infty \frac{1}{(a+n)^s}. \tag{17.154}$$

It is meromorphic in the complex s-plane, its only singularity being a simple pole at $s = 1$ with residue 1. It suffices to consider $\Re(a) \in [0,1]$ since the definition implies

$$\zeta(s,a) = \zeta(s,a+1) + a^{-s}. \tag{17.155}$$

With the first representation in (17.153), one sees that $\zeta_H^{(0)}$ is related to the Hurwitz zeta-function as follows,

$$\zeta_H^{(0)}(s) = \beta\phi_3 \, r_1(s) \left(2(2B)^{\frac{1}{2}-s}\zeta\left(s - \tfrac{1}{2}, \sigma^2/2B\right) - (\sigma^2)^{\frac{1}{2}-s}\right) \qquad (17.156)$$

with $r_1(s)$ in (17.3). In the derivation we used the exact eigenvalues of the Landau problem on a finite torus. Hence, there are no finite size corrections to this formula.

The right-hand side is analytic in a neighborhood of $s = 0$, and its derivative at this point yields the ζ-function regularized determinant. Since the ratio of Γ functions behaves near the origin as $r_1(s) \sim -\sqrt{4\pi}\,s$, the s-derivative at zero and the resulting effective potential are readily evaluated,

$$U_{\text{eff}}^{(0)} = \frac{\sigma^2}{2g^2} + \frac{B\sigma}{4\pi} - \frac{B^{3/2}}{\sqrt{2}\,\pi}\zeta(-\tfrac{1}{2}, \sigma^2/2B). \qquad (17.157)$$

We used that the magnetic flux is $\Phi = BV_s$. With the asymptotic expansion

$$\zeta(s, a \to \infty) \sim \frac{a^{1-s}}{s-1} + \frac{a^{-s}}{2} + \dots, \qquad (17.158)$$

we can extract the zero temperature and density potential for large condensates or for small magnetic fields,

$$U_{\text{eff}}^{(0)} = \frac{\sigma^2}{2g^2} + \frac{\sigma^3}{6\pi} \quad \text{for} \quad B \to 0. \qquad (17.159)$$

This is the well-known result for the GN model without magnetic field. Of course, the limit $B \to 0$ can only be taken (continuously) when the spatial volume is sufficiently large such that the quantized values of B are dense. As we did earlier, we use the minimum σ_0 of the effective potential at zero temperature, density, and magnetic field to set the scale. It follows that the GN coupling is given by $1/g^2 = -\sigma_0/2\pi$.

The finite temperature and density correction $U_{\text{eff}}^{(1)}$ follows readily from the observation that the integral in (17.151) is finite for $s = 0$ and $1/\Gamma(s) \sim s$,

$$U_{\text{eff}} = U_{\text{eff}}^{(0)} + U_{\text{eff}}^{(1)}, \qquad (17.160)$$

$$U_{\text{eff}}^{(0)} = -\frac{\sigma^2}{4\pi} + \frac{B\sigma}{4\pi} - \frac{B^{3/2}}{\sqrt{2}\,\pi}\zeta(-\tfrac{1}{2}, \sigma^2/2B), \qquad (17.161)$$

$$U_{\text{eff}}^{(1)} = \frac{B}{(4\pi)^{3/2}} \int_0^\infty dt\, t^{-\frac{3}{2}} e^{-t\sigma^2} \sum_{n\neq 0} (-1)^n e^{-n^2\beta^2/4t - n\mu\beta} \coth(tB). \qquad (17.162)$$

Here all dimensional quantities are measured in units defined by σ_0. The zero temperature and density contribution depends on the magnetic field, and so does the finite temperature and density correction. Recall that the potential is an even function of B and of σ such that, for example, $B^{3/2}$ means $|B|^{3/2}$.

To rewrite the correction $U_{\text{eff}}^{(1)}$ in a more familiar form, we expand coth in (17.162) in powers of $\exp(-2tB)$ as in (17.153). This leads to a double series the terms of which can be integrated over t with the help of (5.31). This results in

$$U_{\text{eff}}^{(1)} = \frac{B}{4\pi\beta} \sum_{\ell \geq 0} \sum_{n \neq 0} (-1)^n \frac{g_\ell}{|n|} e^{-n\mu\beta - \beta|n|\omega_\ell}, \quad \omega_\ell^2 = \sigma^2 + 2B\ell. \quad (17.163)$$

The alternating geometric series over positive and negative n can be summed,

$$U_{\text{eff}}^{(1)} = -\frac{B}{4\pi\beta} \sum_{\ell \geq 0} g_\ell \left(\log \left(1 + e^{-\beta(\omega_\ell + \mu)}\right) + \log \left(1 + e^{-\beta(\omega_\ell - \mu)}\right) \right). \quad (17.164)$$

Thus, we end up with the two equivalent representations (17.162) and (17.164) for $U_{\text{eff}}^{(1)}$. For $B \to 0$ the values $x_\ell = 2B\ell$ are dense and equidistant on the positive real axis, and the Riemann sum over ℓ can be approximated by a Riemann integral. Inserting $g_\ell = 2$ one recovers the earlier result (17.97).[13]

Catalysis at Zero Temperature and Zero Chemical Potential

The zero-temperature and zero-density effective potential (17.161) contains the Hurwitz zeta-function at $s = -\frac{1}{2}$. For this value of s, neither the integral nor the sum in (17.154) exist. But for a numerical treatment, we need a representation in terms of a convergent integral or series. A stable continuation is obtained by inserting

$$\frac{1}{e^t - 1} = \frac{1}{e^t - 1} - P_t + P_t, \quad P_t = \frac{1}{t} - \frac{1}{2} + \frac{t}{12}, \quad (17.165)$$

where P_t contains the leading terms of the function on the left for small t, into the integral representation (17.154). Subtracting P leads to a convergent integral.[14] The integral with the reinstalled P can be calculated explicitly, and we obtain

$$\zeta(s, a) = \frac{1}{\Gamma(s)} \int dt \left(\frac{1}{e^t - 1} - P_t\right) t^{s-1} e^{-at} + \frac{1}{a^s} \left(\frac{a}{s-1} + \frac{1}{2} + \frac{s}{12\,a}\right). \quad (17.166)$$

In any case, the representation shows the expected pole at $s = 1$ and yields the continuation to the half-plane $\Re(s) > -3$. The continuation is regular in a

---

[13] In the limit it is irrelevant that g_0 is actually 1 and not 2.

[14] One may skip the term $t/12$ in P_t and then the term $s/12a$ in (17.166), and the integral would still exist. But this leads to a numerically less stable representation of the analytic continuation.

Fig. 17.8 The zero temperature and density potential $U_{\text{eff}}^{(0)}$ in three dimensions for four values of the magnetic field. The values at the origin increases with increasing $B \geq 0$, and the potential is non-analytic at the origin. The fields and potential are plotted in units defined by σ_0

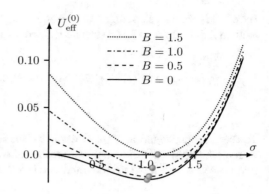

neighborhood of $s = -1/2$ and can be used to rewrite the zero-temperature part as follows:

$$U_{\text{eff}}^{(0)} = -\frac{\sigma_0 \sigma^2}{4\pi} + \frac{\sigma^3}{6\pi}\left(1 + \frac{B^2}{4\sigma^4}\right)$$
$$+ \left(\frac{B}{2\pi}\right)^{3/2} \int_0^\infty dt \left(\frac{1}{e^t - 1} - P_t\right) t^{-3/2} e^{-\sigma^2 t/2B} . \qquad (17.167)$$

We used this integral representation to calculate the effective potential for different values of the magnetic field in Fig. 17.8. Inspecting the plots one can already draw the following qualitative conclusion: at low temperature and low fermion density, the minimum of $U_{\text{eff}}^{(0)}$ (marked by a dot) moves to a larger value of σ with increasing B. In other words the magnetic field amplifies the parity condensate.

Zero Temperature

To find the zero-temperature limit of $U_{\text{eff}}^{(1)}$, we use the representation (17.164). For large $|x|$ we have $\log(1 + e^x) \to x\theta(x)$, such that for $\mu \geq 0$ we have

$$\lim_{T \to 0} U_{\text{eff}}^{(1)} = -\frac{B}{4\pi} \sum_{\ell : \omega_\ell < \mu} g_\ell \cdot (\mu - \omega_\ell), \quad \omega_\ell = \sqrt{\sigma^2 + 2B\ell} . \qquad (17.168)$$

The finite sum (17.168) originates from the discrete Landau levels. Only levels with $\omega_\ell \leq \mu$ or with $\ell \leq \ell_{\text{max}}$ are filled, where

$$\ell_{\text{max}} = \left[\frac{\mu^2 - \sigma^2}{2B}\right] \in \mathbb{N}_0 . \qquad (17.169)$$

Here $[x]$ is the largest integer less or equal to x. If x is negative, then $[x]$ is zero. Actually, the sum in (17.168) can be written as

$$\sum_{\ell:\omega_\ell < \mu} g_\ell (\mu - \omega_\ell) = (\mu - \sigma)\,\theta(\mu - \sigma) + 2\mu\ell_{\max} - 2\sqrt{2B} \sum_{\ell=1}^{\ell_{\max}}(\ell + \sigma^2/2B)^{1/2}.$$

(17.170)

The series representation of the Hurwitz ζ-function in (17.154) reveals that the finite sum on the right-hand side is the difference of two Hurwitz ζ-functions. One of the two cancels in the sum (17.160), and one is left with

$$\lim_{T \to 0} U_{\mathrm{eff}} = -\frac{B}{4\pi} \max\{\mu, \sigma\} - \frac{\mu B}{2\pi}\ell_{\max} - \frac{B^{3/2}}{\sqrt{2\pi}}\zeta\left(-\tfrac{1}{2}, 1 + \sigma^2/2B + \ell_{\max}\right).$$

(17.171)

For the explicit calculation, one uses again the integral representation (17.166) for the Hurwitz function. In terms of

$$X = 1 + \frac{2B}{\sigma^2}\,\ell_{\max} = 1 + \frac{2B}{\sigma^2}\left[\frac{\mu^2 - \sigma^2}{2B}\right] \tag{17.172}$$

the full effective potential of the $3d$ GN model at zero temperature but finite fermion density and magnetic field reads

$$\lim_{T \to 0} U_{\mathrm{eff}} = -\frac{\sigma^2}{4\pi} + \frac{\sigma^3}{6\pi}X^{3/2} + \frac{B}{4\pi}\left(\sigma X^{1/2} - \max\{\mu, \sigma\}\right)$$

$$-\frac{\mu B}{2\pi}\ell_{\max} + \frac{B^2}{24\pi\sigma}X^{-1/2}$$

$$+\left(\frac{B}{2\pi}\right)^{3/2} \int_0^\infty dt \left(\frac{1}{e^t - 1} - P_t\right) t^{-3/2}e^{-(\ell_{\max} + \sigma^2/2B)t}. \tag{17.173}$$

For a vanishing μ, we have $\ell_{\max} = 0$, $X = 1$ and $\max\{\mu, \sigma\} = \sigma$, and we recover the earlier result (17.167).

The zero-temperature potential for $\mu \approx 1$ and small B-fields is depicted in Fig. 17.9. Earlier we have argued that for $B = 0$ and $\mu = 1$, the potential shows a plateau for $\sigma \in [0, 1]$. Switching on a weak magnetic field leads to Landau level oscillations of the order parameter around this plateau [36]. With decreasing B the Landau levels are more dense, and the oscillations have shorter wavelength and smaller amplitude in σ-space. In this parameter region U_{eff} develops many competing local minima, separated by small barriers, and the system shows a cascade of density-driven weakly first-order transitions. For $B \to 0$ these oscillations go away, and one recovers the plateau of the field-free model depicted in Fig. 17.4.

Fig. 17.9 Effective potential
of the 3d Gross-Neveu model
at zero temperature, small B
and $\mu \approx 1$

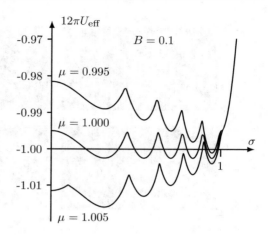

Chiral Condensate in 3d GN Model

In the large-N_f limit, the chiral condensate is proportional to the field σ_{min} which minimizes the effective potential U_{eff} in (17.160),

$$\langle \bar{\psi}\psi \rangle \propto \sigma_{min}, \qquad U_{eff}(\sigma_{min}) \leq U_{eff}(\sigma) . \qquad (17.174)$$

Contour plots of the minimum as function of (μ, T) are depicted in Fig. 17.10 for different values of the magnetic field. To calculate the zero temperature and density contribution, we used the form (17.167) and for the remainder $U_{eff}^{(1)}$ the form (17.164). The color-coding is given aside the last plot for $B = 10$.

For vanishing μ and finite T, we observe magnetic catalysis, i.e., an enhancement of symmetry breaking by the external field. The three-dimensional GN model in the large N_f limit realizes the magnetic catalysis of symmetry breaking in its simplest possible form, similar to other four-Fermi theories [33, 37]. The effect was first observed in the NJL model at $T = 0$ [31–33].

For vanishing T and finite μ, the situation is more involved. For small B-values, the critical μ_c separating the symmetric and the broken phase (with smallest non-zero condensate) increases with increasing B-field, and we observe magnetic catalysis. For example, when B increases from 0 to 0.5, then μ_c increases from 1 to 1.06; see Fig. 17.10. When we further increase B to 1, then μ_c drops to ≈ 0.77. Thus, there is inverse catalysis for intermediate values of B. For larger fields, when the higher Landau levels are empty, we again observe magnetic catalysis. For example, for $B = 10$ the order parameter jumps at $\mu_c \approx 1$. For strong fields the critical μ_c increases monotonically with the field. For an explanation of this effect, see [34].

Considering the plots in Fig. 17.10, we also see that increasing the magnetic field for certain fixed values of T and μ, for example, $T = 0$ and $\mu \approx 0.9$, the system shows a phase transition from the broken into the symmetric phase. If one further increases the field, then the system shows a second transition back into the broken

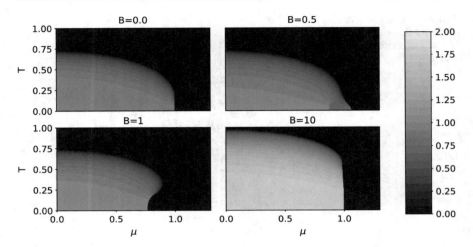

Fig. 17.10 Contour plot of the minimizing σ-field. The temperature-driven transitions at fixed μ are second order, whereas the density-driven transitions at fixed temperature are second and first order for $B > 0$. For $B = 0$ the transition is first order only for $T = 0$ (many thanks to M. Mandl for producing these contour plots)

phase. We see that the third control parameter B gives rise to a complex phase diagram. Symmetry breaking in the large-N_f NJL model in the space of the three control parameters T, μ, and B has been studied in [38].

17.8 Inhomogeneous Condensates in the $2d$ GN Model

Up to now we assumed a condensate homogeneous in space. This assumption seems natural, since the condensate should minimize the free energy of a system with a translation invariant Hamiltonian. But it has been observed that at low temperature and large chemical potential, the minimum of the effective action—which in the large N_f-limit is proportional to the chiral condensate—is not assumed for a homogeneous condensate but for an inhomogeneous one [23–26]. In this region of the phase diagram, the condensate is a periodic function of the spatial coordinate with a wavelength depending on the chemical potential.

17.8.1 The Fermion Determinant for $\sigma = \sigma(x)$

In two dimensions det $\mathscr{D}$ is real, even in σ and even in μ for any field σ; see (17.55). If in addition the field is time-independent, we can factorize the time dependence of the eigenmodes as in (17.61). Since

$$h = i\gamma_* \partial_x + \gamma^0 \sigma = h^\dagger \qquad (17.175)$$

anticommutes with γ_1, the one-particle energies in

$$h\phi_m = \varepsilon_m \phi_m \tag{17.176}$$

come in pairs with opposite signs. Then there is no sign problem because we can use the result (17.63), according to which the determinant is given by

$$(\det \mathscr{D})^2 = \prod_{n,\,m} \left((\omega_n - i\mu)^2 + \varepsilon_m^2 \right). \tag{17.177}$$

Again we shall employ the zeta-function method to regularize this infinite product.

Heat Kernel and Zeta-Function

Whereas for a constant background field the one-particle energies are simple functions of the momenta, we now must solve the spectral problem for h to calculate these energies. Since we didn't specify the one-particle energies in the previous calculation, we can use (17.66), according to which

$$\log \det \mathscr{D} = -\frac{1}{2}\frac{d}{ds}\Big|_{s=0} \zeta_H(s), \quad \zeta_H(s) = \frac{1}{\Gamma(s)} \int dt\, t^{s-1} K_\beta(t)\, e^{-t\varepsilon_m^2}. \tag{17.178}$$

Again we split K_β, given by the alternating series in (17.4), into its zero temperature part plus remainder. The s-derivatives of these contributions at the origin are

$$\frac{d}{ds}\Big|_{s=0} \zeta_H^{(0)}(s) = \frac{\beta}{\sqrt{4\pi}} \frac{d}{ds}\Big|_{s=0} r_1(s)\, \zeta_{h^2}(s - \tfrac{1}{2}), \tag{17.179}$$

$$\frac{d}{ds}\Big|_{s=0} \zeta_H^{(1)}(s) = -2 \sum_{m:\varepsilon_m>0} \log\left((1 + e^{-\beta(\varepsilon_m+\mu)}) + \log(1 + e^{-\beta(\varepsilon_m-\mu)}) \right). \tag{17.180}$$

In writing the sum over only the positive ε_m, we exploited the fact that the one-particle energies come in pairs with opposite signs. Henceforth, we focus on two dimensions, where an inhomogeneous condensate may form, at least in the large-N_f limit. At present it is not known whether inhomogeneous condensates exist in higher dimensions. Variational calculations indicate that this is not the case, at least for the GN model in the continuum limit [39, 40].

In one space dimension $\zeta_{h^2}(s - \tfrac{1}{2})$ has a pole at the origin where r_1 is zero. Hence, one must be careful when evaluating the s-derivative in (17.179). To make progress we subtract and add again the ζ-function of the square of a Dirac Hamiltonian h_0 with constant scalar field $\bar\sigma$. We choose $\bar\sigma$ such that the difference between curly brackets in

$$\zeta_{h^2}(s - \tfrac{1}{2}) = \zeta_{h_0^2}(s - \tfrac{1}{2}) + \left\{ \zeta_{h^2}(s - \tfrac{1}{2}) - \zeta_{h_0^2}(s - \tfrac{1}{2}) \right\} \tag{17.181}$$

is finite at $s = 0$. Later we shall see that this requirement fixes the constant field up to a sign. With this constant field, we can then calculate the derivative in (17.179), since we have already calculated this derivative for constant fields. More precisely, for small s we have $r_1(s) \approx -\sqrt{4\pi}\, s$, such that the derivative is just

$$\frac{d}{ds}\Big|_{s=0} \zeta_H^{(0)}(s) = \frac{d}{ds}\Big|_{s=0} \zeta_{H_0}^{(0)}(s) - \beta \left(\zeta_{h^2}(-\tfrac{1}{2}) - \zeta_{h_0^2}(-\tfrac{1}{2}) \right), \qquad (17.182)$$

where H_0 is the H-operator with constant scalar field $\bar{\sigma}$ and associated Dirac Hamiltonian h_0. For the constant $\bar{\sigma}$, we may use the result (17.82), i.e.,

$$\frac{V}{2g^2}\bar{\sigma}^2 + \frac{1}{2}\frac{d}{ds}\Big|_{s=0} \zeta_{H_0}^{(0)}(s) = V U^{(0)}(\bar{\sigma}), \qquad (17.183)$$

with the zero-temperature and density potential in (17.93). To actually determine $\bar{\sigma}$, we recall that in one (spatial) dimension

$$\zeta_{h^2}(s - \tfrac{1}{2}) = \frac{1}{\sqrt{4\pi}} \frac{1}{\Gamma(s - \tfrac{1}{2})} \frac{a_1}{s} + f(s), \qquad (17.184)$$

with a function f which is smooth at the origin; see, for example, [41]. The integrated heat kernel coefficient a_1 of the square of h in (17.175) is simply

$$a_1 = -\int dx\, \sigma^2(x). \qquad (17.185)$$

It follows that the difference of ζ-functions between brackets in (17.182) is finite if the square of the constant reference field is chosen as follows:

$$\bar{\sigma}^2 = \frac{1}{L} \int_0^L dx\, \sigma^2(x). \qquad (17.186)$$

Denoting the energies of h_0 by $\bar{\varepsilon}_m$ the finite difference reads

$$\zeta_{h^2}(-\tfrac{1}{2}) - \zeta_{h_0^2}(-\tfrac{1}{2}) = \sum_m \left(|\varepsilon_m| - |\bar{\varepsilon}_m| \right) = 2 \sum_{\varepsilon_m > 0} \varepsilon_m - 2 \sum_{\bar{\varepsilon}_m > 0} \bar{\varepsilon}_m. \qquad (17.187)$$

It is proportional to the total energy of the Dirac sea of h minus that of the Dirac sea of h_0. The individual energies of the two Dirac seas are UV divergent, but their difference—properly defined by the first sum—is not.

Inserting our results into (17.182) yields the derivative (17.179). Together with the finite temperature and density correction (17.180), this leads to the ζ-function-defined determinant of H and thus to the effective action density (for static condensates this is the energy per volume or energy density)

$$U_{\text{eff}}[\sigma] = \frac{S_{\text{eff}}[\sigma]}{V} = U_{\text{eff}}^{(0)}(\bar{\sigma}) - \frac{1}{L}\left(\sum_{\varepsilon_m>0} \varepsilon_m - \sum_{\bar{\varepsilon}_m>0} \bar{\varepsilon}_m\right)$$

$$- \frac{1}{V}\sum_{\varepsilon_m>0}\left(\log(1 + e^{-\beta(\varepsilon_m+\mu)}) + (\mu \to -\mu)\right), \quad (17.188)$$

with σ-dependent one-particle energies. At zero temperature the expression for the energy density simplifies,

$$U_{\text{eff}}[\sigma] = U_{\text{eff}}^{(0)}(\bar{\sigma}) - \frac{1}{L}\left(\sum_{\varepsilon_m>\mu} \varepsilon_m - \sum_{\bar{\varepsilon}_m>0} \bar{\varepsilon}_m + \mu N_\mu\right), \quad (17.189)$$

where N_μ counts the number of h-eigenstates with energies $\varepsilon_m \in (0, \mu)$. To find the minima of $U_{\text{eff}}[\sigma]$, we approximate the interval $[-L/2, L/2]$ by a lattice Λ with N equidistant lattice points. Then we decompose an even lattice field into Fourier modes

$$\sigma(x) = \sum_{p \in \Lambda^*} a_p \cos(px), \quad x \in \Lambda \quad (17.190)$$

with momenta on the dual lattice Λ^*. To minimize the effective action on the set of N Fourier coefficients, one may proceed with the following simple iteration:

- Choose an initial set of coefficients $\{a_p\}$ which determine the initial lattice field σ and the average field $\bar{\sigma}$ defined in (17.186). Calculate the positive energies ε_m and $\bar{\varepsilon}_m$ of the corresponding Hamiltonians h and h_0, and calculate the energy density U_{eff} in (17.188).
- Suggest a provisional change of the first Fourier coefficient to a randomly chosen new coefficient. Calculate the energies and energy density for the corresponding trial field.
- If the action decreases then permanently, replace the old coefficient by the new one. Otherwise the old coefficient remains unaltered.
- Proceed with the second coefficient and so on until all coefficients have been tested.
- Typically one sweeps several hundred times through the set $\{a_p\}$ to find a minimum.

We have argued in Sect. 17.5.2 that at zero temperature and for all $\mu \leq 0.707$, the field $\sigma = 1$ minimizes the effective potential (17.93) on the set of constant fields and that $4\pi U_{\text{eff}}(1) = -1$. When searching for a spatially inhomogeneous condensate at zero temperature, we must minimize the action density (17.188). One finds that the minimizing inhomogeneous condensate has lower energy density

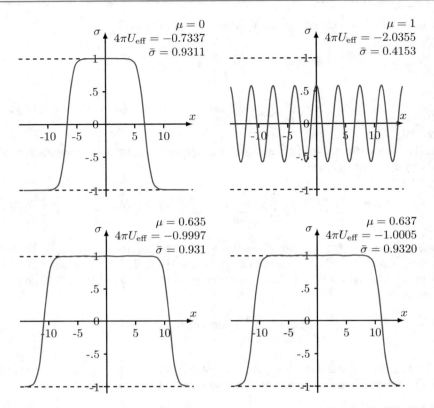

Fig. 17.11 Inhomogeneous condensates at $T = 0$ on a wire with length 30 for different densities. The kink-antikink in the vacuum (top left) is metastable. It has a higher energy than the homogeneous condensate which has $4\pi U_{\text{eff}} = -1$. The top-right figure shows the condensate deep in the inhomogeneous phase. The figures in the bottom row show the condensates near the phase boundary between the homogeneously broken and the inhomogeneous phases. $\bar{\sigma}$ is defined in (17.186)

than the homogeneous condensate for $\mu \gtrsim 0.636 \approx 2/\pi$; see Fig. 17.11. This value is the critical chemical potential which separates the homogeneously and inhomogeneously broken phases. Actually, depending on the initial guess for σ, the iteration process may terminate at different local or global minima. For example, for vanishing T and μ and initial guess $\sigma \propto \cos(2\pi x/L)$, the iteration converges to the metastable kink-antikink configuration depicted in Fig. 17.11 (top left). This configuration represents a baryon-antibaryon system with higher energy than the vacuum configuration $\sigma = 1$,

$$L \cdot \left(U_{\text{eff}}(\text{kink-antikink}) - U_{\text{eff}}(\sigma_0) \right) \approx 0.636\,\sigma_0 \approx \frac{2}{\pi}\,\sigma_0\,, \qquad (17.191)$$

in agreement with the analytic large-N_f prediction by Gross and Neveu [2]. In our numerical algorithm, we used the SLAC derivative introduced on page 404 to calculate the energies of the second-order operator h^2. The high precision of eigenvalues calculated with this derivative was demonstrated in [42]. To get sufficiently accurate values for the minimal energy density, we took a wire of length $L = 30$ (in σ_0-units), discretized by a grid with $N = 201$ lattice points.[15] One checks that for this choice of parameters, finite size effects are negligible, and one is sufficiently close to the continuum. Similar numerical algorithms have been applied to search for an inhomogeneous order parameter in the 3-GN model in the large N_f-limit—either with a Pauli-Villars continuum cutoff or on a lattice with naive fermions to spot one-dimensional structures [40] or by restricting oneself to inhomogeneous condensates with fixed wavelength [39].

To map out the full phase diagram at finite μ and T, one must return to the effective action density (17.188). Applying the iterative minimization to this functional, one obtains the condensates depicted in Fig. 17.12. The figure shows the minimizing configurations for different temperatures at fixed $\mu = 0.7$. For low temperature $T \lesssim 0.21$, the inhomogeneous configurations minimize the effective action density. But when the temperature reaches the value 0.21, the inhomogeneous and the homogeneous minimum $\sigma = 0$ have the same action density ≈ -0.102. For even higher temperature the symmetric phase is favored. The transition from the inhomogeneous to the symmetric phase is smooth since the amplitude of the inhomogeneous condensate approaches zero when T approaches the critical value $T_c \approx 0.21$ from below.

Comparison with Momentum Cutoff

When summing over the Matsubara frequencies only, it makes no difference whether the background is constant or space dependent. Thus, for a static σ one obtains the same intermediate results as in Sect. 17.6.1. In particular, the effective action splits into a zero temperature and density part plus correction term,

$$S_{\text{eff}} = S_{\text{eff}}^{(0)} + S_{\text{eff}}^{(1)}, \quad \text{with} \quad S_{\text{eff}}^{(0)} = \frac{V}{2g_b^2} \bar{\sigma}^2 + V \varepsilon_{\text{cas}}^{(\Lambda)}. \tag{17.192}$$

The correction term $S_{\text{eff}}^{(1)}$ is the second line in (17.188). The zero temperature and density term contains the classical potential energy, which for a static σ is proportional to $\bar{\sigma}^2$, and the regularized Casimir energy density,

$$\varepsilon_{\text{cas}}^{(\Lambda)} = -\frac{1}{L} \sum_{\varepsilon_m > 0} \varepsilon_m. \tag{17.193}$$

We skipped an irrelevant constant counterterm.

---

[15] For periodic boundary conditions and with the SLAC derivative, N must be odd.

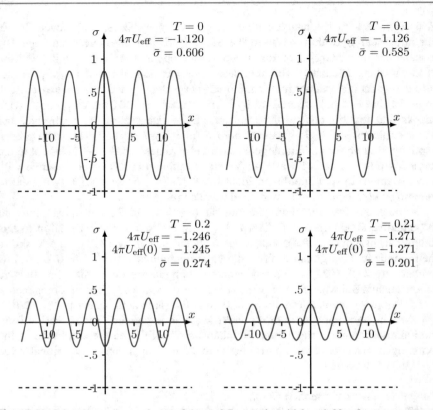

Fig. 17.12 Inhomogeneous condensate for $\mu = 0.7$ on a wire with length 30 at four temperatures. The amplitude of the inhomogeneous condensate decreases with increasing temperature. Slightly above $T = 0.21$, the amplitude goes to zero, and above this temperature the system is in the symmetric phase

To show that with the previously extracted cutoff dependence (17.125) of the bare coupling we can reproduce the ζ-function result, we subtract and add the Casimir energy density $\bar{\varepsilon}_{\text{cas}}^{(\Lambda)}$ for a constant background field $\bar{\sigma}$,

$$\frac{1}{2g_b^2}\bar{\sigma}^2 + \varepsilon_{\text{cas}}^{(\Lambda)} = \varepsilon_{\text{cas}}^{(\Lambda)} - \bar{\varepsilon}_{\text{cas}}^{(\Lambda)} + \left(\bar{\varepsilon}_{\text{cas}}^{(\Lambda)} + \mathcal{V}_B(\bar{\sigma})\right) = \varepsilon_{\text{cas}}^{(\Lambda)} - \bar{\varepsilon}_{\text{cas}}^{(\Lambda)} + U_{\text{eff}}^{(0,\Lambda)}(\bar{\sigma}).$$

$$(17.194)$$

Thus, with $g_b(\Lambda)$ calculated for homogeneous backgrounds, we can safely remove the cutoff, since the regularized effective potential converges to the renormalized potential in (17.127) (without vector field). Furthermore, with the choice (17.186) of the reference field $\bar{\sigma}$, the difference of Casimir energy densities is finite. Thus, we just reproduced the result obtained within the ζ-function regularization.

Gap Equation

We derive the *renormalized* (self-consistent) gap equation in a form which is suitable for a numerical approach. It is just the Euler–Lagrange equation for the renormalized effective action $S_{\text{eff}}[\sigma]$ in (17.188). To calculate the variation of this action, we need the variation of the one-particle energies ε_m of h in (17.175). According to the Hellman-Feynman theorem, they are given by $\delta\varepsilon_m = \langle\psi_m|\delta h|\psi_m\rangle$, where $\{\psi_m\}$ is a set of (adapted) orthonormal eigenstates of h. With $\delta h = \gamma^0\delta\sigma$ we conclude

$$\frac{\delta\varepsilon_m}{\delta\sigma(x)} = \varrho_{mm}(x), \quad \varrho_{mn} = \psi_m^\dagger(x)\gamma^0\psi_n(x). \qquad (17.195)$$

To find the variation of the energies $\bar{\varepsilon}_m$, we observe that $\varrho_{mm}^{(0)} = \bar{\sigma}/\bar{\varepsilon}_m L$ is constant; see Problem 17.9. This is used in the second equation in

$$\frac{\delta\bar{\sigma}}{\delta\sigma(x)} = \frac{\sigma(x)}{\bar{\sigma}L} \implies \frac{\delta\bar{\varepsilon}_m}{\delta\sigma(x)} = \frac{\partial\bar{\varepsilon}_m}{\partial\bar{\sigma}}\frac{\delta\bar{\sigma}}{\delta\sigma(x)} = \frac{\sigma(x)}{\bar{\varepsilon}_m L}. \qquad (17.196)$$

With these relations we can calculate the σ-derivative of the effective action and end up with the renormalized gap equation [43]

$$\frac{\sigma(x)}{2\pi}\log\frac{\bar{\sigma}^2}{\sigma_0^2} - \frac{2\sigma(x)}{\pi}\sum_{n>0}K_0(n\bar{\sigma}L) + \sum_{m:\varepsilon_m<0}\varrho_{mm}(x) - \sigma(x)\sum_{m:\bar{\varepsilon}_m<0}\frac{1}{\bar{\varepsilon}_m L}$$

$$+ \sum_{m:\varepsilon_m>0}\left(\frac{1}{1+e^{\beta(\varepsilon_m+\mu)}} + \frac{1}{1+e^{\beta(\varepsilon_m-\mu)}}\right)\varrho_{mm}(x) = 0. \qquad (17.197)$$

In the derivation we assumed that the energies ε_m are non-degenerate. But as we sum over m, it does not matter which eigenbase of h we choose. We could as well have chosen the adapted eigenbase to derive the Euler–Lagrange equation. This means that (17.197) holds for a Hamiltonian h with degenerate energy levels. Note that the two individual sums over m in the first line are divergent, but the sum of the differences of their terms is convergent.

This renormalized self-consistency equation is a complicated functional equation, the solutions of which have been investigated at various times in the literature. To calculate the chiral condensate at finite temperature and finite density in the large-N_{f} limit, one must simultaneously solve the spectral problem for the σ-dependent Dirac Hamiltonian

$$h\psi_m = \varepsilon_m\psi_m \qquad (17.198)$$

and the gap equation (17.197).

For a homogeneous field, we have $\sigma = \bar\sigma$, and the two sums over m in the first line of the gap equation (17.197) cancel. Then the equation simplifies considerably—in the thermodynamic limit $L \to \infty$ we obtain

$$0 = \frac{\sigma}{2} \log \frac{\sigma^2}{\sigma_0^2} + \sigma \int_0^\infty \frac{\mathrm{d}p}{\varepsilon_p} \left(\frac{1}{1 + e^{\beta(\varepsilon_p + \mu)}} + \frac{1}{1 + e^{\beta(\varepsilon_p - \mu)}} \right). \quad (17.199)$$

There always exists the trivial solution with vanishing condensate field. A nontrivial constant solution satisfied the equation (17.105) which has been solved and discussed in great detail in Sect. 17.5.3. At zero temperature and $\mu \geq 0$, an inhomogeneous solution satisfies

$$\frac{1}{2\pi} \sigma(x) \log \frac{\bar\sigma^2}{\sigma_0^2} + \sum_{m:\varepsilon_m < -\mu} \varrho_{mm}(x) - \sigma(x) \sum_{m:\bar\varepsilon_m < 0} \frac{1}{\bar\varepsilon_m L} = 0. \quad (17.200)$$

For a given (positive) chemical potential, only states in the Dirac sea with energies below $-\mu$ enter the zero-temperature gap equation.

Often the *regularized* gap equation containing the regularized trace of the Green-function with cutoff is considered in the literature [44–47]. Above we used the ζ-function renormalized effective action with physical scale parameter σ_0 such that only finite sums and integrals enter the derivation of the gap equation. At zero temperature and fermion density, Dashen et al. could solve the coupled system for the modes $\psi_m(x)$ and the scalar field $\sigma(x)$ by using powerful inverse scattering methods [44]. They observed that a scalar field could only solve the gap equation if the solutions of the Dirac equation ψ_m are not reflected. Their space-dependent solutions describe n-particle bound states with a filled Dirac sea. Self-consistent solutions at finite temperature and fermion density and in the thermodynamic limit have been constructed much later by Thies et al. [23–26] by some (non-linear) superposition of kink-antikink solutions. They succeeded in constructing periodic solutions $\sigma(x)$ with associated Bloch waves $\psi_m(x)$ of the coupled system (17.198) and (17.197) in a high-density region of the phase diagram. The Bloch waves and the scalar field $\sigma(x)$ are given in terms of Jacobi's elliptic functions. The associated Dirac Hamiltonian has one gap in the spectrum, and the periodic and antiperiodic states at the band edges are given by Jacobi functions. Thus, the property that h shows no reflection for baryon excitations above the vacuum is replaced by the property of having exactly one band gap in the spectrum of the high-density system.

For points in the phase diagram where the inhomogeneous solution has a lower effective action than any homogeneous solution, the system is in a inhomogeneous phase. The correct phase diagram in the large-N_f limit is depicted in Fig. 17.13. Note that the metastable phases and first-order transition line (to guide the eye, this line is kept as dashed line) disappear and are replaced by two second-order transition lines. At low temperature and small chemical potential, there is a homogeneous phase with broken chiral symmetry; at sufficiently high temperature, we are in

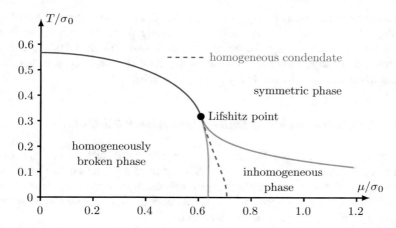

Fig. 17.13 The corrected phase diagram of the large-N_f GN model with homogeneous and inhomogeneous condensates. The first-order line from the Lifshitz point to the line $T = 0$ (dashed line) which exists in the approximation with a homogeneous condensate only, see Fig. 17.5, splits into two-second order lines. In the region with large μ and low T, the condensate is inhomogeneous (many thanks to M. Thies for providing the numbers for the phase diagram and L. Pannullo for producing the diagram)

the homogeneous symmetric phase; and at low temperature and large chemical potential, we are in the inhomogeneous phase with an oscillating chiral condensate.

The wavelength and amplitude of the condensate in the inhomogeneous phase are determined by the chemical potential and the temperature. If one moves within the inhomogeneous phase towards the symmetric phase, then the amplitude of the condensate vanishes. If one moves towards the homogeneously broken phase, then the wavelength of the condensate increases. This is in full agreement with the numerical minimization of the effective action density which, for example, revealed a phase transition at $T = 0$ and $\mu_c \approx 2/\pi$ from the homogeneously broken to the inhomogeneous phase or at $\mu = 0.7$ and $T \approx 2.1$ from the inhomogeneous into the symmetric phase.

17.9 Beyond Large N_f

There exist several alternative approaches to study interacting fermions for a finite number of flavors. For example, one can perform a systematic expansion in $1/N_f$, apply the powerful functional renormalization group method, or perform lattice simulations. We comment briefly on some selected results obtained with these approaches. When comparing with the literature on the three-dimensional Thirring model, one should keep in mind that sometimes N_f refers to the number of four-component reducible spinors and sometimes to the number of two-component irreducible spinors. To distinguish between the two cases, we denote the number of four-component spinors by N'_f and the number of two-component spinors by N_f. Actually, a four-component spinor with mass m is equivalent to two *two*-component spinors, one with mass m and the other with mass $-m$. This explains why a theory with one reducible spinor field has no parity anomaly in three dimensions, in contrast to the theory with two irreducible spinor fields.

17.9.1   Critical Behavior of Three-Dimensional GN Models

The zero-temperature Gross-Neveu models in spacetime dimensions d between the lower critical dimension 2 and the upper critical dimension 4 have a critical coupling g_cr at which the correlation length diverges. In the neighborhood of a critical point, one can define critical exponents as discussed in Sect. 7.3, and these characterize the universality class of the continuous transition.

In this subsection we focus on the three-dimensional GN model and discuss the critical coefficients in the large-N_f limit. The $\mathbb{Z}_2$-symmetric effective potential (17.96) has the small-field expansion

$$U_\mathrm{eff} = U_0 + ta_2\sigma^2 + a_3|\sigma|^3 + \dots , \qquad (17.201)$$

with positive (negative) parameter t in the symmetric (ordered) phase. In the present section this control parameter is related to the coupling g and not to the temperature or chemical potential as in the previous sections. The parameter t is negative for strong couplings $g > g_\mathrm{cr}$ and positive for weak couplings $g < g_\mathrm{cr}$.

U_eff is a Landau free energy for the order parameter σ. It contains the term $|\sigma|^3$ which is not analytic at the origin. This explains why the Landau theory—which assumes an analytical free energy—leads to critical exponents different from the mean-field exponents expected for a $\mathbb{Z}_2$-breaking.

In order to extract the large-N_f critical exponents, we add a source term for the order parameter σ to the effective action, $U_\mathrm{eff} \mapsto U_\mathrm{eff} - h\sigma$. To find the exponent γ, one calculates the minimum $\sigma_0(t, h)$ of U_eff which yields the zero-field susceptibility

$$\chi = \frac{d\sigma_0}{dh}\bigg|_{h=0} = \frac{1}{2a_2t} , \qquad (17.202)$$

and hence $\gamma = 1$. Following the approach outlined in Sect. 7.6, one obtains the exponents $\alpha = \beta = 1$ and $\delta = 2$. Finally, to find the Landau critical exponents $\eta = \nu = 1$, one must determine the subleading term in a gradient expansion of the effective action $S_\mathrm{eff}[\sigma]$; see Sect. 7.6. The mean field critical exponents of the Ising class and of the GN models are given in the second and third columns (with $\Delta = 0$) of Table 17.2.

Table 17.2 MF exponents of the $\mathbb{Z}_2$ Ising class and exponents of the three-dimensional GN model

Exponent	$\mathbb{Z}_2$-Ising	$3d$ GN
β	$\frac{1}{2}$	$1 + \ldots$
γ	1	$1 + 3\Delta + \ldots$
δ	3	$2 + 3\Delta + \ldots$
ν	$\frac{1}{2}$	$1 + \Delta + \ldots$
η	0	$1 - 2\Delta + \ldots$

The corrections of order $1/N'_f$ are proportional to

$$\Delta = \frac{8}{3\pi^2 N'_f}, \quad N'_f = \text{number of 4-component spinors,} \tag{17.203}$$

and have been calculated in [48]. They are contained in the last column of the Table. Actually, within the $1/N_f$-expansion, the GN model was solved in any dimension between the critical dimensions, up to two-loop order, with the fermion anomalous dimension being known even up to three-loop order [48–52]. For example,

$$1/\nu = 1 - \Delta + \left(\frac{9\pi^2}{16} + \frac{79}{6}\right)\Delta^2 + \ldots \tag{17.204}$$

$$\eta = 1 - 2\Delta + \left(\frac{19}{3} - \frac{9\pi^2}{16}\right)\Delta^2 + \ldots \tag{17.205}$$

The exponents have also been computed up to three-loop order within an expansion around the lower critical dimension $d = 2 + \epsilon$ [51, 53] and four-loop order within an expansion around the upper critical dimensions $= 4 - \varepsilon$ [54, 55]. The results available from the $1/N_f$ expansion provide a check for these perturbative calculations. One finds that the four-loop results are in good agreement with the results of the large-N_f expansion. For a comparison of the exponents obtained with different expansions and with the FRG method, the reader may consult [56].

17.9.2 Lattice Simulations of Three-Dimensional GN Models

In various lattice studies, the critical exponents have been extracted from the finite size scaling of thermodynamic potentials and observables [57]. Since the GN models with odd N_f have a sign problem, most simulations were done for the systems with N'_f reducible flavors. Prior to calculating critical exponents, one needs to determine the critical coupling g_{cr} of the phase transition. This can be extracted from the fourth-order cumulant by Binder [58]

$$U_B(L, g) = 1 - \frac{\langle \sigma^4 \rangle}{3\langle \sigma^2 \rangle^2}, \tag{17.206}$$

Table 17.3 Critical
coupling of 3d GN models

$N_f'=$	1	2	4	8
$g_c^2 =$	1.0544	0.8231	0.7318	0.6885

where the powers are taken after summation over the lattice sites. The cumulant
varies between the value 0 in the symmetric weak- coupling regime and the value
2/3 in the broken strong-coupling regime. Only for $g = g_{cr}$ the cumulant is (almost)
independent of the system size L, and this property is exploited to determine the
critical coupling via a finite size analysis of U_B. For example, in the thesis [59] the
critical couplings g_c listed in Table 17.3 have been extracted. These values are used
to evaluate the order parameter and its susceptibility,

$$\Sigma = \langle |\sigma| \rangle \quad \text{and} \quad \chi = V \left(\langle \sigma^2 \rangle - \langle |\sigma| \rangle^2 \right) \tag{17.207}$$

near criticality. From the finite size scaling theory for sufficiently large L, we expect

$$\Sigma \propto L^{-\beta/\nu} \quad \text{and} \quad \chi \propto L^{\gamma/\nu}. \tag{17.208}$$

In a finite system, there is no phase transition in the strict sense, and various
characterizations of a "critical coupling" exist. This ambiguity may be used to check
the stability of the extracted critical exponents. Actually, the exponent $1/\nu$ also
characterizes the slope of the Binder cumulant near criticality as a function of the
coupling, and this allows for an independent evaluation of ν. The critical exponent
β can be estimated directly from a fit to the defining relation

$$\Sigma \propto (g - g_{cr})^\beta. \tag{17.209}$$

The best estimate of critical exponents obtained with chiral SLAC fermions are
listed in Table 17.4. We used the scaling relation

$$\eta_\sigma = \frac{2\beta}{\nu} - 1 \tag{17.210}$$

to calculate the anomalous dimension η_σ of the order parameter from the measured
ratio of exponents β/ν. The hyperscaling relation (11.52) predicts that

$$H = \frac{2\beta}{\nu} + \frac{\gamma}{\nu} - 3 \tag{17.211}$$

Table 17.4 Estimates of
critical exponents for various
N_f' from [59]

N_f'	$1/\nu$	β/ν	γ/ν	η_σ	H
1	1.096(23)	0.824(13)	1.366(26)	0.648	0.01(4)
2	1.07(4)	0.886(16)	1.228(34)	0.772	0.00(5)
4	1.08(5)	0.915(21)	1.17(4))	0.83	0.00(6)
8	1.02(6)	0.960(28)	1.07(6)	0.920	−0.01(8)

should vanish. The values of H are listed in the last column of the Table, and the deviation from 0 measures the quality of the estimated exponents. Recall that for $N_f \to \infty$, all values are 1. For $N_f' = 2$, the critical exponents are in good agreement with previous lattice simulations obtained with staggered fermions [60, 61]. The GN model with 12 reducible flavors has been simulated in [48], and no deviation from the large-N_f prediction for the exponents has been detected. For a careful and exhaustive discussion and comparison with exponents obtained with the large N_f or small ϵ expansions, the bootstrap program, functional methods, and other lattice simulations, see [59].

17.9.3 Lattice Simulations of Three-Dimensional Thirring Models

The two-dimensional massless Thirring model with Lagrangian (17.28) is a conformal field theory without symmetry breaking and thus has no interesting phase structure [16, 17]. But the three-dimensional model with N_f' flavors of reducible spinors could show a spontaneous breaking of the large chiral symmetry,

$$U(2N_f') \to U(N_f') \otimes U(N_f'), \tag{17.212}$$

at least for a small flavor number. Here we focus on the models with *four*-component spinors since they are used to model low-energy properties of condensed matter systems and in addition have no sign problem.

The massless Thirring model with $N_f' = 1/2$ (or $N_f = 1$) is equivalent to the massless GN model with $N_f = 1$ and hence shows spontaneous symmetry breaking [62]. On the other hand, for $N_f \to \infty$ the effective potential can be calculated explicitly. One finds in the infinite volume limit and for $\mu = 0$

$$U_{\text{eff}} = \frac{1}{2g_R^2} v_\nu v^\nu - \frac{T}{2\pi} \int_0^\infty \mathrm{d}p\, p \left(\log\left(1 + \mathrm{e}^{-\beta(p+iv_0)}\right) + \log\left(1 + \mathrm{e}^{-\beta(p-iv_0)}\right) \right).$$

$$\tag{17.213}$$

For a fixed (spatial) momentum cutoff Λ, the bare coupling g relates to the renormalized coupling g_R as

$$g^2 = \frac{4\pi g_R^2}{4\pi + \Lambda g_R^2}. \tag{17.214}$$

At zero temperature U_{eff} is quadratic in the auxiliary vector field v_μ such that there is no phase transition for $N_f \to \infty$. Since there is a broken phase for $N_f' = 1/2$ and none for $N_f' \to \infty$, there exists a critical flavor number N_{fc}' separating the systems with symmetry breaking from those without symmetry breaking.

Early lattice studies were performed with staggered fermions and a small fermion mass to recover the chiral symmetry in the continuum limit. With the help of an

HMC algorithm, simulations with an even N_{f}' and subsequently with non-integer N_{f}' have been presented in [63, 64]. In a more recent lattice study [65] with a similar setup, the authors concluded that the critical flavor number is $N_{\mathrm{fc}}' = 6.6(1)$.

More recent analytic studies and lattice simulations with chiral fermions yield different results—they favor smaller values of N_{fc}'. Based on simulations with massless chiral SLAC fermions which exactly implement all internal symmetries, it has been demonstrated that the $U(2N_{\mathrm{f}}')$ symmetry is never broken for any integer number of *four*-component flavors [66]. In a subsequent publication, the critical flavor number $N_{\mathrm{fc}}' = 0.80(4)$ has been calculated [67].

Independent simulations with *four*-component domain-wall fermions (DWF) in [68] rather point to a N_{fc}' value between 1 and 2. In a follow-up publication, the finding $N_{\mathrm{fc}}' > 1$ has been confirmed [69]. The slight discrepancy of the results obtained with SLAC and DW fermions may be due to uncertainties in the extrapolation to an infinite domain-wall separation. In [69] it was speculated that the two lattice approaches describe different continuum theories and that the bulk DWF formulation more closely conforms to a picture of the strong dynamics in which the auxiliary vector field resembles a gauge field.

Thirring models with an odd number of *irreducible flavors* suffer from a sign problem, and lattice simulations must deal with this problem. But with a combination of analytical and numerical means—based on dualization—the problem can be overcome. It has been demonstrated that systems with odd N_{f} behave differently to those with even N_{f} (which are equivalent to reducible models with $N_{\mathrm{f}}' = N_{\mathrm{f}}/2$) [66]. One finds that the $U(N_{\mathrm{f}})$ symmetry can be broken if N_{f} is odd and $N_{\mathrm{f}} \leq N_{\mathrm{fc}} \approx 9$.

With staggered fermions one ends up with a wrong critical flavor number $N_{\mathrm{fc}}' \approx 6.6$ for the strongly interacting (reducible) Thirring model. We learn that in a field theory with an interacting fixed point, staggered fermions may fail and give rise to a continuum limit in the wrong universality class. The lattice theory with staggered fermions has different global symmetries than the classical continuum model, and the symmetry may not be recovered in the continuum limit—in particular if the quantum system is an asymptotically save theory governed by an interacting fixed point.

In an explorative FRG study with momentum-dependent couplings in combination with the powerful pseudo-spectral method, the critical behavior of four-fermion theories has been reconsidered [70]. While a precise estimate for N_{fc}' remains difficult in these elaborate FRG studies, the new results are compatible with the lattice studies based on chiral fermions.

17.9.4 Inhomogeneous Phases

In Sect. 17.8 we have seen that the two-dimensional GN model in the large N_{f} limit admits an inhomogeneous phase at large chemical potential and low temperature. In this region of the (T, μ) phase diagram, the inhomogeneous solution of the gap equation is thermodynamically favored over the homogeneous one. But this spontaneous breaking of the continuous translation symmetry seems to be in conflict with the celebrated theorem by Mermin and Wagner in statistical mechanics according to which a continuous symmetry cannot be spontaneously broken at finite temperature in one- and two-dimensional statistical systems with short-range interaction. It is also in conflict with the related theorem by S. Coleman for relativistic quantum field theories in $d \leq 2$ dimensions. Indeed, if spontaneous symmetry breaking of a continuous symmetry occurred, as a consequence of the Goldstone theorem, one would expect to find massless Nambu-Goldstone bosons in the particle spectrum. But massless scalars with a relativistic dispersion relation have an infrared divergent correlation function in $d \leq 2$ spacetime dimensions and thus should not exist. However, these no-go theorems do not apply in the limit $N_f \to \infty$. In this limiting case, we have infinitely many degrees of freedom at every point in space, and the system behaves as if it lives in one dimension more.

These arguments seem to imply that inhomogeneous condensates in two-dimensional four-Fermi theories are an artifact of the limit $N_f \to \infty$ and do not exist for $N_f < \infty$. But the situation is not so clear. For example, when proving the Goldstone theorem, one assumes that the theory is Lorentz invariant, the Hilbert space is positive, and a global symmetry group is broken to a subgroup. In non-relativistic systems and for a breaking of spacetime symmetries, the situation is more intricate, since then Goldstone bosons may have unusual dispersion relations which could imply that there are fewer Goldstone bosons than naively anticipated. Possible loopholes in the no-go theorems are discussed in [4].

Thus, it should be of interest to see whether traces of the inhomogeneous phase in the large-N_f system are seen in two-dimensional models with finite N_f. Even more interesting with regard to QCD is the question about inhomogeneous condensates in three- and four-dimensional systems. To answer the first question, two-dimensional GN models with 2, 8, and 16 flavors of chiral SLAC fermions and 8 and 16 flavors of naive fermions[16] have been simulated at finite temperature and density in [4]. First simulations with an HMC algorithm suggest that inhomogeneous structures are present at finite N_f. One observes that at small T and large μ, a generic configuration $\sigma(x)$ exhibits spatial periodic oscillations similar to a cos-function, which might signal an inhomogeneous phase. Actually, the existence of kink-antikink structures in simulations of the model with $N_f = 12$ was observed already many years ago [71]. A typical field configuration for the model with 8 flavors at $(\mu, T) \approx (0.450, 0.030)$ with such periodic oscillations is shown in Fig. 17.14. The configuration is on a lattice with $N_s = 127$ points in spatial direction and lattice constant $a = 0.410$. Similarly as in the large-N_f analysis the condensate σ_0 at zero temperature and density is used to set the scale.

---

[16] Because of fermion doubling $1, 2, 3, \ldots$ naive fermions in two spacetime dimensions describe $4, 8, 12, \ldots$ flavors.

Fig. 17.14 A typical field $\sigma(x)$ generated by the HMC algorithm. The vertical stripes indicate six oscillations in spatial direction (Figure taken from [4])

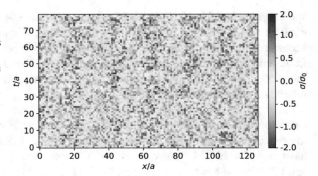

An observable which is able to exhibit information about possibly present inhomogeneous structures is the spatial correlation function of $\sigma(x)$, i.e.,

$$C(x) = \frac{1}{N_t N_s} \sum_{t,y} \langle \sigma(t, y + x) \sigma(t, y) \rangle . \tag{17.215}$$

In the inhomogeneous phase, we can write

$$C(x) = A(x)\, C_{\text{periodic}}(x) , \tag{17.216}$$

where $A(x)$ represents an amplitude function and $C_{\text{periodic}}(x)$ the periodic oscillations. If the system forms a crystal, then $A(x)$ should approach a non-zero constant for sufficiently large separations $|x|$. In case the system has almost long-range order à la Berezinskii, Kosterlitz, and Thouless (BKT) [72, 73], it decreases with an inverse power of $|x|$. Thus, to detect possible deviations of $A(x)$ from the asymptotically constant behavior in the large-N_f limit, the correlation function is calculated on rather large spatial lattices for the smallest possible flavor number $N_f = 2$. Figure 17.15 shows $C(x)$ for up to $N_s = 725$ lattice points corresponding to spatial extents up to $L \approx 297.3$ (in units defined by $1/\sigma_0$). On the whole, 32 statistically significant oscillations in $C(x)$ are observed. The Fourier transform of $C(x)$ shows a pronounced peak at some wave number which we shall refer to as the dominant wave number. The position of the peak is essentially the same for all L demonstrating that the wavelength is independent of the spatial extent.

By fitting the data with one of the functions

$$A(x) = A_{\text{SSB}}(x) \sim \gamma + \alpha e^{-m|x|} + \dots \quad \text{or} \quad A(x) = A_{\text{BKT}}(x) \sim \frac{\alpha}{|x|^\beta} + \dots \tag{17.217}$$

one may hope to decide whether translation invariance is broken spontaneously or whether there is an almost long-range order. But even on a lattice with $N_s = 725$ lattice sites in spatial direction, one cannot rule out either of the two scenarios

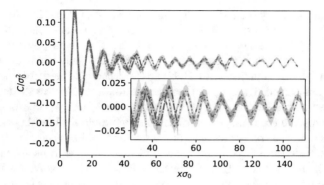

Fig. 17.15 The spatial correlation function $C(x)$ at $(\mu, T) = (0.50, 0.03)$ for *two* flavors and $N_s = 65, 125, 185, 255, 525, 725$. Figure taken from [4]

[4]. Although the answer about the thermodynamic nature of the phase at low temperature and high density remains open, it is clear that many features of the large-N_f solution leave their footprints in systems with finite N_f. Similarly as for the large-N_f solution, one observes oscillatory behavior with μ-dependent wavelength and T-dependent amplitude. If one characterizes the phases of the system —they need not be stable phases in the thermodynamic sense—by the minimal value of $C(x)$ (positive in the homogeneously broken phase, zero in the symmetric phase, and negative in the oscillatory phase), then one finds a phase diagram similar to the one in Fig. 17.13; see [4]. With decreasing N_f, the quantum fluctuations increase, and as a result the broken phase shrinks.

In a subsequent work [74], a strong correlation between the dominant wave number of the spatial inhomogeneities and the baryon number has been observed. This is clear evidence for a region in the phase diagram corresponding to a regime of modulated baryonic matter. The delicate question whether there exists a rigid baryon crystal (as seen in the large-N_f limit) or rather a baryonic liquid [75], where the baryons have a preferred separation locally, but are disordered on large scales, needs further investigations with improved algorithms.

17.9.5 Functional Renormalization Group

In recent years the FRG (see Chap. 12) has been frequently and successfully employed to study the critical behavior of interacting Fermi systems. In the pioneering work [62], the critical exponents $\nu = 0.63$, $\eta_\sigma = 0.31$, and $\eta_\psi = 0.11$ of the three-dimensional $N_f' = 1$ GN model have been presented. A FRG approach based on a partial bosonization with Hubbard-Statonovich field for the GN channel has been developed in [76], and soon a quantitative consensus for the critical regime of these systems has been reached. In subsequent FRG studies, [10] it was convincingly argued that three-dimensional GN models serve as simple realizations of the asymptotic safety scenario in quantum gravity (see Chap. 12). More generally, for relativistic fermion systems in $2 < d < 4$ spacetime dimensions with continuous $U(N_f') \times U(N_f')$ symmetry, the network of interacting fixed point with relevant directions—each of which defines a universality class—has been revealed in [6]. Again a partial bosonization with Hubbard–Stratonovich field for each possible channel (Gross-Neveu, Thirring,...) has been introduced. Beyond structural studies the FRG method has been applied to describe the low-energy properties of various Dirac materials like graphene, He, surface states of topological insulators, or d-wave cuprate superconductors. For reviews on correlated fermion systems, see [77, 78].

Momentum-dependent couplings were included in the flow of the Thirring model and the resulting non-linear differential flow equations solved with powerful pseudo-spectral methods which produce accurate global solutions in [70]. There are indications that the momentum dependencies become more important in Thirring-type models with small N_f' than in Thirring models with large N_f or in GN-type models. With this improvement the authors could show that the critical flavor number of the three-dimensional Thirring models is smaller than predicted by earlier FRG studies, simulations with staggered fermions, and Schwinger-Dyson equations—but in good agreement with recent lattice simulations with chiral fermions [67, 69].

New estimates on critical quantities of the three-dimensional GN model have been obtained by allowing for a continuous number of fermion degrees of freedom and a field-dependent Yukawa coupling in [79]. In [80] more operators were admitted in the flow equation as in earlier investigations. The FRG results agree well with other approaches, for example, lattice simulations or ϵ-expansions. Some representative results are listed in Table 17.5 The FRG results for $N_f' = 1$ and 2 are taken from [80], the lattice MC results from [59], and the results within the $(4 - \epsilon)$ expansion to order ϵ^3 from [54]. The field content of the model with $N_f' = 1/4$ is compatible with that of the $N = 1$ supersymmetric Wess-Zumino model which contains a real scalar field and a Majora spinor field. The FRG numbers quoted for the supersymmetric model are taken from [81].

Table 17.5 Estimates of $1/\nu$ and η_σ for $N_f' = 1$ and 2. The FRG values are from [80]

N_f'	Method	$1/\nu$	η_σ
1	FRG	1.075(4)	0.5506
	MC	1.096(23)	0.648
	ϵ^3	1.166	0.463
2	FRG	0.994(2)	0.7765
	MC	1.07(4)	0.772
	ϵ^3	1.048	0.672
1/4	FRG	1.408	0.180
	MC	1.07(4)	0.772
	ϵ^3	1.419	0.162

17.10 Problems

17.1 (Charge Conjugation) $\mathscr{C}$ is a charge conjugation matrix if

$$\gamma_\mu^T = \eta_c \,\mathscr{C}^{-1}\gamma_\mu\mathscr{C}, \quad \eta_c \in \{-1, 1\}\,.$$

In Table 17.6 you find particular representations for the γ-matrices in $2, 3, 4$, and 5 dimensions with corresponding charge conjugations matrices. In 2 and 4 dimensions, there exist two such matrices, one for each sign of η_c. In *three* dimensions there is one $\mathscr{C}$-matrix with $\eta_c = -1$, and in *five* dimensions there is one with $\eta_c = 1$.

1. Check that the given matrices $\mathscr{C}$ obey the above relation.
2. Try to continue the table to higher dimensions and prove that there are two $\mathscr{C}$ with $\eta_c = \pm 1$ in all even dimensions, one $\mathscr{C}$ with $\eta_c = -1$ in $3 + 4n$ dimensions and one $\mathscr{C}$ with $\eta_c = 1$ in $1 + 4n$ dimensions. This means that the following table, extended to arbitrary dimensions, would have period 4.

17.2 (Symmetries of NJL Model) Prove the transformation rule (17.34) for the isovector and pseudo-scalar field $\bar\psi i\gamma_*\boldsymbol{\sigma}\,\psi$.

17.3 (Parity Matrix) Show that in odd (Euclidean) dimensions there is no parity matrix $\mathscr{P}$ which obeys (17.26) with $\alpha = 1$. Is there a matrix $\mathscr{P}$ with $\alpha = -1$?

Hint 1: Why would (17.27) be in conflict with a solution with $\alpha = 1$?
Hint 2: For $\alpha = -1$ you may choose one of the γ-matrices as $\mathscr{P}$.

17.4 (General Fierz Identities) In this rather lengthy exercise, we discuss Fierz identities in d dimension with arbitrary signature. As a result we shall see how to reshuffle four-Fermi terms.

Table 17.6 Possible representations for the γ-matrices and corresponding $\mathscr{C}$

d	γ^0	γ^1	γ^2	γ^3	γ^5	γ_*	$\mathscr{C}$	η_c	η_c'	η_c''
2	σ_1	σ_3				σ_2	σ_2	-1	-1	-1
2	σ_1	σ_3				σ_2	σ_0	$+1$	-1	$+1$
3	σ_1	σ_2	σ_3			$-\mathbb{1}$	σ_2	-1	$+1$	$+1$
4	$\sigma_1 \otimes \sigma_0$	$\sigma_2 \otimes \sigma_1$	$\sigma_2 \otimes \sigma_2$	$\sigma_2 \otimes \sigma_3$		$\sigma_3 \otimes \sigma_0$	$\gamma^0\gamma^2 = +i\sigma_3 \otimes \sigma_2$	-1	$+1$	$+1$
4	$\sigma_1 \otimes \sigma_0$	$\sigma_2 \otimes \sigma_1$	$\sigma_2 \otimes \sigma_2$	$\sigma_2 \otimes \sigma_3$		$\sigma_3 \otimes \sigma_0$	$\gamma^1\gamma^3 = -i\sigma_0 \otimes \sigma_2$	$+1$	$+1$	-1
5	$\sigma_1 \otimes \sigma_0$	$\sigma_2 \otimes \sigma_1$	$\sigma_2 \otimes \sigma_2$	$\sigma_2 \otimes \sigma_3$	$\sigma_3 \otimes \sigma_0$	$\mathbb{1}$	$\gamma^1\gamma^3 = -i\sigma_0 \otimes \sigma_2$	$+1$	$+1$	-1

1. Consider the linear space of $d_s \times d_s$-matrices, where d_s has been introduced in (17.8), equipped with the scalar product $(M, N) = \operatorname{tr} M^\dagger N$. Show that the matrices

$$\frac{1}{\sqrt{d_s}} \{\mathbb{1}, \gamma_\mu, \gamma_{\mu_1\mu_2}, \dots, \gamma_{\mu_1\dots\mu_D}\} \quad \text{with} \quad \begin{cases} D = d & \text{for even } d \\ D = \tfrac{1}{2}(d-1) & \text{for odd } d, \end{cases}$$

form an orthonormal basis if the indices are ordered, $\mu_1 < \mu_2 < \dots$.

2. For any signature $\gamma^\mu = \gamma_\mu^\dagger$ such that $\gamma_{\mu_1\mu_2\dots\mu_n}^\dagger$ is proportional to $\gamma^{\mu_1\mu_2\dots\mu_n}$. Use this property to show that a d_s-dimensional matrix M can be expanded as

$$M = \frac{1}{d_s} \sum_{n=0}^{D} \frac{1}{n!} (-)^{n(n-1)/2} \gamma_{\mu_1\dots\mu_n} \operatorname{tr}\left(\gamma^{\mu_1\dots\mu_n} M\right),$$

where the sum is over all multi-indices $\mu_1, \dots, \mu_n$—not only the ordered ones.

3. Now we take two spinors ψ and χ with anticommuting components and choose

$$M = \psi\bar\chi \quad \text{or} \quad M_\alpha^{\ \beta} = \psi_\alpha \bar\chi^\beta, \quad \text{such that} \quad \operatorname{tr}\left(\gamma^{\mu_1\dots\mu_n} M\right) = -\bar\chi \gamma^{\mu_1\dots\mu_n} \psi.$$

Show that the above identity implies the general *Fierz-identity*

$$\psi\bar\chi = -\frac{1}{d_s} \sum_n \frac{1}{n!} (-)^{n(n-1)/2} \gamma_{\mu_1\dots\mu_n} \left(\bar\chi \gamma^{\mu_1\dots\mu_n} \psi\right).$$

This means that we can expand the matrix $\psi\bar\chi$ as linear combination of the basis elements $\gamma_{\mu_1\dots\mu_n}$. For Euclidean theories $\gamma_\mu = \gamma^\mu = \gamma_\mu^\dagger$.

4. Show that in *three* dimensions, the identity simplifies to

$$2\psi^a \bar\psi^b = -(\bar\psi^b \psi^a)\mathbb{1} - (\bar\psi^b \gamma_\mu \psi^a)\gamma^\mu$$

and in turn implies the following relation

$$j^\mu j_\mu = -2 \sum_{a,b} (\bar\psi^a \psi^b)(\bar\psi^b \psi^a) - (\bar\psi\psi)^2, \quad \text{where} \quad j^\mu = \sum_a \bar\psi_a \gamma^\mu \psi_a.$$

Hint: Sandwich the first relation between $\bar\psi^c$ and ψ^d, set $b = a$ and $d = c$, and sum.

5. The identities in *two* dimensions follow from those in *three* dimensions if we identify the hermitian γ_2 in $d = 3$ with γ_* in $d = 2$. In particular for one flavor

$$j^\mu j_\mu + (\bar\psi\gamma_*\psi)^2 = -3(\bar\psi\psi)^2.$$

17.5 (Ward Identity) Prove the Ward identity

$$\langle(\bar{\psi}\psi)(x)(\bar{\psi}\psi(y))\rangle = \left(\frac{N_{\rm f}}{g^2}\right)^2 \langle\sigma_x\sigma_y\rangle - \frac{N_{\rm f}}{g^2}\delta(x-y).$$

Hint: The integral of the second derivative of $\exp(-S)$ with respect to $\sigma(x)$ and $\sigma(y)$ vanishes. But to get the above Ward identity, you must consider two more vanishing integrals.

17.6 (Sum Over Bessel Functions) Different regularizations of the fermion determinant give rise to different representations for the renormalized effective potential. For example, to prove that for $L \to \infty$ the sum (17.220) reproduces the integral (17.92), one needs the identity

$$\sum_{n\geq 1}(-1)^n {\rm e}^{-na}\frac{K_{d/2}(bn)}{(bn)^{d/2}} = -\frac{\pi}{b}\frac{|S^{d-2}|}{(2\pi)^{\frac{d}{2}}}\int_0^\infty {\rm d}x\, x^{d-2}\log\left(1+{\rm e}^{-a-b\sqrt{1+x^2}}\right),$$

where $|S^n|$ is the volume of the n-sphere, e.g., $|S^2| = 4\pi$. Try to prove this relation. Hint: The following integral representation for the modified Bessel functions could be useful:

$$K_\nu(z) = \frac{\sqrt{\pi}}{\Gamma(\nu-\frac{1}{2})}\left(\frac{z}{2}\right)^{\nu-1}\int_0^\infty {\rm d}x\, {\rm e}^{-z\sqrt{1+x^2}}x^{2\nu-2}.$$

17.7 (Gap Equation) The transition line where the curvature of the effective potential changes sign at the origin (and thus a local maxima becomes a local minima) is not hard to calculate when one recalls that the right-hand side of (17.104) is proportional to the σ-derivative of $U_{\rm eff}$. Show that the curvature of the potential vanishes at the origin when the integral (17.104) vanishes for $\sigma = 0$. Show that in two dimensions this condition is equivalent to

$$0 = \int_{-\infty}^\infty {\rm d}x\left(\frac{\tanh x}{x - \frac{\beta\mu}{2}} - \frac{1}{\sqrt{x^2 + (\beta\sigma_0/2)^2}}\right),$$

where a principal value prescription is understood. Actually we used this transcendental equation to calculate the second-order transition line left of the Lifshitz point in Fig. 17.5 and the (dashed) line, where a metastable homogeneous symmetric phase develops.
   Hint: Set $\beta p \pm \beta\mu = 2x$, and collect terms with $\pm\mu$.

17.8 (Magnetic Catalysis in $2d$ GN Model) Here we consider the GN model in $1 + 1$ dimensions at finite temperature, finite density, and in the presence of a constant field F_{01}. The relevant fermion operator is

$$\mathscr{D} = \gamma^0(D_0 + \mu) + \gamma^1\partial_x + \sigma = \slashed{D} + \sigma,$$

with constant σ and vector potential giving rise to a constant and quantized F_{01},

$$A_0 = -Ex = -\frac{\Phi}{V}x, \quad \Phi = 2\pi k \quad \text{with} \quad E = -\partial_1 A_0, \quad k \in \mathbb{Z}.$$

1. Calculate the eigenvalues of $\mathcal{D}$ and their degeneracy.
2. Show that for $E > 0$ the ζ-function regularized determinant gives rise to the following effective potential,

$$U_{\text{eff}}(\sigma) = \frac{\sigma^2}{4\pi} \log \frac{2E}{\sigma_0^2} + \frac{E}{4\pi} \log \frac{\sigma^2}{4\pi E} + \frac{E}{2\pi} \log \Gamma(\sigma^2/2E)$$

   which, strangely enough, is independent of μ and T.
3. Prove that for small fields E one recovers the well-known result,

$$U_{\text{eff}}(\sigma, E \to 0) = \frac{\sigma^2}{4\pi}\left(\log\frac{\sigma^2}{\sigma_0^2} - 1\right) + \frac{E^2}{12\pi\sigma^2} + \mathcal{O}(E^4/\sigma^6).$$

4. Show that for strong fields E the potential has the form

$$U_{\text{eff}}(\sigma, E \to \infty) = \frac{E}{4\pi} \log \frac{E}{\pi\sigma^2} + \frac{\sigma^2}{4\pi}\left(\log\frac{2E}{\sigma_0^2} - \gamma\right) + \mathcal{O}(\sigma^4/E).$$

5. We have the peculiar fact that for $E = 0$, the effective potential depends on μ and T, and this dependence completely disappears for any $E \neq 0$. What could be the reason?

Hint: The ζ-function can be written in terms of the Hurwitz ζ-function $\zeta(s, a)$. You will need this function and its s-derivative at $s = 0$, given by

$$\zeta(0, a) = \frac{1}{2} - a, \qquad \zeta'(0, a) = \log \Gamma(x) - \frac{1}{2} \log(2\pi).$$

For small and large fields, you will need the asymptotic formulas

$$\log \Gamma(x) \sim \left(x - \frac{1}{2}\right) \log x - x + \frac{\log 2\pi}{2} + \frac{1}{12x} + \mathcal{O}(1/x^2), \quad x \gg 1,$$

$$\log \Gamma(x) \sim -\log x - \gamma x, \qquad\qquad\qquad\qquad\qquad\qquad x \ll 1.$$

The ζ-function determinant has been calculated in [82], which you may consult in case you have problems.

17.9 (A Useful Relation for the Eigenmodes of h) As in the chapter on inhomogeneous condensates, we define

$$\rho_{mn}(x) = \psi_m^\dagger(x)\gamma^0\psi_n(x)$$

for two orthonormal eigenfunctions of the Dirac Hamiltonian h in (17.175) with energies ε_m and ε_n. Prove the relation

$$\varrho_{mn} = \frac{2\sigma}{\varepsilon_n + \varepsilon_m}\psi_m^\dagger\psi_n + \frac{1}{\varepsilon_n + \varepsilon_m}\partial_x(\psi_m^\dagger\gamma_1\psi_n),$$

and that for a constant σ the ϱ_{mm} are constant as well and given by

$$\varrho_{mm} = \frac{\sigma}{\varepsilon_m L}.$$

Hint: To prove the relation for ϱ_{mn}, one uses the eigenvalue equation for ψ_m. The simple x-dependence of eigenmodes for constant σ is used to prove the last statement.

Appendix: Covariant ζ-Function Regularization

Rather than applying a Poisson resummation formula to the sum over the Matsubara frequencies only, as we did in (17.83), we could resum the Gaussian sum over Matsubara frequencies and spatial momenta at once. With (16.143) this leads to an alternative representation for the ζ-function,

$$\zeta_H(s) = \frac{d_s}{2}\frac{\beta V_s}{(4\pi)^{d/2}}\frac{1}{\Gamma(s)}\int dt\, t^{s-1-\frac{d}{2}}e^{-t\sigma^2}\sum_n(-1)^n e^{-\beta^2 n^2/4t - n\beta\mu}\sum_n e^{-L^2 n^2/4t}.$$

(17.218)

The term with vanishing n and $\boldsymbol{n}$ is identical to the zero-temperature, zero-density, and infinite-volume contribution $\zeta^{(0,\infty)}$ in (17.86). The remaining UV-finite series, denoted by $\zeta^{(2)}$, represents the sum $\zeta^{(0,L)} + \zeta^{(1)}$ of the ζ-functions in (17.86) and (17.90) and comprises the finite temperature, density, and volume corrections. It follows that the first contribution in the associated decomposition of the effective potential

$$U_{\text{eff}} = U_{\text{eff}}^{(0,\infty)} + U_{\text{eff}}^{(2)}$$

(17.219)

is the expressions in (17.87) and (17.89), but without the sums over Bessel functions. Again we use (5.30) to express the remaining alternating series in (17.218) as sum over modified Bessel functions. The derivative with respect to s at $s = 0$ finally yields

$$U_{\text{eff}}^{(2)} = \frac{d_s \sigma^d}{(2\pi)^{d/2}} {\sum_{n,\boldsymbol{n}}}' (-1)^n e^{-n\beta\mu} \frac{K_{d/2}(\beta\sigma\sqrt{n^2 + \tau^2\boldsymbol{n}^2})}{(\beta\sigma\sqrt{n^2 + \tau^2\boldsymbol{n}^2})^{d/2}}, \quad \tau = \frac{L}{\beta}, \quad (17.220)$$

where the prime at the summation symbol means omission of the term with vanishing n and $\boldsymbol{n}$.

In the zero temperature and density limit, only terms with $n = 0$ contribute, and $U_{\text{eff}}^{(2)}$ is equal to the sums over Bessel functions in (17.87) and (17.89). On the other hand, at finite temperature but in the *thermodynamic limit* $L \to \infty$, only terms with vanishing $\boldsymbol{n}$ contribute. Using an integral representation for the Bessel function, one can prove the useful relation given in Problem 17.6. This relation shows that for $L/\beta \to \infty$, the sum (17.220) is identical to the integral (17.92),

$$\lim_{L \to \infty} U_{\text{eff}}^{(2)} = 2d_s \left(\frac{\sigma}{2\pi\beta}\right)^{d/2} \sum_{n \geq 1} (-1)^n \cosh(n\beta\mu) \frac{K_{d/2}(n\beta\sigma)}{n^{d/2}} \quad (17.221)$$

$$= \frac{d_s}{2\beta(2\pi)^{d-1}} \int d\boldsymbol{p} \, \log\left(1 + e^{-\beta(\varepsilon_p + \mu)}\right) + (\mu \to -\mu) = \lim_{L \to \infty} U_{\text{eff}}^{(1)}.$$

More generally, for finite temperature and finite box size $U_{\text{eff}}^{(2)} = U_{\text{eff}}^{(0,L)} + U_{\text{eff}}^{(1)}$.

References

1. W. Thirring, A soluble relativistic field theory. Ann. Phys. **3**, 91 (1978)
2. D.J. Gross, A. Neveu, Dynamical symmetry breaking in asymptotically free field theories. Phys. Rev. **D10**, 3235 (1974)
3. Y. Nambu, G. Jona-Lasinio, Dynamical model of elementary particles based on an analogy with superconductivity I. Phys. Rev. **122**, 345 (1961); Dynamical model of elementary particles based on an analogy with superconductivity II. Phys. Rev. **124**, 246 (1961)
4. J. Lenz, L. Pannullo, M. Wagner, B. Wellegehausen, A. Wipf, Inhomogeneous phases in the Gross-Neveu model in 1+1 dimensions. Phys. Rev. **D102**, 114501 (2020)
5. H. Gies, L. Janssen, UV fixed-point structure of the three-dimensional Thirring model. Phys. Rev. **D82** 085018 (2010)
6. F. Gehring, H. Gies, L. Janssen, Fixed-point structure of low-dimensional relativistic fermion field theories: universality classes and emergent symmetry. Phys. Rev. **D92**, 085046 (2015)
7. G. Parisi, The theory of non-renormalizable interactions: the large N expansion. Nucl. Phys. **B100**, 368 (1975)
8. K. Gawedzki, A. Kupiainen, Renormalizing the nonrenormalizable. Phys. Rev. Lett. **55**, 363 (1985)
9. B. Rosenstein, B. Warr, S.H. Park, Dynamical symmetry breaking in four Fermi interaction models. Phys. Rep. **205**, 59 (1991)

10. J. Braun, H. Gies, D.D. Scherer, Asymptotic safety: a simple example. Phys. Rev. **D83**, 085012 (2011)
11. S. Weinberg, *Critical phenomena for field theorists*, in *Understanding the Fundamental Constituents of Matter*, ed. by A. Zichichi. The Subnuclear Series, vol. 14 (1976), pp. 1–52
12. G.W. Semenoff, Condensed-matter simulation of a three-dimensional anomaly. Phys. Rev. Lett. **53**, 2449 (1984)
13. S. Hands, C. Strouthos, Quantum critical behaviour in a graphene-like model. Phys. Rev. **B78**, 165423 (2008)
14. I.F. Herbut, QED$_3$ theory of underdoped high-temperature superconductors. Phys. Ref. **B66**, 094504 (2002)
15. J.I. Cirac, P. Maraner, J.K. Pachos, Cold atom simulation of interacting relativistic quantum field theories. Phys. Rev. Lett. **105**, 190403 (2010)
16. B. Klaiber, The thirring model, in *Boulder 1967*. Lecture notes in Physics XA (Gordon and Breach, New York, 1968)
17. I. Sachs, A. Wipf, Generalized thirring models. Ann. Phys. **249**, 380 (1996)
18. M. Karbach, G. Müller, Introduction to the Bethe ansatz I. Comput. Phys. **11**, 36 (1997); Introduction to the Bethe ansatz II. **12**, 565 (1998)
19. V.G. Vaks, A.I. Larkin, On the application of the methods of superconductivity theory to the problem of the masses of elementary particles. Sov. Phys. JETP. **13**, 192 (1961)
20. M. Frank, M. Buballa, M. Oertel, Flavor mixing effects on the QCD phase diagram at nonvanishing isospin chemical potential: one or two phase transitions. Phys. Lett. **B562**, 221 (2003)
21. R.L. Stratonovich, On a method of calculating quantum distribution functions. Sov. Phys. Dokl. **2**, 416 (1957); J. Hubbard, Calculation of partition functions. Phys. Rev. Lett. **3**, 77 (1959)
22. H.W. Furry, A symmetry theorem in the positron theory. Phys. Rev. **51**, 12 (1937)
23. M. Thies, K. Urlichs, Revised phase diagram of the Gross-Neveu model. Phys. Rev. **D67**, 125015 (2003)
24. O. Schnetz, M. Thies, K. Urlichs, Phase diagram of the Gross-Neveu model: exact results and condensed matter precursors. Ann. Phys. **314**, 425 (2004)
25. G. Basar, G.V. Dunne, M. Thies, Inhomogeneous condensates in the thermodynamics of the chiral NJL(2) model. Phys. Rev. **D79**, 105012 (2009)
26. M. Buballa, S. Carignano, Inhomogeneous chiral condensates. Prog. Part. Nucl. Phys. **81**, 39 (2015)
27. L. O'Raifeartaigh, A. Wipf, H. Yoneyama, The constraint effective potential. Nucl. Phys. **B271**, 653 (1986)
28. U. Wolff, The phase diagram of the infinite N Gross-Neveu model at finite temperature and chemical potential. Phys. Lett. **157B**, 303 (1985)
29. E.T. Whittaker, G.N. Watson, *A Course of Modern Analysis*, 4th edn. reprinted, section 12.13 (Cambridge University Press, Cambridge, 1963)
30. M. Chamberland, A. Straub, On gamma quotients and infinite products. Adv. Appl. Math. **51**, 546 (2013)
31. S.P. Klevansky, R.H. Lemmer, Chiral symmetry restoration in the Nambu-Jona-Lasinio model with a constant electromagnetic field. Phys. Rev. **D39**, 3478 (1989)
32. K.G. Klimenko, Three-dimensional Gross-Neveu model in an external magnetic field. Theor. Math. Phys. **89**, 1161 (1992)
33. V.P. Gusynin, V.A. Miransky, I.A. Shovkovy, Dynamical flavor symmetry breaking by a magnetic field in 2+1 dimensions. Phys. Rev. **D52**, 4718 (1995)
34. F. Preis, A. Rebhan, A. Schmitt, Inverse magnetic catalysis in field theory and gauge-gravity duality. Lect. Notes Phys. **871**, 51–86 (2013)
35. I.S. Gradshteyn, I.M. Ryzhik, *Table of Integrals, Series and Products* (Academic Press, Amsterdam, 2007)
36. D. Ebert, K.G. Klimenko, M.A. Vdovichenko, A.S. Vshivtsev, Magnetic oscillations in dense cold quark matter with four fermion interactions. Phys. Rev. **D61**, 025005 (2000)
37. I.A. Shovkovy, Magnetic catalysis: a review. Lect. Notes Phys. **871**, 13 (2012)

38. T. Inagaki, D. Kimura, T. Murata, Four fermion interaction model in a constant magnetic field at finite temperature ahd chemical potential. Prog. Theor. Phys. **111**, 371 (2004)
39. R. Narayanan, Phase diagram of the large N Gross-Neveu model in a finite periodic box. Phys. Rev. **D101**, 096001 (2020)
40. M. Buballa, L. Kurth, M. Wagner, M. Winstel, Regulator dependence of inhomogeneous phases in the (2+1)-dimensional Gross-Neveu model. Phys. Rev. **D103**, 034503 (2020)
41. S. Blau, M. Visser, A. Wipf, Zeta functions and the Casimir energy. Nucl. Phys. **B310**, 163 (1988)
42. A. Kirchberg, D. Laenge, A. Wipf, Extended supersymmetries and the Dirac operator. Ann. Phys. **315**, 467 (2005)
43. J. Lenz, L. Pannullo, M. Wagner, B. Wellegehausen, A. Wipf, Inhomogeneous phases in the Gross-Neveu model in 1+1 dimensions. Phys. Rev. **D101**, 094512 (2020)
44. R.F. Dashen, B. Hasslacher, A. Neveu, Semiclassical bound states in an asymptotically free theory. Phys. Rev. **D12**, 2443 (1975)
45. R. Pausch, M. Thies, V.L. Dolman, Solving the Gross-Neveu model with relativistic many body methods. Z. Phys. **A338**, 441 (1991)
46. J. Feinberg, All about the static fermion bags in the Gross-Neveu model. Ann. Phys. **309**, 166 (2004)
47. G. Basar, G.V. Dunne, Self-consistent crystalline condensate inchiral Gross-Neveu and Bogoliubov-de Gennes systems. Phys. Rev. Lett. **100**, 200404 (2008)
48. S. Hands, A. Kocic, J.B. Kogut, Four Fermi theories in fewer than four dimensions. Ann. Phys. **224**, 29 (1993)
49. J. Zinn-Justin, Four fermion interacction near four dimensions. Nucl. Phys. **B367**, 105 (1991)
50. J.A. Gracey, The beta function of the chiral Gross-Neveu model at $O(1/N^2)$. Phys. Rev. **D50**, 2840 (1994)
51. A. Vasiliev, S.E. Derkachov, N. Kivel, A. Stepanenko, The 1/n expansion in the Gross-Neveu model: conformal bootstrap calculation of the index eta in order $1/n^3$. Theor. Math. Phys. **94**, 127 (1993)
52. J. Gracey, Computation of critical exponent eta at $O(1/N^3)$ in the four Fermi model in arbitrary dimensions. Int. J. Mod. Phys. **A9**, 727 (1994)
53. J. Gracey, Four loop MS-bar mass anomalous dimension in the Gross-Neveu model. Nucl. Phys. **B802**, 330 (2008)
54. L.N. Mihaila, N. Zerf, B. Ihring, I.F. Herbut, M.M. Scherer, Gross-Neveu-Yukawa model at three loops and Ising critical behavior of Dirac systems. Phys. Rev. **B96**, 165133 (2017)
55. N. Zerf, L.N. Mihaila, P. Marquard, I.F. Herbut, M.M. Scherer, Four-loop critical exponents for the Gross-Neveu-Yukawa models. Phys. Rev. **D96**, 096010 (2017)
56. L. Janssen, I.F. Herbut, Antiferromagnetic critical point on graphene's honeycomb lattice: a functional renormalization group approach. Phys. Rev. **B89**, 205403 (2014)
57. D. Landau, K. Binder, *A Guide to Monte Carlo Simulations in Statistical Physics*, 3rd edn. (Cambridge University Press, Cambridge, 2009)
58. K. Binder, Finite size scaling analysis of Ising model block distribution functions. Z. Phys. **B43**, 119 (1981)
59. D. Schmidt, Three-dimensional four-fermion theories with exact chiral symmetry on the lattice. Thesis, University Jena. https://doi.org/10.22032
60. L. Karkkainen, R. Lacaze, P. Lacock, B. Petersson, Critical behaviour of the 3d Gross-Neveu and Higgs-Yukawa models. Nucl. Phys. **B415**, 781 (1994)
61. S. Christofi, C. Strouthos, Three dimensional four-fermion models - A Monte Carlo study. J. High Energy Phys. **05**, 088 (2007)
62. F. Hoefling, C. Nowak, C. Wetterich, Phase transition and critical behaviour of the d=3 Gross-Neveu model. Phys. Rev. **B66**, 205111 (2002)
63. L. Del Debbio, S.J. Hands, The three-dimensional Thirring model for $N_f = 4$ and $N_f = 6$. Nucl. Phys. **B552**, 339 (1999)
64. S. Hands, B. Lucini, The phase diagram of the three dimensional Thirring model. Phys. Lett. **B461**, 263 (1999)

65. S. Christofi, S. Hands, C. Strouthos, Critical flavor number in the three dimensional Thirring model. Phys. Rev. **D75**, 101701 (2007)
66. B. Wellegehausen, D. Schmidt, A. Wipf, Critical flavour number of the Thirring model in three dimensions. Phys. Rev. **D96**, 094504 (2017)
67. J.J. Lenz, A. Wipf, B. Wellegehausen, Absence of chiral symmetry breaking in Thirring models in 1+2 dimensions. Phys. Rev. **D100**, 054501 (2019)
68. S. Hands, Critical flavor number in the 2+1D Thirring model. Phys. Rev. **D99**, 034504 (2019)
69. S. Hands, M. Mesiti, J. Worthy, Critical behaviour in the single flavor Thirring model in 2-1d. Phys. Rev. **D102**, 094502 (2020)
70. L. Dabelow, H. Gies, B. Knorr, Momentum dependence of quantum critical Dirac systems. Phys. Rev. **D99**, 125019 (2019)
71. F. Karsch, J.B. Kogut, H.W. Wyld, The Gross-Neveu model at finite temperature and density. Nucl. Phys. **B280**, 289 (1987)
72. V.L. Berezinskii, Destruction of long-range order in one-dimensional and two-dimensional systems having a continuous symmetry group I. Classical systems. Sov. Phys. JETP **32**, 493 (1971)
73. J.M. Kosterlitz, D.J. Thouless, Ordering, metastability and phase transitions in two-dimensional systems. J. Phys. **C6**, 1181 (1973)
74. J. Lenz, L. Pannullo, M. Wagner, B. Wellegehausen, A. Wipf, Baryons in the Gross-Neveu model in 1+1 dimensions at finite number of flavors. Phys. Rev. **D102**, 114501 (2020)
75. D. Chandler, *Introduction to Modern Statistical Mechanics* (Oxford University Press, Oxford 1987)
76. H. Gies, C. Wetterich, Renormalization flow of bound states. Phys. Rev. **D65**, 065001 (2002)
77. T.O. Wehling, A.M. Black-Schaffer, A.V. Balatsky, Dirac materials. Adv. Phys. **63**, 1 (2014)
78. W. Metzner, M. Salmhofer, C. Honerkamp, V. Meden, K. Schonhammer, Functional renormalization group approach to correlated fermion systems. Ref. Mod. Phys. **84**, 299 (2012)
79. G.P. Vacca, L. Zambelli, Multimeson Yukawa interactions at criticality. Phys. Rev. **D91**, 125003 (2015)
80. B. Knorr, Ising and Gross-Neveu model in next-to-leading order. Phys. Rev. **B94**, 245102 (2016)
81. M. Heilmann, T. Hellwig, B. Knorr, M. Ansorg, A. Wipf, Convergence of derivative expansion in supersymmetric functional RG flows. J. High Energy Phys. **02**, 109 (2015)
82. S. Blau, M. Visser, A. Wipf, Analytical results for the effective action. Int. J. Mod. Phys. **A6**, 5408 (1992)

Index

Printed in the United States
by Baker & Taylor Publisher Services